T0155490

# CAMBRIDGE LIBRARY COLLECTION

*Books of enduring scholarly value*

## Life Sciences

Until the nineteenth century, the various subjects now known as the life sciences were regarded either as arcane studies which had little impact on ordinary daily life, or as a genteel hobby for the leisured classes. The increasing academic rigour and systematisation brought to the study of botany, zoology and other disciplines, and their adoption in university curricula, are reflected in the books reissued in this series.

## The Life, Letters and Labours of Francis Galton

A controversial figure, Sir Francis Galton (1822–1911), biostatistician, human geneticist, eugenicist, and first cousin of Charles Darwin, is famed as the father of eugenics. Believing that selective breeding was the only hope for the human race, Galton undertook many investigations of human abilities and devoted the last few years of his life to promoting eugenics. Although he intended his studies to work positively, for eradicating hereditary diseases, his research had a hugely negative impact on the world which subsequently bestowed on Galton a rather sinister reputation. Written by Galton's colleague, eugenicist and statistician Karl Pearson (1857–1936), this four-volume biography pieces together a fascinating life. First published in 1924, Volume 2 focuses on Galton's 'researches of middle life', including his anthropological research and psychological experiments. Pearson himself was later appointed the first Galton professor of eugenics at University College London.

Cambridge University Press has long been a pioneer in the reissuing of out-of-print titles from its own backlist, producing digital reprints of books that are still sought after by scholars and students but could not be reprinted economically using traditional technology. The Cambridge Library Collection extends this activity to a wider range of books which are still of importance to researchers and professionals, either for the source material they contain, or as landmarks in the history of their academic discipline.

Drawing from the world-renowned collections in the Cambridge University Library, and guided by the advice of experts in each subject area, Cambridge University Press is using state-of-the-art scanning machines in its own Printing House to capture the content of each book selected for inclusion. The files are processed to give a consistently clear, crisp image, and the books finished to the high quality standard for which the Press is recognised around the world. The latest print-on-demand technology ensures that the books will remain available indefinitely, and that orders for single or multiple copies can quickly be supplied.

The Cambridge Library Collection will bring back to life books of enduring scholarly value (including out-of-copyright works originally issued by other publishers) across a wide range of disciplines in the humanities and social sciences and in science and technology.

# The Life, Letters and Labours of Francis Galton

VOLUME 2:
RESEARCHES OF MIDDLE LIFE

KARL PEARSON

CAMBRIDGE
UNIVERSITY PRESS

CAMBRIDGE UNIVERSITY PRESS

Cambridge, New York, Melbourne, Madrid, Cape Town,
Singapore, São Paolo, Delhi, Tokyo, Mexico City

Published in the United States of America by Cambridge University Press, New York

www.cambridge.org
Information on this title: www.cambridge.org/9781108072410



This edition first published 1924
This digitally printed version 2011

ISBN 978-1-108-07241-0 Paperback

THE

# LIFE, LETTERS AND LABOURS

OF

# FRANCIS GALTON

Francis Galton, aged 73.

THE

# LIFE, LETTERS AND LABOURS

OF

# FRANCIS GALTON

BY

KARL PEARSON

GALTON PROFESSOR, UNIVERSITY OF LONDON

VOLUME II

RESEARCHES OF MIDDLE LIFE

CAMBRIDGE

AT THE UNIVERSITY PRESS

1924

PRINTED IN GREAT BRITAIN

# PREFACE

THE first volume of this biography appeared in July, 1914, about a month before the outbreak of the Great War. It met with few readers, and failed to repay the cost of production. The war injured the Galton Laboratory in many ways, the chief, perhaps, being that it rendered of small value its publication funds; thus a collection of Galton's published papers, which had been projected before the war, was placed out of the question. Further the relative and the friend of Galton who in 1914 had financed together the first volume were unable to face the excessive post-war costs of printing. In 1919 accordingly, when the pen again replaced ballistics, the tired mind and hand could not seek a legitimate relief in continuing the story of Galton's life. It was only in 1922 that the generous gift of an old schoolfriend, the late Mr Lewis Haslam, M.P., enabled me to face the difficulties of a second volume. I deeply regret that he did not live to see this work in type. But if friends and admirers of Galton find in it anything they value, let them remember the debt they owe to Lewis Haslam.

When it came to planning this second volume the biographer found himself, however, in a very different position from what he had anticipated in 1914. He then considered that the issue of the collected works of Galton would render easy the task of describing Galton's researches. Such issue having ceased to be practicable, a grave problem arose. Many of Galton's papers are now inaccessible, even a record of the original *loci* of publication is not available[1], there is no annotated bibliography to guide men to the memoirs themselves by a suggestion of their contents, and they are scattered, one might almost say, at random not only through the publications of many learned societies and scientific journals, but in the daily, weekly and monthly press, often in magazines which have long ceased to appear. The most striking factor in Galton's work was its pioneer character, he blazed a trail where others have followed with a highway. To grasp his extraordinary suggestiveness—even when his methods are the crude extemporisations of the first settler, ever ready to advance further as others crowd in behind—the reader must study Galton's writings in the mass. But these are in many cases beyond his reach, if not beyond his ken. Thinking the matter out carefully, I determined that this second volume of Galton's biography should to a large extent supply the reader with what the collected works would have done; that the *résumé* of memoirs, books, and articles should be full enough to enable the anthropologist, the geneticist and the statistician to appreciate

[1] The bibliography attached to the *Memories* is very incomplete. Not only do papers fail, but often the description is incorrect either as to volume or as to year, or even as to the title of the journal assigned, while throughout no pages are given.

what Galton had done, and so starting from his suggestions make a more thorough map of a district, where Galton would only claim to have made a chart of the cardinal points. In taking this determination I was soon aware that it meant adding a third volume to this *Life.* I have had to postpone to that volume the discussion of Correlation, the Statistical Theory of Heredity, Personal Identification and Description and Eugenics together with many letters, characteristic of Galton's mentality and of his affectionate disposition. But that volume seems an easy one after the present, for it largely deals with work done after Galton had been recognised as a master and friend.

The multitude of my own tasks from 1880 onwards gave me little leisure to do more than keep in touch with current work; I had small opportunity for considering earlier memoirs, and many of Galton's papers written before I left Cambridge I have only read forty years after their publication. How I now regret that I had not studied them, when with youthful energy still mine I might have pursued further their lines of thought! How many are the suggestions they make for novel and profitable research! I shall indeed be content, if this book of mine opens up to the younger men of to-day that field of inspiration, which Galton provided for some half-dozen of us in the 'eighties. How much one seems to have lost by waiting to explore it fully, until one's *Wanderjahre* were for ever gone!

If this second volume be written essentially to bring the thoughts of a great scientist home to the younger scientists of to-day, to show them the wide regions, practical and theoretical, which Galton opened to the mathematician and statistician, there are still some interludes which appeal to a wider audience, such as the beauty of Galton's friendship for Darwin, the interest of his correspondence with De Candolle, and his brief contact with the "Passionate Statistician." The ingenuity of Galton's mechanisms and the originality of his photographic work will attract others, while in the field of psychology it will be found difficult to refute the claim that he was the first English experimentalist.

If the reader should find Chapter XIII of this work more clumsily worded and carelessly written than those which precede it, he will understand the loss which the biographer incurred by the death of his friend, W. Paton Ker, while the book was passing through the press. Professor Ker's returned proofs, duly loaded with admonition, ejaculation, and humorous chiding, were not only assurance that many of the author's blunders were detected, but led him with delight on more than one occasion to unwonted realms, little sought by votaries of science. Let us rejoice that he has lived,

> "And laugh like him to know in all our nerves
> Beauty, the spirit, scattering dust and turves."

I have again to acknowledge the ready help of Francis Galton's relatives and friends, especially in the matter of portraiture. Even as it is I have had to make a selection from the vast amount of photographic material placed at my disposal, and a portion of that selection is still reserved for the third volume. In contrast to Darwin, Galton was repeatedly photographed, and the result is that we can trace not only the physical changes in his personality from childhood to old age, but I venture to think we can find portraits which emphasise even the individual moods and characteristic phases of his many-sided mentality. This book may help to preserve that play of expression which forms the charm of our memory of a friend, and which is renewed and kept alive by many photographs, until they perish also.

This perishing of photographic portraits, whether negatives or prints, has been sadly impressed upon me not only in the case of photographs of Galton himself—which I have endeavoured to put into a more permanent form—but further in the case of nearly all Galton's own photographic work. Box after box of his negatives as well as the prints from them have perished or are rapidly perishing. I felt strongly the need for preserving at least his hitherto unpublished results in composite portraiture. But to add this number of plates to my volume seemed only possible by curtailing its text. This difficulty was finally overcome by the generosity of Mr Edward Wheler-Galton and by the aid of one who owed much to Sir Francis. In this way it became feasible to give comprehensive illustration of what Galton achieved in composite photography. The exhibit will, I hope, lead to the renewal of this branch of investigation, for I am convinced that its possibilities are by no means exhausted.

I have to acknowledge the great aid I have received from my son Mr Egon S. Pearson in dealing with Galton's photographic material and researches. I have further to thank Major Leonard Darwin and my colleague Miss Ethel M. Elderton for aid in a variety of ways. Lastly I have to place on record a confession. The *Galtoniana* contain a large number of manuscripts and notebooks in Galton's hand; many of these are in pencil, much rubbed, occasionally obliterated. In the earlier chapters of this volume I have constantly used this material. Lately I have been unable, owing to failure of sight, to do so. I may well have missed material which ought to have found its place in these pages. My only apology must be that of what lay in my power to give I have freely given.

KARL PEARSON.

GALTON LABORATORY,
UNIVERSITY COLLEGE, LONDON.
*July* 8, 1924.

Francis Galton in later life. From a sketch by his niece, Miss Eva Biggs (Mrs Ellis).

# CONTENTS OF VOLUME II

# ILLUSTRATIONS TO VOLUME II

Francis Galton
1910

I thought it safer to proceed like the surveyor of a new country, and endeavour to fix in the first instance as truly as I could the position of several cardinal points.

FRANCIS GALTON.

PLATE I

Francis Galton, aged 28, at the end of the "Fallow Years" and before starting for Damaraland.

# CHAPTER VIII

## TRANSITION STUDIES: ART OF TRAVEL, GEOGRAPHY, CLIMATE

### A. ART OF TRAVEL

"The highest minds in the highest races seem to have been those who had the longest boyhood."

FRANCIS GALTON, *Hereditary Talent and Character*, Aug. 1865.

"I have been speculating last night what makes a man a discoverer of undiscovered things; and a most perplexing problem it is. Many men who are very clever—much cleverer than the discoverers—never originate anything. As far as I can conjecture the art consists in habitually searching for the causes and meaning of everything which occurs. This implies sharp observation and requires as much knowledge as possible of the subject investigated. But why I write all this now I hardly know, except out of the fulness of my heart."

CHARLES DARWIN, Letter to his son Horace, Dec. 15, 1871.

WE left Francis Galton at the end of our first volume aged 32, married, with many social friends, an ample competence, and a mind trained both in observation and analysis. His experience had been such that he knew more of mathematics and physics than nine biologists out of ten, more of biology than nineteen mathematicians out of twenty, and more of pathology and physiology than forty-nine out of fifty of the biologists and mathematicians of his day. Added to these advantages he had gained a knowledge of man and his habits in various lands; this gave him additional width of view, if it rendered less obvious to him that field of investigation wherein his powers were ultimately to achieve their most noteworthy successes. Indeed, had Galton been asked in 1854 what was his calling and the nature of his studies, there is little doubt that he would have replied: "I am a traveller by inclination and my study is geography." In his *Memories*[1] Galton tells us that he was

"rather unsettled during a few years, wishing to undertake a fresh bit of geographical exploration, or even to establish myself in some colony; but I mistrusted my powers, for the health that had been much tried had not wholly recovered."

Whether marriage or health was the real source of Galton's 'Wanderlust' being reduced to vacation rambles, it would be hard to say, but we have probably to thank one or the other for his continued presence in England at a time when startling new ideas were to strike upon his receptive mind. Immediately, however, travel, geography, and closely associated therewith, climate, were to occupy his attention; and he did not touch these things without leaving his mark upon them.

"It was not long after my marriage," he writes, "that the character of a piece of work that lay before me was clearly perceived. It was ready to be taken in hand and most suitable to my

[1] pp. 161–2.

powers. It was to aid others in the exploration of the then unknown parts of the world, especially of Africa, of whose total length as much had been seen by me in my two journeys as perhaps by anyone else then living. Being placed on the Council of the Royal Geographical Society[1], I thoroughly utilised that position to fulfil my object."

Galton in the first place set about making travel an art to be learnt, and in the next place he determined to make the traveller a real contributor to geographical discovery.

Those who have watched the development of geography as an academic training in recent years must be struck by the manner in which Galton foresaw its expansion and must, perhaps, occasionally wonder whether its instructors might not still learn something from Galton's early geographical writings. In much the same way Galton's *Art of Travel* first published in 1855—and in the fifth edition of 1872, reaching the most catholic comprehensiveness compatible with pocket dimensions—remains still a treasury not only for the professed traveller, but for the leaders of boy-scouts and girl-guides; nay, there are methods to be learnt in the *Art of Travel* which may bring profit to the ordinary household of to-day[2]. Even what appears on the surface especially intended for the explorer may be found on careful reading to have a home-stayer's application. Thus:

"Travellers are apt to expect too much from their medicines, and to think that savages will hail them as demigods wherever they go. But their patients are generally cripples who want to be made whole in a moment, and other suchlike impracticable cases. Powerful emetics, purgatives and eye washes are the most popular physickings.

The traveller who is sick, away from help, may console himself with the proverb, that 'though there is a great difference between a good physician and a bad one, there is very little between a good one and none at all'." (5th Edn. p. 14.)

or again:

"*Shirt-sleeves*—When you have occasion to tuck up your shirt-sleeves, recollect that the way of doing so is, not to begin by turning the cuffs inside out, but outside in—the sleeves must be rolled up inwards towards the arm, and not the reverse way. In the one case, the sleeves will remain tucked up for hours without being touched; in the other, they become loose every five minutes." (*Ibid.* p. 116.)

It is difficult to give a résumé of the *Art of Travel*. By the fifth edition Galton seems to have thought of most things that any type of traveller would require in any possible climate or country. From outfit and servants to medicine, fire, tents, fuel, food, treatment of animals, signalling, surveying,

[1] Royal Geographical Society: Honorary General Secretary 1857–63, Foreign Secretary 1865–66, Member of the Council either by election or *ex officio* almost continuously from 1854 to 1893, Vice-President 1866–72, 1879–86, 1888–91.

[2] How simple and obvious some of these suggestions are and yet how much might still be learnt from them! Powders for medicinal use should be mixed with coloured flours to distinguish emetics from aperients or to emphasise poisons; where ammonia and the 'blue bag' are not available—on a walk or at a picnic party—oil of nicotine scraped from a pipe, generally at hand, will relieve the pain of a wasp or bee sting. Or, again, such simple directions as those concerning blistered feet where beyond the usual soaping of the insides of the stockings before a long walk, we read "after some hours on the road, when the feet are beginning to be chafed, take off the shoes and change the stockings; putting what was the right stocking on the left foot, and the left stocking on the right foot";—and when a blister is formed pour a little brandy or other spirit into the palm of the hand and drop tallow from a lighted candle into it, rub the feet with this mixture on going to rest, and "on the following morning no blister will exist."

PLATE II

Francis Galton, from a photograph taken after his return from Damaraland, *circa* 1855.

finding the path, etc., even to the conclusion of the journey and the printing of the maps, it is all there, tersely given, with just the needful diagrams and sketches. Many are the mechanical 'dodges,' here given; of such Galton never wearied. He had watched craftsmen in all the lands through which he had travelled and he never tired of experimenting and of model-making[1]. And much of this he has used in his *Art of Travel*. The reader of his South African book will recognise also the individual experiences which gave rise to several of the hints in the present work.

If a few formulae or a small amount of measurement can be thrown in, Galton will gladly provide them. The following is a good illustration:

"THE RUSH OF AN ENRAGED ANIMAL is far more easily avoided than is usually supposed. The way the Spanish bull-fighters play with the bull, is well known; any man can avoid a mere headlong charge. Even the speed of a racer which is undeniably greater than that of any wild quadruped, does not exceed 30 miles an hour, or four times the speed of a man. The speed of an ordinary horse is not more than 24 miles an hour; now even the fastest wild beast is unable to catch an ordinary horse, except by crawling unobserved to his side, and springing upon him; therefore I am convinced that the rush of no wild beast exceeds 24 miles an hour, or three times the speed of a man....It is perfectly easy for a person who is cool, to avoid an animal by dodging to one side or another of a bush. Few animals turn, if the rush be unsuccessful. The buffalo is an exception; he regularly hunts a man, and is therefore peculiarly dangerous. Unthinking persons talk of the fearful rapidity of a lion's or tiger's spring. It is not rapid at all; it is a slow movement, as must be evident from the following consideration. No wild animal can leap ten yards, and they all make a high trajectory in their leaps. Now think of the speed of a ball thrown or rather pitched, with just sufficient force to be caught by a person ten yards off; it is a mere nothing. The catcher can play with it as he likes; he has even time to turn after it, if thrown wide. But the speed of a springing animal is undeniably the same as that of a ball, thrown so as to make a flight of equal length and height in the air. The corollary to all this is that if charged, you must keep cool and watchful, and your chance of escape is far greater than non-sportsmen would imagine." (4th Edn. p. 251.)

While traces of the personality of Galton will be found by those who knew him well on almost every page of the *Art of Travel*, there are passages which mark unconsciously his views and the course of his development from 1853 to 1867. There are omissions also in later editions which tell exactly the stage he had reached.

From Damaraland and Ovampo few if any animals, birds or insects were brought back. Galton then and in the *Art of Travel* considered them from the standpoint of sport and food. His list of instruments contains no microscope or dissecting tools, and of books no work on natural history. The sole reference to the collection of specimens occurs in the last paragraph where a description is given of how to make a specimen box from a flat card (3rd Edition, 1860). There is not a word as to how to observe and record the anthropometric characters, folk-lore or religious customs of savage man; neither callipers, tape, nor colour standards appear in Galton's instrumentarium.

[1] The Galton Laboratory possesses a whole series of rough models in card, wood or glass; 'Galton's Toys,' as we call them. Of the purpose of many we know absolutely nothing; others were initial attempts at Galton's hyperscope, heliostat, etc. Besides these 'Toys' are quite a number of instruments chiefly optical made by practical instrument makers to Galton's plans, but in certain cases it has so far been impossible to determine for what purposes they were intended.

These are very noteworthy omissions in a man who was among the foremost anthropologists in this country later to study both the psychic and physical characters in man. In the *Art of Travel* Galton had essentially the needs of the geographer—in the narrower sense—in view; the physical country is more important than its inhabitants. It is possible that this is a general rule in life; in youth it is the novelty of the physical environment, but at a later age it is the novelty of new organic types that forms the intense pleasure of travel[1].

Even as late as 1878 when Galton edited the fourth edition of *Hints to Travellers*, a useful compendium for travellers issued by the Royal Geographical Society, while we find a section on the "Collection of Objects of Natural History" there is no reference to man, and the sole approach to any "Hint" of an anthropological nature in the work is a brief note of 17 lines on p. 71 by the Rev. F. W. Holland describing how paper squeezes may be taken of inscriptions. Even the article on photography does not refer to the photography of the natives, or their habitations and occupations. The book is excellent as a guide to the instruments and processes needful to the mapmaker; it lacks all that would give the local human colour to the environmental description. We are not criticising the book from the standpoint of modern academic geography, which does consider man in relation to the physical environment it depicts. We are merely emphasising that Galton in the period we are discussing had not yet discovered his real *métier*—anthropology in its broadest sense. He was doing yeoman service for geography, but the study of man's development, its knowable past and probable future, had not yet fascinated him, still less did it dominate his activities.

The *Art of Travel* shows us indirectly also how undeveloped Galton's mind was in another direction even in 1860. In the third edition of this book we read:

"The method of obtaining fire by rubbing sticks together was at one time nearly universal. It seems remarkable that the time of discovery of the art of fire-making is not recorded in the Bible. We may easily imagine that our first parents obtained their fires from natural sources; of which, some parts of the Caucasus at least, abound in examples. But when Cain was sent an outcast, how did *he* obtain fire? It is remarkable that his descendants are precisely those who invented metallurgy, and arts requiring fire. We might almost theorise to the effect that he or they discovered the art of fire-making, and pushed the discovery into its applications." (p. 27.)

Then follows the well-known passage from Pliny's *Natural History* on the best wood for fire-sticks. In the fourth edition of the *Art of Travel* of 1867 this passage as to our "first parents" and as to Cain as the inventor of fire-sticks has disappeared. Between 1860 and 1867 Galton had read and assimilated Darwin's *Origin of Species*, and in Galton's own words that book had formed "a real crisis in my life" and had driven away "the con-

[1] The rule is of course not invariable. The present writer spent much time in the Austrian Tyrol in his youth, and was on one occasion asked to write a handbook to the Tyrol as one of a series of guidebooks to the Alps. Nothing came of the proposal, because he replied that it must in the first place be a guide to the folk-lore, history, art and institutions of the Tyrolese themselves, and in the second place only a route book to valleys, passes and peaks.

straint of my old superstition as if it had been a nightmare, and was the first to give me freedom of thought." (See Galton's letter to Darwin, Plate II of Vol. I.)

In the rapid growth of our knowledge of the wonderful process of human development, extending now over nearly a quarter of a million years, and with our present certainty that man has used fire for a great portion of that period, the suggestion that the discovery of how to make fire was a product of those last few thousand years which biblical folk-lore endeavours to cover, may well raise a smile. But does the modern reader realise when he smiles at and criticises the mid-Victorians that it was they—Darwin, Galton, Huxley, Clifford and others—who worked their way from such ignorance to insight and gave him the power to smile at it[1]?

To turn to a lighter matter before we leave the *Art of Travel* for good, we may find, even in such a work cram full of detail and technique, sure traces of Galton's sense of humour. Thus, having remarked that asses to kick must put their head to the ground and to bray must raise their tail, and described how the head can be kept up and the tail kept down, he remarks:

"In hostile neighbourhoods, where silence and concealment are sought, it might be well to adopt this rather absurd treatment [lashing the tail to a heavy stone]. An ass who was being schooled according to the method of this and the preceding paragraph, both at the same time, would be worthy of an artist's sketch." (4th Edn. p. 61.)

Again, talking of *Duck-Shooting*, Galton remarks:

"It is convenient to sink a large barrel into the flat marsh or mud, as a dry place to stand or sit in, when waiting for the birds to come. A lady suggests to me, that if the sportsman took a bottle of hot water to put under his feet, it would be a great comfort to him, and in this I quite agree; I would take a keg of hot water, when about it." (*Ibid.* p. 253.)

Talking about *Natives' Wives* as members of a party, Galton commends them as giving great life to a party and as being invaluable in picking up gossip, which will give clues of importance, otherwise often missed. He considers in a special paragraph the *Strength of Women*, which he finds adequate for the march, and adds:

"It is the nature of women to be fond of carrying weights; you may see them in omnibusses (!) and carriages, always preferring to hold their baskets or their babies on their knees, to setting them down on the seats by their sides. A woman, whose modern dress includes, I know not how many cubic feet of space, has hardly ever pockets of a sufficient size to carry small articles, for she prefers to load her hands with a bag or other weighty object." (*Ibid.* p. 8.)

Lastly while Galton admitted that men without independent means could turn travel to excellent account as in opening up new countries, finding natural history specimens or hunting for ivory, there is no doubt that he

[1] Nay, does he realise how widespread is still the ignorance of human history in the apparently 'educated' classes? During the few months that the cases containing objects bearing on man's development have been on view in the little museum of the Galton Laboratory we have received more than one remonstrance against the dating of a neolithic skeleton at 8000 years and of palaeolithic man at over 50,000 years, as incompatible with the 'well-known date' of the creation of the world!

thought, as most will hold too, that the ideal traveller was a man like himself:

"If you have health, a great craving for adventure, at least a moderate fortune, and can set your heart on a definite object, which old travellers do not think impracticable, then—travel by all means. If, in addition, you have scientific taste and knowledge, I believe that no career, in time of peace, can offer to you more advantages than that of a traveller." (4th Edn. p. 1.)

Such then is the *Art of Travel* planned as Galton himself states during his South African exploration of 1850–51. It deserves a new edition, brought up in substance and illustration to date—if the all-round knowledge, such as Galton had, still has its representative[1].

We cannot, however, leave the subject of travel without referring to two or three other enterprises in which Galton had a hand. Notable among these is the *Vacation Tourists and Notes of Travel*, of which he was the originator and editor. It was to be an annual volume and issues appeared for 1860, 1861 and 1862–63. The work Galton tells us just paid its way, and the idea certainly was, and might still be, a good one. The 1860 volume contained papers by W. G. Clark, Leslie Stephen, John Tyndall and others, besides one by Galton himself[2]. Much of the matter can be read with pleasure and profit to-day. How wonderfully wise in the light of recent experience seems W. G. Clark's talk with the Frenchman in Genoa over the latter's view that "there will be no secure and lasting peace for Europe until its political system is based upon the principle of nationalities." How this search for a definition of a 'nationality' might have warned President Wilson of the difficulties and the danger of the creed he was to propound 60 years later at Versailles! The accounts of early Alpine ascents, the Allelein Horn by Leslie Stephen, the Eggischhorn by Tyndall and the attempt on the Matterhorn by Hawkins are all still worthy of perusal. Galton contributed to the first volume a paper on Spain and the total eclipse of June 1860. He went out with a party in charge of the Astronomer-Royal in H.M.S. *Himalaya* and saw the eclipse from La Guardia. This was his first visit to Spain and he saw a good deal of the country, staying in the Pyrenees after the eclipse, and "here that remarkable madness of mountain climbing, to which every healthy man is liable at some period of his life, and which I had always believed myself to have gone through once for all in a mitigated form, began to attack me with extreme severity[3]." But while he gives us little account of his mountaineering, he takes up very seriously the question of sleeping or camping out at great altitudes, and gives a very full description of suitable rations for a six-days' outing, and above all of the knapsack sheepskin sleeping bags[4] of the French 'douaniers' or the frontier

[1] A great mass of material for a new edition was collected for and sent to Galton by the late Mr Howard Collins. It will be found in the Galton Laboratory Archives.

[2] Galton himself in his list of published papers, Appendix to *Memories*, p. 324, says the *Vacation Tourists* contained *two* memoirs by himself. I have failed to find more than one.

[3] *Vacation Tourists*, 1860, p. 446. Galton became from this date even to the end of his days a frequent visitor to the Pyrenees.

[4] With his usual desire to test practical efficiency, Galton carried one of these bags 1000 feet above Luchon and spent the night in it during a terrific thunderstorm! While familiar to

watchers against smugglers. With the Basque districts of Spain Galton was delighted. He writes:

"Every act of the people was original—their gait, their implements, their way of setting to work. I looked into many shops—such as tinkers', blacksmiths', potters', and so forth—and came to the conclusion, speaking very broadly, that if any of their patterns were introduced into England, or that if any of ours were made to replace theirs, the change would involve decided incongruity, and lead to questionable improvement. Another subject which struck me at once, and with which up to the last moment of my stay in Spain, I became no less charmed, was the graceful, supple and decorous movement of every Spanish woman[1]. It was a constant pleasure to me to watch their walk, their dress, and their manner, as it is a constant jar to all my notions of beauty to see the vulgar gait, ugly outlines, mean faces, bad millinery and ill-assorted colours of the vast majority of the female population that one passes in an English thoroughfare."

Galton contrasts the peasantry, especially the Basque peasantry of Spain, with the inferiority of physique, manner and address of the upper classes of Madrid society, and with conditions in England, where he tells us that "the higher classes, speaking generally, have the higher make of body and mind and by far the nobler social tone." But the peasantry in almost every land, if it has been long on the soil, appears to the visitor harmonious and even beautiful—think only of the Italian, the Austrian, the upland Baden and the Norwegian tillers of the earth, each admirable in their own way and each suited by centuries of selection to their own environment. The grace of an autochthonous peasantry, the suitability of their dwellings to their climate, of their clothing to their habits, and their artefacts to their domestic and agricultural needs, impresses us in the same way as the grace of a wild animal, adapted in every instinct and habitude to its native haunts, impresses us if we observe it unawares in its own surroundings.

The reader of Galton's paper will realise how he was beginning in 1860 to turn his thoughts more to man, and this also may be read between the lines of the account he gives of the public baby-dandle in Logroño:

"In the afternoon, the military were paraded, and the bands played in the square. Of course all the spare population went to see them; but what amused us especially, was the part taken by the nurses and the children, both here and at Vittoria. They came in hundreds, scattered among the crowd. The instant the music began, every nurse elevated her charge, sitting on her hand, at half-arm's length into the air, and they all kept time to the music by tossing the babies in unison, and slowly rotating them, in *azimuth* (to speak astronomically) at each successive toss. The babies looked passive and rather bored, but the energy and enthusiasm of the nurses was glorious. At each great bang of the drummers a vast flight of babies was simultaneously projected to the utmost arms' length. It was ludicrous beyond expression." (*V.T.* 1860, p. 436.)

Another feature of this travel paper is Galton's increased interest in meteorology and generally in climate. There is even a touch of it in his description of the corona during the eclipse; he is inclined to treat the

Arctic travellers, sleeping bags had not up to that date been used by Alpine climbers, and Galton at a dinner of the English Alpine Club was toasted as the greatest 'bagman' in Europe. *Memories*, p. 190.

[1] In a letter to his mother (July 19, 1860) Galton writes: "I cannot tell you how I enjoy Spain. The people are so civil and nice and *clean*. Italy won't bear comparison on the score of cleanliness with Spain. Everybody is happy and graceful and well-to-do." This letter contains an account of the eclipse and a rather brilliant pen and ink drawing of the corona.

corona as a mock halo. His job at the eclipse was to have been the taking of observations with an actinometer[1], but on the day before the eclipse, when the instrument was unpacked, it was found to be broken.

"I candidly confess that a rising feeling of exultation accompanied this discovery; I was not now necessarily obliged to spend the precious three minutes of the eclipse in poring on an ascending column of blue fluid in a graduated stem, and noting down the results by feeble lamplight, but I was free to enjoy to the full the whole glory of the eclipse." (*V.T.* 1860, p. 437.)

Galton decided to sketch the corona and to determine from its effect on colours the exact colour of the eclipse light about which there had been controversy. His account[2] of the eclipse is worth reproduction in part, if only for the originality of his views on the corona.

"2 hrs. 50 m. Indian yellow, cobalt and emerald green are lower in tone. I can distinguish all twelve colours perfectly. Light much fainter. 55 m. Light far fainter. I made a hole in a paper screen, and watched the crescentic image of the speck of sunlight that shone through it on the floor. The shadows were very dark and sharp. Air cold. 58 m. The numerous pigeons of the place began to fly home, fluttering about hurriedly, taking shelter wherever they could. There was something of a hush in the crowd.

At about 3 h.—I forgot to note the exact watch time, I am sorry to say—totality came on in great beauty. The Corona very rapidly formed itself into all its perfectness. It did not appear to me to *grow*, but to stand out ready formed, as the brilliant edge of the sun became masked. I do not know to what I can justly compare it, on account of the peculiar whiteness of its light, and of the definition of its shape as combined with a remarkable tenderness of outline. There was firmness but no hardness. In its general form, it was well balanced, but larger on one side than the other. It reminded me of some brilliant decoration or order, made of diamonds and exquisitely designed. There was nothing to impress terror in the sight of the blotted-out sun; on the contrary the general effect of the spectacle on my mind was one of unmixed wonder and delight.......The Corona-light sufficed abundantly for writing rough notes and for seeing my colours. Oddly enough, the *burnt sienna* and the *vermillion* alone ceased to be distinguishable from each other. Indian yellow had greatly lost brilliancy. I made a rough sketch of the Corona—it was too manifold in its details and too beautiful in its proportions for me, bad artist as I am, to do justice to it in the short time the spectacle lasted—yet the drawing which I made and which is given here [see Fig. 1, p. 9], is to my mind a fair diagram of this splendid meteor[3]. I drew it without taking any measurements to guide me, but simply as I would sketch any ordinary object. The uppermost part is that which was uppermost when I drew it. I used no lantern and required none; there was a sufficiency of light. The principal facts were, firstly, that the long arms of the Corona [see Fig. 2, p. 9] do not radiate strictly from the centre, neither are they always bounded by straight lines[4]. The upper edge of *a* was truly tangential, that of *d* and others nearly so; *c* was remarkably curved, and so was the lower edge of *b*, though less abruptly; it was like a finch's beak, and remarkably defined. Secondly the shape of the Corona was not absolutely constant; speaking generally, it was so; but in small details it appeared to vary continually, by a slow diorama-like change. There was no pulsation or variation of intensity, visible in its light. I was particularly impressed by its solemn steadiness.

[1] Galton had previously to his departure been instructed by Sir John Herschel in the use of this invention of his.

[2] His sketches and other details appear in the Royal Astronomical Society's *Memoirs*, Vol. XLI, pp. 563–4, as well as in *Vacation Tourists*.

[3] Elsewhere in this paper (p. 423 "each phenomenon of that strange and magnificent meteor") Galton uses this word in the sense of an unusual atmospheric appearance.

[4] Even Sir George Airy doubted the curvature Galton gave to some of his rays, but photographs of subsequent eclipses have confirmed the curved rays. There was no photograph of this eclipse, the first probably at which photography was possible, although a photographer was present. He inserted his slide and exposed, but had forgotten to put a *plate into his slide!*

It seemed scarcely possible to believe that the light of the Corona was other than the rays from the sun made visible in some incomprehensible manner round the edge of the moon, the appearance being eminently suggestive of a brilliant glistening body, hidden behind a screen. The nearest resemblance I can think of, to express my meaning (not that I am to be understood as supposing the remotest analogy between the causes of the two appearances), is the effect of a jet of water, playing from behind against some obstacle, and throwing an irregular halo of spray around it, on all sides. That a reasonable foundation may exist for ascribing the Corona to some diversion of the ordinary rays of the sun, however unintelligible the cause of this diversion may be, and not to a luminous atmosphere surrounding the sun, was powerfully impressed on me by certain appearances that were observed when totality was passed: they were these. Four or five minutes after the reappearance of the sun, Mr Atwood called attention to remarkable luminous radiations, like sunbeams slanting through a cloud, and proceeding in narrow but long brushes from the cusps of the sun. They changed their angular directions and even their shapes with such rapidity, that I was almost bewildered in a first attempt to draw them. If I looked down on my paper to draw a few strokes, the appearance had become changed, when

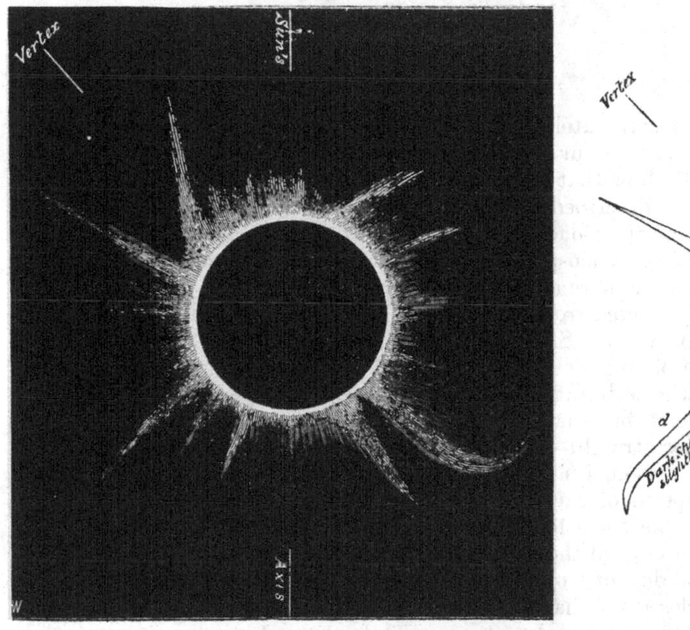


Fig. 1.

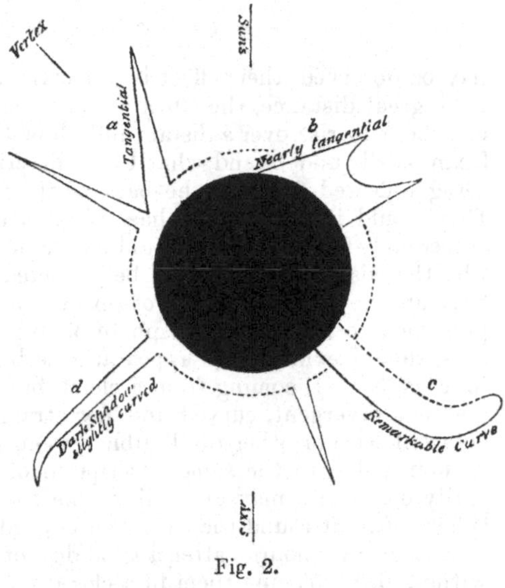


Fig. 2.

I again raised my head. Nevertheless between 3 h. 11 m. and 3 h. 13 m. I managed to make three sketches; the two that were most characteristic are here very fairly represented. After 3 h. 13 m. the light of the emerging sun was too strong to admit of further observations. The brushes were perfectly distinct and unmistakeable, they were best seen by holding up the hand so as to mask the sun, and they were perfectly visible through the telescope when it was so turned as to exclude the sun. There was no mistake whatever about their existence. I trust that the attention of observers of future eclipses will be directed to them, both before and after totality. Now whatever may have been the cause of the brushes, would also, I should guess, be competent to create the greater part of the Corona; the two appearances being of identically the same genus. It will be observed that the brushes in Fig. 3[1] enclose an angle of about 130°, on the side of the emergent sun, and that the same angle had changed to about 195° in Fig. 4, to say nothing of the appearance of a central bar of light. The angular change of the brushes was continuous, so long as I had an opportunity of looking continuously at them.

I have since often looked for, and have only just seen (Sunday Feb. 10th[2]), an almost precise

[1] See Figs. 3 and 4, p. 10. [2] Presumably 1861.

representation of these appearances, in the case of small black snow-laden clouds sailing before the sun. When the clouds are in any way transparent, though some indications of these brushes

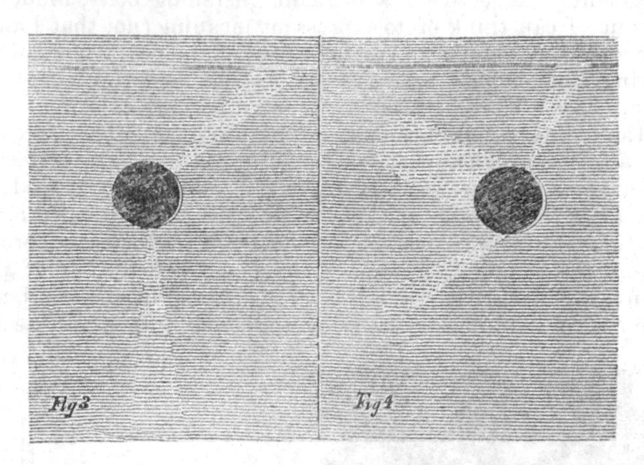


may be observed, their effect is proportionately feeble, and if the sun be masked by an object at no great distance, the effect does not occur at all. The common artist representations of the sun about to rise over a distant hill, show that these appearances are generally recognised. Now I can hardly understand what I have described, on any other supposition than that of sun-beams being reflected from off the back of the cloud at a very acute angle athwart the line of sight. They would illuminate the haze of the atmosphere through which they passed, and being seen exceedingly foreshortened, would be the more apparent. But here I stop. I do not comprehend why the wisps of light should be projected from the cusps of the uncovering sun, and therefore have an apparent movement of revolution. Still less can I understand why the moon, which is presumed to have no atmosphere of any description[1], capable of being illumined by passing rays, should exhibit this appearance so beautifully. When I shall have seen wisps of light, as in Figs. 3 or 4, coming from a cloud, but shaped in any way like those of Fig. 1—convergent and not divergent, curved and not straight—whether owing to irregular distribution of the adjacent haze or other intelligible reason, I shall hardly resist feeling satisfied that the Corona is mainly due to the same description of cause that produces them, whatever that cause may really be. There may, in addition, be some luminous effect produced by an enveloping atmosphere of light round the sun, seen beyond the edges of the eclipsing moon......

As to my colours: after a good deal of trouble, I find I can reproduce the exact effect that I witnessed, by placing them in a closed box having a dark ceiling, and admitting a faint white light at a low angle. I then view the colours also at a low angle through a piece of dull yellow glass. All these details seem essential to effect: they are in some sort, the equivalents to a yellow sky near the horizon, and gloom above head." (*V.T.* 1860, pp. 440–4.)

We have given this long extract from Galton's paper because it shows not only the working of Galton's mind at the time, but is very characteristic of the general manner in which he approached problems. He thought and reasoned about things for himself even when they might lead him astray. His *curved* corona rays have been confirmed. He alone noted that cusp rays were still visible when the crescent was masked by the hand. Galton's observations (but not his inferences from them) will be found in Ranyard's "Observations made during Total Solar Eclipses" (*Memoirs of R. Astron. Soc.* Vol. XLI, 1879), where his sketches are reproduced (pp. 563–4) in more finished form.

[1] The brushes according to Galton's sketch extended to three times the moon's diameter.

PLATE III

42, Rutland Gate, S.W. Galton's home from 1857 to 1911. (House with *white* door.)

The trusty Gifi, Galton's Swiss Servant. ("Mr Galton is at home, Sir, and has been expecting you.")

Galton himself says[1] of the *Vacation Tourists*, that excision was often an unwelcome duty, and illustrates it by the statement that among the contributions offered for one volume were thirteen separate descriptions of sea-sickness! Yet the volumes have something of the charm of leisurely mid-Victorian journalism; and should not be allowed to pass into complete oblivion.

In 1864 Galton wrote for John Murray a knapsack guide for Switzerland[2], which just deserves mention under the heading of travel. It reached a second edition, the one I have examined. The late Mr John Murray paid Galton £150 for the copyright. It was one of a series of four (Switzerland, Italy, Norway and Tyrol) which have since passed out of sight. The general plan is very much what we now associate with Baedeker, and the hints as to hotels and the character of landlords were more or less original in those days. Galton's name is not associated with the work and there is little to identify it with his personality. He does not mention it in the *Memories*, it is omitted in lists of his books and memoirs, and the present writer never heard him refer to it. It is nevertheless a substantial piece of work. How and when did Galton obtain his knowledge of Switzerland? The answer may be found in the brief yearly records of "Frank's Life" and "Louisa's Life" on opposite pages from 1830 to 1853, and then carried on in common by Mrs Galton until her death in 1897, which year is written by Francis Galton himself. From this *Record*[3] we find that not only was a considerable portion of the wedding tour devoted to Switzerland (1853), but in 1856 the Galtons were in Switzerland and the Tyrol. In 1857, 1861, 1862 both were again in Switzerland, and in 1863, Francis Galton, probably to complete the knapsack guide, was alone in Switzerland. Thus his experience was fairly ample for a guide which was intended not for the high-peak climber, but for the 'Thalbummler,' or for the tourist in the broader sense.

The autumn travel often extended to two or three months, and the visits to Mrs Galton's family at Gayton or Julian Hill, or to Francis Galton's relatives at Leamington, Claverdon or Hadzor were a constant feature of the Galtons' life; they consumed much time, but had no doubt compensating advantages especially as the health of both was at times indifferent. Beyond these travels and visits social life is often referred to, and the names of the Russell Gurneys, the Gassiots, the Norths and of Spottiswoode begin to appear in the diary. After their return to England in 1853, the Galtons had occupied lodgings in Portugal Street; then they lived at 55 Victoria Street, Westminster, and finally in 1857 they took possession of the house in Rutland Gate, which remained Galton's home till his death in 1911, and is the environment with which most of his surviving friends will chiefly associate him. The light and airy, white enamelled drawing-room, with its furniture of many periods and styles; the long dining-room with its bookcase at the back, Galton's working table[4] in the front window, and on the walls the prints

[1] *Memories*, p. 187.

[2] *The Knapsack Guide for Travellers in Switzerland.* New edition revised, 1867. "Frank busy editing Murray's Handbook." *L. G.'s Record*, 1864.

[3] In future to be cited as *L. G.'s Record.*

[4] Now in the *Galtoniana* of the Galton Laboratory with his writing chair.

of Galton's friends—Darwin, Grove, Hooker, Brodrick, Spencer, Spottiswoode[1], etc.; the dark back room with its shelves loaded with pamphlet cases filled with letters and manuscripts, the boxes of models, and the notes of a long lifetime of collecting, mostly indexed by Galton himself, all these will be familiar memories to his friends, and formed a singularly unique environment, very characteristic of the man. A great reference library, one of the principal rooms of the house devoted to a study, reproductions of modern or medieval art, these were not essential needs of Galton's nature. Among his books were no long series of foreign journals or transactions, and but few fundamental treatises on anthropology or natural history. His library consisted chiefly of books which their writers presented to him—such as Darwin's works—or of offprints and papers sent to Galton when his name had become known. There are scarcely two dozen books in Galton's library as we now have it, which we can assert he must have purchased to forward his work. There are masses of measurements and observations of Galton's own, but unlike Darwin he did not start by analysing published material. He collected afresh either directly or through others and formed his conclusions *de novo.* I do not think he ever studied Laplace or Poisson; I am confident that he had never considered the original papers of Gauss; even while Galton's work seems to flow naturally from that of Quetelet[2], I am very doubtful how far he owed much to a close reading of the great Belgian statistician. He formed no collection of his books, and the few references to Quetelet in Galton's writings are such as might easily arise from indirect sources. Galton took up his problems one after another and worked at them largely disregarding their past history, when indeed they had one. This is only possible in a man of great insight and brimming over with suggestive ideas and novel processes. But the method has some drawbacks, when adopted by lesser men, and even, as in his above-cited account of his observations on the Corona, we may find it open to criticism in Galton himself. But unless my readers grasp this characteristic of Galton's nature they will fail to understand how, with all his travel and the social and executive calls on his time, he was yet able to accomplish so much. It is dangerous advice to give to *every* scientific worker, but it is the only useful advice to give to a young man of genius: Find a little trodden path, and explore it rapidly alone, without regard to the work of others; many precious hours will be wasted if you follow up the spoor of each one who has passed athwart your path before. Galton took some time to find his individual track, but having found it, he went ahead without much regard to forerunners or even to those working on parallel lines. It was this individuality of method which impressed itself on his environment and rendered him so independent of the usual appurtenances of the scholar. He thought and he worked with the simplest of tools, and these mostly of his own making.

[1] Hung at the present time on the walls of the Committee Room in the Galton Laboratory. The family portraits which hung in the drawing-room and Galton's bedroom may still be seen on our walls, and even the quaint stunted cupboard wardrobe from Galton's dressing-room now serves as a store for mechanical calculators—a use which would have delighted his heart!

[2] In a card index prepared by Galton himself to his books and pamphlets the name of Quetelet does not appear.

I have interpolated this paragraph into our history of the man himself, because while those travels, that social life and that scientific executive work[1] went on continuously, I shall only refer to them again very incidentally, they serve merely as a back-ground to the intellectual life of the man with which we shall in this second volume be principally concerned.

The reader will perhaps have observed that the year 1855 does not occur on our p. 11 as one in which Galton visited Switzerland. It was the year of the Crimean War, a year of grave depression for all those who had the national welfare at heart. The crass ignorance which rules in high places, the criminal want of preparation characteristic of nearly all British executive bodies showed themselves in the general breakdown of 1855, as intensely as they did in 1914, or in the Boer War of 1900. The shame of 1855 is almost forgotten now in the light of more recent impressions. But it led the patriot, man or woman, of those days to cry: "How can I aid this helpless, foolish country of mine? What can I contribute that it lacks? How is this brainless executive to be pushed on to firmer ground?" And men and women stepped out of their seclusion and their studies in 1855 in their tens, as they did in the last war in their hundreds, and demonstrated that the nation's real strength lay in its reserve of brain-power, and not in its political leaders and the paid servants of the government. The world rung with the glorious work—the almost Joan d'Arc task of Florence Nightingale; the inner circle might know that her greatest services to the nation were not those which caught the public imagination; but the public were right in identifying its ideal with a definite personality, above all with such a marked one as that of the 'Lady of the Lamp.' But at the time of the Crimean War as in recent years there were others also who asked themselves what is my *métier* and how can I supply in one way or another what the nation lacks? Among these self-questioners was Francis Galton.

He heard of the terrible sufferings of our soldiers in the trenches, due in the first place to the ignorance of their officers, men who in the majority of cases had had no experience of bivouac and camp. Galton realised the need of our armies in one way as Florence Nightingale did in another. His own words best express the situation:

"The outbreak of the Crimean War showed the helplessness of our soldiers in the most elementary matters of camp-life. Believing that something could be done by myself towards

[1] Council of the Royal Geographical Society 1854–93, Secretary 1857–63; General Secretary to the British Association (at first with W. Hopkins and then with T. Archer Hirst) 1863–67, President of the Geographical Section 1862 and 1872, President of the Anthropological Section 1877 and 1885; he was twice invited to be President of the Association, 1890 and 1905, but declined on the ground of health and strength. ["They wanted to nominate me as President of the British Association for 1893, but I have definitely declined, as I did for 1891 [?1890], being out of my element in dining out day after day, and making speeches, which I detest. Besides I am too deaf to do the ordinary presidential duties well." *Letter to Sister Bessie*, Feb. 13, 1892.] He was Chairman of the B. A. Anthropometric and of the Local Societies Committees in 1883; elected Fellow of the Royal Society 1860; Member of Council 1865–6, 1870–2, 1876–7, 1882–4, and Vice-President in the last three series, Member of the Kew Observatory Committee 1858, Chairman on the death of De la Rue 1889 until 1901. Meteorological Committee 1855. Meteorological Council 1901. President of the Royal Anthropological Institute 1885–88; Chairman of the Royal Society Committee on Evolution 1896, etc. etc.

removing this extraordinary and culpable ignorance, I offered to give lectures on the subject; gratuitously, at the then newly-founded camp at Aldershot. As may be imagined from what is otherwise known of the confusion of the War Office at that time[1], no answer at all was sent to my letters, until I ventured to apply personally to the then Premier, Lord Palmerston, who at once caused me to be installed. It is evident from my old notebooks that I worked very hard to frame a suitable course of practical instruction and of lectures for those who cared to profit by them[2]."

[1] Rather at all times. The War Office should have inscribed over its portals: Crimea, 1855; South Africa, 1900; Flanders, 1914; with a blank space for its future achievements.

[2] *Memories*, pp. 163–4. I have found a printed document in our *Galtoniana*, marked "Printed for Private Circulation only," which is probably the actual form of Galton's request to the War Office. It is dated 4th May, 1855 and is entitled *Ways and Means of Campaigning.* It begins:

"The helplessness of our soldiers, when they are thrown for awhile upon their own resources, has been so frequently insisted on and deplored in the Evidence taken before the Sebastopol Committee, and in speeches in both Houses of Parliament that, while it becomes impossible to doubt it as a fact, there arises a serious question whether, and in what way, we should attempt to remedy it.

"Now as matters bearing upon this question have been my special study in extended travel, so far as to have induced me to write upon them, quite irrespectively of the present war, I thought myself justified in communicating to the military authorities a scheme that I had matured to meet the present occasion, and, whilst my proposal remains under consideration, I embrace the opportunity of putting what had been scattered over many pages of writing, at different times, into the present condensed and legible form......I have offered my gratuitous services in organising a SCHOOL OF INSTRUCTION in such of the ways and means of campaigning as fall under the following heads: 1st The best of those MAKESHIFTS AND CONTRIVANCES which those people adopt who have been thrown on their own resources in all parts of the world. 2nd The elements of those HANDICRAFTS which experience has shown to be most needful in those circumstances.

"My wish is to reduce the teaching of these matters to a regular system. I am quite convinced it can be done, and that an interesting and very useful course in them could be afforded to the army generally, at a very small cost and without clashing with their regular duties. I seek for an opportunity of proving this practically. If I succeed in doing so to the satisfaction of our military rulers, they might extend the system as widely as they pleased, and my classes would have instructed a number of persons who would afterwards be qualified to teach. As yet the matter is a novelty; no one can point to experience and say: 'These things are the best to be taught—this the best way of teaching them—such is the time required for a good practical instruction—such the likelihood of the course being a popular one. But if the experiment has been once set on foot, even so short a period as two months, would go far towards deciding these points, and affording sound ground for future plannings."

Then follows the suggestion of beginning at Aldershot and a summary of what the contents of the course might be, and the paper ends with a sincere hope that the Military Authorities may think fit to countenance a scheme which requires a mere trifle of material support.

The *Times* of September 25, 1855 devoted two columns to Galton's work on travel and campaigning in a very sympathetic spirit. "The camp of Aldershot" it wrote "at this moment is the scene of a remarkable experiment. Mr Galton, a gentleman of considerable experience in the shifts and contrivances available for travelling in wild countries, has obtained the permission of Lords Panmure and Hardinge to tenant two huts with an enclosure adjoining for the purpose of communicating his experience to the British soldier. His services are rendered gratuitously at present, for his arts are untested and his teaching is a novelty. But the day will perhaps come when their value will be recognised and he himself be duly installed as our first Professor of Odology [? Hodology]. *A priori* we should say that a traveller may be competent to supply a void in military education. A soldier, like a traveller, may be thrown upon his own resources, but without having acquired a traveller's self-dependence. In such circumstances he may eventually obtain the knowledge which is prized as the qualification of an old campaigner, but necessity will be his teacher and his apprenticeship will be cruel......*Au reste*, whatever his success may be in his immediate undertaking Mr Galton has the merit of having

General Knowles then in command gave Galton all the aid in his power, two huts were placed at his disposal and he took a small house, Oriel Cottage, two miles away and walked to and fro daily to his work. He started in July and, except for a short visit to Paris in September, continued to lecture until the following spring.

The syllabus of Galton's lectures in 1856 lies before me, and on p. 16 is shown the first page. The Lectures were to be illustrated by Pictures, Models and Experiments.

"*Water* deals with methods of finding as well as using (purifying and filtering) and carrying, greased canvas bags, skins, etc. *Fire* with 'Lucifer matches,' burning glasses, fire-sticks, sulphur matches, etc. *Bivouac* with natural and artificial screens, sleeping bags from blankets sewn along edges, sleeping when in urgent danger, etc. *Food* with proper proportions of fat and meat, cooking, grinding, preserving, fishing, game, etc. *The March*, watching, hearing, tracking, scouting, prisoners, sore feet and drying clothes. *Rivers and Bad Roads* with temporary bridges, fords, swimming cattle and waggons, rafts, rough boats, steep pitches and waggon brakes, etc. *Crafts and Mechanics* with felling timber, seasoning, making axles, lathes, bending wood, case hardening, fuel, turning, soldering, rude capstans, pulleys, knots, etc. *Animal Products* with bones, horn, catgut, bladders, hair, shells, hides, charcoal, glue, oil, candles, soap, etc. Next we learn about writing materials, substitutes for paper, and ink, secret writing, inscriptions for secret information, conveying letters, etc. *Animals of Draught and Burden* deals with hobbling and tethering, watering of cattle, nosebags, pack saddles, saddlery, waggon harness and waggon-mending, accidents to waggons and animals, etc. *Tents and Hutting* describes various kinds of tents and material, best place for pitching, action of rain and dew on tents, huts rude and more elaborate, whitewash and plaster, seats and tables, windows, floors, etc."

Much of this syllabus, which I have given in very abbreviated form, will be familiar to the reader of the *Art of Travel*, but the matter was adapted to the special needs of the army and to a special audience.

Galton writes in his *Memories* (p. 164) with undue modesty of the attempt he made in 1855–6 to teach the soldier the art of campaigning which at that time appeared to form no part of the military curriculum[1]. The sluggish War Office at last seems to have recognised its value, for two years after the war was over it caused to be constructed and distributed to various centres ten sets of cases of models and specimens illustrative of Galton's lectures to be prepared after the design of a set which he himself had made and presented to Woolwich. These cases of models were accompanied by a catalogue of 20 pages prepared by Galton himself, this being an enlargement of the above-mentioned syllabus. In the preface he writes:

"It is trusted that they may not only serve to interest and instruct the soldier, but especially to suggest to officers, who take an active part in educating their men, the precise subjects on which practical classes in the Arts of Camp Life may most usefully be employed......An old

framed a code of rules for situations in which Englishmen of all men are most apt to find themselves. The most popular type of an Englishman ever conceived is that of an isolated self-relying unit; and for this individual Mr Galton has formed a manual, which he might properly term 'Robinson Crusoe made Easy.' He offers a consistent rule of life to the vagabond, and settled principles to the restless wanderer. In this sense he converts the savage into the sage and makes the whole wilderness blossom with his red handbook [i.e. *The Art of Travel*]."

[1] I have before me the list of officers and others who attended his lectures in July, August and September 1855, and if the lectures after the first were not crowded, the total number who attended was not insignificant. Galton had never had the training that an academic teacher acquires, and his formal lectures, at any rate in his later life, were not as good as his talks and demonstrations.

# ARTS OF TRAVELLING AND CAMPAIGNING

THE FOLLOWING
COURSE
OF
PUBLIC LECTURES
WILL BE DELIVERED AT
THE CAMP AT ALDERSHOT,
BY

## FRANCIS GALTON Esq., F.R.G.S.

*Author of the "Art of Travel" and of "Explorations in Tropical South Africa"*

The Lectures will commence at HALF-PAST SEVEN, and will be delivered on WEDNESDAY Evenings, as follows:—

| *Page* | *Subject of Lecture* | *Day* |
|---|---|---|
| 3, | Water for drinking . . . . . . . . | Jan. 16 |
| 4, 5 | Fire. Bivouac . . . . . . . . | Jan. 23 |
| 6, | Food . . . . . . . . . . | Jan. 30 |
| 7, 8, 9 | The March. Rivers and bad Roads . . . . | Feb. 6 |
| 10 | Crafts and Mechanics . . . . . . . | Feb. 13 |
| 11, 12, 13 | Bush Manufactures . . . . . . . | Feb. 20 |
| 14 | Animals of Draught and Burden . . . . . | Feb. 27* |
| 15 | Tents and Hutting . . . . . . . | March 6 |

Mr Galton will also be present in the Lecture Room, from 10½ to 12 o'clock, on the THURSDAY Mornings, where he will be happy to explain any matters connected with the Lecture of the preceding Evening; and to repeat it at 11 o'clock, in the event of there being a sufficient attendance.

LONDON
PRINTED BY T. BRETTELL, RUPERT STREET, HAYMARKET

1856

* Had Galton overlooked that 1856 was Leap Year?

campaigner's acquirements consist partly in knowledge and partly in handiness. Field lectures, illustrated by experiments, may convey the first to an intelligent novice, and these models will explain what kind of things must be made by his hands, before he can acquire the latter.

FRANCIS GALTON.

42 Rutland Gate[1],
*April* 5, 1858."

Only one of Galton's Aldershot lectures, the inaugural one "on the opening of his Museum and Laboratory in the South Camp, V. Nos. 18 and 20," was, I believe, printed. It was issued by John Murray in 1855 as "Arts of Campaigning, an Inaugural Lecture delivered at Aldershot." From this lecture it would appear that one of the huts was turned into a museum, illustrating by sketches and models and a small library the arts of campaigning; it was open from 1.30 to 6.30; the second hut was a workshop, and a place for storing tools.

"Next as regards teaching the *hand*. I am collecting a motley stock of very simple tools and raw materials, planks, logs, twigs, canvas, cloths, and everything necessary for making with the hand those very things that you will see pictured in the museum; I urge you to come and make use of them. In the palisadoed plot of ground, between the huts, you can sit and work just as roughly as you would in the Crimea, and you will from time to time have intelligent workmen to assist you in your difficulties, and explain the use of the tools you work with....There is no habitable country so wild and so inhospitable as not frequently to afford ample materials for making each thing I have mentioned. But unless we learn to draw our supplies from nature, and not through the medium of manufactories, we may sit with our hands folded in unwilling idleness, and complaining of want when we are really in the midst of abundance, and surrounded by opportunities of using them......I hope that these huts may be looked upon more as a laboratory where learners may teach themselves, which is the best kind of learning,—rather than as a place where they are formally taught. I wish to make it a kind of head-quarters of the knowledge of those shifts, contrivances, and handicrafts that are available in camp life; and I call upon you to help me with your assistance. Write to your friends from the Crimea, or from the bush, who take an interest in these things, get hints of original experiences from them, and communicate them to me; they will not lie idle, but will at once be turned to account in increasing a store already large, and will remain recorded in pictures or in models for the good of ourselves and all who follow us."

Throughout Galton exhibits his innate modesty; asking for help rather than offering to teach, he proposes expeditions to distant points of the heath to illustrate camp contrivances. He endeavours to give a thoroughly practical turn to his instruction, avoiding scrupulously all that was simply fanciful.

About the same time as these lectures, but at a date unknown to me, Galton gave a lecture on the Art of Campaigning at the United Services Institution in Whitehall Yard. On this occasion according to the *Times* report (cutting without date) "there were present many wounded officers from the Crimea, and the gallery was filled as usual with non-commissioned officers from the Guards, Artillery, Household Cavalry and other troops forming the garrison of London and Woolwich." Galton showed experimentally how a tree might be cut down and turned into a pole without tools and a hole dug in the hardest ground for it without a spade or other tool than a small stick or iron ramrod. He lashed a common clasp-knife to a

[1] This is the first published paper dated from Rutland Gate.

piece of wood and made a spokeshave and so obtained shavings to make a very comfortable bed. Further he explained and demonstrated experimentally how a tent peg might be fixed in loose sand drift to give a resistance of 70 to 90 lbs., etc. He was reported as having been listened to with great attention.

In 1912 one of Galton's cabinets of "Models illustrative of the Arts of Camp Life," which had found its way to the South Kensington Museum, was transferred to the Royal Geographical Society, where it still is, an object of interest in the museum, if hardly of study; whether any of the others have survived I do not know. Probably Galton's activities to some extent helped to relieve the situation in the Crimea, and doubtless had his work been done in 1917 instead of 1856, he would have been offered an O.B.E.

Two memoirs by Galton probably arose from this association with military campaigning. The first is that describing his invention of the Hand Heliostat, an instrument for the purpose of flashing sun signals[1], and the second "On a New Principle for the Protection of Riflemen[2]." The latter would have been more effectual with the spherical bullets and the lower muzzle velocities of those days than with modern trajectories; it depends on the well-known fact that owing to the resistance of the air, the trajectory of a shot is not symmetrical about its highest point, and accordingly a shot fired from $A$ to $B$, and another from $B$ to $A$, do not follow the same curve. It is accordingly possible to intercept one of these and not the other by a properly placed screen.

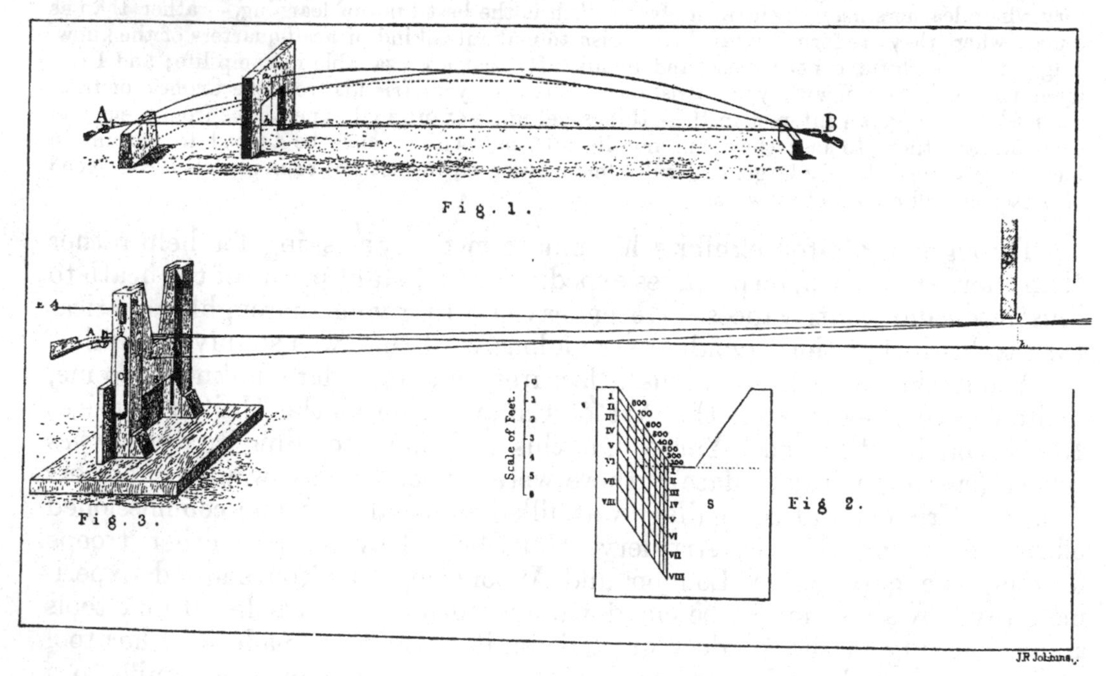


[1] *British Association Report*, 1858, part II, pp. 15–17; also with diagram in *The Engineer*, Oct. 15, 1858, p. 292; *R. Geog. Soc. Proc.* IV, 1860, pp. 14–19.

[2] *United Services Journal*, 1861, Vol. IV, pp. 393–6.

Fig. 1 (p. 18) explains, of course without the details, Galton's idea. I do not know whether it was ever dealt with experimentally; the extremely flat trajectories of modern bullets, the fact that *A* would have to protect himself not against a single rifleman at a definite range *B*, but against a number of riflemen in different and only vaguely known positions, tells against the method. Even in the case of a single opponent *B*, the latter might, by shifting his position somewhat, actually use *A*'s protection as a protection for himself. Still it is conceivable that the method might be of service in trench warfare, when the ground in front had been accurately ranged, and the danger of drawing artillery fire by the screen had been if possible overcome.

Galton's other paper, namely on the heliostat, is of more direct interest as indicating the mechanical bent of his mind. He had been interested in heliostatic work since his Long Vacation reading party at Keswick in 1841 when he

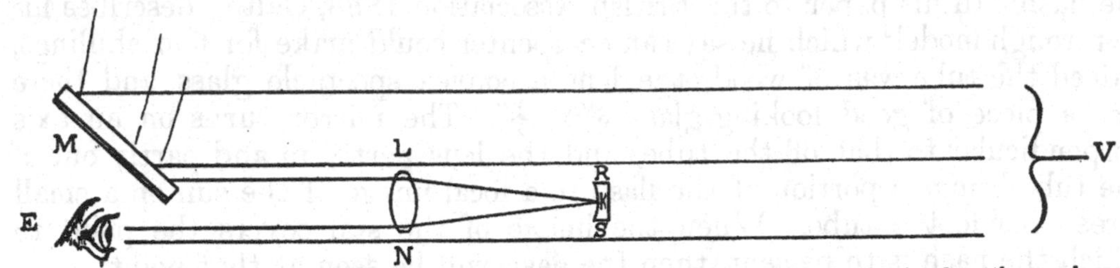


**Fig. A. From Galton's original drawings for his heliostat. Diagrammatic figure indicating how the mock-sun is formed and seen covering a portion of the field of view at *V*. The small screen *K* only intercepts a small portion of the field of view. Cf. Fig. 1, p. 20.**

**Fig. B. From Galton's original drawings for his heliostat. Field of view of the telescope with the mock-sun covering the point of a promontory to which the instrument would flash light.**

found on the top of Scawfell a party of ordnance surveyors endeavouring to get into touch with Snowdon and obtain its bearing by aid of a heliostat[1]. Galton's ambition was to construct a pocket heliostat and he spent much time in preparing models, of which the Galton Laboratory possesses almost

[1] *Memories*, p. 61.

as many as it does of his 'hyperscope,' a forerunner of the modern periscope[1]. Galton's hand-heliostat while described in 1858 in a form which he himself carried in a large waistcoat pocket and which he considered efficient up to ten miles and said would on many occasions have been most valuable to him in Damaraland was, in a larger size and with a stand, ultimately manufactured by Messrs Troughton and Simms under the name of *Galton's Sun-Signal.* According to the former Admiralty Hydrographer, Sir William Wharton, it was used quite recently in nautical surveys to enable shore parties to make their exact whereabouts visible to those on the ship[2]. The principle on which Galton based his heliostat is a fairly simple one. He intercepted a small part of the flash and by aid of it created an image of the sun in the field of view of a telescope; this image was then thrown on any required point of the landscape, and when so thrown any one at that point would see the flash. In his paper to the British Association 1858, Galton describes his own rough model[3] which he says any carpenter could make for four shillings, indeed the tube was of wood, the lens a convex spectacle glass, and there was a piece of good looking-glass $3'' \times 4\frac{1}{2}''$. The mirror turns on an axis perpendicular to that of the tube, and the lens partly in and partly out of the tube brings a portion of the flash to a focal image of the sun on a small screen inside the tube. When the image of the sun covers the point to which the flash is to be sent, then the flash will be seen at that point.

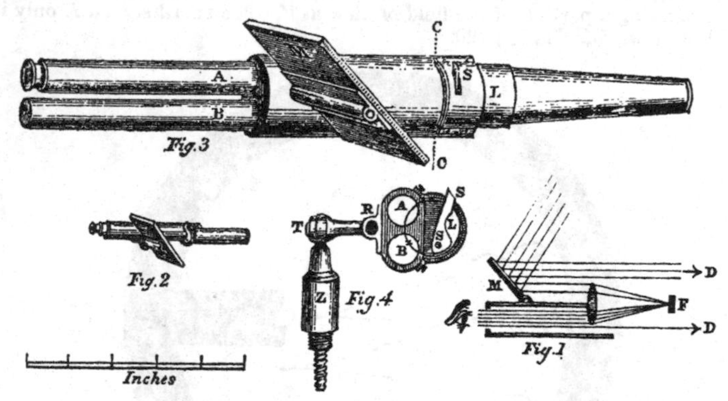


Fig. 1 explains the working; *M* is the mirror with the sun's rays falling on it and reflected in direction *D*, *F* is the screen at the focus of the lens, which is seen by the eye as superposed on the object at *D*. Fig. 2 is a simple pocket form; Fig. 3 a more elaborate form, which has a theodolite telescope *A*, and a plain tube *B* as a finder. Fig. 4 shows the section of Fig. 3 at *C* with a holder which can be screwed on to a camera tripod. A fairly

[1] Both are arrangements of parallel mirrors set at an angle of 45° to the axis of a square tube (generally of card!) with a hole in the opposite walls facing each mirror. Galton designed them in order to see a ceremony over the heads of a crowd or to inspect what lay beyond a high wall. The instrument has also been called the altiscope, and the principle of the modern periscope of the submarine is identically the same. Hyperscopes, probably not under this name or with any knowledge of Galton's early work, were used in the trenches in the course of the recent war.

[2] *Memories*, p. 165.

[3] In the Galton Laboratory.

full account of the instrument is given in the paper of November 28, 1859, on "Sun-Signals for the use of Travellers," and ample directions for its use are provided in the instructions for Galton's Sun-Signal in the larger size which accompany the instrument as it was made by Messrs Troughton and Simms.

Two other researches belonging to this period (1860–62) exist among Galton's voluminous papers. I do not think they were ever printed. In the first he considers the bulk of gold in the world—i.e. in currency, ornaments, etc. This in 1800 was estimated at £225,000,000. Galton computes the volume of pure gold therein and concludes that it would occupy 3053 cubic feet. "Hence my room without extra window space, but disregarding curve at corners of cornice, would hold more gold than was extant in 1800 by 94 cubic feet."

The second paper contains a suggestion of how to reach a decimal coinage for England by the introduction of two new coins, the 'mite' or 'quint,' a fifth part of a penny and the 'cent' or 'groat' of 12 quints. The groat would thus be $\frac{1}{100}$ of a pound, and the florin = 10 groats the intermediate link. The object of the mite or quint was to get the existing coins in easy terms of groats and quints. Thus:

| one penny | threepence | sixpence | one shilling | two shillings |
|---|---|---|---|---|
| = 5 mites | = 15 mites | = 30 mites | = 60 mites | = 120 mites |
| | = $1\frac{1}{4}$ cents | = $2\frac{1}{2}$ cents | = 5 cents | = 10 cents |

Galton put his scheme before Archibald Smith whose gravest objection against it seems to have been that the smallest ultimate divisions were not but ought to be binary.

### B. GEOGRAPHY

While Galton was thus turning his travel experience and his study of the works of travellers to national use, he did not overlook geographical research. In 1855 he had contributed a paper of thirty pages entitled "Notes on Modern Geography" to a work issued by Messrs Parker and Son entitled *Cambridge Essays contributed by Members of the University.* I have not succeeded in discovering the real origin of this work, but if all the essays came up to Galton's in suggestiveness, it must be a matter of regret that it has passed out of recollection. Galton's aims with regard to geography were of a three-fold character, namely: (*a*) to encourage geographical research by travel, and to make it easier by suggestion of methods and instruments to travellers, (*b*) to make geography a school and academic study, and (*c*) to revolutionise and humanise maps. The object of the essay just referred to was undoubtedly to popularise modern geography, but throughout his interest in the improvement of maps is dominant.

"There is usually" he writes[1] "as great a difference in geographical value between an ordnance map and, it may be, a beautifully engraved, popular one, as there is in poetical merit between a copy of Shakespeare and a gorgeously bound volume of the vilest trash that was ever published by aid of titled interest and half-extorted subscriptions."

[1] *Loc. cit.* p. 91. The *Cambridge Essays* were issued from 1855 to 1858; their pages excluded 'scientific' subjects according to the preface (? written by the publisher) so that they might

But Galton wanted more than the accuracy of the ordnance map, he wanted a pictorial map, a bird's-eye view of a coloured model.

"It is hardly to be expected that travellers should always find it advisable to draw up for publication large pictorial charts of the routes they have travelled, but duplicates of their sketches and surveys would be a very valuable acquisition to the records of Geographical Societies, where they could be studied by map-makers, who wished to compile a pictorial chart of the country in which they lay. It would, I should think, be a very interesting task to endeavour to map a district on this method, and the result would be sure to be a gratifying one, if the traveller had the eye and the touch of an artist[1]. The strictly accurate, but meagre information that is afforded to a student by ordinary maps is more tantalising than satisfactory. A blind man fingering a model could learn as much from his sense of touch alone, as they convey to our eyes. They are little more than an abstraction, or a ghost of the vivid recollections with which the memory of the traveller is stored, not that these recollections are very varied or shifting—one image succeeds another in rapid changes—but that the somewhat stereotyped survey which the mind recalls when it attempts to image to itself the features of a once-visited country, is a matter of colour and blaze of sunshine, and dancing waters and quaint crags or well-marked headlands, and here and there stretches of level land clothed with russet forests or lying open in tawny plains. It is surely not too much to expect that at least some allusion to these features—which are everything to the memory, which are precisely what every traveller whom we address is mentally referring to as *his* map, whilst he answers our questions—should find a legitimate place even in the highest and driest system of topography[2]."

In short Galton wanted geological and vegetational information added to the maps then in vogue, and he thought it possible to combine a graphic picture with a sufficiently faithful ground-plan. He had great hopes from the art of colour lithography then being rapidly developed[3]. Galton's senses were keenly alive when travelling and he remarks in this paper that France, Switzerland, Germany and almost every European country has its pervading smell, and its pervading sounds, all widely alien to the experiences of our own mother-country. It was something of the impressions of all this local colouring which Galton found so painfully missing in maps.

"be made intelligible and interesting to the general public of educated men" (!). In the first volume, besides papers by Galton's friends Charles Astor Bristed, the American (see *Memories*, p. 77), and Charles Buxton (*Ibid.* p. 69), there were papers by Liveing, Fitzjames Stephen, and W. G. Clark (*Ibid.* p. 70), also a close intimate of Galton. The *Essays* thus were the product of Galton's close contemporaries, if they did not actually spring from his entourage. I have failed to find who really set them going.

[1] Galton refers in this matter to popular coloured bird's-eye views of the Crimea and Baltic, poor in execution, but supplying a distinct want. He notes also Ziegler's geological maps.

[2] *Loc. cit.* p. 97. Eighteen years later a letter to George Darwin shows that Galton's thoughts were still working on the same lines. After referring to projecting mouldings on maps to represent mountain chains, modelling from successive contours, Galton continues:

"I have often thought of procuring a really artistically made and coloured globe [elsewhere he suggests one of 9 feet diameter] and once had much correspondence about it. Ruskin wrote a very good letter. It seems to me that one might set to work by making a spherical shell, cutting it up *into convenient parts* like a puzzle map, and mount the parts that were temporarily wanted to be consulted on a convex table. These could be multiplied by casts, also by electrotype." (British Association, Bradford, Sept. 24, 1873.) The last sentence shows that Galton intended his globe to be a model of the world's *surface*, not a mere map.

[3] Seventeen years later Galton proposed at a council meeting of the Royal Geographical Society that the interest on the Murchison Fund be expended this year (? 1872) in procuring specimens of, and a report on, the various styles of cartographic representation now in use both in England and abroad, as regards shading, colours, symbols, and method and cost of production, but not as regards projection, and that a committee should be appointed to arrange particulars.

Galton speaks highly of geographical societies in the help they provide for qualified travellers, and then doubtless comes a touch of his own experience, and

"their moral influence is not to be disregarded, by which they sustain the courage and perseverance of a traveller, whose special tastes find little countenance and sympathy from the associates whom the accidents of birth and neighbourhood have made nearest to him[1]."

He kept frequently in his mind the discovery of any means of easing the path of the travelling geographer. Thus we have his "Table for Rough Triangulation without the usual Instruments and without Calculation" of March 1860. This was a simplified form of measuring the distance of an object (breadth of river, etc.) lying off the traveller's path first suggested with more rigorous calculations by Sir George Everest, formerly Surveyor-General of India[2]. Galton appears to have issued his table on a single leaf, probably for the use of travellers, and it was afterwards incorporated in the *Hints for Travellers* he edited for the Royal Geographical Society. The idea is an exceedingly simple one. We proceed thus:

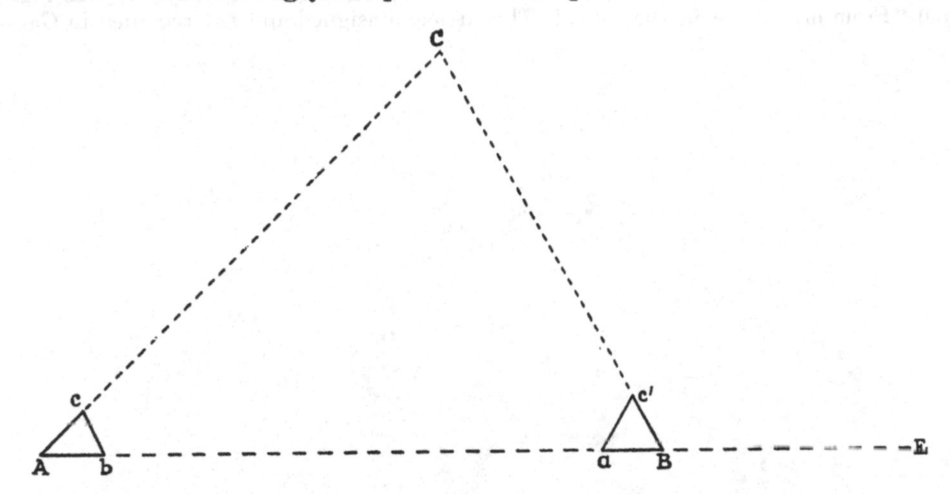


$C$ is the inaccessible object assumed to be either in the same horizontal plane as our base line $AB$, or else to be the projection of an object on that plane. We walk ten paces $Ac$ from a peg at $A$ towards $C$ and insert a peg at $c$. We then walk ten paces towards an accessible object at $B$ and set a peg at $b$. We pace $cb$ to the nearest quarter pace. Then we walk 100 paces from $A$ to $B$ and in order to maintain a straight line always look at a distant object $E$ behind $B$; at $a$ which is 90 paces from $A$ we insert a peg and then do ten more paces to $B$ (peg). Next we step ten paces $Bc'$ from $B$ towards $C$, and finally pace $c'a$ to the nearest quarter pace. The number of paces in $cb$ and $c'a$ are the two arguments by which we enter horizontally and vertically Galton's Table, and under them we find $AC$ given in paces. If we enter with $c'a$ and $cb$ horizontally and vertically we find $BC$. If $AB$

[1] *Loc. cit.* p. 100.

[2] *Journal of the Royal Geographical Society*, 1860, pp. 321 *et seq.*

be a north and south line, the angle $cAb$ is the *bearing* of $C$ and this is given in the first vertical column of the table. There is no instrument needful and no calculation. Of course the method is rough, but if care be taken that no angle of the triangle be less than 30°, practically valuable results may be obtained.

Another paper with the same intention of aiding the geographer was that of 1858 on "The Exploration of Arid Countries[1]." Galton states his problem as follows:

"I suppose an '*exploring*' party, as few in numbers as is consistent with efficiency, to be aided by a '*supporting*' party, who may be divided into two or more sections. The duty of this supporting party is to carry provisions, partly to be eaten on the way out, and partly to be '*câched*,' or buried in the ground, in order to supply the wants of a homeward journey. After a certain distance from camp had been reached, and the loads of one '*section*' of the supporting

[1] *Proc. R. Geographical Society*, Vol. II, pp. 60–77. Galton was much interested in the difficulties of exploring the arid centre of Australia. Among his papers I found a little map of Australia indicating by different shading the settled and squatting districts, and with the desert routes of the Gregories, Stuart and of Burke's cross-continental fatal journey. It was marked in pencil "From my article in the ——." This article, unsigned and not recorded in Galton's

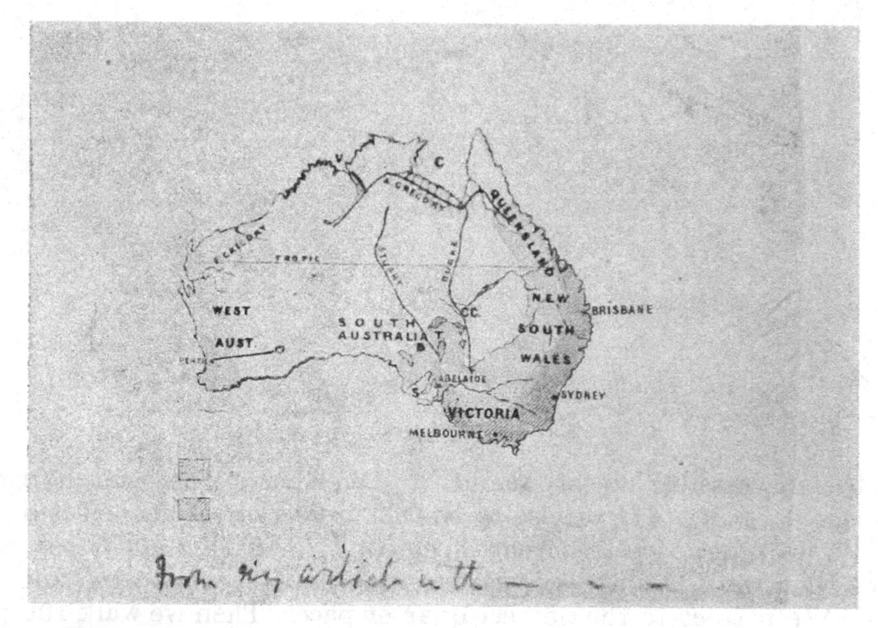


list of published papers, was ultimately run to ground in the *Cornhill* for 1862, pp. 354–64. Its title is "Recent Discoveries in Australia" and it describes the work of the Gregories, Babbage, McDonal, Stuart, Burke and King. It might be read to-day by anyone desiring to get an interest in Australian discovery. One wonders whether Burke's life would have been saved if Galton's system of câches had been fully adopted. One passage may be cited: "It appears hopeless to ascertain the habitable qualities of any district of Australia by seeing it only once. The arid plains after a month's soaking rains are wholly altered. An unexpected fact still remains, it is that wherever a sheep station is by any means established, the country becomes rapidly improved by its influence. It is a subject for Darwinian speculation. Grazing improves grasses, occupier dams up creeks and deepens water-hole. Perhaps the grasses and bushes flourish through the moisture. Their roots will then form a natural matting that checks evaporation while long fibres of the roots encourage more water to enter deeply into the soil."

party had become exhausted in furnishing meals and câches to the entire expedition, this section would separate from its companions and return home. A second '*section*' would subsequently act as the first had done, and afterward a third and even a fourth, according to their original number. Finally the explorers would be left by themselves at some days' journey in advance of the farthest known watering place, with their own loads of provisions untouched, and with other provisions stored in câches, fully sufficient for their return, and in every respect as capable of further exploration as if it was from their own camp, and not from a spot in the heart of the desert, whence they were about to take their departure.

Doubtless the same general idea must often have occurred to other travellers besides myself; but whether it is because the details have been found puzzling and difficult to work out, or because the necessary vessels for carrying water were not to be met with when wanted, no traveller in arid countries has ever availed himself of the great power which this method of exploration affords[1]."

Galton starts with a table of the weight of water and food, needed as rations by horse or mule, ox, and man per day, and also the total weights which each can drag or take on its back. The problem is then to determine at what distances each section of the party is to return so as to leave the ultimate exploring party with full weights of rations and full câches for the return journey. It was exactly the sort of problem which delighted Galton; there was a little of mathematics[2], a little of statistics and considerable amount of ingenuity required, and the whole had a practical bearing. He adopted the *binary* system by which half the remaining party returned at the end of each stage. It would not be fitting here to discuss at greater length Galton's tables and results. He had chiefly in view the then unexplored regions of Australia.

As Secretary of the Royal Geographical Society Galton came into touch with many famous travellers—Burton, Speke, Grant, Stanley, etc. Galton himself drafted the instructions for the Burton-Speke expedition of 1856, which led to the discovery of Tanganyika and the Victoria Nyanza lakes, a discovery made at the painful cost of a quarrel between Burton and Speke. Galton had an all-round admiration for Speke and Grant, and a respect for the eccentric genius of Burton. Of Stanley he thought less favourably as of a man inclined to sacrifice the scientific aspect of geography for what the younger generation would term journalistic 'stunts.' The letters of both Burton and Speke to Galton give evidence of the difficult position of the latter in his relation to the two former as Secretary of the Royal Geographical Society[3].

[1] *Loc. cit.* p. 61.

[2] There is a misprint in Eqn. (3) of p. 65 where in the value for $s$, the section that turns back first, the numerators of the two fractions should of course be $a^{r-1}$ and $a^{r-2}$ respectively.

[3] Galton was first among those who worked for the Speke Obelisk in Kensington Gardens, and he desired above all things that a joint memorial to Speke, Burton, Grant, Baker, Stanley and Livingstone should be arranged near the Speke Obelisk as a reminder to later generations of what our nation has done for African discovery.—We might well add Galton himself to the list.—The time has, perhaps, come now when the smaller, if very human, side of these men might be forgotten under a common monument. Galton wrote a letter to the *Times*, May 25, 1904, advocating such an African memorial to include the names of earlier travellers—Bruce, Mungo Park, Lander, Clapperton and Barth (who was subsidised by England). He suggested a massive block of stone with a map of Africa in bold coloured mosaic on its curved top, even as Africa would appear on a five foot globe. The memorial was to be surrounded by such African trees, shrubs and flowers as will grow in this country. But although Burton and Speke were both dead, the wound was not yet healed and nothing came of Galton's plan.

We can here only afford space for one letter of interest, that from Speke of February 26th, 1863; it indicates the growing feeling between Speke and Burton and at the same time the difficulties that Galton had to encounter.

"{My reports will be sent from Khartoum as soon as we arrive there.} GONDOKORO 26 *February* /63
(to be posted on arrival at [Khartoum[1].]
14° 30′ N. Lat. Head winds keep us back.)

27 *March.*

MY DEAR GALTON

Petherick has shown me a paper of the R. G. Society by which I infer you wrote me a letter suspecting the V. N'yanza to be the source of the Congo or perhaps one of Du Chaillu's rivers because a river was made to run both in and out of it. I fear you did not receive the letter I wrote from Madeira after reading Burton's journal in the Society's volumes, else you could not have supposed so, for in addition to the fact that *every* Arab knew the 'Kiuiva' river ran out of the Lake and told us they supposed the Lake to be the source of the Jub, every Arab had heard of the vessels on the Nile though Burton tried to hide these matters from the public; I suppose to excuse himself for not visiting the N'yanza. I can only say it is a pity my geographical papers *read* before the Society were not put into the Society's Journal in preference to Burton's papers, which were *not read* and therefore not commented on, for that alone has put everybody wrong. Burton's geography was merely a copy of my unfinished original maps, left open until I reached England for further information. Burton wanted me to instruct him, acknowledging that he knew nothing of the typographical features of a country. He could not have written one word unless I had instructed him, but he gave up his lessons too soon, imagined largely on the nucleus I gave him and fell into error accordingly.—You will find all the information you require upon this journey in my reports, so I will now open a new project to you for crossing Africa from East to West following as close as possible upon the line of the Equator; for unless I do it, it will not be done this century. It can be done easily enough on a large scale and with a power of money, but not as I have been travelling at the beck and call of every chief that falls in the way. This is the sum total of my requirements, provided the Govt. is enlightened enough to accept it, which is doubtful we know: Four men of science as captains to 400 negroes, half Crue men from the West Coast, and the other half from Zanzibar, all hands to be furnished with carbines. I should then want a vessel to visit Venice and pick up boats, pass round the West Coast for Crue men, and continue to Zanzibar where the vessel would wait until I commenced the march and then return by the Cape to the mouth of the Congo, where it would await my arrival and convey the 400 men to their respective homes. But this is not all, for I should require another vessel to go up the Nile and form a depôt at Gondokoro. The rest you can imagine. With one word more, I will close my letter and I tell it to you as an overseer to the Society—I firmly believe I should have reached this one year ago and at $\frac{1}{3}$ less expense, if my projects for the journey had been promptly attended to. I asked for leave and money 12 months before starting in order that I might form two Depôts in the interior, but I neither got my leave nor the money until 2 months or so before I started, and therefore could only form one Depôt in advance. That has been the root of my disasters and delay—but 'all's well that ends well,' and there is an end of it, only let the warning be a caution for the future. How I should have rejoiced to receive your letter, but nothing has reached me, not even a letter of advice from Rigby, which announced the departure of some letters and a host of delicacies sent by kind friend Rigby. And now old Galton with Grant's best wishes and my own to yourself and Wife, believe me

Yours ever sincerely H. SPEKE.

P.S. I have sent a map and several papers as I shall not be home in time to contribute to this year's Journal and I fancy it important this should have an early issue.

Let us pity "old Galton" as he read of those 400 men, each to be furnished with a carbine! That great journey, which would have antedated

[1] Deleted.

Stanley's, never came off. In his *Memories* (pp. 201–2) Galton vividly describes the scene at the Bath Meeting of the British Association in 1864, where Burton was to read a paper attacking Speke, and Speke shooting in the neighbourhood had been invited to reply. In the Committee of the Geographical Section (which meets before the open session), at which the President Sir Roderick Murchison, Sir James Alexander, Captain Burton, Galton and others were present, a letter (in the course of a discussion whether the Council of the Association should be requested to bring Speke's services to the notice of the Government) went round the table. It was to announce that Speke had accidentally shot himself dead in drawing his gun after him, while getting over a hedge. Thus ended the life of a man—whom Galton described as "a thorough Briton, conventional, solid and resolute...a fine manly fellow"—in a tragedy, which, one might have hoped—but would have hoped in vain—must stop controversy and bitterness. Burton, Galton tells us, "had many great and endearing qualities with others of which perhaps the most curious was his pleasure in dressing himself, so to speak, in wolf's clothing, in order to give an idea that he was worse than he really was."

I have not dwelt more at length on this painful controversy as it only indirectly concerns Galton, but it made a very deep and lasting impression on his mind. I am not at all sure it was not the origin of his very strong dislike in later life of all forms of controversy, so that he would let a criticism pass without reply which is not always the most effective manner of fostering the growth of a new and therefore reluctantly accepted branch of science. No reply is too often taken by a thoughtless public to be identical with an admission of error. When the Biometric Laboratory started the series of papers *Questions of the Day and of the Fray* in 1906, Galton expressed his grief at what was not indeed an offensive but in many cases a too long delayed defensive.

Besides preparing plans of travel for various discoverers[1], Galton took, as his manuscript notes show, a large part in the executive work of the Royal Geographical Society. Thus we find about 1858 numerous plans for a meeting-room, the main outline of which he adopted from the old debating-room of the Cambridge Union Society. This, I suppose, had impressed him in Cambridge days as an excellent speaking-chamber, and he wrote to Montagu Butler for a plan and details[2]. Galton further started the movement for increased interest in geography in schools, and it resulted in the Society

[1] Cf. for example the paper "Additional Instrumental Instructions for Mr Consul Petherick" by F. Galton, *Proc. R. Geog. Soc.* Jan. 28, 1861, p. 96, which is a model of what such instructions should be. And again, we may note the "Report on African Explorations," *Proc. R. Geog. Soc.* May 26, 1862, p. 175, as indicating how Galton kept in touch with African exploration and how fully he carried out his duties of Secretary to the Society. In this paper his condemnation of uncontrolled trade on the White Nile "mostly in the hands of reckless adventurers and lawless crews" is characteristic of the man, who was later greatly revolted by Stanley's proceedings.

[2] The Secretary informs me that the minutes of the Royal Geographical Society make no reference to these plans, but that the House Committee considered in 1857 proposals for extending the premises. Galton is not named as a member of this committee, but he was probably *ex officio* one of their number, and prepared the plans for their consideration.

offering an annual gold medal to be competed for by public school boys. He afterwards took a considerable part in the agitation which ended in the recognition of geography as an academic study.

Galton was active in many other ways for the cause of geography. In 1861 he was asked to give the Church Missionary Society some information as to Zanzibar as a possible centre of missionary enterprise, having regard to its climate, physical features and the moral and social condition of the people. Galton read a paper to the society on June 1st and it is published in their journal *The Mission Field*[1]. In the paper he points out the dominant Arab and Moslem influence which radiated from Zanzibar, not only all along the coast of the mainland but far into the continent, perhaps one-third across. Galton gave his information from manuscript notes of Burton and from photographs of Grant lent by Speke. Galton on the whole spoke well of the Arabs, but ill of the negro natives of the mainland, thus following Burton rather than Speke. He concluded as follows:

"The natives are most assuredly no inquiring race, open to influence, but the very contrary. Again their countries are intersected by commercial routes through which a tide of Moslem ideas is constantly flowing, and could a handful of missionaries, looking at past and present history to guide us in our speculations, be supposed to avail against it? It strikes me, too, as something not quite generous to avail ourselves of the courtesy and the unusual tolerance of a Moslem power to sow seeds of a certain harvest of discord. What we find in Zanzibar is a far-reaching and far-influencing, but not a strong power; anxious to do well, seeking to consolidate itself, amenable to a good English influence, but above all things, the *sine quâ non* of its existence is that it should be Moslem. With our very limited missionary agency, it seems to me that we should divert its current to healthier and more hopeful fields than Zanzibar, and that England, so far as she may interfere at all, whether through her representative or by any other agency, should try to effect the following results: To relieve the Sultan, by means of our moral support, from the embarrassment of foreign pressure; to promote safe lines of legitimate and civilising traffic into the far interior of Africa; and to open better communication between Zanzibar and the more civilised world than now exists. This is the schedule of what England is actually doing, and I further believe it is all she ought, for the present, to undertake in Zanzibar[2]."

This is not the first, nor the last, occasion on which Galton[3] emphasised the possibly superior civilising effect of Moslems over Christians on barbarous races. Of course he speaks here of the state of affairs in 1861, before the medical work (India and China) or the craft-school factor (Nigeria) had been added to the purely religious activities of the Christian missionaries. There is a characteristic table of the Zanzibar climate on p. 124, detailing the wind, the rainy, cold and hot seasons and the seasonal healthiness; the paper probably has now fallen much behind the present state of knowledge.

In 1862 Galton took Sir Roderick Murchison's place, who fell ill just before the meeting, as President of the Geographical Section of the British Association. If he gave any opening address, it was certainly a makeshift effort and has not been published. Mrs Galton merely notes that her husband was at Cambridge for the Association[4]. Ten years later, however, 1872, Galton was again President of the same section and gave the cus-

[1] Vol. VI, No. 66, pp. 121–30. [2] *Loc. cit.* p. 130. [3] See Vol. I, p. 207.

[4] Galton recounts an amusing incident of the meeting in his *Memories*, pp. 208–9.

tomary opening address. In this address Galton emphasises the important relations of climate to geography and remarks how human agency can influence both.

"We are beginning to look on our heritage of the earth much as a youth might look upon a large ancestral possession, long allowed to run waste, visited recently by him for the first time, whose boundaries he was learning, and whose capabilities he was beginning to appreciate. There are tracts in Africa, Australia, and at the Poles, not yet accessible to geographers, and wonders may be contained in them; but the region of the absolutely unknown is narrowing, and the career of the explorer, though still brilliant, is inevitably coming to an end. The geographical work of the future is to obtain a truer knowledge of the world. I do not mean by accumulating masses of petty details, which subserve no common end, but by just and clear generalisations. We want to know all that constitutes the individuality, so to speak, of every geographical district, and to define and illustrate it in a way easily to be understood; and we have to use that knowledge to show how the efforts of our human race may best conform to the geographical conditions of the stage on which we live and labour."

Galton finally turns to maps; he does not refer as in the *Cambridge Essays* to the ultimate goal of the geographer—a map of the world with ordnance map accuracy—but he does return to the idea of combining geographer and artist.

"The facility of multiplying coloured drawings will probably lead to a closer union than heretofore between geography and art. There is no reason now why 'bird's-eye views' of large tracts of country should not be delicately drawn, accurately coloured, and cheaply produced. ......It is therefore to be hoped, that the art of designing the so-called 'bird's-eye views' may become studied and that real artists should engage in it[1]."

Galton finally concentrates on two practical map proposals. He notes how difficult it is to procure ordnance maps and how much their format hinders their popular use. He suggests that pocket forms of Government maps should be issued, and what is more be sold at every head post-office. We have had to wait many years for anything like the carrying out of Galton's proposal. Why should we, he asks, not succeed in getting "the beautiful maps for which we, as tax-payers, have paid, but only copies or reductions of them, not cheaper than the original and of very inferior workmanship and accuracy"? Galton's second proposal was that the Government should be petitioned to issue a five mile to the inch map of the country "to serve as an accurate route-map and to fulfil the demand to which the coarse county maps, which are so largely sold, are a sufficient testimony." Yet 1872 antedated the cycling and by much more the motoring tourist![2]

[1] Galton's suggestions were not carried very far, but has not the time come for resuscitating two of his ideas by means of air-plane *colour* photography and the use of the stereoscope? So far I have not seen any attempts at air-plane colour photography, nor using air-plane stereoscopic cameras, but such may exist. There seems a real field for experiment in this direction, which might possibly fulfil some of Francis Galton's fond hopes. Stereoscopic air-plane photography might also have military value. [Since writing these lines I have learnt that the latter aspect of the matter has been considered by the Air Ministry.]

[2] Galton's criticism did not wholly fail. The General Committee of the British Association resolved: "That Sir Henry Rawlinson, F. Galton, Admiral Ommanney, J. Hawkshaw, Bramwell, W. De la Rue, Godwin Austin be a Committee (with power to add to their number) for the purpose of representing to the Government the advisability of an issue of the one-inch ordnance maps, printed on strong thin paper, each sheet having a portion of an index map impressed on the outside to show its contents and those of the adjacent sheets and their numbers. Also that

Galton's presidency of Section E at Brighton was marked by the reading of a paper by Stanley concerning his Congo travels. The ex-Emperor Napoleon and the Empress Eugénie were present, and it was feared Stanley might use the occasion in an inappropriate way[1]. The meeting passed off, however, with only one interference on the President's part, but with some tension. For those who would realise Galton's strong feelings about Stanley's proceedings the following extract from an article—really by Galton—in the *Edinburgh Review* for January 1878 (pp. 166–91) will be indicative.

"The exploration of Africa has been conducted of late on a new system. The routes of the earlier travellers passed either through parts of the continent where the population is sparse, as in Caffre-land or in the Sahara, or in those where it is organised into large kingdoms, such as lie between Ashanti and Wadai, and which are much too powerful to admit of any traveller forcing his way against the will of their rulers. The older explorers were therefore content to travel with small retinues, conciliating the natives of the larger kingdoms by patient persistence and feeling their way. But of recent years all this has been changed. The progress of discovery has transferred the outposts of knowledge and the starting-points of exploration to places where the population is far more abundant than that which is met with in either the northern or the southern portions of Africa, yet where it is for the most part divided into tribes. Hence modern explorers have found the necessity for travelling with large and strongly armed retinues. This new method has been frequently adopted in the upper basin of the White Nile, which has been the scene of many military expeditions sent by the Egyptian Government to force a way into the Soudan, including that commanded by Sir Samuel Baker. So, in the south, Livingstone's comparatively small band of determined Caffres, placed at his disposal by a chief whose confidence he had gained, enabled him to cross the Continent in the latitude of the Zambesi. Subsequently other travellers like Burton, Speke, Grant and Cameron, starting from Zanzibar, have adopted a similar plan. Their forces were large enough to enable them to pass as they pleased through regions where the tribes were small, they were sufficiently powerful to make larger tribes fear to attack them, and as they invariably adopted a conciliatory policy with the latter, they never came into serious collision with the natives. Mr Stanley has adopted the plan of travelling with an armed retinue on a much larger scale than any of those we have named, and he has certainly carried, by these means, a great expedition successfully through

these maps should be sold in all important towns and, if possible, at the several post-offices, and that Mr F. Galton be the Secretary." Mr Ayrton, the minister under whose control the Ordnance Survey Office was, saw the deputation but it is not very clear that any definite impression was made on him. On the other hand, Major-General Sir Henry James, Superintendent of the Ordnance Survey Office, wrote a number of appreciative letters, and maps were prepared in accordance with Galton's suggestions. James writes: "As to the difficulty of getting our maps it arises from the fact that we have four agents only for England and Wales instead of at least 250 or one in every tolerably sized town, and these agents are all in London and receive $33\frac{1}{3}$ per cent. instead of 25 per cent. But few and dear as the agents are, they seem to have been selected because they are themselves map-makers and sellers, and sell to the public bad copies of the ordnance maps taken by robbery! This was done against my earnest protest and the Government is losing some £3000 a year by the arrangement, and the public are everywhere dissatisfied." Henry Fawcett wrote a sympathetic letter and regretted that the matter had not also been brought before the Committee of Section F, of which he was a member.

Galton's index map, however, is now a commonplace of many map publications.

[1] "Mr Stanley had other interests than geography. He was essentially a journalist aiming at producing sensational articles." *Memories*, p. 207. What vexed Galton peculiarly was that Stanley had made no proper positional observations, and Galton ventured to utter the words "sensational geography." Stanley in his letters used violent language about the Royal Geog. Society, and about Markham and Galton. He was no doubt excited by the inquiries as to his birth, from which that Society had not and could not entirely disassociate itself. He did not meet that question straightforwardly and fearlessly as he might well have done. (Letters of Stanley and others in *Galtoniana.*)

Africa. Thus he states, 'I led 2,280 men across hostile Unyoro' on an expedition intended to cross the Albert Nyanza. Again, when he leaves Nyangwe on his final expedition down the Lualaba, he starts with a body of 500 fighting men. Thus with a larger military force than hitherto employed, and making a determined use of it, Mr Stanley has conducted a geographical raid across the middle of Africa, which has led him into scenes of bloodshed and slaughter, beginning at the Victoria Nyanza, and not ending until he arrived in the neighbourhood of the Western Coast. [This achievement undoubtedly places Mr Stanley in the foremost rank of African discoverers and ensures to him a hardly-earned and lasting fame[1].] The question will no doubt be hotly discussed how far a private individual, travelling as a newspaper correspondent, has a right to assume such a warlike attitude, and to force his way through native tribes regardless of their rights, whatever those may be. A man who does so acts in defiance of the laws that are supposed to bind private individuals. He assumes sovereign privileges, and punishes with death the natives who oppose his way. He voluntarily puts himself into a position from which there is no escape, except by battle and bloodshed; and it is a question, which we shall not argue here, whether such conduct does not come under the head of filibustering. Nations are above laws, and may and do decide what expeditions they may care to launch, but the assumption of such a right of private individuals is certainly open to abuse, and seems hard to defend. It is impossible to speak of Mr Stanley's journey without noticing this exceptional character of it. At the same time it is not our present object to discuss the morality of his proceedings, but to occupy ourselves with his discoveries, which are unquestionably of the highest geographical importance, and may lead to consequences in comparison with which the death of a few hundred barbarians, ever ready to fight and kill, and many of whom are professed cannibals, will perhaps be regarded as a small matter[2]."

Galton next proceeds to discuss Stanley's geographical discoveries in relation to those of Schweinfurth and of Barth. Then he turns to the trade products of Africa and its means of transport. Under the latter heading he certainly did not anticipate the modern railway developments nor the rapid increase in mineral and agricultural exports. The general sense of his papers seems to be that the trade is scarcely worth the European's while and is best undertaken by the Arab. But the most valuable and interesting part of the paper which indicates surely the change which had occurred in Galton's outlook—his advance towards anthropology—is the long account he gives of the physical and mental characters of the negro. His judgment is not favourable:

"By picking and choosing out of a multitude of negroes, we could obtain a very decent body of labourers and artizans; but if we took the same number of them just as they came, without any process of selection, their productive power, whether as regards the results of toilsome labour or of manual dexterity, would be very small." (p. 180.)

"Leaving for a moment out of consideration the combative, marauding, cruel and superstitious parts of his nature, and all that is connected with the satisfaction of his grosser bodily needs, his supreme happiness consists in idling and in gossip, in palavers and in petty markets. He has no high aspirations......He loses more of that which is of value to him in consequence of his labour than he gains by what his labour produces. He has little care for those objects of luxury or for that aesthetic life which men of a more highly endowed race labour hard to attain. His coarse pleasures, vigorous physique, and indolent moods, as compared with those of Europeans, bear some analogy to the corresponding qualities in the African buffalo, long since acclimatised in Italy, as compared with those of the cattle of Europe. Most of us have observed in the

[1] In Galton's own copy these words are enclosed with an ink border and against them is written "an editorial insertion—not mine. F. G." Thus does the smaller intelligence, editing the broader mind, make nonsense of Galton's meaning. If "Barabbas was a publisher," of a surety the railing malefactor was an interpolating editor!

[2] Were it not for the 'perhaps' we might suspect the last few lines to be another editorial interpolation.

Campagna of Rome the ways of that ferocious, powerful and yet indolent brute. We may have seen him plunged stationary for hours in mud and marsh, in gross contentment under a blazing sun; at other times we may have noticed some outbreak of stupid, stubborn ferocity; at others we may have seen him firmly yoked to the rudest of carts, doing powerful service under the persistent goad of his driver. The buffalo is of value for coarse, heavy and occasional work, being of strong constitution and thriving on the rankest herbage; else he would not be preserved and bred in Italy. But he must be treated in a determined sort of way, by herdsmen who understand his disposition, or no work will be got out of him, and besides that, he is ferocious and sufficiently powerful to do a great deal of mischief." (p. 180.)

For Galton in 1872, as one race of animals differs from another, so all races of men are not equal. He has started on his recognition of hereditary superiorities. Because the negro belongs to an inferior race, the Arabs, who coalesce with the natives, inter-marry, and do not look upon a converted negro as an inferior, have done far more than Christian missionaries to educate and civilise the negro.

"Of Mohammedanism and Christianity—we do not speak here or elsewhere as to their essential doctrines, but as they are practically conveyed by example and precept to the negro—the former has the advantage in simplicity. It exacts a decorous and cleanly ritual that pervades the daily life, frequent prayers, ablutions and abstinence, reverence towards an awful name, and pilgrimage to a holy shrine, while the combative instincts of the negro's nature are allowed free play in warring against the paganism and idolatry he has learned to loathe and hate. The whole of this code is easily intelligible, and is obviously self-consistent. It is not so with Christianity, as practised by white men and taught by example and precept to the negro. The most prominent of its aggressions against his every-day customs are those against polygamy and slavery. The negro, on referring to the sacred book of the European, to which appeal is made for the truth of all doctrines, finds no edict against either the one or the other, but he reads that the wisest of men had a larger harem than any modern African potentate, and that slave-holding was the established custom in the ancient world. The next most prominent of its doctrines are social equality, submission to injury, disregard of wealth, and the propriety of taking no thought of to-morrow. He, however, finds the practice of the white race, from whom his instructions come, to be exceedingly different from this. He discovers very soon that they absolutely refuse to consider him as their equal; that they are by no means tame under insult, but the very reverse of it; that the chief aim of their lives is to acquire wealth; and that one of the most despised characteristics among them is that of heedlessness and want of thrift. Far be it from us to say that the modern practice in these matters may not be justified, but it appears to require more subtlety of reasoning than the negro can comprehend, or, perhaps, even than the missionary can command, to show their conformity with Bible teaching." (p. 187.)

Galton's anthropological sense was rapidly developing; he was firmly grasping the relativity of religions, how they have no absolute validity, but the suitability of any creed depends on the stage of the mental and even the physical development of a given race. Shortly, a religion must be in harmony with the habits and culture of a given people at a given time, or it will fail to fulfil its purpose—which, from the anthropological side, is to strengthen and to stabilise the social purposes and gregarious instincts of a definite group of men. As Galton realised, Christianity has built up no negro kingdoms, but Mohammedanism has done so, and the negro converts erect mosques, maintain religious services, and conduct their schools without external support.

That Galton was not wholly content to leave Africa to Negro and Arab is evidenced by his letter to the *Times* of June 6, 1873. In that letter he recognised that much of Africa cannot be occupied by the European, that

PLATE IV

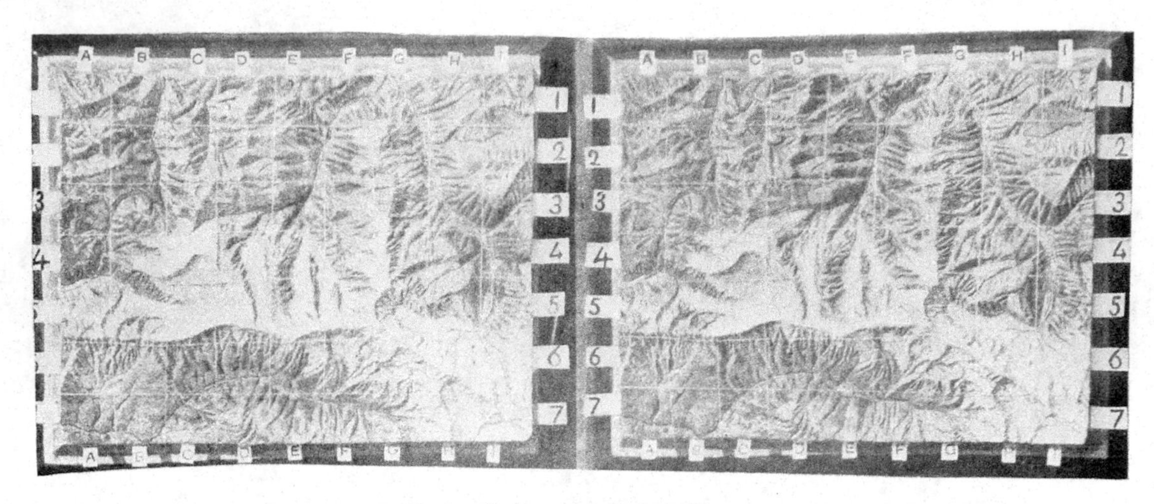


Ortler Spitze and Stelvio Pass.

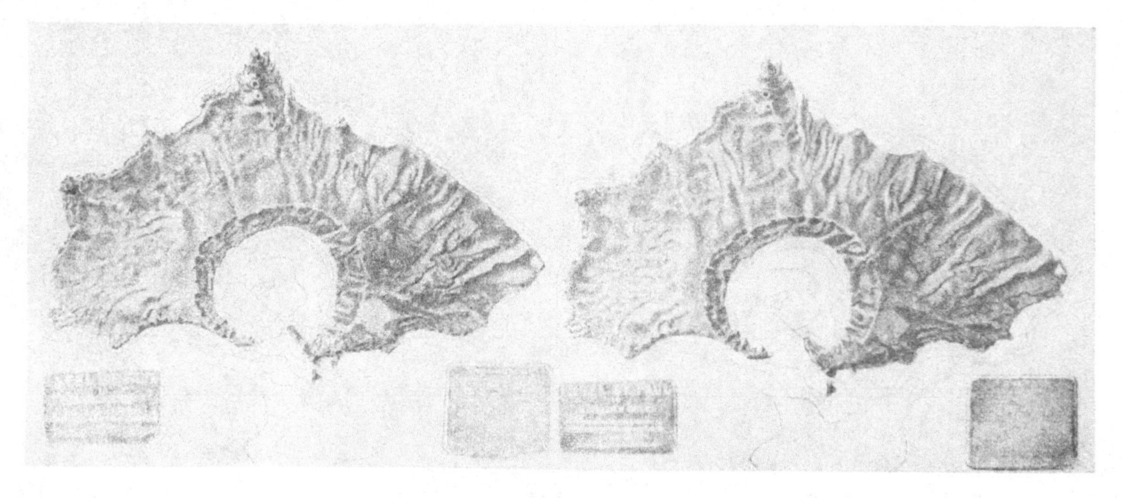

Island of St Paul.

Stereoscopic views of Geographical Models (should be examined with a stereoscopic lens-doublet.)

"average negroes possess too little intellect, self-reliance and self-control to make it possible for them to sustain the burden of any respectable form of civilisation without a large measure of external guidance and support. The Chinaman is a being of another kind, who is endowed with a remarkable aptitude for a high material civilisation. He is seen to the least advantage in his own country, where a temporary dark age still prevails, which has not sapped the genius of the race, though it has stunted the development of each member of it, by the rigid enforcement of an effete system of classical education, which treats originality as a social crime[1]......The natural capacity of the Chinaman shows itself by the success with which, notwithstanding his timidity, he competes with strangers, wherever he may reside. The Chinese emigrants possess an extraordinary instinct for political and social organisation; they contrive to establish for themselves a police and internal government, and they give no trouble to their rulers so long as they are left to manage these matters for themselves."

"The history of the world tells a tale of the continual displacement of populations, each by a worthier successor, and humanity gains thereby. We ourselves are no descendants of the aborigines of Britain, and our colonists were invaders of the regions they now occupy as their lawful home. But the countries into which the Anglo-Saxon race can be transfused are restricted to those where the climate is temperate. The tropics are not for us to inhabit permanently; the greater part of Africa is the heritage of a people differently constituted to ourselves. On that continent, as elsewhere, one population continually drives out another. We note how Arab, Tuarick, Fellatah, Negroes of uncounted varieties, Caffre and Hottentot surge and reel to and fro in the struggle for existence. It is into this free fight among all present that I wish to see a new competitor introduced—namely the Chinaman. The gain would be immense to the whole civilised world if he were to outbreed and finally displace the negro, as completely as the latter has displaced the aborigines of the West Indies. The magnitude of the gain may be partly estimated by making the converse supposition—namely the loss that would ensue if China were somehow to be depopulated and restocked by negroes."

Whatever opinion we may hold of Galton's views on the Chinaman, there is no doubt that this passage marks not only his full acceptance of the doctrine of the survival of the fitter race as applied to man, but further his opinion that civilised man could himself directly expedite the processes of evolution.

A few further memoirs having a bearing on geographical or allied topics may be noted here. We have already referred to his views on maps. In 1865 the idea occurred to Galton that, as maps so conspicuously fail to give us the leading features of a mountainous country and are indeed so incapable of representing crags and cliffs successfully, a stereoscopic photograph of a model would be of extreme value[2]. Indeed a coloured model on this plan with reproduction by colour photography might go a long way to satisfy Galton's craving for something more illustrative of the floral and geological environment than an ordinary map can provide. He suggests what might be done in this way by photography of the models of the English Lakes at Keswick, of the Pyrenees at Luchon, and of the Alps at Berne, Zurich, Lucerne and Geneva. With the assistance of Mr R. Cameron Galton he was able to obtain and exhibit stereoscopic photographs of the following models:

(1) Island of St Paul in the Indian Ocean[3], from an Austrian bronze model.

[1] Was this sentence a thrust at another race, and was Galton thinking of his own bitter experience? See Vol. I, pp. 12, 142.

[2] "On Stereoscopic Maps taken from Models of Mountainous Countries," *R. Geog. Soc. Journal*, 1865, pp. 99–106.

[3] Midway on Mercator's chart between Melbourne and Cape Town.

(2) The Ortler group of mountains, from an Austrian model.
(3) Mount Blanc district, from Bauerkeller's relief map.
(4) Cape Town and Table Mountain, from a coloured model.
(5) Abyssinia, from a rude model.
(6) The Isle of Wight, from a rude model.

The paper itself is accompanied only by a photograph of (1)[1], somewhat confusing as it illustrates also the proposal of Galton to build up large maps in stereoscopic sections.

The Galton Laboratory possesses stereoscopic slides of (1) giving the whole island, and of (2)—(6) inclusive, and also of a seventh slide-part of the Ile de Porquerolles in the Mediterranean off Toulon. Our photographs have faded in the course of nearly sixty years but they show still with extraordinary effect the success of Galton's idea. The Stelvio stands out in a way that no map can compete with, and hardly a bird's-eye view from the Spitz itself could give such a good conception of the 'lie of the land.' It is a grievous pity that the stereoscopic idea of map-models has been forgotten, and we might hope for its resuscitation in association with the air-plane as already suggested. We provide in the accompanying plate copies of the faded photographs of (1) and (2) which nevertheless will suffice—if the reader be lucky enough to possess a pair of stereoscopic lenses—to justify this statement.

Another paper of this same year is entitled: "Spectacles for Divers and the Vision of Amphibious Animals." In this paper Galton states that if water is *in contact* with the human eye[2] a double convex lens of flint glass, each of whose surfaces has a radius of 0·48 inch, will correct the concave water-lens. It will require to be supplemented by another of moderate power according to the convexity of the individual eye and refractive power of the different kinds of flint glass. Galton found, however, that even with a lens of this kind under water the eye had not much power of accommodating itself to different distances, and his own distinct vision was restricted to a range of about eight feet. He considered, however, the glasses he used only provisional[3]. He thought such spectacles might be useful to divers in pearl and sponge fisheries, or to sailors examining the bottoms of ships. The paper suggests that amphibious animals must have a power of adjusting their sight, i.e. seals, otters, diving birds, etc., but does not enter into the *modus operandi.* Here again as in the case of stereoscopic maps I think an interesting question has failed to be carried further.

As late as 1881 Galton still maintained some interest in geographical research, but his main work was directed into other and more congenial channels. In the *British Association Report*, 1881[4], there is a brief comparison by Galton of the equipment of exploring expeditions in 1830 and 1880. He notes the progress that has been made in certain instruments:

[1] Search at the Royal Geographical Society having failed to discover the originals, a further hunt among the negatives of the *Galtoniana* has brought to light the originals—too late for reproduction here.

[2] *B. A. Report*, Vol. xxxv, 1865 (Sect.), pp. 10–11, not as when the diving helmet is used.

[3] Still extant in the *Galtoniana*. [4] pp. 738–40.

e.g. mercurial horizons, thermometers and barometers, the binocular glasses, steel and stylographic pens; the progress in clothing, flannel, peacoats and macintoshes; progress in preserved foods; progress in the *personnel*, the educated classes are physically better developed, which Galton attributes to their leading healthier lives, owing to the heavy eating and drinking having ceased, to the better ventilated bedrooms and proper holidays. Lastly he notes the greater ease and quickness with which an explorer can reach the starting-point of his wanderings. The idea suggested in the last sentence probably led Galton to what, I think, was his last contribution to geographical science. In the same year[1] he constructed an "isochronic passage chart for travellers." It consists of a map of the world on Mercator's projection indicating by five colours in two shades the number of days required to reach from London all parts of the world. The map might easily be a little more detailed as the unit of time ten days is rather large, extending from London to Jerusalem, Peru, and Hammerfest, but not be it noted in those days to New York. A similar map made to-day would be of much interest, especially in view of the great development in forty years of trans-continental railways and fast steamships. Galton took, as his authorities, time tables of steamship companies and railways, with public and private post-office information.

It cannot be denied by those who study Galton's memoirs on geography that they mark a continuous development. He remains to the end keen on the mechanical 'dodges' and graphical artifices which had delighted the boy at Atwood's[2] and the youth at Cambridge[3]; but travel for novelty soon became for him travel for a knowledge of physical environment; in this stage Galton was a pure geographer, but then very rapidly the important part of this environment became for him its relation to man and Galton, without realising the full meaning of the change, had passed from the geographer to the anthropologist[4]. Even by the 'seventies' geography had become a secondary study.

The last of Galton's writings that touches on exploration was his graceful preface to W. E. Oswell's *William Cotton Oswell, Hunter and Explorer* of 1900. In this Galton claims justice for Oswell as the first explorer to reach Lake Ngami; Livingstone simply went with Oswell and Murray as a guest, but Livingstone's later fame and Oswell's reticence led to a retrospective credit being given to the former for this first great journey.

### C. CLIMATE

Parallel with Galton's geographical research we find a correlated study—that of meteorology. The services he rendered to this science have been only occasionally recognised at their full value, and much that he has suggested would be worthy of reconsideration and adaptation to the modern state of meteorological knowledge.

[1] "On the Construction of Isochronic Passage Charts," *British Association Report*, 1881, pp. 740–41; *Roy. Geog. Soc. Proc.* 1881, pp. 657–58. [2] Vol. I, p. 77. [3] Vol. I, p. 148.

[4] A very valuable letter of Galton's, advocating the adequate representation of Geography and Anthropology in the 'Proposed Imperial Institutes,' will be found in the *Times*, October 6, 1886 (p. 8).

Galton's first considerations on climate sprang directly from his geographical work and were closely associated with the relation of the health of the explorer to the climate[1]. Thus we have already referred to his little table of climate in Zanzibar in the lecture of 1861. Two years later he prepared a table giving the "Climate of the countries bordering Lake Nyanza, 1861–2." He published it in the *Royal Geographical Society Proceedings*, Vol. VII, 1863 (pp. 225–8), in a paper entitled "On the Climate of Lake Nyanza": it gives with the exception of the middle fortnight of November meteorological details for every 'week' of every month[2]. The mean temperature for the week, the maximum and minimum and the extreme ranges are provided. The rainfall in inches is given where available, the number of rainy days per week, the number of days per month of rain sufficient to be measured, and the total number of days per month of rain and slight showers, also the prevalent wind for each month. The whole of the material was due to Speke and Grant, but it obviously required much 'dressing, i.e. smoothing and interpolation, etc. It was based on observations taken at Karagwè (5100 ft.), Uganda (3400 ft.), Unyoro (3200 ft.) and in a camp 3400 ft. above sea-level, so that there is considerable heterogeneity in the data. Speke's original log is among the *Galtoniana*.

The tables thus formed led Galton to consider the possibility of maps combining at a single glance much meteorological data. This occupied his mind largely in 1861–2, and there is no doubt that Galton was the first to publish meteorological maps of Europe, possibly of any country at all. He induced his friend W. Spottiswoode, the head of the great printing firm, to cast movable types which were used in Galton's first maps. These represented by shaded rectangles Rain, Snow, Dull and Overcast, Overcast, Mostly clouded, Half-clouded, A few clouds, Clear blue sky; the direction and force of the wind were given by another series of symbols, and finally the height of the barometer and the temperature—ordinary and wet bulb—were printed in figures. Thus a rectangle 8·5 mm. high by 5·5 mm. broad contained the information as to rain and cloudiness, intensity and direction of wind, state of barometer and thermometers at a given meteorological station, and these rectangles were placed centrally to each station on a map of the stations at which observations had been made. Galton's first map[3] is

[1] The effect of climate on the traveller had been brought home to Galton very emphatically by his own experience of the after-effects of his travels in Syria and tropical Africa.

[2] 1st to 7th, 8th to 15th, 16th to 23rd, 24th to end. Thus the first 'week' was 7, the second and third 'weeks' each 8, and the fourth 'week' 7 or 8, or in the case of February only 5 days.

[3] Among Galton's papers is another "Weather Map of the British Isles for Tuesday, Sept. 3, 9 a.m." No year is stated, but I should think it was more probably Tuesday, Sept. 3, 1861, than Tuesday, Sept. 3, 1867. In this map five conditions of wind intensity, five conditions of sky (clear, detached cloud, overcast or fog, showers, rain) and three pressure conditions (barometer falling, stationary, rising) are indicated by no less than 75 circular 'stamps.' The direction of the wind is given by the direction of the arrows which measure wind intensity and the circular stamps are rotated to give this direction. A map is printed with the names of the 60 to 70 recording stations, and underneath these names the appropriate stamp is affixed with the right orientation. 25 copies of these attachable stamps for the case of rising barometer

PLATE V

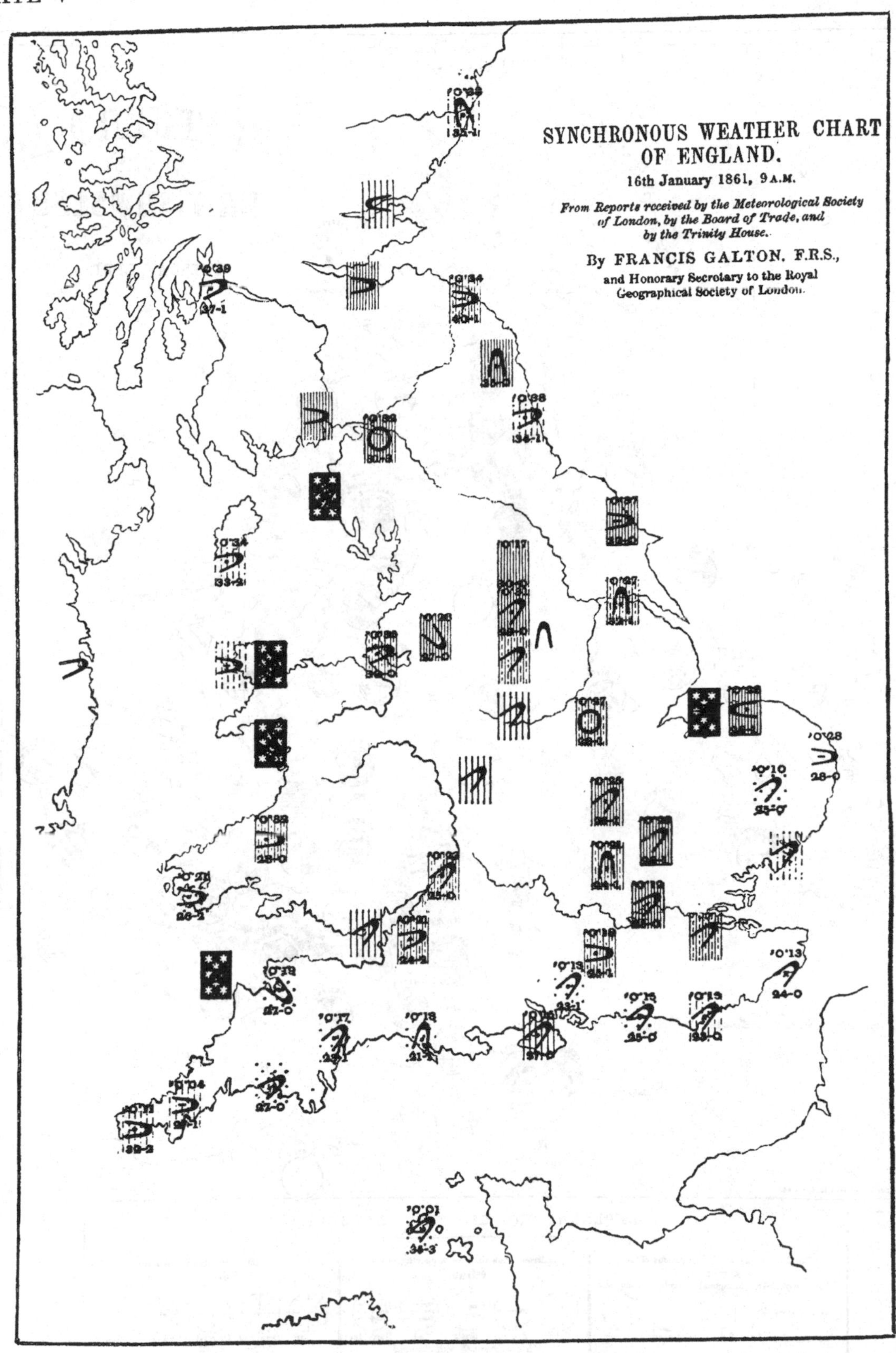


One of Galton's earliest synchronous weather maps, issued with his circular concerning European weather in 1861: see our p. 38.

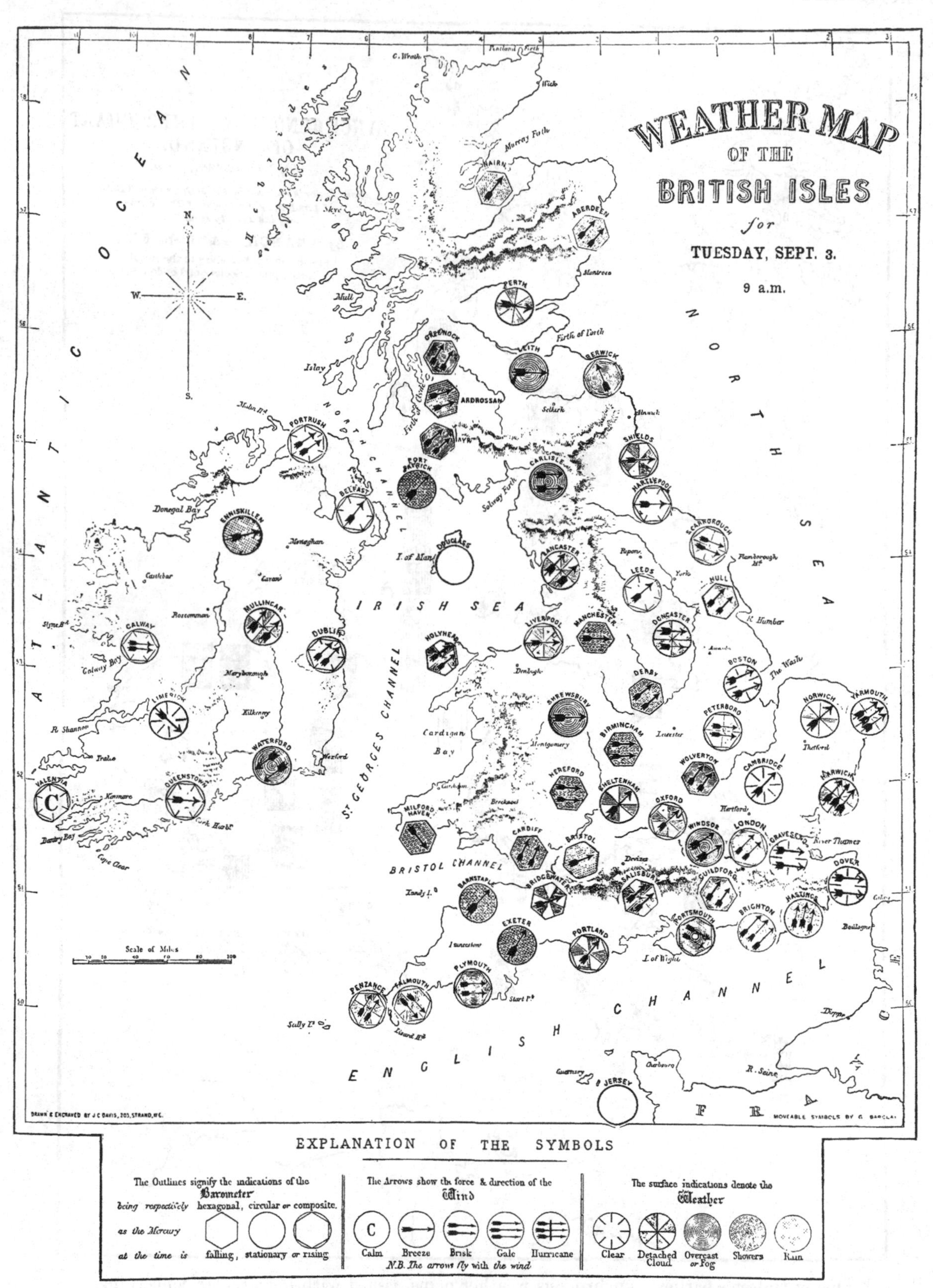


One of Galton's earliest synchronous weather maps, probably for Sept. 3, 1861, showing the use of his circular stamps to indicate direction of wind and nature of barometric change.

printed only in two colours, red outline for England and black shading for rain and cloudiness. It is entitled: "English Weather Data, Feb. 9, 1861, 9 h. a.m.," and is part of a circular issued from 42 Rutland Gate and dated June 12, 1861. "The accompanying sheet has been printed as an experiment, by means of movable types which I have had cast for meteorological purposes." To save confusion of figures, barometric heights were not inserted on the map, but lines of equal pressure having been deduced, the places where the isobars of each $\frac{1}{10}$ of an inch cut the right- and left-hand borders of the map were marked, and a straight line joining any pair of corresponding figures was taken to be approximately the corresponding isobar. These isobars were not given on the map[1].

In July 1861 Galton issued another circular, this time addressed to European meteorologists and printed in English, French and German[2]. He appeals to them to provide synchronous meteorological data for a series of aërial charts of Northern Europe (latitudes, 42° 25′ on the south, including all France and Perugia, to 61° on the north, including Shetland, Bergen and Christiania; from the westernmost limit of the British Isles to Königsberg, Warsaw and Budapest). The data were to be for the whole month of

have survived. When the proper stamps have been attached the map is ready for photography or engraving. I do not know why Galton replaced these circular stamps by the oblong blocks of his later maps, possibly because the oblongs were easier to set up in a press and actually print on to the map. This map looks more graceful than those of the circulars, but contains somewhat less information. The two maps can be compared in the accompanying plates.

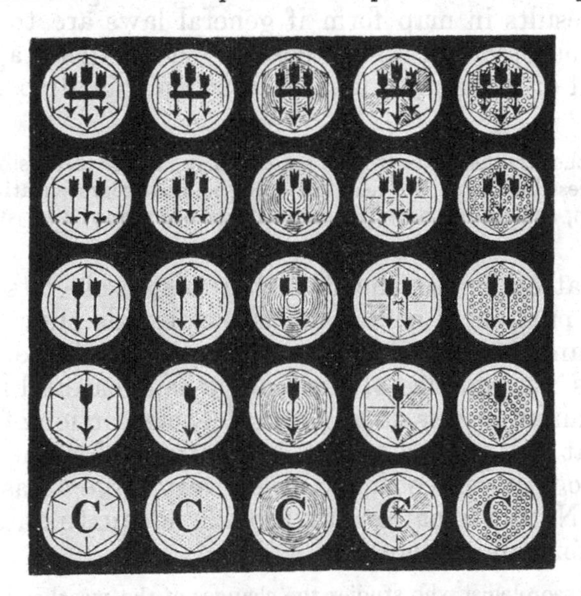


Specimens of Galton's circular stamps for attaching to maps and so forming synchronous weather charts. See footnote 3, p. 36.

[1] This circular almost in the same words appears as an article entitled "Meteorological Charts" in the *Philosophical Magazine*, Vol. XXII, 1861, pp. 34–5.

[2] This is, I believe, Galton's first appeal by circular for the filling in of schedules, a practice considerably developed by him later.

December 1861, and synchronously at 9 a.m., 3 p.m. and 9 p.m. for each day. The specimen map sent with the chart is a considerable improvement on that of the June circular. It is entitled: "*Synchronous Weather Chart of England, 16th January* 9 A.M. From Reports received by the Meteorological Society of London, by the Board of Trade and by the Trinity House." About 50 stations were used. It is printed in three colours—the outline map in green, the rain and cloudiness rectangles in brick red, and the wind symbols and figures for barometer and thermometers in black. The circular itself gives the most minute directions for observations, and even rules as to postal dispatch. Finally also a blank schedule was sent on which the desired data would be written in together with printed tables for reducing Centigrade and Réaumur to Fahrenheit, and millimetres, Paris lines and Russian lines to English inches for the barometer. As a return for assistance Galton promised a copy of his publication to contributors.

We reproduce here Galton's map of 1861 in a single colour and reduced to the size of our page, and also one of the *Meteorographica* maps of 1863.

The materials obtained by Galton's circular were somewhat disappointing, yet Galton proceeded to reduce them; his book or better atlas: *Meteorographica or Methods of Mapping the Weather; illustrated by upwards of* 600 *printed and lithographed diagrams referring to the Weather of a large part of Europe during the month of December* 1861, was published by Macmillan and printed by Eyre and Spottiswoode. In the text of this work Galton insists not only on the need of tabulating observations, but on representing the results in map form if general laws are to be drawn from them. Maps are as essential to meteorology as to geography. I believe Galton was the first or among the first to insist on this almost obvious truth. But, alas!—

"A scientific study of the weather on a worthy scale seems to me an impossibility at the present time from want of accessible data. We need meteorographical representations of large areas, as facts to reason upon, as urgently as experimental data are required by students of physical philosophy."

Galton draws attention to the fact that meteorologists are strangely behindhand in the practice of combining the materials they possess. While there are more than 300 skilled observers recording thrice daily with excellent instruments, the practice of combining their material is absent. "No means exist of obtaining access to any considerable portion of these observations without great cost, delay and uncertainty[1]." For the most valuable results in meteorology it is needful to study very large areas, or indeed the world as a whole. No single nation can provide adequate data spread over a wide enough area for valid conclusions.

"The labour of a meteorologist who studies the changes of the weather is enormous before he can get his materials in hand and arrive at the starting-point of his investigations. In the

[1] Thirty or more years later the biographer found the same difficulty still in existence, when correlating barometric heights across the Atlantic, eastward from Hammerfest to Cape Town, westward from Halifax to the Falklands; the required data existed in manuscript, but were very costly to get copied.

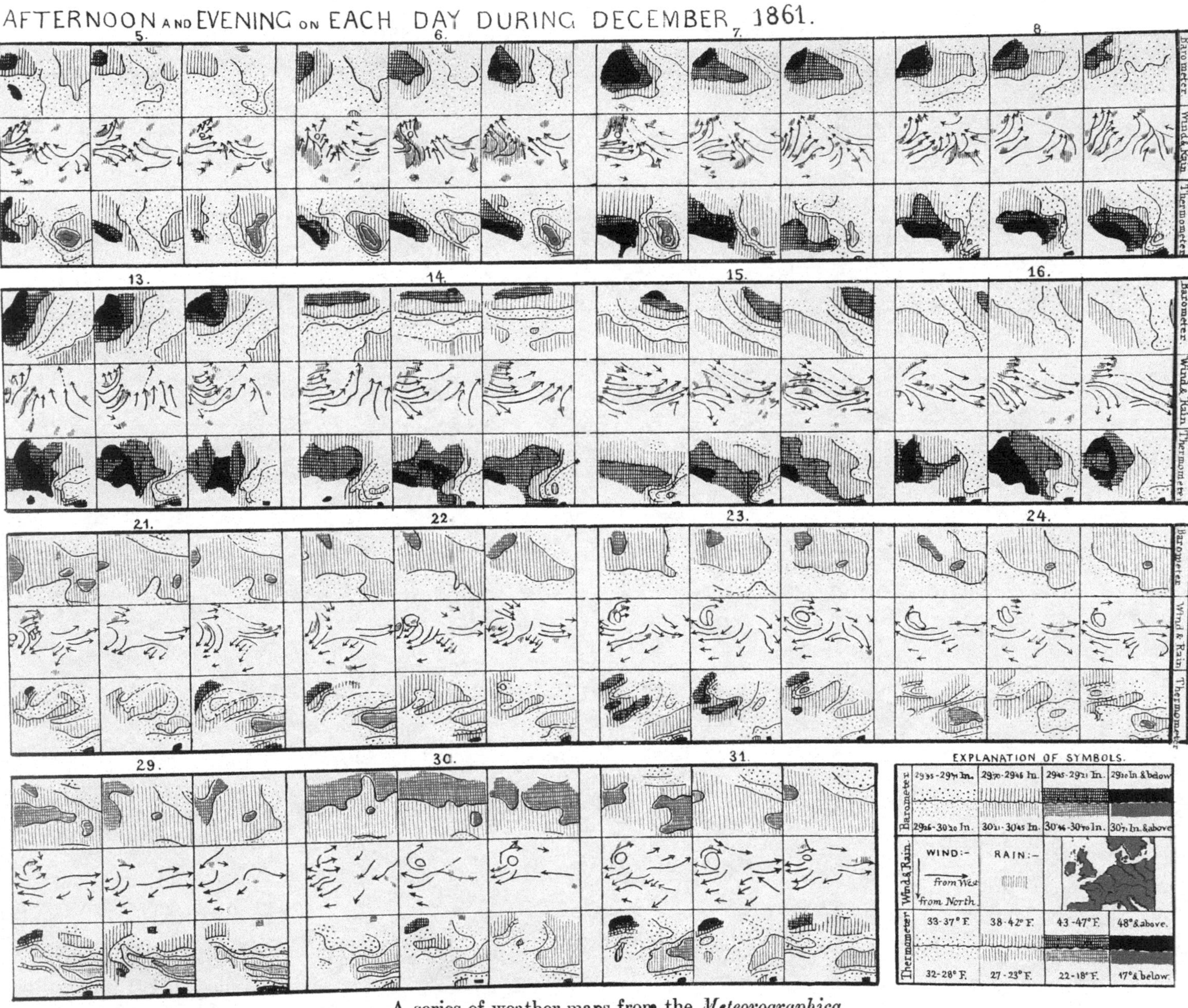


A series of weather maps from the *Meteorographica.*

ordinary course he has to apply, with doubtful chance of success, to upwards of 10 Meteorological Institutes in Britain and Europe, for the favour of access to the original documents received by them, and to fully 30 individuals besides. He has next to procure copies, then to reduce the barometer and thermometer readings to a common measure, and finally to protract on a map. I feel that all this dry, laborious, and costly work, which has to be undergone independently by every real student before he can venture a step into scientific work, is precisely that which should be undertaken by Institutes established for the advance of Meteorology." (p. 3, col. i.)

Galton's own list of failures is considerable:

"There was no central Institute in Switzerland......neither was there any recognised Institute in Denmark or Norway. Whether by accident or misunderstanding, several promised communications from Denmark have never reached me, to my great regret, for its weather was closely linked with our own. From Sweden I could obtain nothing, from France next to nothing[1], from Bavaria only the valuable observations made at Munich. From Italy I had considerable hopes held out to me, but little fruit. The interior of Ireland is wretchedly represented, and would have presented a gap, like France, were it not for two eminent astronomers and some chance assistance besides." (p. 4, col. ii.)

The bulk of Galton's data came from Belgium (with the aid of Quetelet), Holland (with the help of Buys Ballot), Austria (from Kreil) and Berlin (from Dove). To the three former Galton tenders his special thanks. Then comes Galton's excuse for his publication of a work based on admittedly inadequate data:

"Entertaining the views I have expressed on the necessity of meteorological charts and maps, and feeling confident that no representation of what *might* be done would influence meteorologists to execute what I have described, so strongly as a practical proof that it could be done, I determined to make a trial by myself, and to chart the entire area of Europe, so far as meteorological stations extend, during one entire month, and I now publish my results." (p. 3, col. ii.)

A most important discovery was made by Galton as soon as he had begun plotting his wind and pressure charts. While Dove had recognised that centres of low pressure in the northern hemisphere were associated with counter-clockwise directions of the wind round a centre of calms, and termed this system a cyclone, Galton noted that centres of high pressure are associated with clockwise directions of the wind round a centre of calms. Galton termed this system an *anticyclone*, and the name rapidly came into general use, and is very familiar now although few who use it remember that Galton first noticed the system and coined the name[2].

When one studies Galton's tiny charts of pressure and wind for the thirty-one days of December 1861, each chart extending over the whole of Central Europe, and thinks of the paucity of his data, one cannot but wonder at the inspiration which led him to his conclusions. Luckily December 1861 was a month of contrasts, the first half of the month marked a series of cyclones—

[1] Appeal to France for scientific information is even after the war nearly always in vain; letters remain unanswered, and presents of memoirs unacknowledged. From both Germany and Austria, even at the present day, one is fairly certain of a full and courteous reply, and almost any German University Library will still lend a book inaccessible in this country. Narrow nationalism in science is a crime against our common humanity.

[2] "A Development of the Theory of Cyclones." Received Dec. 25, 1862. *Royal Society Proceedings*, Vol. XII, 1863, pp. 385–6.

the black areas of low pressure on the barometer charts corresponding to a whole series of counter-clockwise running arrows on the wind charts; and the second half of the month marked a series of anticyclones—the red areas of high pressure on the barometric charts corresponding to a whole series of clockwise running arrows on the wind charts. About the middle of the month we have the transition from black to red areas on the barometric charts, and here sure enough are two systems of arrows on the wind charts one counter-clockwise and one clockwise. *But* it is very clear that the broad band from the Skelligs to Königsberg, west and east, and from Siena to Christiania, south and north, was largely inadequate to exhibit the 'cores' of a cyclone and

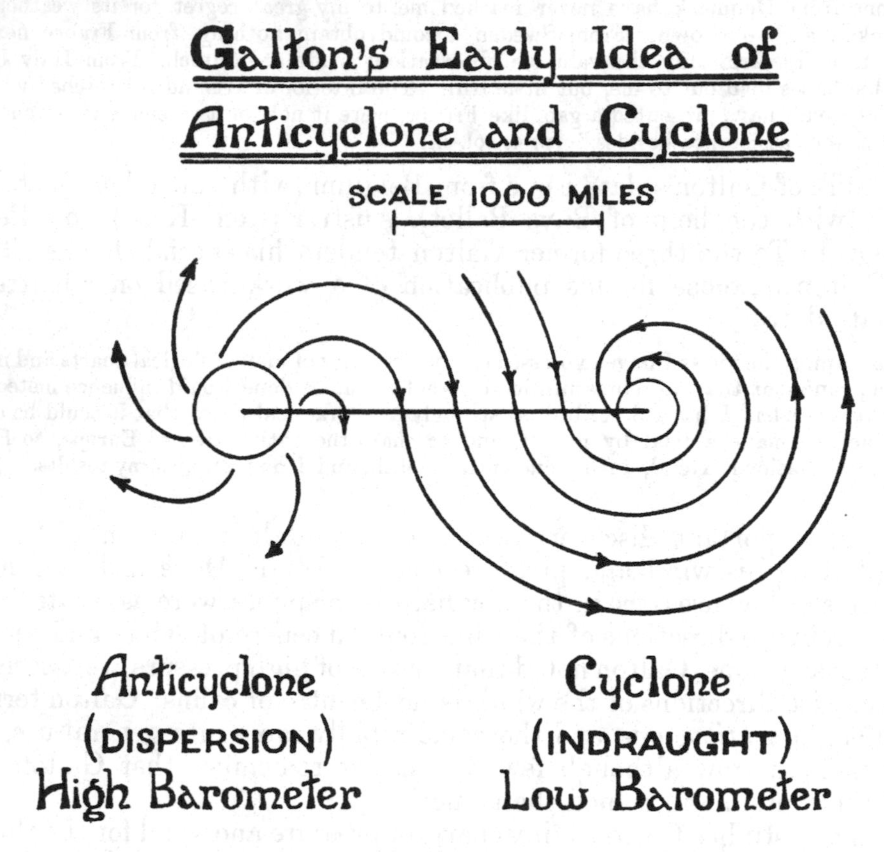


anticyclone on the same chart. The cores of one or other or even of both lay outside the large area for which Galton was plotting simultaneous observations. As I have already remarked, a single continent is scarcely sufficient for the study of meteorological observations. Such is one of the main lessons of the *Meteorographica*, and one doubts if it had been realised before that publication. Yet Galton recognised that if an observer in the northern hemisphere supposed himself standing at the core of an anticyclone—i.e. a centre of high pressure—and facing towards the core of a cyclone—i.e. a centre of low pressure—the winds would pass from his left to his right hand. If we term the line of his sight a bi-cyclonic line, Galton in his Royal Society paper of December 1862

Francis Galton, aged 38, from a photograph of 1860.

supposed that the wind would cross this bi-cyclonic line at an angle of 45°. In his *Meteorographica* he had modified this statement. He writes:

"Many meteorologists will refer with eagerness to these wind charts, to see how far they may confirm or oppose the theory of cyclones. I deduce from them......that they testify to the existence, not only of cyclones, but of what I ventured to call 'anticyclones.' If the lines of wind currents, in the black and red lithographs, are compared with the barometrical charts immediately above them, one universal fact will be found throughout the entire month. It is that on a line being drawn from the locus of highest to the locus of lowest barometer, it will invariably be cut more or less at right angles by the wind; and especially, that the wind will be found to strike the *left* side of the line, as drawn *from* the locus of highest barometer. In short, as by the ordinary well-known theory, the wind (in our hemisphere) when indraughted *to* an area of light ascending currents, whirls round in a *contrary* direction to the movements of the hand of a watch, so, conversely, when the wind disperses itself *from* a central area of dense descending currents, or of heaped up atmosphere, it whirls round in the *same* direction as the hands of a watch. I confidently appeal to these maps, and especially to the original MSS. whence these charts have been reduced, to confirm the theory." (*Meteorographica*, p. 7, col. i.)

From the temperature charts Galton did not draw conclusions as epoch-making as from the pressure charts, perhaps he laid overmuch stress on the *direction* of the wind as the chief source of hot and cold areas; but when we persist beyond that first feeling of repugnance which the crudely hatched masses of red and black on his charts excite in our minds, we catch glimpses of broad generalisations, or if the reader prefers suggestions, of what might flow from the more accurate synchronous data plotted by similar methods for still more extended areas.

"The areas of barometric elevation and depression are enormous, and in their main features are very regular. They are easily recognised by the lithographic maps, in black and red. There is no case in which the Charts include the whole breadth or length of any one of these areas, and there are cases where clearly not one-half of them is included, yet the map is about 1,200 geographical miles in height and 1,500 in breadth. They do not move with regularity, ridge behind ridge, like waves of the sea, but they are ever changing their contours and their sections. They also vary in the speed and directions of their movement of translation." (p. 7, col. ii.)

Galton had seen that Great Britain was not a large enough area for meteorological inquiry; he then attempted what might be learnt from what he terms an "enormous area," only again to realise that 2000 miles is hardly adequate to exhibit at the same time a cyclonic and an anticyclonic system. He thus prepared the way for that world meteorology on which modern forecasting essentially depends, and which is now-a-days a commonplace of our daily papers.

Nay, it is to Galton himself that we owe those little weather charts which form a familiar item of our morning news, e.g. in the *Times* newspaper. There is a little series of maps in the *Galtoniana* of the Galton Laboratory of which the diagram on page 42 is a reproduction, under which Galton has written "First attempt made for *Times* by a drill pantagraph in plaster and a stereo taken from it, my proposal." The maps are for December 10, Evening, and show by different types of shading the areas which lie between certain ranges of the meteorological characters[1]. It is interesting to compare

[1] Galton does not on these first maps state what characters were represented by the two systems, probably pressure and cloudiness; there are no indications of wind direction and no printed figures.

Galton's suggestions with the isobar maps of to-day, still giving wind by arrows, recording temperature by figures and state of the heavens in words. This original meteorological map for the *Times* must have been at a later date than we are now considering, perhaps about 1869, as the drill pantagraph must have been previously constructed.

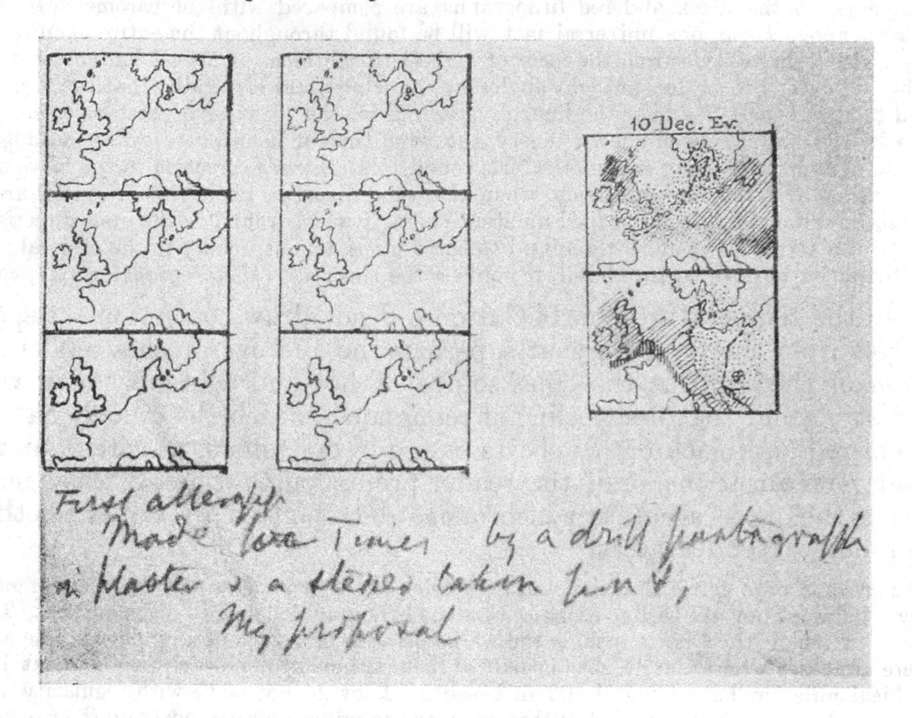


But I cannot find that it was ever published. The first issue of a weather-map in the *Times* was on April 1, 1875, and we give on p. 43 a reproduction of it. An account of the matter was published in *Nature*, April 15, 1875. We give a few sentences from it:

"The method of preparation of the chart seems simple enough at present, but it has been the fruit of much thought, as the problem of producing in the space of an hour a stereotype fit for use in a Walter machine has not been solved without many and troublesome experiments."

Then follows a brief description of the material, the drill pantagraph (see our p. 46) and the engraving of the block.

"The initiative in this new method of weather illustration is due to Mr Francis Galton....... It is hardly necessary to allude to the value of such charts as these as a means of leading the public to gain some idea of the laws which govern some of our weather changes."

The *Shipping Gazette* started publishing on January 4, 1871 a daily chart for the winds round the coast of the British Isles on the basis of reports telegraphed to the Meteorological Office. It states in its issue that "this new system of showing the direction and force of the wind by movable types etc. has been entered at Stationers' Hall." After Galton's maps of 1861 and 1863, it is difficult to see why the system should be called 'new.'

The publication of the *Meteorographica* placed Galton at once among the leading English meteorologists. The history of weather forecasting in England starts from Admiral R. Fitzroy of 'storm cone' fame[1]. By his exertions the English Meteorological Office was founded in 1854. Fitzroy had more enthusiasm than science. On his death in 1865 the Board of Trade appointed a small departmental committee to consider the whole subject. It consisted of Mr (afterwards Lord) Farrer, then permanent secretary of the

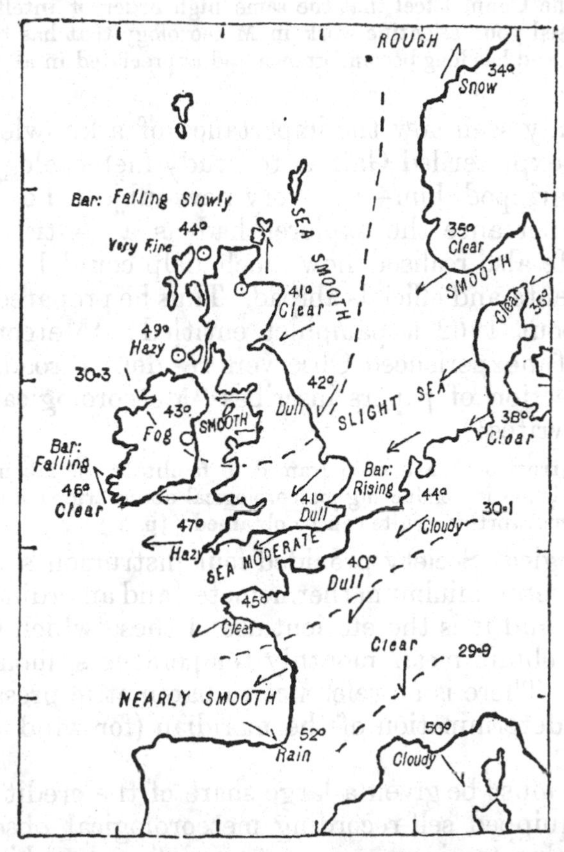


**The dotted lines indicate the gradations of barometrical pressure, the figures at the end showing the height, with the words "Rising," "Falling" &c., as required. The temperature at the principal stations is marked by figures, the state of the sea and sky by words. The direction and force of the wind are shown by arrows, barbed and feathered according to its force. ⊙ denotes calm.**

Galton's Weather Map, *The Times*, April 1, 1875.

Board of Trade, Captain Frederick Evans, the Hydrographer, and Francis Galton. They reported in 1866, and as a result of their report the Meteorological Committee was appointed in 1868 with Galton as a member. This committee worked for some years, but it was felt that a wider scope of action was desirable, and after a second Government committee appointed by the Board of Trade and Treasury conjointly, Galton again being a member, it emerged as the Meteorological Council, and of this Galton was a member until 1901.

[1] Better known to some of our readers as the Captain Fitzroy of the *Beagle*, the surveying ship on which Charles Darwin sailed as Naturalist.

Thus for nearly forty years Galton was intimately associated with both the theory and practice of meteorology in this country. In a letter to Galton on his resignation from the Meteorological Council in 1901, the then chairman, Sir Richard Strachey, wrote:

"It is no exaggeration to say that almost every room in the Office and all its records give unmistakable evidence of the active share you have always taken in the direction of the operations of the Office. The Council feel that the same high order of intelligence and inventive faculty has characterised your scientific work in Meteorology that has been so conspicuous in many other directions, and has long become known and appreciated in all centres of intellectual activity[1]."

We have already seen how the importance of a knowledge of climate to the traveller and explorer led Galton to study meteorology; but as soon as this subject had 'gripped' him—as every new subject he attacked did—he recognised the importance the explorer had as a contributor to meteorological science. He also realised how much help could be obtained for this science from residents and officials abroad. Thus he prepared for the Meteorological Society about 1862 a pamphlet entitled: "Meteorological Instructions for the use of inexperienced Observers resident abroad." This pamphlet Galton in his collection of papers inscribes "Meteorological Instructions for *Travellers*." He writes:

"The following instructions have been framed to facilitate the labours of those who have little leisure and experience in conducting meteorological observations, and show the minimum of effort with which trustworthy results can be obtained." (p. 2.)

The Meteorological Society provided four instruments at a small cost,—maximum thermometer, minimum thermometer, and an ordinary thermometer, with a rain-gauge, and it is the efficient use of these which Galton describes. The object was to obtain mean monthly temperatures, monthly ranges, rain and wind returns. There is no reference to barometric pressure. Geographical position and a determination of the meridian (for wind observations) are also referred to.

To Galton also must be given a large share of the credit for devising and organising well-equipped self-recording meteorological observatories. Continuous photographic tracings were arranged for the chief meteorological instruments. These are very familiar now, but they required much time and thought in those early days of meteorology and photography[2]. When these 'tracings' were obtained they were not in a form for reproduction and publication, and the difficulty, which meets the editor of every journal, was encountered by Galton, namely: How can diagrams be reproduced so as

[1] Letter of May 9, 1901 from Meteorological Office. It would be impossible to enumerate here all Galton's work for the Meteorological Committee. The index to its *Minutes* must be consulted by those desiring further information. His plans from anemometers and pantagraphs to methods of "weighting" ship-logs, of lithographing and charting are scattered broadcast through these *Minutes*. Galton devised a "Torsion anemometer" and a "Hand anemometer" for use on ships. The latter may still be seen in the Science Museum, South Kensington. See *Catalogue of Meteorology*, pp. 53, 61, 1922.

[2] "I had the satisfaction in its [i.e. Meteorological Council's] early days, when new instruments and methods were frequently called for, of being able to do my full share of the work." *Memories*, p. 234.

to reduce the scales in two directions at right angles in any desired and *different* ratios? A photograph, or with more labour a pantagraph, will reduce in *both* directions in the *same* ratio, but this is not what is needed. A contributor of a memoir rarely pays any attention to the proportions of the page in which he desires his paper to appear, and then it is a mere chance whether his diagrams however neatly constructed can be used without redraughting. The ideal remedy would be a photographic process of bi-directional

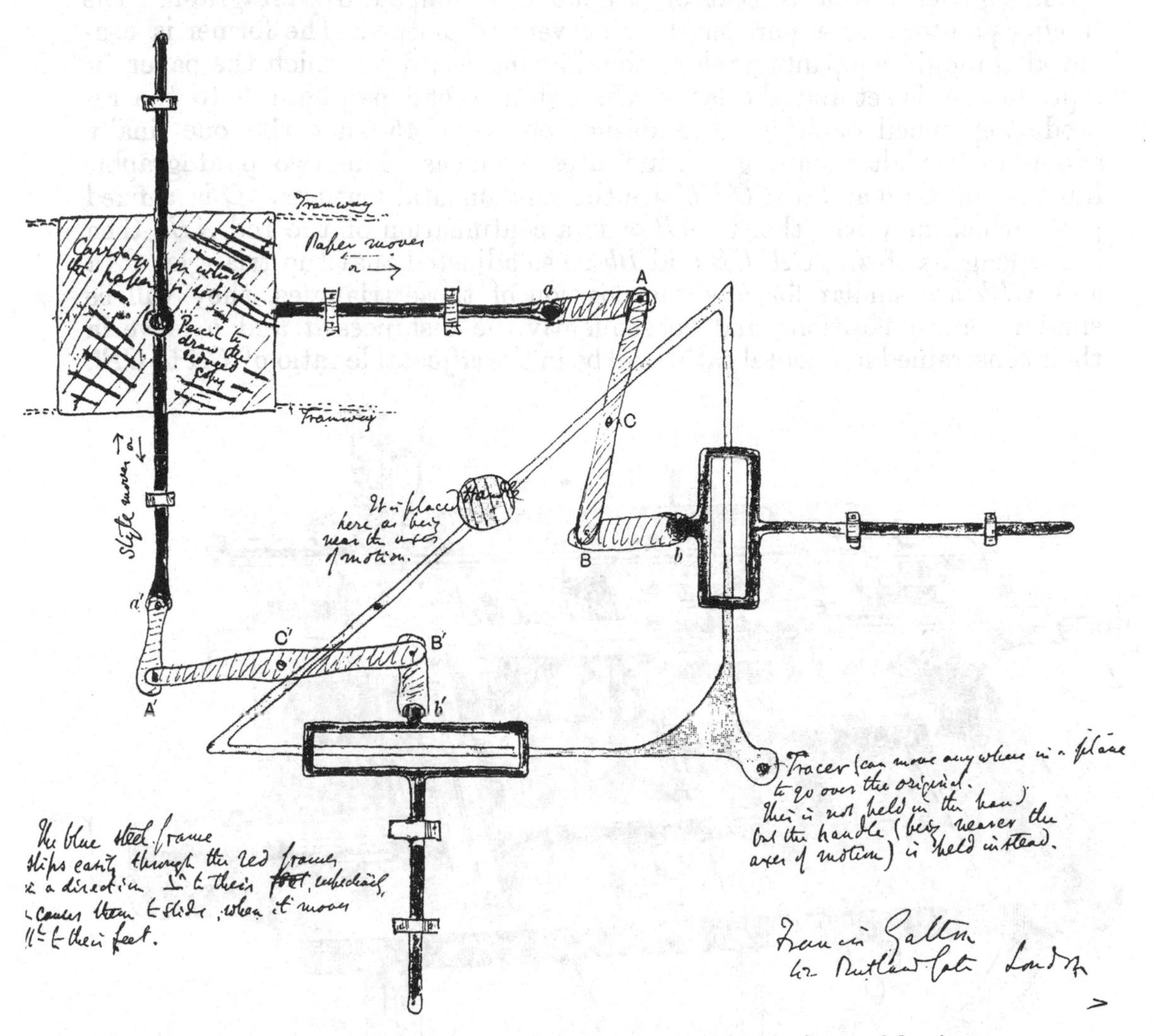


One of Galton's original designs for double pantagraph, coloured in the actual drawing.

reduction, because photography is so much shorter and cheaper than pantagraphic work. The difficulties as to distortion of lettering would be overcome by pasting on the printed lettering to the reduced photograph, instead of to the original drawing. But although this topic more than once formed the subject of long talks of the present writer with Francis Galton, no practical photographic scheme was then evolved[1]. The difficulty of bi-projection met Galton at a very early stage of his meteorological work, and he solved it by the construction of a compound pantagraph. He gave a great deal of thought to the subject, and his papers contain numerous devices and suggestions for an instrument of this character. His work upon it began in 1867, but was not completed till 1869, when the first compound pantagraph constructed by Mr C. Beck was placed in the Meteorological Office.

The general idea is that of a double or compound pantagraph. The tracing pointer has a horizontal and a vertical motion. The former is conveyed through one pantagraph to the drawing-board on which the paper for reproduction is set and the latter through a second pantagraph to the reproducing pencil or style. The design on page 45—not the one finally adopted—in Galton's autograph indicates his ideas. The two pantagraphic linkages $aACBb$ and $a'A'C'B'b'$ are the fundamental features. $C$ is a fixed pivot which may be either in $AB$ or in a continuation of the rod $AB$; then if the lengths of $aA$, $CA$, $CB$ and $Bb$ be so adjusted that the triangles $CAa$ and $CBb$ are similar for any one position of those triangles, they will be similar for all positions, and consequently the distances $a$ and $b$ move in their constrained horizontal paths will be in the adjustable ratio of $CA$ to $CB$[2].

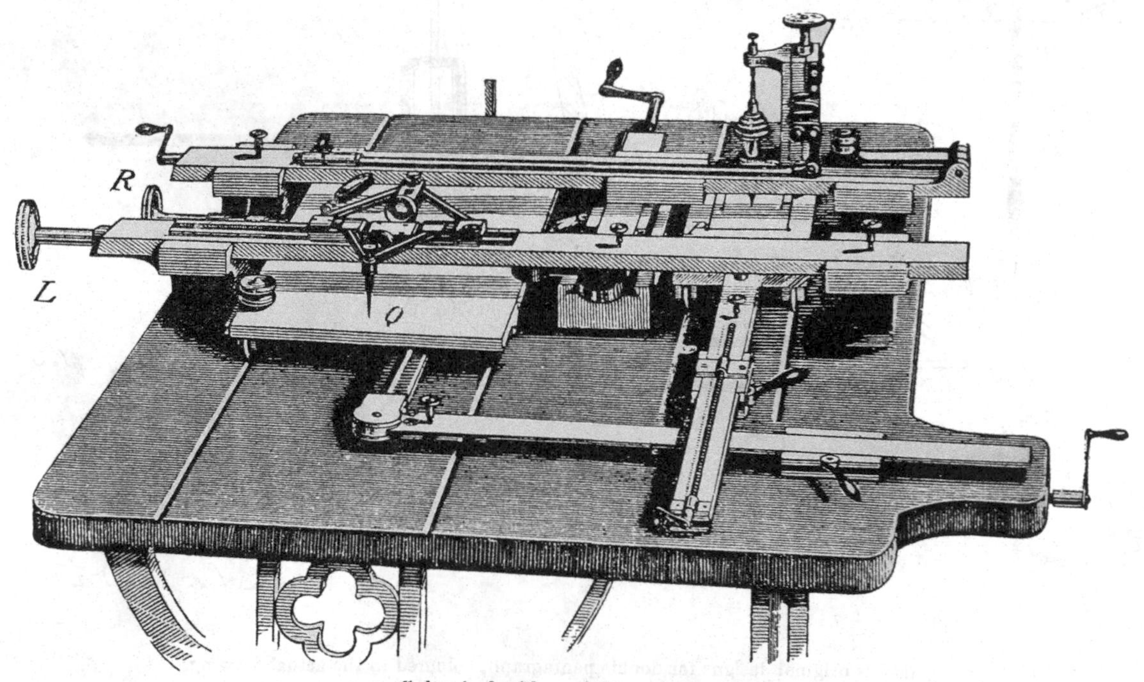


Galton's double or drill pantagraph.

[1] We shall see later that Galton actually solved the problem, but did not publish his solution. A simple 'bi-projector' was made for the writer by Mr Horace Darwin many years ago for drawing parabolae, ellipses, probability curves, etc., but it involved the cutting of a definite metal template for each type of curve; it might possibly be adapted for reducing drawings.

[2] This fundamental principle of Galton's compound pantagraph is discussed by him with proof in a letter of July 15, 1869, but he was inquiring for a maker even in May of 1869.

Perhaps a simpler arrangement would be to replace the linkage by pins at $a$ and $b$ running in a slotted bar turning about a pivot $C$. We reproduce (page 46) an illustration of the final apparatus; the pencil or style could be replaced by a drill[1]. With this instrument for twelve years the *continuous* automatic weather records for seven stations (velocity and direction of wind, dry and wet bulb thermometers, barometer, vapour-tension and rain) were reduced to manageable dimensions and published. Of this publication Galton remarks:

"It surprises me that meteorologists have not made much more use than they have of these comprehensive volumes. But there is no foretelling what aspect of meteorology will be taken up by the very few earnest and capable men who work at it. Each of them wants voluminous data arranged in the form most convenient for his own particular inquiry[2]."

Probably the use has not been made of these graphical charts that might well have been made; but Galton's own results indicate that we need *simultaneous* data for a far wider range than Great Britain, and further modern methods of multiple correlation, which seem likely to be most productive of result in present day meteorology, demand numerical values, and these are hard to obtain from the graphs; not only can they scarcely be read off with the requisite accuracy, but to reconvert the graphs into any numbers whatever is in itself a most arduous task.

Galton's compound pantagraph has indeed a far wider field of usefulness than reducing automatic weather returns. The difficulty is that it is not made commercially and procurable at a moderate cost.

A second instrument devised by Galton about this time will be found described in the *Report of the Meteorological Committee*, 1871 (p. 30). It was devised for obtaining mechanically the vapour-tension curve from the curves of dry and wet bulb thermometers, but again it can be used to serve a much more general purpose, namely to obtain the curve of a variate whose ordinate is a given function of the ordinates of two other curves—all three curves having the same abscissa. The machine depends upon the construction of a surface corresponding to the function the variate is of the two other ordinates (i.e. in Galton's case the vapour-tension in terms of wet and dry bulb thermometer readings). By fine screw adjustment the cross-hairs in two microscopes are brought into accordance with the tops of the ordinates in the two curves, but the screw which adjusts one microscope moves the surface parallel to one axis, and the screw which adjusts the other microscope moves the surface perpendicular to this direction. Thus a vertical style resting on the surface raises to an adequate height a scriber which marks the ordinate or function-value of the compound variate[3]. It would be out of place here to give a more complete account of the instrument, but my more mechanically minded readers will grasp the general idea from the

[1] The theory is fully described in the *Minutes of the Meteorological Committee*, 1869, p. 9. It is also figured in the *Katalog mathematischer Modelle, Apparate und Instrumente*, of the Deutsche Mathematiker-Vereinigung, 1892, p. 232.

[2] *Memories*, p. 236.

[3] In Galton's actual instrument (see our p. 48) the required curve was recorded on a zinc plate (partly removed in figure to show scriber $R$). The scriber received when adjusted a blow from the hammer $H$ worked by the action of the operator's foot on a treadle.

accompanying figure. Galton constructed his surface from a table of 400 values of the vapour-tension, 400 holes being bored into a solid rectangular block to these 400 values spaced properly apart, and then the remainder cut away, filed and smoothed. The construction at that time did not cost more than £6. Here again it is easy to think of many purposes to which a machine of this kind could be put, but as it has never been made as a commercial article, it has never come into general use. Perhaps this brief notice may remind investigators of the existence of Galton's design.

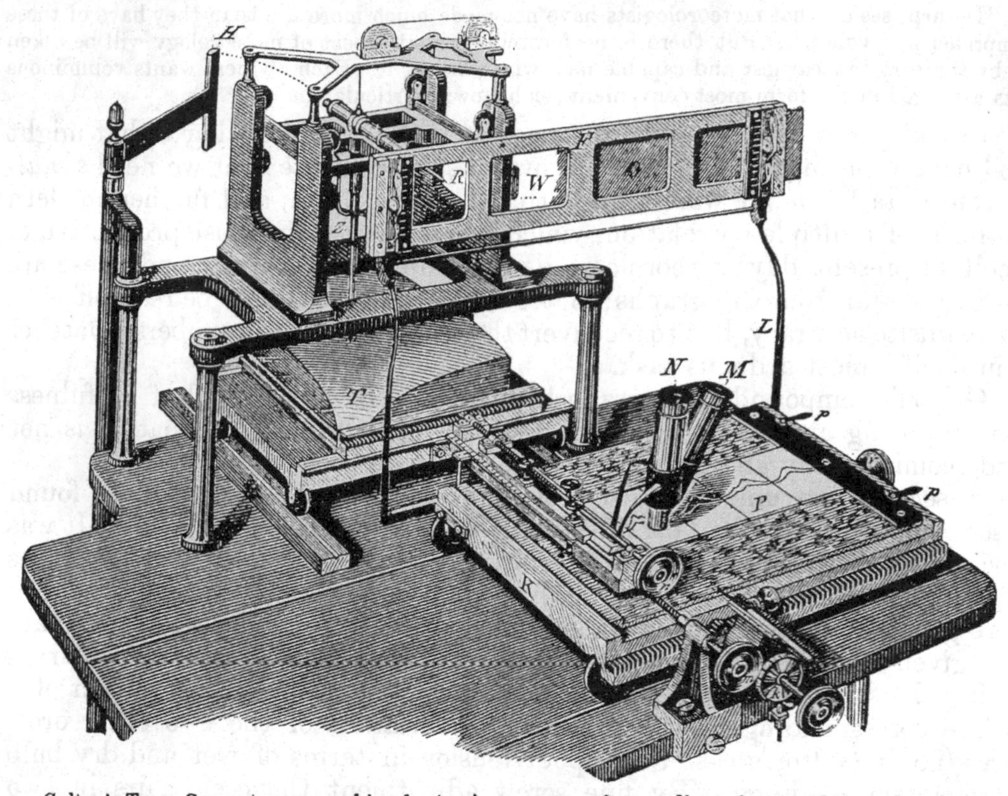


**Galton's Trace Computer—a machine for tracing a curve, whose ordinate is any arbitrary function of two other variate values at the same abscissa or time.**

A third instrument designed by Galton a little earlier (1867) never came in being, owing probably to a discouraging letter from Balfour Stewart at that time at the Kew Observatory, who laid great stress on comparison of pairs of automatic meteorological records at different intervals. Galton was easily discouraged and was apt to treat the judgments of the really able people whom he consulted as sure to be better than his own. It certainly was a pity that in this case he was put off completing his model. It was of the following nature: a map is mounted horizontally on, let us say, a metal plate; then holes are drilled at each meteorological station, and a rod of a convenient length is free to move vertically up and down in the hole. Templates are now cut to the continuous automatic records of any meteorological character for these stations and are fixed in vertical planes running east and west on a carriage which runs east and west on rails under the map. Let us suppose the top of the rods to rest on the templates; then if the templates be adjusted by shifting east and west, so that the rods all rest on the points of the templates corresponding to the same instant of time, their tops will mark the contemporaneous value of the chosen variate at that time; and, if the stations be fairly numerous, will indicate a sort of surface of the variate. Let now the carriage be moved along, and the surface will change with the time, and the eye will recognise how the fall in one area is accompanied by a rise in another. For example we should actually see a cyclone or anticyclone passing along.

As a matter of fact Galton linked up his vertical rods with his templates by a system of levers, and this might be needful for one or two stations absolutely in the same latitude, but the cheapest construction would be fairly light rods ending with knife-edges to rest on the corresponding templates. He proposed also to convert the up and down motion of the wind curves into the angular motion of an arrow turning round a vertical axis at the station on the map. Galton's drawing of his apparatus is dated April 6, 1867, Sorrento, Italy, and his description of it April 11, 1867[1]. Mrs Galton's diary says that they travelled to Italy at the end of January 1867, "staid chiefly in Rome and Naples and the neighbourhood of Venice, then by S. Tyrol to St Moritz, where the cure did wonders for me, but did not suit Frank." Then the Galtons went to Heidelberg and Bavaria, reaching England in October, where after a round of visits they settled in London by the end of November. Such were the conditions under which Galton had largely to do his work! One is forced to believe that he walked and thought, and his pocket notebooks suggest that he jotted down his diagrams and rough calculations at odd moments.

### D. OTHER MECHANISMS

We have already referred to 'Galton's Toys,' models made up of strings, pieces of wood, lenses, or often of card and bits of glass only, which it is now practically impossible to interpret. But these 'Toys' are not all, there are constantly diagrams and schemes for instruments or the improvement of instruments among Galton's papers. I take almost at random a bundle with papers dating from 1858 to early in the 'seventies; this contains *inter alia* the following packets:

(*a*) One entitled "Examination of Sextants." It appears that at Galton's suggestion in 1858 the General Committee of the British Association passed the following resolution:

"That the consideration of the Kew Committee be requested to the best means of removing the difficulty which is now experienced by officers proceeding on Government Expeditions, and by other scientific travellers in procuring instruments for determination of geographical positions, of the most approved portable construction, and properly verified. That the interest of geographical science would be materially advanced by similar measures being taken by the Kew Committee in respect to such instruments to those which have proved so beneficial in the case of magnetical and meteorological instruments."

[1] Balfour Stewart's letter is dated Kew Observatory, May 1, 1867.

This resolution led to Galton being placed on the Kew Committee, and to an endeavour being made to raise the standard of angle-measuring instruments in this country, and their comparison with those of foreign make. The Royal Geographical Society was approached in the matter and that Society passed a resolution to offer a prize of £50 or a gold medal to the "Designer or maker of the most serviceable Reflecting Instrument for the Measurement of Angles"—doubtless at the instance of its Honorary Secretary. There exists a whole series of letters to Galton on the point. Sir Edward Sabine in a letter of Feb. 16 refers to both resolutions as Galton's. The latter proposed a Kew certificate for sextants, and a study of errors due to special forms of mercurial horizon, as well as of those peculiar to the prismatic compass.

(*b*) But Galton did not confine his attention to the above instruments. In 1864 Casella brought out a pocket 'altazimuth,' "improved and modified by the kind assistance of Francis Galton Esq. F.R.S." It could be used in two positions, in one as a good azimuth compass, and in another as a weighted disc for altitudes. It could also be used as an ordinary compass or as an ordinary clinometer[1].

(*c*) Galton also designed for Casella a small pocket instrument termed a *Zeometer*[2] (from Greek ζέω, boil), by means of which with an ounce of water and drachm of spirit the height of any mountain could be obtained and index correction of the aneroid determined; Galton provided a table of corrections for cases in which there was a considerable portion of the mercury in the stem of the thermometer outside the vessel containing the boiling water, and this table accompanied the directions for the use of the Zeometer.

(*d*) There is the design of another instrument to show by the action of a piece of catgut or of whalebone strips on the motion of a clock the number of hours per day in which humidity has exceeded a datum value. Galton considered that a similar arrangement could be made for temperature.

(*e*) Details of a linkage for determining the conjugate foci of a lens mechanically[3].

(*f*) A note on lighthouse signals. Galton notes that the period of a complete breath is very nearly and very regularly four seconds. That this four seconds as a period recognisable by everyone should be taken as the base unit for lighthouse flashing signals.

(*g*) The original design of the hand-heliostat, with diagrams of its working and a water colour illustration of the field of view with the mock-sun covering the requisite flash point of the landscape. See p. 19 above.

(*h*) An instrument termed the "Tactor" machine. The diagram shows it to consist of two levers each with a tooth working on one of two complicated eccentrics on the same axis and apparently causing certain blocks to rise, fall and grip. I have no idea for what purpose the "Tactor" machine

[1] There exists still Galton's determination by aid of it of the latitude of Rutland Gate!

[2] There are very full details for the construction of this instrument, apparently in the draft of a letter to Casella.

[3] This occupied Galton again later, when he was busy with photographic change of scales, and in conjunction with Mr (now Sir) Horace Darwin a very reasonable linkage was devised to keep object and focal plane image at their proper distances from the optical centre of the objective.

was designed, and there is nothing but the design, no accompanying description to explain matters.

(*i*) A very detailed account of the "Wave Engine" in two notebooks dated 1871–2. Galton was very busy during these years with an endeavour to invent an engine by which the energy of waves might be rendered available for useful work, and in particular for the propulsion of ships. Galton's attention was probably first drawn to the matter by the difficulty there is in getting from an open boat on board a vessel at sea.

"Those who in rough weather have had occasion to get on board a vessel at sea are well aware of the large and rapid changes of relative position between the boat and the vessel. At one moment the boat has to be fended off from the sides of the companion ladder against which it is violently dashed, at another it is lying many feet below its lowermost steps. No ordinary activity and presence of mind are required in a person unaccustomed to the rhythm of these changes to seize the exact moment when it is possible to jump onto the ladder without accident. Even if the waves be so short compared to the length of the vessel that she rests in perfect steadiness while the boat is tossing about, the difficulty of embarkation is still very great, for the rise and fall of the boat is 4 feet in moderate weather in a roadstead like Spithead[1] (of course it is much more in the open sea) and it will be repeated perhaps 12 times in the minute. It is clear that this energy might be made to do work, if the boat were secured to the end of an arm, moving vertically up and down like a pump handle, that handle might be connected with suitable mechanism and caused to perform useful work."

The simplest conception is that of a buoy attached to a lever with fixed fulcrum; the up and down motion of the lever may be turned to useful work. Galton calculates a table of wave energy, measured in horse-power per ton of surface water. Thus for a wave of 5 ft. height (from trough to crest) with a period of 5 seconds, and a vessel displacing 1 ton of water, the horse-power would be 1·3. Galton next takes two vessels $V$ and $W$ and he proposes to link them together in such wise that they have complete liberty within the range of the slide which forms part of the "link."

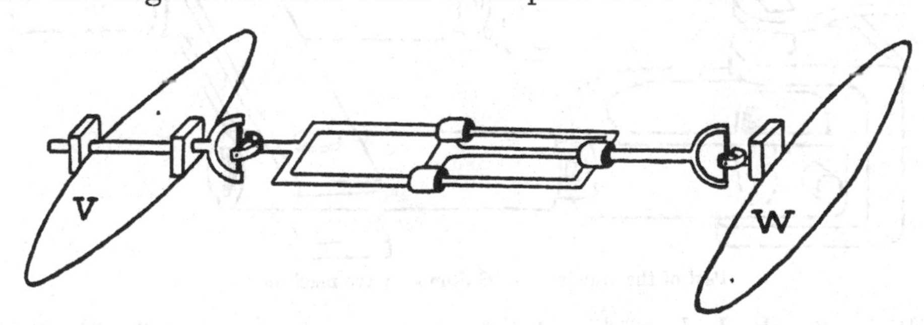


"The link consists of a Hooke's joint at the side of $W$, which allows $W$ to roll and to yaw,—it will be obvious that the same movement which permits rolling obviously includes heaving. An

[1] A few extracts from *L. G.'s Record* throw light on these years: "Frank gave up his rabbit-breeding and took to machine inventing....We were at Southsea enjoying the Dockyard at Portsmouth, and the sight of the great ships of war. Captain Hall took us about in his steam launch. We went over the Wellington, the Victory, the St Vincent training ship, the Queen's Yacht the Enchantress, and the Monarch and Devastation, the great ironclads, also the Trafalgar previous to its sailing next day....Frank taken up with spiritualism and attended meetings at Mr Crookes's and Mr Cox's. We went to Brighton for the British Association, and Emma [Francis Galton's sister] joined us on the 10th [August]. Frank President of the Geographical Section. Stanley made himself most conspicuous and obnoxious" [see our p. 30].

axle passing across *V* allows the relative pitching and tossing of the two vessels. This axle is connected by a Hooke's joint which allows exactly the same movements of rolling (inclusive of heaving) and yawing to *V* that the first-mentioned joint did to *W*. And lastly the two Hooke's joints are connected by a sliding arrangement, which permits the vessels to approach or separate from one another within the range of the slide."

"In the case I am about to consider, I will suppose the three motions consisting of (1) the relative pitching of the two vessels, (2) the rolling of *V* and (3) the yawing of *V* to be transferred to a wave engine' on *V*, and the other three motions consisting of (1) the relative separation (or approach) of the two vessels, (2) the rolling of *W* and (3) the yawing of *W* to be transferred to the 'wave engine' on *W*."

The bulk of Galton's paper is then concerned with the mechanical arrangements by which every phase of the relative motion of the parts of the "link" can be applied to producing rotatory motion on *V* and *W*. Such mechanical arrangements constitute the "wave engine." To describe them would take us beyond our proper limits, but they exhibit all Galton's ingenuity from the mechanical side. I do not know whether any one had considered previously the possibility of using the relative tossing and pitching of two hulks as a source of power.

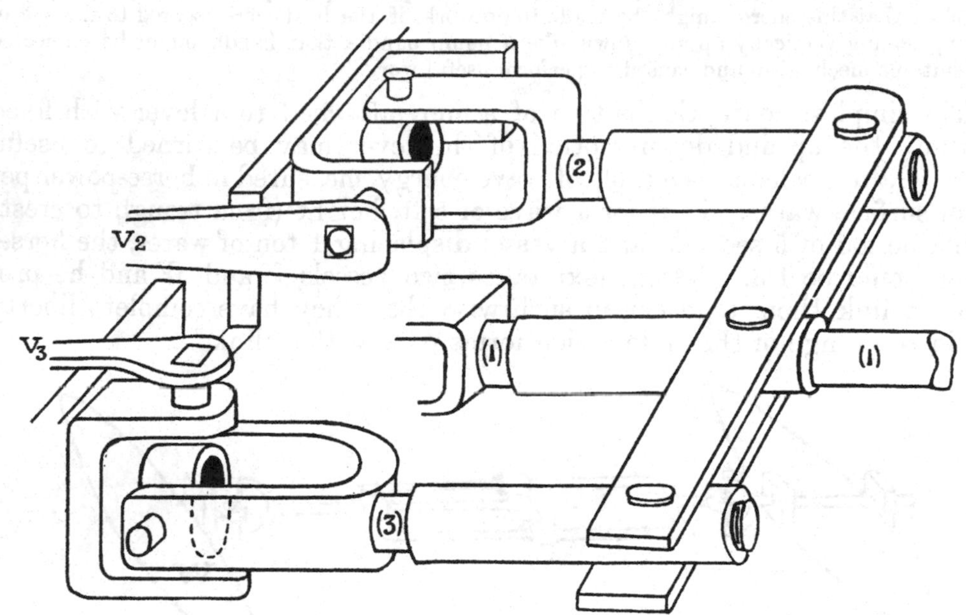


Part of the drawings for Galton's "wave machine."

Galton consulted three friends about his "wave machine"—Mr C. W. Merrifield, the Rev. H. W. Watson the mathematician, and Mr George Darwin. Merrifield considered the matter at considerable length with regard to the horse-power available, the actual mass in motion and the friction. He sums up as follows:

"My theoretical conclusion is therefore against the machine being of practical utility, by reason of its probable efficiency not being adequate to its cost and its inconvenience. I consider, however, that both the idea and the machinery are ingenious in a very high degree; and I should be sorry if you allowed one adverse opinion (coming from myself) to discourage you from trial."

Galton seems at first to have had an idea of utilising his relative motions to work some form of air-engine, and it was about this phase of his invention that he wrote to Mr Watson, who replied as follows:

"I am truly rejoiced to find that you are so sanguine. I am confident you have hit upon something real, and not a chimaera, and only hope you may be able to bring it to some practical end. I am not inclined to think you could utilise the power you have discovered in the way you suggested."

Galton had also suggested that motive power for a double ship might be obtained from the relative motion of its twin parts, and this point is taken up by George Darwin:

"I will keep your secret strictly. I am glad to hear that you are going to patent it, as it sounds as if it ought to be a great mercantile invention

Will it be possible to unyoke your ships? If not they would be rather unmanageable in rivers and harbours. Will not the danger of collisions be much increased by the great width and what will happen when the helm has to be turned hard to avoid anything? If one of the ships got at all out of hand, it would be rather an awkward combination wouldn't it? My father is very incredulous *in re* the spirits. I am sorry to hear that Miss F. is to have her familiars with her as 3 conjurors could combine to do their tricks without much chance of being found out[1]."

Whether it was Merrifield's criticism or George Darwin's irony[2] which led Galton to abandon his scheme, I cannot say; a last letter from Merrifield indicates that in April 1872, Galton was proposing to employ his apparatus to measure the energy of a sea-disturbance. Galton's idea, which must of course be distinguished from the use of tidal energy, seems to possess much originality. As our coal and oil supplies run short, possibly men will turn again to Galton's suggestion of harnessing the waves.

### E. CLIMATE (*continued*)

A meteorological paper of August 1866[3], read before the British Association in that year[4], deserves a passing notice. In this paper Galton criticises the statistical methods of the old Meteorological Office—otherwise of Admiral Fitzroy. It is entitled: "On an Error in the usual method of obtaining Meteorological Statistics of the Ocean." He points out that the

[1] See our p. 51 ftn. Galton's investigation of spiritualism interested Charles Darwin and will be referred to again later.

[2] Major L. Darwin assures me that the irony would be quite unconscious on Sir George Darwin's part.

[3] *B. A. Report*, Vol. XXXVI, 1866 (Sect.), pp. 16–17. Also *Athenaeum*, Sept. 1, 1866. We may just mention in this footnote that in the previous year (October 1865) Galton wrote a long notice in the *Edinburgh Review*, pp. 422–55 of J. F. Campbell's *Frost and Fire*. There was much in this book to interest Galton and excite his criticism and suggestion; thus he explains from close observation (p. 433) that trees do not as Campbell suggests 'bend to the wind,' they bend under the weight of branches, which can only flourish on the lee-side of the trees.

[4] Read by the Secretary of the Section as Galton was ill at the time. He had gone to the British Association at Nottingham, but had been "done up and obliged to leave." The Galtons then went to Leamington where "Dr Jephson prescribed for Frank, he grew very weak under the treatment. End of September returned home and remained six weeks, then went to the Norths, and took lodgings at Hastings in Breeds Place; stayed there till near end of January 1867. Frank rode constantly." *L. G.'s Record*. It was this illness which probably led Galton to resign the Secretaryship of the British Association and spend much of 1867 in travel.

ocean being divided into areas of 5° angle in longitude and latitude, and the ship returning *all* its observations, subject to the sole condition of an interval of eight hours between observation and observation, a ship will give more observations when the wind is unfavourable than when it is favourable; accordingly there will be an error produced—since favourable and unfavourable winds are peculiar to certain areas, and ships outward and inward bound follow different courses—in taking not only the mean direction of the wind for certain areas but also in other meteorological variates highly correlated with the wind, such as temperature and dampness. The remedy would be to enforce not only an interval in time, but an interval in distance of the positions of successive observations.

Galton's criticism is of less importance now that steamships have replaced sailing vessels, but the paper is of interest as marking probably the first occasion on which Galton exhibited publicly his fine instinct for the discovery of statistical fallacies.

The reader will not appreciate Galton's work at this period unless he remembers that Galton's earliest travels were associated with *sailing* ships; it was in such a vessel, the *Dalhousie*, that he sailed for Africa; and he thought for many years of his life in terms of wind and not steam as a motive power[1]. Thus it came about that when Galton turned his study of meteorology in the direction of ocean travel, he thought in terms of sailing vessels. The wind had for Galton a singular fascination, and for him the problem always was: What can we learn from the wind, how can we make it of greatest service?

Three or four of his papers touch on wind problems, and these we will now briefly consider.

The first one that may be referred to is entitled: "Barometric Predictions of Weather," and the paper was read at the British Association Meeting in 1870[2]. Galton's paper is suggestive, because, what he is actually seeking for in his linear prediction formula of the velocity of the wind in terms of barometric height, temperature and damp is what is now familiar to statisticians as a multiple regression formula. Galton very properly saw that the relation of barometric height to wind-velocity did not depend upon the instantaneous wind, and he accordingly experimented with average wind-velocity for a series of two, three, etc. hours. He came to the conclusion that the best period for the average was about twelve hours. He considered that twelve hour averages should also be taken for temperature and damp. Galton easily found his averages from the automatic record of continuous temperature, wind-velocity and damp. He explains clearly why he takes an average, namely the barometric pressure acts in sympathy with a much larger wind-velocity area, than that immediately in its own neighbourhood. The pressure (as in the case of water) is affected some time

[1] I think this is true even as late as the early 'seventies when Galton was busy with his "wave engine" (see p. 51). Such an engine as a propulsor would hardly have occurred to one who had grown up in an era of steam vessels.

[2] *Brit. Assoc. Report*, 1870, *Trans. Sections*, pp. 31–33; *Nature*, Vol. II, Oct. 20, 1870, pp. 501–3.

before the arrival of the centre of greatest disturbance. Accordingly Galton reaches a formula of the form

$$h_1 - h_2 = m\{v_1(12) - v_2(12)\} + p\{t_1(12) - t_2(12)\} + q\{d_1(12) - d_2(12)\},$$

where $h$ = pressure, $v(12)$ equal average wind-velocity, $t(12)$ equal average temperature and $d(12)$ equal average damp for 12 hours round an epoch, and the subscripts 1 and 2 represent epochs of time at a few hours interval. Galton then determines in rough figures the values of $m$, $p$, $q$ from observations at Falmouth. So far he might—by very crude methods indeed—be determining a multiple regression formula. But the next step he takes is erroneous; he transfers what amounts to $v_2$ to the other side of his equation, and proceeds to predict $v_2$ from barometric height, etc. It was not till much later that Galton realised that in the simplest case the prediction formula of $v$ from $h$ is not the same formula as that of $h$ from $v$. Hence although his conclusion that average wind-velocity cannot be closely predicted from barometric height is true, his method really failed to demonstrate it rigidly.

"The barometer when consulted by itself, without a knowledge of the weather at adjacent stations, can claim but one merit, namely, to guide us in a form of storm which does not occur once a year in the British Isles, of a fall in the mercury outstripping in an extraordinary degree the increasing severity of the weather; and I believe it to be on account of this rare phenomenon here, and of the reports of sailors from hurricane latitudes, where it is much more frequent, that the fame of the instrument has been so widely spread."

With his usual instinct Galton had reached a true conclusion, although his method was at fault. For us the interest of his paper lies in the evidence that he was feeling his way towards 'correlation'[1].

A series of three papers must now be considered in conjunction. The earliest of these is entitled: "On the Conversion of Wind Charts into Passage Charts." It was read in Section A of the British Association, 1866 (*Trans. Sections*, pp. 17–20), and published also in the *Philosophical Magazine*, Vol. XXXII, pp. 345–8, 1866. Galton explains his purpose in the following words:

"The most direct line between two points of the ocean is seldom the quickest route for sailing vessels. A compromise has always to be made between directness of route on the one hand, and the best chance of propitious winds and currents on the other. Hence it is justly argued that an inquiry into the distribution of the winds over all parts of the ocean is of high national importance to a seafaring people like ourselves. A knowledge of the distribution of the winds would clearly enable a calculation to be made which would show the most suitable passage in any given case[2]. But as a matter of fact, no calculations have yet been made upon this base; much less have charts been contrived to enable a navigator to estimate by simple measurements the probable duration of a proposed voyage. The wind charts compiled by the Meteorological Department of the Board of Trade are seldom used by navigators; for they do

[1] Galton's paper led to a long correspondence with G. H. Darwin (afterwards Professor Sir George Darwin), chiefly noteworthy because from this date an intimate correspondence sprung up and touched many other problems that Galton was considering in later years.

[2] Galton was clearly endeavouring to replace the straight lines and loxodromes of Mercator's Chart by a modern theory which should take account of the variations of the wind—less suited indeed to the examination room, but of more practical value. How few Cambridge mathematical examiners appear to have realised even since Galton's time the futility of loxodromics, which pay no attention to the individuality of the ship or the local characteristics of the wind!

not afford the results that seamen principally require; they are only data from which those results might be calculated by some hitherto unexplained process, which, we can easily foresee, must be an exceedingly tedious one."

It is this process which Galton proceeds to unfold for moderate winds in the case of a "merchantman of the class that usually navigates the Atlantic." To carry it out we require to know: (1) the proportionate time (or the relative frequency) that the wind blows in a given area from each of say eight points of the compass, (2) the number of miles that the particular ship will make in an hour at each angle to the wind. Combining these two results we can measure for the average of the winds in that area the average progress of the ship towards each point of the compass in an hour. If the distance reached[1] in an hour be plotted from a centre in the arc, we obtain a closed curve whose radius vector measures the efficiency of the ship in that particular area for a particular course. If now the chart be divided up into areas and in each area be placed the corresponding polar diagram, we have converted a wind chart diagram into a passage chart diagram. A navigator now plots his proposed course across these areas, and sets off with his compasses the distance run per hour in the direction of the course from the nearest polar diagram. In this way he is able to calculate the average time on the proposed course and can compare it with the time on other courses.

"He will thus be able to select the quickest out of any number of routes that may be suggested to him, and to determine, on the most trustworthy of existing data, what is the best course to adopt in sailing from one part of the ocean to another."

Galton suggests the modification of the polar diagram when (*a*) force of wind and (*b*) current are taken into account.

The next paper on this subject was published in the Minutes of the Meteorological Council for December 2, 1872[2]. In this communication Galton advances a considerable stage further. The Meteorological Office had sorted out the whole of the data for direction and force of wind and for current into "single degree squares." Thus the resultant direction and strength of current, the average force of the wind and its proportional directions were more or less accurately known for each area, for each month of the year. Galton now terms the polar diagrams of his earlier paper "isodic curves" or briefly "isods." He calculates them for the month of January for "2° squares" from Longitude 0° to 10° N. and from Latitude 20° to 30° W., allowing for current, and force of wind as well as direction, and taking as his standard type the "Beaufort ship." The rays now represent the average space run in 8 hours, and Galton enters into details of how to construct 'isodic charts and passages He seems, however, to have been in some doubt as to whether his name 'isod' was appropriate. In his own copies of this paper, he questions in pencil whether the word should not be 'ishodic.' But another doubt must have arisen in his mind; his isods did not represent equal paths, but the paths

[1] Not the distance traversed, because to reach a given point the ship will generally have to tack.

[2] Presented to Sir Edward Sabine, the Chairman, Mr Galton, Major-General Smythe and Sir Charles Wheatstone.

run in *equal times*. Thus a suitable name would have been 'hodogram' had not something like that word been already appropriated in another and rather unfitting sense by Sir William Hamilton. Galton in his third paper, published in the *Royal Society Proceedings*[1], and entitled: "On the Employment of Meteorological Statistics in determining the best course for a

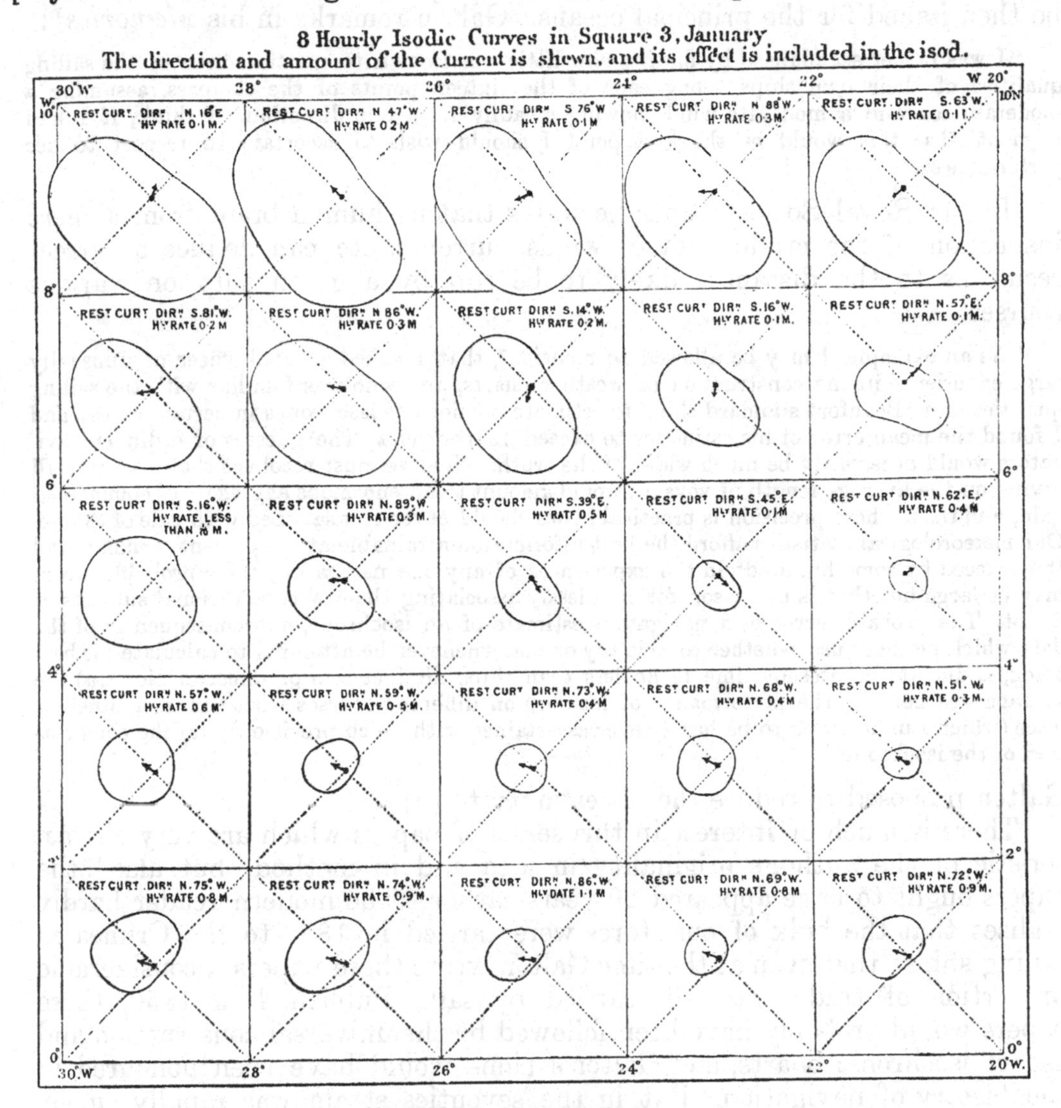


Note: REST CURT = RESULTANT CURRENT.

**Specimen of Galton's Isodic Curves from Minutes of Meteorological Council, 1872.**

ship whose sailing qualities are known," terms these hodograms "isochronous curves" or simply "isochrones." The discussion of the construction of isochrones, if somewhat fuller, follows here the lines of that in the *Meteorological Council Minutes*. A new feature is the description of a somewhat elaborate

[1] Vol. xxi, pp. 263–74, April 1873.

machine for plotting isochrones; the individuality of the ship is represented by zinc templates cut to its sailing qualities under each force of wind, and each template corresponding to the number of points the course of the ship lies off the wind. Galton considered that such templates could be cut for some moderate number of classes of ships, and charts of isochrones for such classes be then issued for the principal oceans. Galton remarks in his *Memories*[1]:

"I was rather scandalised by finding how little was known to nautical men of the sailing qualities of their own ships, along each of the sixteen points of the compass, assuming a moderate sea and a moderate wind blowing steadily from one direction. I think, if I had a yacht, that this would be the first point I should wish to ascertain in respect to her performances."

In his Royal Society paper he states that no human brain from a mere inspection of the crude data of winds, currents, etc. can deduce a correct result as to the distances likely to be run by a given ship on various courses.

"As an example, I may be allowed to mention[2], that I asked a naval officer of unusually large experience in the construction of weather charts, and who was familiar with the sailing qualities of a 'Beaufort standard ship,' to estimate portions of isochrones in certain cases; and I found the mean error of his estimates to exceed 15 per cent. The guesses of ordinary navigators would necessarily be much wider of the truth. Now we must recollect that a very small saving on the average length of voyages would amount to an enormous aggregate of commercial gain, and that, where precision is practicable, we should never rest satisfied with rule of thumb. Our meteorological statistics afford the best information attainable at the present moment, and they exceed by some hundredfold the experiences of any one navigator; their probable errors may be large but that is no reason for needlessly associating them with additional subjects of doubt. The probable error of a navigator's estimate of an isochrone, and consequently of the data which he must use, whether consciously or not, whenever he attempts to calculate his best track, is due at the present time to no less than three distinct sets of uncertainties: (*a*) the average weather; (*b*) the performance of his ship on different courses with winds of different force (which I understand to be hardly ever ascertained with much precision); (*c*) the computation of the isochrone."

Galton proposed to reduce the uncertainty to (*a*).

There is much of interest in this series of papers which are very characteristic of the author's originality in idea and in method, but alas! the papers ought to have appeared 20 years earlier. The modern reader hardly realises that the bulk of our stores were carried in 1857 to the Crimea in sailing ships; that even at the time Galton wrote these papers a considerable proportion of trade was still carried by sail. Published in 1850, these papers would probably have been followed by the universal construction and use of isochronic charts, and Galton's name would have been honoured in the history of navigation. But in the 'seventies steam was rapidly superseding sails, and sails were practically discarded before the Meteorological Office had time to collect the more ample and trustworthy data of ocean statistics, on the publication of which Galton's charts depended. Each mode of transit is succeeded by another, the railways killed canals, as motor traffic is killing the railways. It is hard on the discoverer and inventor to be working at a period of transition on a method of transit which has not

[1] p. 240.

[2] *Roy. Soc. Proc.* Vol. xxi, p. 267.

yet been recognised as moribund. His work, however good, perishes with its subject. The greatest tragedy in the history of discovery is the invention of a great improvement on some existing process, which process itself is then in a brief time completely replaced by some novel and wide-reaching development.

Closely allied to Galton's meteorological work was his association with the Kew Observatory. The Kew Observatory, constructed for George III's amusement, had been handed over by the Government at the suggestion of the Council of the British Association (1842) as a centre for testing scientific instruments, and it ultimately fell under the control of the Royal Society (1872). We have already seen that Galton was placed on the Managing Committee in 1858 as a result of the movement set going by him for the standardisation of sextants and other portable angle-instruments. On this Committee Galton made or strengthened several scientific friendships, notably those with Sir Edward Sabine, who largely influenced Galton's scientific career, with J. P. Gassiot[1], and with Warren De la Rue. Galton succeeded De la Rue as Chairman of the Kew Committee in 1889 and held that post till 1901, when the Committee ceased to exist as an independent body on the constitution of the National Physical Observatory. Sabine had made Kew a central magnetic observatory for the world. Galton busied himself mostly with apparatus for the testing and standardisation of instruments of all kinds. Sextants, thermometers, watches, telescopes, field-glasses, photographic lenses were all tested at Kew, and in many of these cases it was Galton on whom fell the chief responsibility for selecting the methods and instruments used in the tests. We have already referred to Galton's first proposal to test sextants[2] by heliostatic processes, i.e. by flashing light from the Observatory to distant fixed mirrors, which would reflect the light for angular measurement back to the Observatory. This method was discarded owing to its dependence on suitable weather; it was succeeded by a system of collimators. Next, an instrument for standardising thermometers devised by Galton with the aid of suggestions by De la Rue was made by Mr R. Munro, and set up at Kew in 1875. After two years service, which suggested certain modifications, the instrument and its method of use were described by Galton in a paper entitled: "Description of the Process of Verifying Thermometers at the Kew Observatory," read at the Royal Society, March 15, 1877[3]. The apparatus reveals Galton's characteristic ingenuity, but is of too specialised a nature to be described here[4]. In 1890 a pamphlet entitled: *Tests and Certificates of the Kew Observatory.*

[1] The Gassiots are frequently mentioned in *L. G.'s Record*, as present on social occasions and as joining the Galtons when on travel.

[2] Even as late as 1889, if we exclude thermometers, sextants stood second only to Navy binoculars, 292 to 341, in the statistics of instruments tested at Kew. In 1912 over 1000 sextants a year were being examined.

[3] *Proc. Roy. Soc.* Vol. XXVI, pp. 84–9, 1877. See also *Phil. Mag.* 1877, pp. 226–31.

[4] In 1912 it was still in use at Kew and was familiarly called "The Galton." That it should have survived nearly forty years service is a strong testimonial to its inventor's instrumental thoroughness.

*Issued by the Kew Committee of the Royal Society*, was published. Galton includes it both in his list of memoirs and in the bound volumes of his papers, so that it was doubtless compiled by him. It gives information as to the history of the Observatory, the wide range of instruments tested by the staff, the nature of many of the tests and the charges for testing. The Committee[1] of which Galton was then Chairman was indeed a strong one, and the general progress made in thirty years very noteworthy.

But Galton was not only interested in the methods of testing, but also in the convenience of the building itself and of its environment. General Strachey coming out one day from the Observatory noticed that the Mid-Surrey Golf Club had established a green immediately in front of the Observatory, and thinking how the matter might develop held that some means must be taken to secure a protected area round the building. But the institution possessed no funds for such an expenditure; accordingly Francis Galton (1893) generously and quietly provided the money, between £300 and £400, for placing a fence enclosing about six acres of ground round the Observatory. Dr Chree, the Superintendent, writing to me in 1912, said:

"Sir Francis' interests according to my recollections were more with instruments and their verification than with observational work. He usually professed to regard himself as a poor man of business and finance, but I think this was partly a pretence intended to form an excuse for leaving financial matters largely to General Strachey,—a very great friend of Sir Francis'—who liked to deal with matters of that kind....The Kew Committee used to meet once a month with a long vacation in summer—and generally Sir Francis got me to go up to Rutland Gate before each meeting and go through the business with him. His long experience of the Observatory rendered him so familiar with the work that he used to get along wonderfully well as Chairman, notwithstanding his deafness."

An amusing anecdote may be told to illustrate Galton's kindliness of spirit. With the increase of the testing work the Royal Society officers decided that the then existing system of Kew Observatory accounts—which was of General Strachey's arranging, somewhat primitive, and not requiring any special financial training in the Observatory officials—must be altered, and the Royal Society's auditor proposed a scheme of the complexity naturally dear to the professional mind. General Strachey was much hurt and Galton said privately that something must be devised to soothe General Strachey. This proved easier than might have been anticipated, for the non-financially trained, on close scrutiny of the accounts, discerned that the Royal Society had been recovering income tax and inadvertently not paying it over to the Kew Committee! That Committee was accordingly able to extract a substantial sum from the Royal Society and General Strachey was thus led to feel he was a match for the financial experts of the Society!

One or two miscellaneous papers may be fitly touched on in this chapter because they illustrate either Galton's instrumental ingenuity, or have more or less relation to the subjects here discussed. About 1877 Galton sent a letter to the *Field* newspaper suggesting a very simple speedometer for bicycles. This was a small sand-glass and all the rider had to do was to

[1] Abney, Grylls Adams, Creak, Carey Foster, Admiral Richards, the Earl of Rosse, Rücker, R. H. Scott, Generals Strachey and Walker, and W. T. L. Wharton.

count the number of strokes he made with the foot on one treadle, while the sand-glass run down. This provided the number of miles per hour at which he was moving. For example a sand-glass running out in 6 secs. is appropriate to a wheel of 2 ft. 9½ inches diameter, while one of 10 secs. corresponds to one of 4 ft. 8 inches diameter[1]. I am not aware that these sand-glasses ever came into use; with the differential gearing of the modern cycle, and with the free-wheel, several modifications would be needful.

A last meteorological paper by Galton was read at the British Association meeting 1880[2]. It is entitled: "On determining the heights and distances of clouds by their reflexions in a low pond of water and in a mercurial horizon."

"The calm surface of a sheet of water," Galton writes, "may be made to serve the purpose of a huge mirror in a gigantic vertical range-finder, whereby a sufficiently large parallax may be obtained for the effective measurement of clouds. The observation of the heights and thicknesses of the different strata of clouds, and of their rates of movement, is at the present time perhaps the most promising, as it is the least explored branch of meteorology. As there are comparatively few places in England where the two conditions are found of a pool of water well screened from wind, and a station situated many feet in height above it, the author hopes by the publication of this memoir to induce some qualified persons who have access to favourable stations to interest themselves in the subject, and to make observations."

The observations were to be made with a sextant and mercurial horizon, and demand a knowledge of, or a discovery of, the following quantities:

(*a*) the difference of level between the surfaces of the mercury and of the pool ($d$);

(*b*) the angle between the reflection of a part of the cloud in the mercury and in the pool ($p_1 = n$ minutes of angle, say);

(*c*) the angle between the portion of the cloud and its reflection in the mercury ($2\alpha$, Galton identifies $\alpha$ with the altitude of the cloud and suggests that it may be measured directly by a pocket altazimuth: see our p. 50).

Galton gives the approximation:

$$\text{Vertical height of cloud} = \frac{d}{n}\, 6875{\cdot}5 \cos(\alpha + p) \sin \alpha$$

and tabulates the factor by which $d/n$ must be multiplied for various values of $\alpha + p$ and $p$, or $n$, to obtain this vertical height. He mounted his horizon on a bar attached to a camera tripod, so that the reflection from the pool was seen *under* the mercury[3].

In this chapter of Galton's life I have endeavoured to indicate the chief scope of his activities during the ten years which followed his South African travels and his marriage. On his return home he came into touch with men like Sir Edward Sabine and Sir Roderick Murchison whose enthusiastic spirit caused Galton's labours to be directed in their own specialised directions, and the impulses thus given led to phases of study the ramifications

[1] Only those who remember the cycles of the 'seventies will appreciate this diameter.

[2] *Report*, pp. 459–61.

[3] I have checked Galton's formula of which he gives no proof on the assumption that the observer may be assumed to have his eye at the mercury, but I have had no opportunity of testing whether the method is fairly easy of application. Pools and cliffs are innumerable, but few of them are associated. The ideal spot would be a disused and flooded quarry.

of which lasted through many years of a long life. But Galton, while he maintained a keen interest in these topics, at least till 1900, grew less and less actively productive. A new and wider aspect of the cosmos was opening up for him, and evidence of an entirely different intellectual influence becomes apparent even in the 'sixties. His thoughts had begun to turn from the study of physical environment to the study of the organic contents of that environment, or in a narrower sense from cosmography to biology—from geography and meteorology to anthropology and psychology. There can be little doubt that the incentive in these directions came from his growing friendship with Charles Darwin, and the appearance of Herbert Spencer and Huxley in the circle of his acquaintances.

### F. SPIRITUALISM AND JOURNALISM

I have attempted in this chapter to give a more or less complete account of those labours of Galton which deal with the physical and the mechanical rather than the human side of his studies. In case some of my readers may have found this account tedious, for not everyone can have understanding and sympathy for the catholicism of Galton's pursuit of knowledge, I will conclude this chapter with brief accounts of two other matters of more general interest, which occupied a good deal of Galton's time in the period under discussion. The man of science, who with the history of the world before him finds it impossible to accept a primitive folk's account of man's creation and its purpose, is tempted to consider whether the methods in which he puts his trust for solving problems of the phenomenal universe may not be adequate as instruments of research in the unknown vast of the hyper-phenomenal[1]. Such a man of science, possibly owing to a lack of epistemological study, forgets that his senses have been developed to grasp physical phenomena, that his concepts are deductions from his sensuous perceptions, and that neither his sensuous nor mental outfits are adapted for sensating, perceiving and conceptualising the hyper-phenomenal. Some men grasp this truth by the logic of reasoning, others by the logic of experience, others by a healthy instinctive appreciation, and some never grasp it at all. To the first group we may, perhaps, say Huxley belonged, to the second Galton, to the third Darwin, and to the fourth Crookes and Alfred Russel Wallace.

Galton at any rate thought in 1872 that that branch of the 'supernatural' which we term spiritualism was at least worthy of inquiry. He endeavoured in a series of letters to interest Charles Darwin in his inquiries, and the latter appears to have been willing to give the matter a trial, but I have not been able to trace his letters; one at any rate went to Crookes, and another to Home, but in a letter of Darwin's son George to Galton there is a report of his father's incredulity as to the doings of Miss F.[2] It seems clear from the letters that Home had no great inclination to exhibit his powers to

[1] The good word 'supernatural' has become vulgarised until it signifies little more than something of which the user has inadequate previous experience occurring as a phenomenon, i.e. in the 'natural' world.

[2] See above, p. 53.

Darwin and Galton alone. It is probable that this reluctance led Galton to a less agnostic attitude towards spiritualism. If there were any other experiences which led to his final rejection of spiritualist claims, I have not found traces of them in his correspondence. It is possible that he became acquainted with Robert Browning's poem, *Mr Sludge, "The Medium,"* and he would be certain to read Huxley's report to Darwin on the trickery of another medium. The letters and paragraphs in letters to Charles Darwin on this subject are as follows:

5 Bertie Terrace, Leamington. *March* 28, '72.

MY DEAR DARWIN I enclose the revised statement about the curious trick[1] in Dr X's family. I questioned his widow only a fortnight before her death, all his 7 children, his son's wife and her 2 nurses. There is no contradictory evidence whatever.

Now about Mr Crookes, I have been twice at his house in séance with Miss F. who puts her powers as a friend entirely at his disposal, and once at a noisy but curious séance at Sergeant Cox's. I can only say, as yet, that I am utterly confounded with the results, and am very disinclined to discredit them. Crookes is working deliberately and well. There is not the slightest excitement during the sittings, but they are conducted in a chatty easy way; and though a large part of what occurs might be done, *if the medium were free*, yet I don't see how it can be done when they are held hand and foot as is the case. I shall go on with the matter as far as I can, but I see it is no use to try to enquire thoroughly unless you have (as Crookes says) complete possession of a first class medium. The whole rubbish of spiritualism seems to me to stand and fall together. All orders are given by raps,—levitation, luminous appearances, hands, writings and the like are all part of one complete system.

The following is confidential at present. What will interest you very much, is that Crookes has needles (of some material not yet divulged) which he hangs *in vacuo* in little bulbs of glass. When the finger is *approached* the needle moves, sometimes (?) by attraction, sometimes by repulsion. It is not affected at all when the operator is jaded but it moves most rapidly when he is bright, and warm and comfortable, after dinner. Now different people have different power over the needle and Miss F. has extraordinary power. I moved it myself and saw Crookes move it, but I did not see Miss F. (*even* the warmth of the hand cannot radiate through glass). Crookes believes he has hold of quite a grand discovery and told me and showed me what I have described quite confidentially, but I asked him if I might say something about it to you and he gave permission[2].

I can't write at length to describe more particularly the extraordinary things of my last séance on Monday. I had hold in one of my hands of *both* hands of Miss F.'s companion who also rested *both* her feet on my instep and Crookes had equally firm possession of Miss F. The other people present were his wife and her mother and all hands were joined. Yet paper went skimming in the dark about the room and after the word 'Listen' was rapped out the pencil was heard (in the complete darkness) to be writing at a furious rate under the table, between Crookes and his wife and when that was over and we were told (rapped) to light up, the paper was written over—all the side of a bit of *marked* note paper (marked for the occasion and therefore known to be blank when we began) with very respectable platitudes—rather above the level of Martin Tupper's compositions and signed "Benjamin Franklin"[3]! The absurdity on

[1] An hereditary habit of rather violently stroking the nose, while asleep, so that the thumbnail occasionally lacerated that organ. I have just ascertained that it has been transmitted to a generation born since 1872.

[2] It is not clear from this passage how far either Crookes or Galton originally associated Crookes' radiometer with mediumistic powers.

[3] As to Franklin: "......yourself
Explained the case so well last Sunday, sir,
When we had summoned Franklin to clear up
A point about those shares i' the telegraph:"

*Mr Sludge, "The Medium,"* R. Browning's Works, Vol. VII, p. 183, 1889.

*Sludge* was Home, and Browning although, unlike Galton, he was able to convict him of fraud

the one hand and the extraordinary character of the thing on the other, quite staggers me; wondering what I shall yet see and learn I remain at present quite passive with my eyes and ears open. Very sincerely yours, FRANCIS GALTON.

5 Bertie Terrace, Leamington. *March* 31st, '72.

MY DEAR DARWIN Your letter will be a great encouragement to Crookes and I have forwarded it to him to read, telling him what I had written.

About the 'female'—I hesitated a full 10 minutes before inserting the word 'it' on the ground that the subject of the story might be identified in after life and that the knowledge of the trick might damage her marrying value! I do not know if I am over fastidious. It is purely my own idea—no objection was raised by any of the family. So do entirely as you like[1]. Very sincerely yours, FRANCIS GALTON.

42 Rutland Gate. *April* 19/72.

MY DEAR DARWIN I have only had one séance since I wrote, but that was with Home in full gas-light. The playing of the accordion, held by *its base* by one hand under the table and again, away from the table and behind the chair was extraordinary. The playing was remarkably good and sweet[2]. It played in Sergeant Cox's hands, but not in mine, although it shoved itself, or was shoved under the table, into them. There were other things nearly as extraordinary. What surprises me, is the perfect apparent openness of Miss F. and Home. They let you do whatever you like, within certain reasonable limits, their limits not interfering with adequate investigation. I really believe the truth of what they allege, that people who come as men of science are usually so disagreeable, opinionated and obstructive and have so little patience, that the séances rarely succeed with them. It is curious to observe the entire absence of excitement or tension about people at a séance. Familiarity has bred contempt of the strange things witnessed, and the people find it as pleasant a way of passing an idle evening, by sitting round a table and wondering what will turn up, as in any other way. Crookes, I am sure, so far as it is just for me to give an opinion, is thoroughly scientific in his procedure. I am convinced, the affair is no matter of vulgar legerdemain and believe it well worth going into, on the understanding that a *first rate medium* (and I hear there are only 3 such) puts himself at your disposal.

Now considering that the evenings involve no strain, but are *a repose*, like the smallest of occasional gossip; considering that there is much possibility of the affair being in many strange respects true; considering that Home will, bonâ fide, put himself at our disposal for a sufficient time (I assume this from Crookes' letter and believe it, because it would be bad for Home's reputation, if after offering he drew back; but of course this must be made clear); considering,

(Griffin and Minchen, *Life of Robert Browning*, p. 203, 1910) only did so publicly in this poem, which so strangely echoes Galton's account of the séances.

[1] This last paragraph refers to an entry in the pedigree of the nose-stroking family.

[2] So Browning again:

"All was not cheating, sir, I'm positive
I don't know if I move your hand sometimes
When the spontaneous writing spreads so far,
If my knee lifts the table all that height,
Why the inkstand don't fall off the desk a-tilt,
Why the accordion plays a prettier waltz
Than I can pick out on the piano-forte,
Why I speak so much more than I intend
Describe so many things I never saw.
I tell you, sir, in one sense I believe
Nothing at all,—that everybody can,
Will, and does cheat; but in another sense
I'm ready to believe my very self—
That every cheat's inspired, and every lie
Quick with a germ of truth."

*Mr Sludge*, "*The Medium*," *loc. cit.* p. 236.

I say, all these things, will you go in for it, and allow me to join? Home is a restless man, as regards his movements, and could be induced to go to and fro. I am sure I could—if I could *ensure a dozen séances*, at which only our two selves and Home were together. (Others might be in the room, if you liked, but, I would say, not present within reach.)

It is impossible, I see, to prearrange experiments. One must take what comes, and seize upon momentary means of checking results. Home *encourages* going under the table and peering everywhere (I did so and held his feet while the table moved), so I am sure you need not feel like a spectator in the boxes while a conjuror is performing on the stage. He and Miss F. just want civil treatment and a show of interest. Of course, while one is civil and obliging, it is perfectly easy to be wary. Pray tell me what you think of the proposal in Crookes' letter. Very sincerely yours, F. GALTON.

42 Rutland Gate, S.W. *May* 26, '72.

MY DEAR DARWIN I feel perfectly ashamed to apply again to you in my recurring rabbit difficulty[1], which is this: I have (after some losses) got 3 does and a buck of the stock you so kindly took charge of cross-circulated, and so have means of protracting the experiments to another generation of breeding from them and seeing if their young show any signs of mongrelism. They do not thrive over well in London, also we could not keep them during summer at our house, because the servants in charge when we leave could not be troubled with them. Is it possible that any of your men could take charge of them and let them breed, seeing if the young show any colour, then killing the litter and breeding afresh, 2 or 3 times over? I would most gladly pay even a large sum—many times the cost of their maintenance—to any man who would really attend to them. Can you help me?

As regards spiritualism nothing new *that I have seen* since I wrote, for Home and Miss F. have been both absent. I wrote a letter of overtures to Home when I enclosed yours, but got no reply. I have kept up communication with Crookes, and am satisfied that he has the investigation thoroughly in hand, and delays publication on grounds of desiring a little more completeness of data. He is a most industrious taker of notes.

How very kind your letter was about Home. It grieved me much that you had to speak in such terms about your health. Ever sincerely yours, FRANCIS GALTON.

Three days later, to a letter arranging to lunch at Down, Galton adds the postscript: "The spiritualists have given me up, I fear. I can't get another invite to a séance." Darwin evidently wrote about this time to Galton asking the latter to introduce a friend to the spiritualistic fraternity, for on June 7, 1872, Galton writes from Rutland Gate:

MY DEAR DARWIN I did not reply yesterday about the Spiritualists as I expected that day and this to have heard from Mr Home, and Crookes is out of Town. It will give me great pleasure to do what I can for X.Y. but I rather doubt whether I shall have power to do much. I can't myself get to these séances as often as I like—indeed I have had no opportunity for a long time past. The fact is, that first class mediums are very few in number and are always acting. Also that Crookes and others are working their very best at the subject and entertain a full belief that they will be able to establish something important and lastly what, I see, is a real difficulty with them, the introduction of a stranger always disturbs the séances. I say all this to excuse me in your eyes, if I don't fulfil your wishes as you would like; but I will do my best and write—whenever I have anything to say to X.Y. as you propose.

The person most likely to help would, I think, be Lord L.

I wonder if I have offended Home by my last letter to him—he has never replied and I hear incidentally there is to be an important séance this very night! Alas for me! Ever yours sincerely, FRANCIS GALTON.

The last letter to Darwin concerning Home was in November of 1872. In the concluding paragraph of a further communication regarding the nose-stroking family Galton writes:

[1] This refers to a continuation of the 'Pangenesis' experiments after the publication of the memoir of 1871 to be discussed later.

"Crookes wrote to me that Home's presence was very important, for the experiments were far more successful when he was the medium, than when anyone else was, and he is now in Russia and will not return until May. So I will wait."

One almost imagines that Home fled[1] before the courteous manner but scrutinising eye of Galton! Our truthseeker did not immediately give up the pursuit of the hyper-phenomenal. In February of the following year (1873) there is a letter to Huggins about the psychology of the latter's dog 'Kepler[2],' and it ends with a few remarks on a suggestion of Huggins' that a medium who untied himself in a room—presumably through spirit assistance—should be watched through an aperture. Galton fears the room would be too dark, but says that he will suggest to Crookes that previous experiment be made to see if it is.

Certain letters given in Francis Darwin's *Life of Charles Darwin*, Vol. III, p. 187, probably indicate the end of Darwin's and Galton's inquiries into spiritualism.

"Spiritualism was making a great stir at this time. During a visit to Erasmus Darwin's in January 1874, a *séance* was arranged with Mr X., a paid medium, to conduct it. We were a largish party sitting round a dining-table, including Mr and Mrs G. H. Lewes (George Eliot). Mr Lewes, I remember, was troublesome and inclined to make jokes and not play the game fairly and sit in the dark in silence. The usual manifestations occurred, sparks, wind blowing, and some rappings and movings of furniture. Spiritualism made but little effect on my mother's mind [Mrs Charles Darwin] and she maintained an attitude of neither belief nor unbelief." *A Century of Family Letters*, 1904, Vol. II, p. 269.

Darwin himself wrote [Jan. 18, 1874] about this séance:

"We had great fun, one afternoon, for George hired a medium, who made the chairs, a flute, a bell, and candlestick, and fiery points jump about in my brother's dining-room, in a manner that astounded everyone, and took away all their breaths. It was in the dark, but George and Hensleigh Wedgwood held the medium's hands and feet on both sides all the time. I found it so hot and tiring that I went away before all these astounding miracles, or jugglery, took place. How the man could possibly do what was done passes my understanding. I came downstairs and saw all the chairs, etc., on the table, which had been lifted over the heads of those sitting

[1] As Browning puts it:

"What? If I told you all about the tricks?
Upon my soul—the whole truth and nought else,
And how there's been some falsehood—for your part,
Will you engage to pay my passage out,
And hold your tongue until I'm safe on board?
.......................... Begin elsewhere anew!
Boston's a hole, the herring-pond is wide,
V-notes are something, liberty still more,
Beside, is he the only fool in the world?"

*Mr Sludge, "The Medium," loc. cit.* pp. 184, 245. Browning's contact with Home appears to have been in 1857 or 1858. Sutherland Orr, *Life and Letters of R. Browning*, p. 216, 1891. *Dramatis personae* containing *Mr Sludge, "The Medium"* dates from 1864. The Galton-Darwin letters from 1872. Did Galton chance to read Browning? Home's habit of slipping across the 'herring-pond' when the environment was growing difficult for him seems to have been characteristic.

[2] 'Kepler' was one of a family of dogs that feared a butcher's shop and were furious at butchers. Galton writes "What you say about dogs' reasoning reminds me of a phrase used by the master of some performing dogs: 'Dogs, sir, do a deal of *pondering*'." See *Nature*, Vol. VII, p. 281, 1873.

PLATE IX

Francis Galton, aged 42, from photographs of 1864.
Co-editor with Herbert Spencer and Norman Lockyer of *The Reader*.

round it. The Lord have mercy on us all, if we have to believe in such rubbish. F. Galton was there, and says it was a good *séance*......

Such as it was it led to a second smaller and more carefully organised one with Huxley present, who reported to Darwin."

That report is printed in Huxley's *Life and Letters*, Vol. II, pp. 144–9. His conclusion was that the medium was "a cheat and an impostor." It produced the following letter from Darwin:

DOWN, *January* 29 [1874]

MY DEAR HUXLEY.—It was very good of you to write so long an account. Though the séance did tire you so much it was, I think, really worth the exertion, as the same sort of things are done at all the séances, even at ———'s; and now to my mind an enormous weight of evidence would be requisite to make one believe in anything beyond mere trickery....I am pleased to think that I declared to all my family, the day before yesterday, that the more I thought of all that I had heard happened at Queen Anne St, the more convinced I was it was all imposture....My theory was that [the medium] managed to get the two men on each side of him to hold each other's hands instead of his, and that he was thus free to perform his antics. I am very glad that I issued my ukase to you to attend.

Yours affectionately, CHARLES DARWIN.

Probably Galton also saw Huxley's report and concurred in his judgment. At any rate he very soon became a despiser of 'spiritualistic' séances.

Such are the last traces I can find of Galton's investigations into spiritualism. Some thirty-five years later Galton knew that the present writer had been invited to attend a séance by one who had sought aid from spiritualism in what formed for different reasons a crisis in the lives of all three. From his few written words on that occasion I know that Galton must long and definitely have been convinced of the futility of any light reaching human affairs from that strange medley of self-deception, chicanery and credulity which passes under the name of spiritualism. But I have no clue to the events or mental processes by which his attitude passed from the stage of agnosticism to that of complete rejection.

I have already indicated elsewhere that Galton was young till his death. Even between forty and fifty he was a boy who must try his powers on all things that came his way; it is true that he had had for some years experience of editing the Royal Geographical Society's *Journal*, but in 1865, amid all his other projects and work, Galton took upon his shoulders a very considerable share of the editorial duties of a weekly journal—*The Reader*. Galton himself says it was an amusing experience, and indicates that the loss of the guaranteed £100 was more than compensated by the gain of an unexpected view of the seamy side of journalistic enterprise[1]. The attempt was a brave one, and one of the committee of three, Spencer, Galton and Lockyer, which was appointed to make the preliminary arrangements, very shortly after made a marked success with a somewhat similar journal—*Nature*.

*The Reader* had been established in January 1863 as a journal of Literature, Science and Art, and when purchased towards the end of 1864, the programme of its future aims was propounded as follows:

"The very inadequate manner in which the progress of Science and the labour and opinions

[1] *Memories*, pp. 167–8.

of our scientific men have been recorded in the weekly press, and the want of a weekly organ which would afford scientific men a means of communication between themselves and with the public, have been long felt."

The aim of *The Reader*, without neglecting Literature, Art, Music and the Drama, was to supply this need. The prospectus then goes on to say that "the scientific arrangements of *The Reader* have the support and approval of"—and then follow 75 names, which cover practically all the men who created mid-Victorian science: Darwin, Galton, Grove, Hooker, Huxley, Lubbock, Lyell, Murchison, Sabine, Spottiswoode, Tyndall and Wallace; Adams, Balfour Stewart, Cayley, Crookes, De la Rue, Frankland, Glaisher, Hind, Hirst, Hofmann, Maskelyne, Odling, Roscoe, Stokes, Tait, William Thomson (Lord Kelvin) and Williamson; Babington, G. Bentham, G. Busk, John Evans, W. H. Flower, Andrew Ramsay, Sclater, Sharpey and Woodward, with many other names familiar enough to the scientific world of the third quarter of the nineteenth century. It was a tremendous force to bring together, and, all because there was no one man who would devote his whole life and whole energy to the projected task, *The Reader* came to nought.

The original shareholders in the company were G. Burges, J. E. Cairns, Rev. Ll. Davies, Galton, Gassiot, Huth, T. Hughes, Huxley, Lubbock, Lockyer, Robins, Roget, Spottiswoode, Spencer and Tyndall!

The first meeting was held in the rooms of Tom Hughes[1] in Lincoln's Inn Fields on Nov. 15, 1864, and the rough notes of the proceedings are in Galton's handwriting. £2250 were to be paid for the paper, plant and lease. Cairns was to take charge of the Political Economy, Galton of Travel and Ethnology, Huxley of Biology, Lewes of Fiction and Poetry, Spencer and Bowen of Philosophy, Psychology and Theology, while Seeley was to be asked to take charge of Classics and Philology. There were to be ten pages of Literature, three of Miscellanea, eight of Science, two of Art, and two of Music and the Drama. Four thousand copies were to be printed at a weekly cost of £110 including printing, paper, publication and office expenses. The returns were modestly estimated at sales 2000 copies £25 and advertisements £65, so that an initial loss weekly of £20 was anticipated. It made a brave show on paper—Tom Hughes' familiar legal blue 'opinion' paper—but the outcome was a little different. Herbert Spencer wasted the time of the committee in discussing 'first principles'; the powerful scientific support failed when it was pressed for reviews and articles, the paid sub-editor, the only man with 'real journalistic experience,' rather got on the nerves of the managing committee through his methods of procuring advertisements. Learned but illegible contributors sternly remonstrated with the editors about the inadequate and imaginative efforts of the proof-readers. The reviewers knew in some cases more of the subject than the authors of the books reviewed, and as a consequence the latter wrote long and angry letters to *The Reader*. Notably Burton, within a few months of Speke's death, replying to a review of his own *Nile Basin*, presumably by Galton, sent a truculent letter carrying on *post-mortem* hostilities. The critic, Burton tells us, ought to have known that his theory was one of the

[1] The author of *Tom Brown's Schooldays*.

most ridiculous ever put forth by man and was cobbled up in the map-room of the Royal Geographical Society. It would have its merit in the eyes of those who collect "romantic geography." A friend of Speke might wonder whether publication or non-publication was the wiser course. Poor Galton, endeavouring to still the fight and be fair to *both* men, had indeed his Scylla and Charybdis to steer between! The trials of an editor are manifold, but the trials of an editorial committee must be computed by multiplication not by division. The ship had too many first rank commodores aboard, and no one whose livelihood depended on a successful voyage. It is small wonder that it never reached port. *Pereat Lector, Natura resurgat*[1].

Many years later Galton was again an editor. In 1901 he consented to be "Consulting Editor" of *Biometrika*, a post, I think, he appreciated[2] though the acting editors did not trouble him much. The 'pink sheets' with résumés of the conclusions reached in the papers of each part, which were features of the earlier volumes, were undertaken at his suggestion[3].

The contents of this chapter will probably lead the reader to think we must have exhausted Galton's activities and labours during the twenty years that followed his marriage; on the contrary we have hardly considered a moiety of them, and those which remain to be discussed are of the greater importance.

[1] The *Reader* expired in 1866; *Nature* with an almost identical science programme appeared in November 1869, with Norman Lockyer as sole editor. But the introduction was by Huxley ("half a century hence curious readers will probably look at our *best*, not without a smile"), and Galton, Wallace, Darwin, G. H. Lewes, Sir William Thomson, Tylor, Balfour Stewart, Roscoe, etc., all the crew of the old *Reader* manned the new vessel and helped to steer its course into smooth waters.

[2] He inserts it as an item in his list of memoirs (*Memories*, p. 330), and included it in a privately printed list of "Biographical Events."

[3] Galton's journalistic suggestions were often of surprising originality when they were made, but will now seem commonplaces. Thus his idea of weather charts in the daily press, unthought of when he made it (1868); the idea that foreign and *colonial* books especially were not, but ought to be, adequately noticed in the English press; that new maps ought to be reviewed and criticised; that as to "Blue Books, no notices of them were published except in a list at the beginning and end of the session or very rarely at other times although 50 volumes appeared a year, but they ought to be continually reviewed"; that a list of new publications ought to be issued weekly under a suitable classification (1864, I cite from Galton's suggestions for *The Reader*); these ideas were practically novelties when Galton propounded them. Like forks and brooms they are such commonplaces of our traditional culture to-day, that not one person in a hundred feels any gratitude to the unknown originator.

# CHAPTER IX

## EARLY ANTHROPOLOGICAL RESEARCHES

### A. THE PASSAGE FROM GEOGRAPHY TO ANTHROPOLOGY AND RACE-IMPROVEMENT

"About the time of the appearance of Darwin's *Origin of Species* I had begun to interest myself in the Human side of Geography, and was in a way prepared to appreciate his view. I am sure I assimilated it with far more readiness than most people, absorbing it almost at once, and my afterthoughts were permanently tinged by it. Some ideas I had about Human Heredity were set fermenting and I wrote *Hereditary Genius*. In working this out I forced myself to become familiar with the higher branches of Statistics, and, conscious of the power they gave in dealing with populations as a whole, I availed myself of them largely."

*Manuscript Note* of Francis Galton in the handwriting of Mrs Galton found among his papers.

I HAVE indicated in the preceding chapter how Galton's interests were turning from man's environment to man himself—not only to his physical but to his psychical characters. One of the most conspicuously interesting facts in Galton's development is that in 1865 he had reached, we might almost say had planned out, the main conception of his work on Man. It is not possible to say from the dates of issue which of Galton's anthropological papers of this year, namely "The first steps towards the Domestication of Animals"[1] or "Hereditary Talent and Character"[2], was the earlier, because the date of publication is not necessarily that of writing the paper. Mrs Galton's 'Record,' however, shows that both papers antedate 1865:

1863. Returned by the Riviera road [from Switzerland] and home in November. Frank appointed Secretary to the British Association. Wrote paper on Domestication. Visited the Norths, etc.

1864. Very cold beginning of year. Went to Leamington at Easter [to visit Galton's mother]....Emma and Milly to us in May. Went abroad in July to Switzerland, Peiden, Grindelwald, St Luc. Returned for British Association at Bath. I at Julian Hill meanwhile [Mrs Galton's mother's house]. Visited Hadzor [home of the Howard Galtons] and Leamington. ...Went to the Norths in December. Frank writing Hereditary Talent....Frank busy editing Murray's Handbook. Christmas at home and alone."

The "Domestication" paper is chiefly of value as showing the transition of Galton's thoughts. Examining the accounts travellers give of savage races and his own experiences, Galton propounds the view that wild animals were tamed as pets or even kept for religious purposes[3] before they were

[1] *Transactions of the Ethnological Society of London*, Vol. III, pp. 122–38, 1865. I have no record of when it was read.

[2] *Macmillan's Magazine*, 1st Paper, June 1865, 2nd Paper, August 1865, Vol. XII, pp. 157–66, 318–27.

[3] Several of the cases cited by Galton, e.g. that of the kites from Shark's Bay, Australia, and the serpents at Whydah in Africa, suggest that totemism even might be the ultimate source of domestication.

domesticated for food or transport. He cites many—but relatively few out of the statements he had collected—of natives keeping animals as pets and even of native women feeding the young of wild animals from their own breasts, Australian women puppies, presumably young dingoes, New Guinea women young pigs, and Indian women of North America bear cubs. Galton considers that the value of domestication as a source of food was only found out incidentally as a result of taming animals for pets[1].

"Have," he writes, "extraordinary geniuses arisen who severally taught their contemporaries to tame and domesticate the dog, the ox, the sheep, the hog, the fowl, the llama, the reindeer and the rest? Or again: Is it possible that the ordinary habits of rude races, combined with the qualities of the animals in question, have sufficed to originate every instance of established domestication? The conclusion to which I have arrived is entirely in favour of the last hypothesis."

Because all savages maintain pet animals, because many tribes have *sacred* ones, and because kings of ancient states had *imported* animals on a vast scale from their barbarian neighbours, Galton holds that every animal of any pretension has been held in captivity over and over again and had numerous chances of becoming domesticated. We have no more domesticated animals than exist, because there are no others suited for domestication. Suitability for domestication depends upon an animal (i) being hardy, (ii) having an inborn liking for man, (iii) being comfort-loving, (iv) being useful to man, (v) breeding freely in captivity, (vi) being gregarious in its nature. These conditions Galton illustrates and states the exceptions. He gives due place to continual selection after domestication.

"To conclude. I see no reason to suppose that the first domestication of any animal, except the elephant, implies a high civilisation among the people who established it. I cannot believe it to have been the result of a preconceived intention, followed by elaborate trials, to administer to the comfort of man. Neither can I think it arose from one successful effort made by an individual, who might therefore justly claim the title of benefactor to his race; but on the contrary, that a vast number of half-unconscious attempts have been made throughout the course of ages, and that ultimately, by slow degrees, after many relapses, and continued selection, our several domestic breeds became firmly established." (p. 138.)

We know much more of the history of man now than was known in 1863. We realise the long history of man, and how he knew the elephant, the reindeer and the horse as sources of food long before he tamed them. It is, indeed, doubtful whether palaeolithic man ever domesticated any form of animal. His art shows no trace of the pet, and there is only one and that a very doubtful case of a possible bridle—the horse is merely 'game,' as the reindeer or earlier the mammoth.

[1] There was probably some correspondence between Galton and Darwin as to savages' pets I am unable to date the following letter, but it probably belonged to this period.

DOWN, BROMLEY, KENT, *July 7th.*

MY DEAR GALTON, I return the enclosed signed with great pleasure. Many thanks for information about Dr Barth's work, which I will read. I continue much interested about all domestic animals of all savage nations, though I shall not take up cattle in detail. If on reading I shall have anything to ask I will accept your kind offer and ask. Anything about savages taking any the least pains in breeding or crossing their domestic animals is of particular interest to me. With kind remembrances to Mrs Galton, Pray believe me, Yours very sincerely, CH. DARWIN.

When we pass from the palaeolithic to the neolithic stage we are not, unfortunately, able to trace the pet developing into the domesticated animal. What does seem to loom through the mist of prehistory is a series of races each with a more or less completed culture breaking up an older culture, a race which has domesticated cattle, a race which has domesticated horses, a race living with dogs or with reindeer; they loom upon us from the unknown, replacing less effective cultures. This does not exclude the domesticated animal arising from the pet, but it does suggest that either environment or religious belief led to pets of a certain type, and that only certain groups had the inspiration to turn their pets to the service of the group. Galton states that when he travelled in Damaraland the chiefs took pleasure in their herds of cattle rather for their stateliness and colour than for their beef—they were as the deer of an English squire.

"An Ox was almost a sacred beast in Damaraland, not to be killed except on momentous occasions, and then as a sort of sacrificial feast, in which all bystanders shared. The payment of two oxen was hush money for the life of a man. I was considerably embarrassed by finding that I had the greatest trouble in buying oxen for my own use, with the ordinary articles of barter. The possessors would hardly part with them for any remuneration; they would never sell their handsomest beasts." (p. 135.)

The possibility that the pet was a totem, or an animal of religious character, might throw light on the association of special domestic animals with definite races and their cultures.

A paper which may be considered in relation to that on "Domestication" may be fitly referred to here, as it also arose from Galton's travel-experience. It is entitled: "Gregariousness in Cattle and in Men." It was published in *Macmillan's Magazine* for 1872[1]. The theme of this article is a remarkable one, namely that our remote ancestors lived in herds or packs and that this gregarious or herd instinct is the source of many of man's intellectual weaknesses in his advanced civilisation[2]. The same idea of the instinct of the herd in man appears to have occurred to a number of writers recently, but they do not seem to have known or at any rate do not acknowledge Galton's priority of idea. The most interesting point of the article lies not only in the fact that Galton here assumes that he has in the earlier memoirs, which will shortly be discussed, proved the inheritance of the mental and moral characters in man, but in the fact that he stands definitely on the Darwinian platform and considers what heredity and selection have made of man. He opens his paper with the following description of its aims:

"I propose, in these pages, to discuss a curious and apparently anomalous group of base moral instincts and intellectual deficiencies, to trace their analogies in the world of brutes, and to examine the conditions, through which they have been evolved. I speak of the slavish aptitudes, from which the leaders of men, and the heroes and the prophets, are exempt, but which are irrepressible elements in the disposition of average men. I refer to the natural tendency of the vast majority of our race to shrink from the responsibility of standing and acting alone, to their exaltation of the *vox populi*, even when they know it to be the utterance of a mob of nobodies, into the *vox Dei*, to their willing servitude to tradition, authority and custom. Also,

[1] Vol. XXIII, pp. 353–7.

[2] Galton does not say that it may also be the source of some of his higher altruistic sympathies and social habits, but he had recognised this in his paper of 1865.

I refer to the intellectual deficiencies corresponding to these moral flaws, shown by the rareness with which men are endowed with the power of free and original thought, as compared with the abundance of their respective faculties and their aptitude for culture. I shall endeavour to prove that the slavish aptitudes, whose expression in man I have faintly but sufficiently traced, are the direct consequence of his gregarious nature, which, itself, is a result both of his primaeval barbarism and of his subsequent forms of civilisation. My argument will be that gregarious animals possess a want of self-reliance in a marked degree, that the conditions of the lives of these animals have made gregarious instincts a necessity to them, and therefore by the law of natural selection, these instincts and their accompanying slavish aptitudes have gradually become evolved. Then I shall argue, that our remote ancestors have lived under parallel circumstances, and that we have inherited the gregarious instincts and slavish aptitudes which were developed under those circumstances, although in our advanced civilisation they are of more harm than good to the race." (p. 353.)

Galton points out how in earlier life he had gained an intimate knowledge of certain types of gregarious animals. First he had found the camel's need for companionship a never exhausted topic of curious admiration in his tedious days of travel across North African deserts (see our Vol. I, pp. 199–205). Secondly and chiefly he had spent more than a year in close association with the semi-wild cattle of Damaraland. He had travelled an entire journey on the back of one of them with others at his side either as wagon or pack cattle for which nearly a hundred were broken in, or wholly unbroken and serving the purpose of an itinerant larder. He had often spent the night in their midst while the cries of prowling carnivora sounded around.

"These opportunities of studying the disposition of such peculiar cattle were not wasted upon me. I had only too much leisure to think about them, and the habits of the animals strongly attracted my curiosity. The better I understood them, the more complex and worthy of study did their minds appear to me." (p. 354.)

Galton then gives us a very striking account of the psychology of the herd.

One of the difficulties in breaking in wild cattle is to obtain 'fore-oxen for the team; these must be those who are of an exceptional disposition—born pioneers and leaders.

"Men who break in wild cattle for harness watch assiduously for those who show a self-reliant nature, by grazing apart or ahead of the rest, and these they break-in for fore-oxen. The other cattle may be indifferently devoted to ordinary harness purposes, or to slaughter; but the born leaders are far too rare to be used for any less distinguished service than that which they alone are capable of fulfilling." (p. 354.)

Galton considers that the law of "deviation from an average"—about which he had recently been writing (see our Chapter X)—would certainly be applicable to independence of character in cattle. He found every degree of it from the ox that could be ridden even at a trot apart from his fellows down to the ox that exhibits every sign of mental agony when segregated from the herd. The herd, with its mutual defence, its 'fore-oxen' as material for leaders, and its own leader, is the product of a country infested by large carnivora. A crouching lion fears oxen who turn boldly upon him, and does so with reason. The 'fore-oxen,' who are self-reliant, tend to be destroyed.

"Natural selection tends to give but one leader to a herd....Looking at the matter in a broad way we may justly assert that wild beasts trim and prune every herd into compactness, and tend to reduce it into a closely united body with a single well-protected leader. The development of independence of character in cattle is thus suppressed far below its healthy natural standard by the influence of wild beasts, as is shown by the greater display of self-reliance among cattle whose ancestry for some generations have not been exposed to such danger." (p. 357.)

It would be impossible in a résumé like this to cite all Galton's acute observations on the cattle herds of Damaraland, but the paper is well worth reading even to-day. He then proceeds to apply its lesson with certain modifications to savage man, but he insists

"on a close resemblance in the particular circumstance that most savages are so unamiable and morose as to have hardly any object in associating together besides that of mutual support."

As in the case of cattle herds there is a definite size in a given environment which is best suited to the human herd. A very large tribe is deficient in centralisation or is straitened for food and falls to pieces; a small tribe is sure to be overrun, slaughtered or driven into slavery. The law of natural selection

"must discourage every race of barbarians which supplies self-reliant individuals in such large numbers as to cause their tribe to lose its blind desire of aggregation. It must equally discourage a breed that is incompetent to supply such men, in a sufficiently abundant ratio to the rest of the population, to ensure the existence of tribes of not too large a size." (p. 357.)

Galton now proceeds to his 'moral': All through primaeval times, the steady influence of social condition summed up in the clan, the tribe, the petty kingdom tended to exterminate a superfluity of self-reliant men.

"I hold that the blind instincts evolved under those long-continued conditions have been deeply ingrained into our breed, and that they are a bar to our enjoying the freedom which the forms of modern civilisation could otherwise give us. A really intelligent nation might be held together by far stronger forces than are derived from the purely gregarious instincts. It would not be a mob of slaves, clinging together, incapable of self-government, and begging to be led; but it would consist of vigorous, self-reliant men, knit to one another by innumerable attractions, into a strong, tense and elastic organisation. Our present natural dispositions make it simply impossible for us to attain this ideal standard, and therefore the slavishness of the mass of men, in morals and intellect, must be an admitted fact in all schemes of regenerative policy. The hereditary taint due to the primaeval barbarism of our race, and maintained by later influences, will have to be bred out of it before our descendants can rise to the position of free members of a free and intelligent society: and I may add, that the most likely nest, at the present time, for self-reliant natures, is to be found in the States founded and maintained by emigrants." (p. 357.)

Wonderful, is it not, how Darwinism had already gripped Galton? How he thought in terms of heredity and natural selection and was ready to apply them to the past history of man in order to explain its present and suggest its future! The notion that it is necessary for human progress to breed out the men of slavish morals and intelligence—the essential foundation of eugenics—is already a truth to him.

Democracy—moral and intellectual progress—is impossible while man is burdened with the heritage of his past history. It has bound mankind to a few great leaders; it has produced a mass of servile intelligences; and only man's insight—man breeding man as his domesticated animal—can free mankind. This was Galton's view. Possibly the historian of man in the dim future may grasp that man in the age of nations was as much a product of natural selection as man in the age of tribes; and that a nation was not stable when it produced too many self-reliant 'fore-oxen,' or, worse still, when each ruminant and stolid ox no longer considered the common determination of the herd as binding on his conscience. He might even cite as illustration the Ireland of the twentieth century!

The world has seen numerous travellers, many men of mechanical genius, and not a few students of nature who grasped the evolution of human societies. But Galton the Cambridge mathematician, Galton the ox-rider, Galton of the wave-machine, and Galton the eugenist, seem at first sight so widely incongruous, and yet rightly estimated are necessary features of that all-round individuality—observant, constructive, calculating, and enthusiastic—of Galton the anthropologist, using that term in its widest sense, who by originality of method, wide experience of men and ripe judgment of affairs influenced the development of many younger men in the last quarter of the nineteenth century.

The paper just discussed was taken somewhat out of its proper order because it springs so directly from Galton's travel-experience, and because it indicates so clearly the growing tendencies of Galton's mind. But the reader must remember that Galton did not suddenly rush to the conviction, that from this time onward dominated his view of life, namely that the psychical characters in man, and also in the lower animals, are hereditary. He had been working on this subject for at least six or seven years. The best evidence of this is the paper written in 1864 on "Hereditary Talent and Character" (see our p. 70). It is singular how this foundation stone of Galton's anthropological work—the equal inheritance of the psychical and physical characters—has been disregarded even by some of his professed followers. As for the psychologists by calling they at first left this fundamental problem to others, and later, instead of observing and experimenting themselves, wasted energy in futile criticisms. Few men are willing to admit that their folly on the one hand is inbred, or that their talent which has led to success is not a product of their own free industry. Even men of quite reasonable intelligence daily confuse the possession of knowledge with mental endowment, and, as a result of their confusion, assert that psychical characters are chiefly the result of training. Another very common argument is of the following kind: A dictionary of biography is appealed to and it is found that far more distinguished men are sons of mediocre parents than of distinguished parents. It is then asserted that talent cannot be inherited. The fallacy is fairly flagrant, if examined, but is sufficiently plausible to be often repeated. Let us suppose that one parent in a thousand is distinguished, and, the rate of reproduction being the same, one offspring in ten to distinguished parents is distinguished, but only one offspring to the hundred in the case of non-distinguished parents. Then in a community of 10 distinguished and 10,000 non-distinguished parents we shall have one distinguished individual born of distinguished parents and 100 distinguished individuals born of mediocrity.

The fallacy consists in emphasising the 100 against the unit, and overlooking the fact that the distinguished parent produces distinction at ten times the rate of the mediocre parent. Galton is very careful in this paper to compare rates and not totals, and he realises that if we could increase the fertility of the able and check that of mediocrity, we should effectually alter the intellectual standard of our race. It is difficult for the mind with even a small modicum of statistical intelligence to appreciate how Galton's thesis could possibly be upset by showing that a larger *total* of talented men were born from mediocre than from able parents; that is only to proclaim the enormous prevalence of mediocrity!

"The power of man over animal life, in producing whatever varieties of form he pleases, is enormously great. It would seem as though the physical structure of future generations was almost as plastic as clay, under the control of the breeder's will. It is my desire to show, more pointedly than—so far as I am aware—has been attempted before, that mental qualities are equally under control.

A remarkable mis-apprehension appears to be current as to the fact of the transmission of talent by inheritance. It is commonly asserted that the children of eminent men are stupid; that, where great power of intellect seems to have been inherited, it has descended through the mother's side; and that one son commonly runs away with the talent of a whole family. My own inquiries have led me to diametrically opposite conclusions. I find that talent is transmitted by inheritance in a very remarkable degree; that the mother has by no means the monopoly of its transmission; and that whole families of persons of talent are more common than those in which one member only is possessed of it. I justify my conclusions by the statistics I now proceed to adduce, which I believe are amply sufficient to command conviction. They are only a part of much material I have collected, for a future volume on this subject[1]; all of which points in the same direction." (p. 157.)

Galton writing in 1864 points out that while we are perfectly certain of the inheritance of qualities in the brute world and breeders have learnt many empirical rules by experience, we have not advanced even to this limited extent in the case of man. It appears to have been nobody's business to study heredity in man, and the facts that only two generations are likely to be born in the lifetime of any observer, and that each individual rarely marries more than once, render the study harder in man than in the animals. Still nobody has doubted that the *physical* characters of man are equally transmissible with those of brutes. Galton then notes that as far as he is aware no one has bred animals for intelligence, but only for qualities which are useful to man. He suggests that instead of breeding dogs for special aptitudes and points we should try breeding them for intelligence:

"It would be a most interesting occupation for a country philosopher to pick up the cleverest dogs he could hear of, and mate them together, generation after generation—breeding purely for intellectual power, and disregarding shape, size and every other quality. (p. 158.)

The reader would have to remember that the wise dog would not always be most sympathetic to man; he might be a dexterous thief, a sad hypocrite or show marked contempt for humanity. As no one has bred for intelligence in animals, so no one has considered the possibility of breeding for intelligence in man. Galton's aim is to show that it is feasible because talent and character are inherited. His method is precisely that of his first great book *Hereditary Genius* (1869), only therein it is more fully developed. He estimates that of the men who receive a fair education about 1 in 3000 reach

[1] *Hereditary Genius*, 1869.

distinction; it would not have affected the validity of his argument had he taken 1 in 1000. He then takes from biographical dictionaries and other sources the number of distinguished men who have had distinguished fathers, sons or other relatives, and shows that they have far more distinguished sons than could possibly be anticipated in a like number of the general population. For example, taking 391 painters from Bryan's Dictionary there are 33 cases of sons who have renown as artists. Now supposing each painter had on an average 3 sons, we have 33 distinguished in a group of 1173, or about 1 in 36. Galton's statistics show something of this order—when allowance is made for size of family—for the frequency of distinction of all kinds in the sons of a population of fathers of distinction. But in the general population of the educated[1] distinction is only of the order 1 in 3000—or 1 in 1000 if the reader prefer. This rough method—ample enough for its purpose—was Galton's first application of statistics to the problem of heredity. It is the way he convinced himself that the mental characters in man were transmissible. But Galton was not content with merely reaching a truth. His next step was to consider what its relation to race betterment might be, and then—in 1864—we suddenly find the whole doctrine of eugenics as the salvation of mankind developed half-a-century too early!

"As we cannot doubt that the transmission of talent is as much through the side of the mother as through that of the father, how vastly would the offspring be improved, supposing distinguished women to be commonly married to distinguished men, generation after generation, their qualities being in harmony and not in contrast, according to rules, of which we are now ignorant, but which a study of the subject would be sure to evolve!" (p. 163.)

Galton next meets the "great and common mistake" of supposing that high intellectual powers are generally associated with puny frames and small physical strength. He says that men of remarkable eminence are almost always men of vast powers of work. He notes how even sedentary workers astonish their friends when on vacation rambles, and how frequently men of literary and scientific distinction have been the strongest and most daring of alpine climbers.

"Most notabilities have been great eaters and excellent digesters, on literally the same principle that the furnace that can raise more steam than is usual for one of its size must burn more freely and well than is common. Most great men are vigorous animals, with exuberant powers and an extreme devotion to a cause. There is no reason to suppose that in breeding for the highest order of intelligence, we should produce a sterile or a feeble race." (p. 164.)

Galton condemns the civilisation of the Middle Ages that enrolled so many youths of genius in the ranks of a celibate clergy; and he condemns the costly tone of society to-day which also forces genius to be celibate during the best period of manhood. He finds that very great men are not averse to the other sex, for many have been noted for their illicit intercourses, and in this respect he especially blames great lawyers. But science does not escape his censure; he takes the commoners who have been Presidents of the British Association as a fair list of leaders in science of the present day,

[1] Galton limits his field because of the handicap on the uneducated, however talented they may be.

and finds that only one-third of them have been married and had children. After drawing attention to the ages which thought it quite natural that the strongest lance should win the fairest lady in the tournament, Galton concludes the first part of his lecture half humorously, half earnestly as follows:

"Let us, then, give reins to our fancy and imagine a Utopia—or a Laputa if you will—in which a system of competitive examination for girls, as well as for youths, had been so developed as to embrace every important quality of mind and body, and where a considerable sum was yearly allotted to the endowment of such marriages as promised to yield children who would grow into eminent servants of the State. We may picture to ourselves an annual ceremony in that Utopia or Laputa, in which the Senior Trustee of the Endowment Fund would address ten deeply-blushing young men, all twenty-five years old, in the following terms: 'Gentlemen, I have to announce the results of a public examination, conducted on established principles; which show that you occupy the foremost places in your year, in respect to those qualities of talent, character, and bodily vigour, which are proved, on the whole, to do most honour and best service to our race. An examination has also been conducted on established principles among all the young ladies of this country who are now of the age of twenty-one, and I need hardly remind you, that this examination takes note of grace, beauty, health, good temper, accomplished housewifery and disengaged affections, in addition to noble qualities of heart and brain. By a careful investigation of the marks you have severally obtained, and a comparison of them, always on established principles, with those obtained by the most distinguished among the young ladies, we have been able to select ten of their names with special reference to your individual qualities. It appears that marriages between you and these ten ladies, according to the list I hold in my hand, would offer the probability of unusual happiness to yourselves, and what is of paramount interest to the State, would probably result in an extraordinarily talented issue. Under these circumstances if any or all of these marriages should be agreed upon the Sovereign herself will give away the brides, at a high and solemn festival six months hence in Westminster Abbey. We on our part are prepared, in each case, to assign 5,000*l.* as a wedding present, and to defray the cost of maintaining and educating your children, out of the ample funds entrusted to our disposal by the State.'

If a twentieth part of the cost and pains were spent in measures for the improvement of the human race that are spent in the improvement of the breed of horses and cattle, what a galaxy of genius might we not create! We might introduce prophets and high priests of civilisation into the world, as surely as we can propagate idiots by mating *crétins*. Men and women of the present day are, to those we might hope to bring into existence, what the pariah dogs of the streets of an Eastern town are to our own highly-bred varieties.

The feeble nations of the world are necessarily giving way before the nobler varieties of mankind; and even the best of these, so far as we know them, seem unequal to their work. The average culture of mankind is become so much higher than it was, and the branches of knowledge and history so various and extended, that few are capable even of comprehending the exigencies of our modern civilisation; much less of fulfilling them. We are living in a sort of intellectual anarchy, for the want of master-minds. The general intellectual capacity of our leaders requires to be raised, and also to be differentiated. We want abler commanders, statesmen, thinkers, inventors, and artists. The natural qualifications of our race are no greater than they used to be in semi-barbarous times, though the conditions amid which we are born are vastly more complex than of old. The foremost minds of the present day seem to stagger and halt under an intellectual load too heavy for their powers." (pp. 165–6.)

Here was Galton fifty years ago calling out for the 'superman,' much as the younger men of to-day are doing. But he differed from them in that he saw a reasoned way of producing the superman, while they do not seem to get further than devoutly hoping that either by a lucky 'sport' or an adequate exercise of will power he will one day appear!

One point—possibly the tragedy of his own life—Galton overlooked: you may be a man of wide intelligence, of a gifted stock with fertile parents, you may marry a woman of gifted stock and fertile parents also, and yet have no need to ask the 'Senior Trustee of the Endowment Fund' to maintain and educate children, if they are denied you. There are not only the physical and mental, but the physiological harmonies to be considered and of these the 'Senior Trustee' does not give us a hint. In later years Galton modified his views; he would, I think, have been content to grade physically and mentally mankind, and have urged that marriage within your own grade was a religious duty for those of high grade or caste.

In the second part of his paper Galton adds a number of interesting considerations and meets probable criticisms. Thus he starts with the statement that out of a hundred sons of men highly distinguished in the open professions eight are found to have rivalled their fathers in eminence[1]. But Galton considers that the mother has in most of these cases been selected 'at haphazard.' He points out that, where even *both* parents are of eminence, it would be absurd to expect their children to be on the average equal to them in natural endowment, because beyond the parents they would necessarily have much 'mongrel' ancestry.

"No one, I think," he writes, "can doubt, from the facts and analogies I have brought forward, that if talented men were mated with talented women of the same mental and physical characters as themselves, generation after generation, we might produce a highly bred human race, with no more tendency to revert to meaner ancestral types than is shown by our long-established breeds of race-horses and fox-hounds." (p. 319.)

In this passage we see Galton feeling towards the effect of (what I later termed) 'assortative mating,' and pointing to the 'mongrelism' of previous ancestry as the true source of his own 'law of regression.'

Galton next indicates that while marriage within the like intellectual grade would tend to differentiate society into two grades or castes objection may be raised that it would not tend to elevate society as a whole[2]. He suggests that (what I have later termed) 'reproductive selection' would or should be called into play. In the first place natural selection, he considers, would powerfully assist in the substitution of the higher caste *A* for the lower caste *B* by pressing heavily on the minority of weakly and incapable men. He did not at that time seem to have recognised that, while in the 'sixties the fertilities of the two castes were—what they no longer are—very nearly equal, the whole course of modern social evolution has been to suspend the action of natural selection. Galton did see, however, that a differential fertility has to be brought about, and he suggested that if intermarriage between *A* and *B* be looked upon with strong disapproval, so the early marriage of *A* and the discouragement or postponement of that of *B* would be "agencies amply sufficient to eliminate *B* in a few generations."

[1] I am not clear that Galton here accurately expresses the result deducible from his figures. I cannot find that he has allowed for size of adult family—at most say an average of $2\frac{1}{2}$ sons. If so, then we ought to say out of the 100 sons of distinguished men a little over three on the average would be distinguished, which statement would still be ample to prove Galton's point.

[2] It would tend to produce more and stronger leaders for the nation which adopted it, which after all may be more important than elevating society as a whole, especially if we lay stress on Galton's herd analogies.

I venture to think that Galton hardly gave due weight to such a primary human instinct as that of mating, if he really thought that the matings of *B* could be effectively discouraged or retarded by any existing agencies short of another primal force—such as natural selection—of equal intensity. The possibility of separating marriage from birth, better realised to-day than in the 'sixties, is of course a factor of great importance, but the general influence of birth-control so far has been little short of disastrous from the eugenic standpoint; it has tended to decrease the fertility of the intelligent relatively to that of the unintelligent caste in the community. That a wider knowledge of birth control will produce the desired 'reproductive selection,' as some eugenists apparently hold, is to assume that Caste *B* is intelligent and social enough to adopt control, and Caste *A* is altruistic enough to discard it, even at the cost of that social eminence which is the natural reward of its superior intelligence. Galton held that the improvement of the breed of mankind presents no insuperable difficulty.

"If everybody were to agree on the improvement of the race of man being a matter of the very utmost importance, and if the theory of hereditary transmission of qualities in men was as thoroughly understood as it is in the case of our domestic animals, I see no absurdity in supposing that, in some way or other, the improvement would be carried into effect." (p. 320.)

May we not answer to that proposition: Undoubtedly true, but how bring every one, nay even a majority of Caste *B*, to agree? To think it possible is to assume they belong to Caste *A*! Galton himself when he returned to Eugenics forty years later seems more fully to have recognised these difficulties, to have appreciated that, important as the problem is, its solution will be long and difficult. That we must be content to create a 'religious' feeling on the subject, to endeavour legislatively to strengthen the economic position of Caste *A* so that it may multiply, and legislatively to restrict the propagation of the worst members of Caste *B*, its minority, the mentally defective, the deaf and dumb, the blind and the deformed, when, as is mostly the case, these characters are of the hereditary type. Not one all-embracing remedy—certainly not birth-control—is the solution of the eugenic problem, but a steady examination of all social schemes, philanthropic and legislative, from the eugenic standpoint, and the bringing of an enlightened public opinion to bear upon them, so that the main idea of Galton, a differential fertility in favour of Caste *A*, is little by little brought into existence[1].

Galton in the paper under discussion emphasises the fact that he is not dealing solely with ability; he is thinking of all "mental aptitudes" as well as of "general intellectual power." He cites even mental and physical pathological states as certainly hereditary; cravings for drink, gambling, strong sexual passion, proclivities to fraud, pauperism or crimes of violence, longevity and premature death go by descent; many forms of insanity, gout, tendency to tuberculosis, heart disease, diseases of brain, liver and kidneys, of ear and of eye, etc. In fact Galton outlines the vast field of hereditary

[1] I have here (as Galton does) only spoken of two divisions, Castes *A* and *B*, to show his argument. Actually society is made of an infinity of grades for any character, and this Galton fully recognised.

research, which science is only slowly, if surely, investigating forty years later.

To illustrate what he means by "mental aptitudes" Galton refers to those of the American Indian and of the Negro; both have been reared under the most different environments from the North to the South of the world, and again under the most diverse social and political institutions, yet in all their essential mental characteristics they remain Red Man and Negro. Nature, as Galton later expressed it, is ever dominant over nurture. The Red Man has everywhere great patience, great reticence, great dignity, yet he has the minimum of affectionate and social qualities compatible with the continuance of his race.

"The Negro has strong impulsive passions, and neither patience, reticence nor dignity. He is warmhearted, loving towards his master's children and idolised by the children in return. He is eminently gregarious, for he is always jabbering, quarrelling, tom-tom-ing and dancing. He is remarkably domestic, and is endowed with such constitutional vigour, and is so prolific that his race is irrepressible." (p. 321.)

The characterisation Galton gives of Red Man and Negro—briefly resumed above—has been equalled, if scarcely bettered, by other anthropologists, but it remained for him to draw the essential conclusions that if the Negro is more unlike the Red Man in his mind than in his body, and this holds for all the environments in which you find them, then a race is a race because mental and moral characteristics are hereditary, and heredity will maintain these features dominating the slight, we might almost say superficial, effects of the most varied environment.

"Our bodies, minds and capabilities of development have been derived from them [our forefathers]. Everything we possess at our birth is a heritage from our ancestors." (p. 321.)

Galton next turns to the question whether habits acquired by the parents can be inherited by their offspring, and discusses it at length.

"I cannot ascertain that the son of an old soldier learns his drill more quickly than the son of an artizan. I am assured that the sons of fishermen, whose ancestors have pursued the same calling time out of mind, are just as sea-sick as the sons of landsmen when they first go to sea."

Galton rejects the inheritance of acquired characters whether mental or physical. Then, if in vague language, he propounds a doctrine probably for the first time in the history of science, which amounts to the theory of the continuity of the germ plasm. He boldly asserts that there is nothing in the embryo of an individual that was not in the embryos of its parents; that all the parental life from embryo to adult age, and from that to senility, has contributed nothing to the offspring embryo.

"We shall therefore take an approximately correct view of the origin of our life, if we consider our own embryos to have sprung immediately from those embryos whence our parents were developed, and these from the embryos of their parents, and so on for ever. We should in this way look on the nature of mankind, and perhaps on that of the whole animated creation, as one continuous system, ever pushing out new branches in all directions, that variously interlace, and that bud into separate lives at every point of interlacement."

"This simile does not at all express the popular notion of life. Most persons seem to have a vague idea that a new element, specially fashioned in heaven and not transmitted by simple descent, is introduced into the body of every newly-born infant. Such a notion is unfitted to stand upon any scientific basis with which we are acquainted. It is impossible it should be true unless there exists some property or quality in man that is not transmissible by descent. But the terms *talent* and *character* are exhaustive; they include the whole of man's spiritual nature so far as we are able to understand it. No other class of qualities is known to exist, that we might suppose to have been interpolated from on high. Moreover the idea is improbable from *à priori* considerations, because there is no other instance in which creative power operates under our own observation at the present day, except it may be in the freedom in action of our own wills. Wherever else we turn our eyes, we see nothing but law and order, and effect following cause." (pp. 322–3.)

The reader will now grasp how necessary it is to appreciate that the inheritance of mental and moral characters in man was the fundamental concept in Galton's life and work. It led him to all his later quantitative investigations on heredity; it led him to his conception of the 'stirp,' or as it was later termed the principle of the continuity of the germ plasm, but it led him also to his rejection of the doctrine of an implanted 'soul'—"*talent* and *character* are exhaustive; they include the whole of man's spiritual nature so far as we are able to understand it." Galton's free thought was the product of his views on heredity, and Darwin's *Origin of Species* had led Galton directly to the study of heredity in man. This is very obvious in the present paper. He admits variation between the embryos due to the same parents, although we know not the how of that variation. He applies selection directly to man. It is Nature's

"fiat that the natural tendencies of animals should never disaccord long and widely with the conditions under which they are placed. Every animal before it is of an age to bear offspring has to undergo frequent stern examination before the board of Nature, under the law of natural selection; where to be 'plucked' is not necessarily disgrace, but is certainly death." (p. 323.)

Then Galton proceeds to question whether the moral character is not also selected in words which Huxley would have done well to ponder on before he gave many years later his much misguided—for so I must venture to characterise it—Romanes Lecture.

"In strength, agility, and other physical qualities, Darwin's law of natural selection acts with unimpassioned merciless severity, the weakly die in the battle for life; the stronger and more capable individuals are alone permitted to survive, and to bequeath their constitutional vigour to future generations. Is there any corresponding rule in respect to moral character? I believe there is, and I have already hinted at it when speaking of the American Indians. I am prepared to maintain that its action, by insuring a certain fundamental unity in the quality of the affections, enables men and the higher order of animals to sympathise in some degree with each other, and also that this law forms the broad basis of our religious sentiments[1]." (p. 323.)

Galton then goes on to point out that animal life in all but the lowest classes depends on at least one and more commonly on all of the following types of affection: sexual, parental, filial and social.

"The absence of any one would be a serious hindrance if not a bar to the continuance of any race. Those who possess all of them, in the strongest measure, would, speaking generally, have an advantage in the struggle for existence. Without sexual affection, there would be no marriages, and no children; without parental affection, the children would be abandoned; without filial affection, they would stray and perish; and without social, each individual would be

[1] I think Galton is here applying the word 'religious' in its correct philological sense to the sentiments which unite society or bind men together; the theistic sentiment (p. 324) is hardly known to savages; religion in the earliest development is a social or tribal bond.

single-handed against rivals who were capable of banding themselves into tribes. Affection for others as well as for self, is therefore a necessary part of animal character. Disinterestedness is as essential to a brute's well-being as selfishness. No animal lives for itself alone, but also, at least occasionally, for its parent, its mate, its offspring and its fellow. Companionship is frequently more grateful to an animal than abundant food......But disinterested feelings are more necessary to man than to any other animal, because of the long period of his dependent childhood, and also because of his great social needs, due to his physical helplessness. Darwin's law of natural selection would therefore be expected to develop these sentiments among men, even among the lowest barbarians, to a greater degree than among animals. I believe that our religious sentiments spring primarily from these four sources." (pp. 323–4.)

"In the same way as I showed in my previous paper that by selecting men and women for rare and similar talent, and mating them together, generation after generation, an extraordinarily gifted race might be developed; so a yet more rigid selection, having regard to their moral nature, would, I believe, result in a no less marked improvement of their natural disposition[1]." (p. 325.)

In short Galton puts forth as his faith, that morality and the religious sentiments so far from being inexplicable on the basis of natural selection, as Huxley thought them, are its direct products. He thought that until a society has developed under natural selection a morality, religious sentiments and an instinct of continuous steady labour it would never be stable, and these thoughts suggested his later researches into social stability[2]. Indeed power of continuous steady work, prolonged or late development and tameness[3] of disposition are the three features which differentiate for Galton civilised man from the savage. He goes on to consider some of the effects of civilisation in diminishing the rigour of natural selection. It preserves weakly lives that would perish in barbarous lands. Above all he emphasises the ill-effects of inherited wealth.

"'The sickly children of a wealthy family have a better chance of living and rearing offspring than the stalwart children of a poor one.' 'Poverty is more adverse to early marriages than is natural bad temper or inferiority of intellect.' 'Scrofula and madness are naturalised among us by wealth; short-sightedness is becoming so.' 'There seems no limit to the morbific tendencies of body or mind that might accumulate in a land where primogeniture was general, and where riches were more esteemed than personal qualities.'"

Such are a few of Galton's aphorisms. He again and again points out how little value there is in a 'noble' descent, for generally the 'nobility' of a family is represented by a few slender rills amid a superfluity of non-noble sources. Nor is there, he holds, any limit

"to the intellectual and moral grandeur of nature that might be introduced into aristocratic families, if their representatives, who have such rare privilege in winning wives that please them best, should invariably, generation after generation, marry with a view of transmitting these noble qualities to their descendants." (p. 326.)

[1] Galton cites as an illustration of the alteration of natural disposition the evolution of the North American people, the selection of the emigrants from the most restless, combative and rebellious classes of Europe. "If we estimate the moral character of Americans from their present social state we shall find it to be just what we might expect from such a parentage!" I do not cite Galton's estimate because I think it truer when he wrote than to-day, partly owing to the change in the nature of emigrants, and partly owing to the same sort of natural selection within that nation itself.

[2] His schedule on this subject will be referred to in a later chapter.

[3] By 'tameness of disposition' Galton denotes the opposite to wild and irregular disposition, the untameable restlessness which is innate in the savage and to some extent in the gypsy.

And then follows Galton's first enunciation of the Law of Ancestral Heredity. He writes:

"The share a man retains in the constitution of his remote descendants is inconceivably small. The father transmits, on an average, one-half of his nature, the grandfather one-fourth, the great-grandfather one-eighth; the share decreasing step by step in a geometrical ratio, with great rapidity." (p. 326.)

Galton is clearly on the right track, but the numbers he gives would only be correct if he used parental generation, grandparental generation, great-grandparental generation, instead of father, grandfather, great-grandfather. He has overlooked the mother, and overlooked the multiplicity of the ancestral individuals. The numbers as he gives them in a later publication are one-fourth for the parent, one-sixteenth for the grandparent and one-sixty-fourth for the great-grandparent, etc. Galton's view was that this rule held for either blended or alternative inheritance[1]. Thus in 1865 Galton had already in mind this law of ancestral heredity, although by an obvious oversight he gave the wrong proportions. It has been suggested by certain over-confident geneticists that Mendelism has for ever done away with Galton's Law. I doubt the validity of this conclusion, as I doubt whether hybridisation will explain variation, or mutation account for the origin of species. The subject is, however, too controversial and technical to be discussed here, although it must be referred to as long as the tendency is to belittle Galton's contributions to heredity[2].

Galton draws from his law the conclusion that there is nothing much to be proud of in descent in a single line from a Norman baron for it would contribute on the average only $(\frac{1}{4})^{26}$, if 26 generations intervened, to the individual's natural aptitudes,

"an amount ludicrously disproportioned to the value popularly ascribed to ancient descent. As a stroke of policy, I question if the head of a great family, or a prince, would not give more strength to his position, by marrying a wife who would bear him talented sons, than one who would merely bring him the support of high family connections." (p. 327.)

[1] See his *Natural Inheritance* (pp. 138 and 151) where he applies it both to stature and to eye colour; in the one case to deduce the average stature of the offspring, and in the other case to deduce the distribution of eye colour in the offspring.

[2] If Galton in 1865 had applied his law to the following scheme of the crossing of two pure races $(DD)$ and $(RR)$ to produce the hybrid $(DR)$—to use Mendelian notation—let us see what it would have given him for the offspring of $(DR)$ mated with $(DR)$. The two parents are $(DR)$ and $(DR)$, the four grandparents are $(DD)$, $(RR)$, $(DD)$, $(RR)$, the eight great-grandparents are four $(DD)$ and four $(RR)$ and so on. Hence according to Galton's Law applied to alternative inheritance, the constitution of the offspring would be for a family of $4f$

$$\begin{aligned} 4f\{\tfrac{1}{4}(DR)+\tfrac{1}{4}(DR)+\tfrac{1}{16}[2(DD)+2(RR)]+\tfrac{1}{64}[4(DD)+4(RR)]+\tfrac{1}{256}[8(DD)+8(RR)]+\ldots\} \\ =f\{2(DR)+(\tfrac{1}{2}+\tfrac{1}{4}+\tfrac{1}{8}+\ldots)[(DD)+(RR)]\} \\ =f\{(DD)+2(DR)+(RR)\}. \end{aligned}$$

Or, this simple application of his law would have led Galton to the fundamental equation of Mendelian hybridisation. On the other hand had Galton applied it to $(DD)$ mated with $(DR)$—or to a pure race mated with a hybrid—he would have obtained different results according to the origin of the hybrid, i.e. whether it was an immediate result of crossing pure races or the product of two hybrids, or the product of hybrid and pure race, i.e. whether it was not or was an 'extracted' hybrid. And who shall say that he would not have been right? My own experience certainly leads me to doubt whether all hybrids, 'extracted' or not, are of equal germinal valency.

Galton was an ardent democrat, if such means to refuse birth any privilege if birth be not accompanied by mental superiority; he was a thorough-going aristocrat, if such be involved in a denial of the equality of all men; he would have graded mankind by their natural aptitudes, and have done his best to check the reproduction of the lower grades. The last paragraphs of the paper deal in a very novel manner with the theological problem of the sense of original sin. We have already noted that Galton held that conscience and with it the religious sentiments were developed in man by natural selection, they were the highest form of the herd instinct, and the tribe in which they were developed had greater social stability than any group that did not possess them. But man was barbarous but yesterday, and many of his native qualities are not yet moulded into harmony with his recent advance. Even our Anglo-Saxon civilisation is but skin deep, and the majority of English were the merest boors at a much later date than the Norman conquest. We are still barbarians in a large part of our nature; our no very distant ancestry grubbed with their hands for food, and dug out pitfalls for their game, and holes for their hut-poles and palisades with their fingers as tools. We see it all in the pleasure which the most delicately reared children take in dabbling and digging in the dirt, an inheritance from barbarian forefathers, akin to that of the pet dog who runs away from its mistress to sniff at any roadside refuse in the instinct to find the lost pack. The whole moral nature of man is tainted with 'sin,' which prevents him following his conscience, his social sense. From the Darwinian view the development of our religious sentiment has advanced—at any rate in certain members of the community—more rapidly than the elimination of the savage instincts of past stages of culture. The more recent the barbarism the more conscious the race is of the inadequacy of its nature to its moral needs.

"The conscience of a negro is aghast at his own wild, impulsive nature, and is easily stirred by a preacher, but it is scarcely possible to ruffle the self-complacency of a steady-going Chinaman." (p. 327.)

The revivalist meets with the greater success, the more degraded and less cultured is the population he works on.

"The sense of original sin would show according to my theory, not that man was fallen from a high estate, but that he was rapidly rising from a low one. It would therefore confirm the conclusion that has been arrived at by every independent line of ethnological research—that our forefathers were utter savages from the beginning, and that after myriad years of barbarism, our race has but very recently grown to be civilised and religious." (p. 327.)

Thus on the basis of Darwin's law of natural selection, and on the theory that natural aptitudes are not at the same time harmoniously developed, or eradicated, Galton accounts for the conflict in human nature summed up in the doctrine of 'original sin.' It will not be cleared away by any atonement, but solely by breeding out the uneradicated and hereditary savagery of human nature still dominating civilised man. What an illustration of his views Galton might have drawn from the events of the decade which followed his death!

I have given what the reader may consider undue space to this one magazine article, until he comes to see its relation to Galton's later views. It is really an epitome of the great bulk of Galton's work for the rest of his life; in fact all his labours on heredity, anthropometry, psychology and statistical method seem to take their roots in the ideas of this paper. It might almost have been written as a résumé of his labours after they were completed, rather than as a prologue to the yet to be accomplished. It is not only that Galton here gives us clearly his religious creed—religion has ceased for him to have a supernatural and taken on a purely anthropological value—but he formulates the work he intends to do, and actually did do in the remaining forty-five years of his life. Few realise that Galton was already in 1864 a thorough-going eugenist, that here in the prime of his life—in his 42nd year—he stood free of all the old beliefs which he implicitly accepted ten years earlier[1]. He acknowledges that his freedom was due to Darwin. But he does not hint that he had stept out beyond Darwin. For Darwin wrote:

"You ask whether I shall discuss 'man.' I think I shall avoid the subject, as so surrounded with prejudices, though I fully admit it is the highest and most interesting problem for the naturalist[2]."

and Galton said:

"I shall treat of man and see what the theory of heredity of variations and the principle of natural selection mean when applied to man."

So he came to sketch out his future work and whither he thought it would lead him in the course of years. Reading this article we see that his researches in heredity, in anthropometry, in psychometry and statistics were not independent studies, they were all auxiliary to his main object—the improvement in the race of man. Those who, ignoring what Galton and others have done, would cast doubt on the inheritance of the mental and moral characters, at once withdraw the foundation stone of Galton's life-work. That principle was essential to his views on the past evolution of man, was the mainstay of his religious belief, and the rock on which he built his scheme for man's future progress. For him the chief difference between barbarous and civilised man lay not in their physical qualities but in their mental or moral aptitudes, and all recent progress has been made by the action of natural selection on these hereditary characteristics.

It was by furthering this work of selection, by, in a broad sense, the further domestication of man, that Galton hoped to produce supermen. And, however desirable later writers, ignoring Galton, have proclaimed this end to be, they have provided no rational and scientific means, such as he did, of attaining it. Natural, albeit idle curiosity would like to know how Galton's orthodox friends and clerical relations met this bolt from the blue. The only letter, however, that has reached me from 1865 is one of May 31st

[1] See my account of his *Art of Travel*, p. 4.

[2] *Letter of Darwin to Wallace*, 1857. Between 1857 and 1871 Darwin's views of these prejudices changed. I venture to think Galton's voice crying in the wilderness had aided in the preparation of public opinion.

from Frank Buckland, to whom, I think, Galton must have sent an advanced copy of his paper, for Buckland says that he cannot thank Galton sufficiently for the copy of *Macmillan's Magazine*.

"Your theory is most excellent, and I shall endeavour to collect facts for you with a view to its elucidation."

And some facts Buckland does give, especially with regard to his experience of soldiers, but they are scarcely scientific observations. One point may be noted, because it carries us back into the age in which Galton was working.

"I have heard," Buckland writes, "that when a fine-looking Englishman travels in the Southern States the slave-owners offer him the best-looking girls, as a cross between a tall strong Englishman and a fine made Black girl produces a good useful slave worth money."

The son of the Dean makes no comment, however, on the originality and heterodoxy of Galton's standpoint; it is probable that he had not yet seen the second half of the paper[1]. While Galton always dated his letters fully, Darwin rarely, if ever, did; and it has taken a good deal of consideration and labour to place these letters in approximate order. I cannot, however, find out that Galton sent a copy of this paper to Darwin, although he sent most of his publications. It would, indeed, be of interest to have seen Darwin's comments upon it, if he made any. What Galton himself wrote in 1908 is indeed the best general comment:

"I published my views as long ago as 1865, in two articles written in *Macmillan's Magazine* while preparing materials for my book, *Hereditary Genius*. But I did not then realise, as now, the powerful influence of Small Causes upon statistical results. I was too much disposed to think of marriage under some regulation, and not enough of the effects of self-interest and of social and religious sentiment. Popular feeling was not then ripe to accept even the elementary truths of hereditary talent and character, upon which the possibility of Race Improvement depends. Still less was it prepared to consider dispassionately any proposals for practical action. So I laid the subject wholly to one side for many years. Now I see my way better, and an appreciative audience is at last to be had, though it be small[2]."

Galton laid it aside as propaganda, but as I have said it is the key to nearly the whole of his work for twenty years.

### B. *HEREDITARY GENIUS*, 1869 (Second Edn. 1892)

We now turn to the first step Galton took in the scientific demonstration of his creed—the study of the heredity of the mental and moral characters as a basis for Race Improvement. It was the first of four fundamental treatises in whole or great part devoted to the inheritance of the mental aptitudes in man. The other three are: *English Men of Science, their Nature and Nurture*, 1874; *Human Faculty*, 1883 and *Natural Inheritance*, 1889, and round these four greater works a whole swarm of memoirs and minor researches group themselves like flotillas of destroyers about a battle-fleet. A bio-

[1] Of some interest for the history of journalism is Buckland's statement that "in order to give the public numerous facts connected with Natural History I propose to start a new paper of my own to be called *The Land and the Water*."

[2] *Memories*, p. 310.

grapher indeed needs courage, when he starts upon the task of conveying to his readers even a moiety of the original ideas—both as to methods and conclusions—offered by this mass of material. It would need undoubtedly a robust conscience to realise that not for many years, possibly never again, will one individual read through practically the whole of Galton's published and unpublished writings, and then be confident that no idea of ripe suggestiveness, which might have developed in the minds of others into a noble scientific or social growth, has escaped his record! And yet the biography of a man of such a productive mentality and of such a lengthy activity as Galton should not only describe the many-sidedness of its subject, but enable readers of many tastes to find out what is of special interest to them in his writings. Galton's biographer has to provide an index to a veritable encyclopaedia, as well as trace the evolution of an original mind. The general scheme of *Hereditary Genius* was outlined in the first part of "Hereditary Talent and Character," but Galton's more complete demonstration of the heredity of mental aptitudes took five further years of work[1].

Galton's book is written with more gravity and less suggestiveness than his preliminary magazine article, and this is fitting.

"I propose to show in this book that a man's natural abilities are derived by inheritance, under exactly the same limitations as are the form and physical features of the whole organic world. Consequently, as it is easy, notwithstanding these limitations, to obtain by careful selection a permanent breed of dogs or horses gifted with peculiar powers of running, or of doing anything else, so it would be quite practicable to produce a highly-gifted race of men by judicious marriages during several consecutive generations. I shall show that social agencies of an ordinary character, whose influences are little suspected, are at this moment working towards the degradation of human nature, and that others are working towards its improvement. I conclude that each generation has enormous power over the natural gifts of those that follow, and maintain that it is a duty that we owe to humanity to investigate the range of that power, and to exercise it in a way that, without being unwise towards ourselves, shall be most advantageous to future inhabitants of the earth." (p. 1.)

In his preface Galton tells us that he was drawn to the subject of hereditary genius in the course of a purely ethnological inquiry into the mental peculiarities of different races.

As the quotation at the head of this chapter indicates, Galton had been led from Geography to Man, but when he came to examine the peculiar characteristics of human races, he found their psychical characteristics as marked and as permanent as their physical characteristics. Such a result

[1] In *L. G.'s Record* these years are represented by continual poor health in both Husband and Wife. They are also years of long continental travels and many home visits. We read under 1869 for example: "My health very troublesome till June and a great hindrance to my doing much. Frank in good health and able to dine out again. Went to Bertie Terrace [Francis Galton's mother's] at Easter and was not the better for it. Emma ['Sister Emma'] came to us in March and June. Lucy Wheler [Mrs Studdy] in May. Started in July for the Tyrol and Bavaria, Venice and home by the Sprügen, returned Oct. 4. Met the R. Gurneys and Mrs Bather at Basle. Went to Julian Hill [Mrs Galton's mother's] and Bertie Terrace before settling down. Dear Mr North died October 26 at Hastings. Frank's book 'Hereditary Genius' published in November, but not well received, but liked by Darwin and men of note. He began his experiments in Transfusion and became a member of the Zoological and Royal Institution. I attended Tyndall's Lectures after Easter. Spent Christmas at home and alone."

Mrs Francis Galton, from a portrait in the Galton Laboratory.

could only be reached by admitting the heredity of mental and moral aptitudes. Such was the evolution of Galton's theory and when he began to apply it to his own contemporaries at school, at college and in after-life he realised the truth that ability goes by descent.

"The theory of hereditary genius, though usually scouted, has been advocated by a few writers in past as in modern times. But I may claim to be the first to treat the subject in a statistical manner, to arrive at numerical results, and to introduce the 'law of deviation from an average' into discussions on heredity." (p. vi.)

We have here Galton's first direct appeal to statistical method and the text itself shows that Downes' translation of the *Letters on Probabilities* by Quetelet (London, 1849) was Galton's first introduction to the Laplace-Gaussian or normal curve of deviations, which was later to play such a large part in Galton's anthropometric work.

Galton's general plan is first to justify, in the case of men of great ability, the measurement of their ability by their reputation. Men *reputed* as endowed by nature with extraordinary genius are taken in the default of better evidence to be of surpassing ability, and the correlation is probably so high that little error in the highest grades of intellect will be introduced by this identification. Probably the identification is somewhat looser in Galton's second and third grades though the correlation must be something considerable here. Galton runs through various methods of appreciating 'eminence,' and comes to the conclusion that they indicate very approximately the same result:

"When I speak of an eminent man I mean one who has achieved a position that is attained by only 250 persons in each million of men, or by one person in each 4000.... The mass of those with whom I deal are far more rigidly selected—many are as one in a million, and not a few as one of many millions. I use the term 'illustrious' when speaking of these. They are men whom the whole intelligent part of the nation mourns when they die; who have, or deserve to have, a public funeral; and who rank in future ages as historical characters." (p. 11.)

It was at this time that Galton, I think, first realised the great principle that while between men of moderate ability there is scarcely any difference, between 'illustrious' and even 'eminent' men there are extraordinary differences. Galton (p. 19) illustrates this on the marks of the men in two years of the Mathematical Tripos at Cambridge, the total marks being 17,000.

| Number of Marks | Under 500 | 500—1000 | 1000—1500 | 1500—2000 | 2000—2500 | 2500—3000 | 3000—3500 | 3500—4000 | 4000—4500 | 4500—5000 | 5000—5500 | 5500—6000 | 6000—6500 | 6500—7000 | 7000—7500 | 7500—8000 | Total |
|---|---|---|---|---|---|---|---|---|---|---|---|---|---|---|---|---|---|
| Number of Candidates | 24 | 74 | 38 | 21 | 11 | 8 | 11 | 5 | 2 | 1 | 3 | 1 | 0 | 0 | 0 | 1 | 200 |

Here the difference between the first and second man is upwards of 2000 marks, but 157 of the mediocre men fall within the same range of 2000 marks. This difference between the extreme men, whether on the able or the stupid side, had great interest for Galton and he recurred to the problem many years afterwards, as I shall indicate later in this work.

Having reached series like this Galton considered how he should represent them theoretically, and he came to the conclusion that the proper method would be to use the normal curve, or curve of errors of the astronomers. This had already been used by Quetelet for *physical* measurements and he cites as illustrations the distributions of measurements of 5738 Scottish soldiers for chest size and of 100,000 French soldiers for stature. These results from Quetelet are from our present standpoint not very convincing; but supposing they do show that physical measurements may be approximately described in this manner, it does not follow that psychical measurements will also follow this distribution. The only real evidence Galton gives (p. 33) on this point is to show the marks obtained by 72 Civil Service Candidates in fact and in theory. Tested by modern methods the theory fits the facts to the extent that if the theory were true one sample in six would give results more divergent from the theory than the observed facts are. It cannot therefore be said that Galton demonstrates that intellectual ability is distributed according to the normal law of deviations. We are not even certain of that to-day. But demonstration was not really needful for Galton's purpose; he could legitimately classify human intelligence by applying a 'normal scale' to it, and he would still have his eight classes $A, B, \dots F, G$ and $X$ above and $a, b, \dots f, g$ and $x$ below the average, but he could not claim that these grades were separated by equal "amounts of intelligence," although more recent experience—i.e. with quantitative mental-tests—suggests that it is approximately correct.

Galton's classes: $F$, 1 in 4300, $G$, 1 in 79,000 and $X$, 1 in 1,000,000, correspond roughly to his three highest grades of intelligence. Galton then gives the following more popular description of his classification:

"It will be seen that more than half of each million is contained in the two mediocre classes $a$ and $A$; the four mediocre classes $a, b, A, B$ more than four-fifths and the six mediocre classes more than nineteen-twentieths of the entire population. Thus the rarity of commanding ability, and the vast abundance of mediocrity is no accident, but follows of necessity, from the very nature of these things.

The meaning of the word 'mediocrity' admits of no doubt. It defines the standard of intellectual power found in most provincial gatherings, because the attractions of a more stirring life in the metropolis and elsewhere are apt to draw away the abler classes of men, and the silly and the imbecile do not take a part in the gatherings. Hence, the residuum that forms the bulk of the general society of small provincial places, is commonly very pure in its mediocrity. The class $C$ possesses abilities a trifle higher than those commonly possessed by the foreman of an ordinary jury. $D$ includes the mass of men who obtain the ordinary prizes of life. $E$ is the stage higher. Then we reach $F$ the lowest of those yet superior classes of intellect, with which this volume is chiefly concerned.

On descending the scale, we find by the time we have reached $f$, that we are already among the idiots and imbeciles. We have seen...that there are 400 idiots and imbeciles, to every million of persons living in this country; but that 30 per cent. of their number appear to be light cases, to whom the name of idiot is inappropriate. There will remain 280 true idiots and imbeciles to every million of our population. This ratio coincides very closely with the requirements of class $f$.........

Hence we arrive at the undeniable, but unexpected conclusion, that eminently gifted men are raised as much above mediocrity as idiots are depressed below it; a fact that is calculated to considerably enlarge our ideas of the enormous differences of intellectual gifts between man and man." (pp. 35–6).

The "undeniable conclusion" is really based on assuming the normal distribution of intelligence. In the light of more recent investigations we may say that such a distribution is at any rate a rough approximation to the state of affairs, and Galton's conclusion within broad limits is correct. This and, in a more proven, if experimental way, the evidence from Tripos marks convinced Galton that men, like races of men, are not of equal natural ability.

"I have no patience with the hypothesis occasionally expressed, and often implied, especially in tales written to teach children to be good, that babies are born pretty much alike, and that the sole agencies in creating differences between boy and boy, and man and man, are steady application and moral effort. It is in the most unqualified manner that I object to pretensions of natural equality. The experiences of the nursery, the school, the university, and of professional careers, are a chain of proofs to the contrary. I acknowledge freely the great power of education and social influences in developing the active powers of the mind, just as I acknowledge the effect of use in developing the muscles of a blacksmith's arm, and no further. Let the blacksmith labour as he will, he will find there are certain feats beyond his power that are well within the strength of a man of herculean make, even although the latter may have led a sedentary life[1]." (p. 14.)

In his chapter on "The Comparison of the Two Classifications," in which Galton treats of how far a man's success and reputation is a measure of his natural power of intellect, he explains the method of his selection; he did not regard high social or official position—but "reputation in the opinion of contemporaries, revised by posterity."

"I speak of the reputation of a leader of opinion, of an originator, of a man to whom the world deliberately acknowledges itself largely indebted." (p. 37.)

Galton analyses the qualities, which lead a man to eminence, into capacity, zeal and adequate power of doing a great deal of very laborious work. He holds that men who achieve eminence and those who are naturally capable are to a large extent identical. By genius Galton understands the

"nature, which, when left to itself, will, urged by an inherent stimulus, climb that path that leads to eminence, and has strength to reach the summit—one which, if hindered or thwarted, will fret and strive until the hindrance is overcome, and it is again free to follow its labour-loving instinct. It is almost a contradiction in terms, to doubt that such men will generally become eminent" (p. 38)

and there are few men who reach eminence who do not possess this combination of powers. A boy who is carefully educated learns little useful information at school, he is taught the art of learning; the man who overcomes hindrances learns the same art in adversity. If the hindrances due

[1] Galton illustrates this by a case in which trained Highlanders challenged all England to compete with them in their games of strength. They were beaten in the foot-race by a youth, a pure Cockney, and clerk to a London banker. Perhaps I may be permitted to cite another illustration, from an occurrence at 'varsity sports over 40 years ago. The high jump had been won by a highly trained athlete, and the rod had been replaced at the last half inch he had failed to surmount; a non-combatant, a somewhat sedentary mathematician in every day costume, stepped from among the spectators, leapt the rod to the astonishment of the onlookers, and disappeared again into the crowd.

to humble rank were so severe as they are sometimes described, then all those who surmount them would be prodigies of genius; on the contrary we find many who have risen from the ranks have no claim to 'eminence.' Hindrances of rank only repress mediocre men, and perhaps some men of pretty fair powers—men of classes below $D$. Many of $D$ and a great many of $E$ ability rise from the ranks, and Galton holds the very large majority of the intelligences above $E$[1].

"If a man is gifted with vast intellectual ability, eagerness to work, and power of working, I cannot comprehend how such a man should be repressed. The world is always tormented with difficulties waiting to be solved—struggling with ideas and feelings to which it can give no adequate expression. If, then, there exists a man capable of solving those difficulties, or of giving voice to those pent up feelings, he is sure to be welcomed with universal acclamation." (p. 39.)

Galton undoubtedly did not believe in any large frequency of "mute inglorious Miltons[2]": he felt convinced

"that no man can achieve a very high reputation without being gifted with very high abilities; and I trust I have shown reason to believe, that few who possess these very high abilities can fail in achieving eminence." (p. 49.)

Having made these postulates Galton then proceeds to discuss his material. His method is precisely that of the paper on "Hereditary Talent and Character"; that is to say he makes no attempt to measure in any way the intensity of heredity. He takes the high grades of ability and measures the frequency of their appearance; he then measures the frequency of the appearance in the limited population of kinsmen of the eminent in some special degree, and finding this much greater than in the general population he argues that it can only be because the special talent runs in families. The whole argument is drawn on statistical lines, but, perhaps, it is not more convincing than the pedigrees themselves of illustrious men, many of which Galton gives in part, and which might easily be amplified and brought up to date.

One of the difficulties of Galton's task is the discovery or appreciation of the number of relatives in each grade of important individuals, and his values, or rather appreciations, are open at times to question. Thus he credits the judges on an average with only one son each, say with a family of two, i.e. one son and one daughter. But he makes the judge to have on an average $1\frac{1}{2}$ brothers and $2\frac{1}{2}$ sisters, or to spring from a family of five. In both cases

[1] Galton cites America as a country of more widely spread culture and education than Great Britain, where the hindrances to rising from the ranks are smaller, and yet their men eminent in science and literature are fewer than ours. Hence he argues that if our hindrances were lessened we should not become materially richer in highly eminent men. The footnote which follows seems to justify this opinion stated in 1869.

[2] A great educationalist recently put to the writer this question: We have had millions of children in London alone through our primary schools, we select the best annually and send them to the secondary schools and the best of these again go to the universities, why have we not yet found a single Darwin, Newton or Milton? The failure is, I think, explicable by the fact that a selection of Galton's $F$, $G$ and $X$ stocks had been going on for centuries before the County Councils took the net in hand to fish, possibly a trifle crudely, in nearly exhausted waters.

the son or brother is supposed to reach an age of 50 before he can be tested for eminence. Now it is rather difficult to accept the statement that the average offspring of a judge is two and of his father five[1]; notwithstanding the discredit Galton casts elsewhere on the moral conduct of lawyers[2]. I believe that owing to the difficulty of getting accurate information, although Galton did not go beyond the ordinary sources—for example to herald's visitations, etc.—it would have been best to take average values from pedigrees of the period. For example Galton gives 36 °/₀ of eminent sons to the judges on the basis of one son apiece, but 14·4 °/₀ on the basis of 2·5 sons would have been equally effective for his purpose, which was to show that a judge being one barrister in a hundred, or, since as he remarks barristers are highly selected, one man possibly in 4000, the chances are enormously against judges having 14·4 °/₀ of legally eminent sons on the assumption of a pure chance occurrence.

Galton's chapter on the judges—his most complete and detailed section—is a very fine piece of work, and might be used as the starting-point for still more complete pedigree work on the heredity of legal ability[3].

The next chapter deals with 'Statesmen,' and Galton admits the difficulty of steering between first the acceptance of mere official position or notoriety as equivalent to a more discriminative reputation, and secondly a selection with an unconscious bias towards facts favourable to inheritance. Thus he writes:

"It would not be a judicious plan to take for our select list the names of privy counsellors, or even of Cabinet ministers; for though some of them are illustriously gifted, and many are eminently so, yet others belong to a decidedly lower natural grade. For instance, it seemed in late years to have become a mere incident to the position of a great territorial duke to have a seat in the Cabinet as a minister of the Crown. No doubt some few of the dukes are highly gifted, but it may be affirmed, with equal assurance, that the abilities of the large majority are very far indeed from justifying such an appointment." (p. 104.)

Galton is indeed a democrat in his views on the nobility:

"A man who has no able ancestor nearer in blood to him than a great-grandparent, is unappreciably better off in the chance of being himself gifted with ability, than if he had been taken out of the general mass of men. An old peerage is a valueless title to natural gifts, except so far as it may have been furbished up by a succession of wise intermarriages. When, however, as is often the case, the direct line has become extinct and the title has passed to a distant relative, who had not been reared in the family traditions, the sentiment that is attached to its possession is utterly unreasonable. I cannot think of any claim to respect, put forward in modern days, that is so entirely an imposture, as that made by a peer on the ground of descent, who has neither been nobly educated, nor has any eminent kinsman, within three degrees." (p. 87.)

What would Galton have said had he written fifty years later when peerages appear to be given away, not for noble education, eminent kinsmen, or distinguished public service, but apparently on the ground of men being

[1] Of course the judge may have no offspring and his father must have had *one* at least.

[2] "Hereditary Talent and Character" (p. 164). "Great lawyers are especially to be blamed in this [illicit intercourse followed by a corresponding amount of illegitimate issue], even more than poets, artists or great commanders."

[3] There is a considerable correspondence with E. B. Denison in the *Galtoniana* letters for 1869 about the ability and fertility of judges and peers.

successful tradesmen! And yet Galton is above all an aristocrat. When we read his 'Judges' and his 'Statesmen' we see him almost swept off his feet when he discovers for the first time from his own reading the characteristic ability of the Montagus and Norths, or of the Temples and Wyndhams. There was almost a simplicity about his adoration of ability and he positively gloated over it, if it took an unusual and individual turn. Many very able men scarcely appreciate high ability in others, because, as in the matter of wealth, a man is apt to judge relatively to his own holding. Not so Galton; had he used himself as a standard measure, I fear his modesty would have led him to revise more than one of his estimates.

"A collection of living magnates in various branches of intellectual achievement is always a feast to my eyes; being as they generally are such massive, vigorous, capable-looking animals[1]". (p. 332.)

Galton had an immense veneration for genius as he defines it; not only like Carlyle would he have made his heroes rulers of the mediocre, but unlike Carlyle he would have had his heroes steadily and surely replace the latter. That men of genius are unhealthy puny beings—all brain and no muscle—weaksighted, and generally of poor constitutions, Galton will not accept for a moment.

"I think most of my readers would be surprised at the statures and physical frames of the heroes of history, who fill my pages, if they could be assembled together in a hall. I would undertake to pick out of any group of them, even out of that of the Divines, an 'eleven' who should compete in any physical feats whatever, against similar selections from groups of twice or thrice their numbers, taken at haphazard from equally well-fed classes." (p. 331.)

Perhaps Galton laid too great stress on the high wranglers and classics of his own day who had been ''varsity blues'; or again on the big-headed men on the front benches at the Royal Society meetings in the early 'seventies[2].

One characteristic, but an all-important one, Galton admits both his 'Judges' and 'Statesmen' did not possess; the power of being prolific. It will be obvious that if men of ability are unprolific, as they are often supposed to be, then the families of great men will be apt to die out, and Galton's project for creating a race of 'supermen' must be defeated. This point—whether or no a breed of men gifted above the average could maintain itself during an indefinite number of generations—is so important that Galton devotes a special chapter to the subject. Turning to the 'Judges' he first cites Lord Campbell's statement that when he first became acquainted with the English Bar, one-half of the Judges had married their mistresses,

[1] "One comfort is that Great Men taken up in any way are profitable company. We cannot look, however imperfectly, upon a Great Man, without gaining something by him. He is the living light-fountain, which it is good and pleasant to be near." *Lectures on Heroes*, p. 2.

[2] He was very unhappy about the low correlations I found between intelligence and size of head, and would cite against me those 'front benches'; it was one of the few instances I noticed when impressions seemed to have more weight with him than measurements. It is possible, however, that between his day and mine science changed its recruiting fields, and 'eminence' became less common.

"the understanding being that when a barrister was elevated to the Bench, he should marry his mistress, or put her away." Galton describes this statement as 'extraordinary,' but no effort of memory is needed to recall at least one illustration of it in the case of those as old as the present biographer. The advanced period of their lives, if Lord Campbell's statement be correct, at which they would marry, might account for a reduced number of children, but even under this disadvantage Galton asserts that the Judges were by no means an unfertile race (p. 131). Still it may be that their families die out. He finds that out of 31 judges who became peers before the close of George IV's reign, 12 are no longer represented in the peerage. Galton then inquired into the particulars of the marriages of these law-lords, their children and grandchildren.

"I found a very simple, adequate, and novel explanation, of the common extinction of peerages, stare me in the face. It appeared, in the first instance, that a considerable portion of the new peers and of their sons had married heiresses....But my statistical lists showed, with unmistakable emphasis, that these marriages are peculiarly unprolific. We might, indeed, have expected that an heiress, who is the sole issue of a marriage, would not be so fertile as a woman who has many brothers and sisters. Comparative infertility must be hereditary in the same way as other physical attributes, and I am assured it is so in the case of domestic animals. Consequently the issue of a peer's marriage with an heiress frequently fails, and his title is brought to an end." (p. 132.)

This generalisation of Galton's, most brilliant in its suggestiveness, he extends to the 'Statesmen' and to a considerable portion of the peerage. Everywhere with the same result, that in a large proportion of the cases in which peerages become extinct there have been marriages with heiresses. Is Galton's conclusion, however, that heiresses come of sterile families the correct explanation? I venture to think it is not. In the first place there is in man some, but no very important, correlation between fertility in mother and daughter, and this result has been confirmed for mice and horses. The inheritance of grades of fertility seems hardly adequate to account for such rapid extinction as Galton records. I think explanation may be found in two other directions. The son of a peer starting with an assured position spends a life of ease and pleasure, which is often synonymous with a life which ruins health and squanders wealth. The fortune of the family has then to be retrieved, and the solution is marriage with an heiress. And here the words of Dr Erasmus Darwin are appropriate and he also was a keen observer:

"As many families become gradually extinct by hereditary diseases, as by scrofula, consumption, epilepsy, mania, it is often hazardous to marry an heiress, as she is frequently the last of a diseased family[1]".

The fertility of a libertine and a woman of decadent stock is likely to be much below the normal both in quantity and survival value. Only a very detailed investigation of a long series of cases would allow us to determine whether Erasmus Darwin or his grandson has taken the more correct view;

[1] *Temple of Nature* (Additional notes), p. 45, 1803, cited by Galton on the interleaf of his copy of *Hereditary Genius*, 1869.

and the investigation would be well worth making. Galton himself realised that there might be other points than the sterility of heiresses:

"The reason I have gone so far is simply to show that, although many men of eminent ability (I do not speak of illustrious or prodigious genius) have not left descendants behind them, it is by no means always because they are sterile, but because they are apt to marry sterile women, in order to support the peerages with which their merits have been rewarded. I look upon the peerage as a disastrous institution, owing to its destructive effects on our valuable races. The most highly gifted men are ennobled; their elder sons are tempted to marry heiresses, and their younger sons not to marry at all, for these have not enough fortune to support both a family and an aristocratical position. So the side-shoots of the genealogical tree are hacked off, and the leading shoot is blighted, and the breed is lost for ever. It is with much satisfaction that I have traced and, I hope finally disposed of, an important cause why families are apt to become extinct in proportion to their dignity—chiefly so, on account of my desire to show that able races are not necessarily sterile, and secondarily because it may put an end to the wild and ludicrous hypotheses that are frequently started to account for their destruction." (pp. 139–40.)

The following chapters on 'Commanders,' 'Literary Men,' 'Men of Science,' 'Poets,' 'Painters,' 'Musicians' and 'Divines' and 'Senior Classics' we must pass over more briefly; they follow the same general lines as those on 'Judges' and 'Statesmen,' and the same general criticisms apply. Perhaps the most important of these is that, up to and including 'Men of Science,' Galton still appears to consider that 100 eminent men have only 100 sons who reach adult life. If the wickedness of Judges and the heiress-hunting of Statesmen could justify such a fertility, it certainly does not seem reasonable in the case of other classes; although as I have pointed out it does not really affect Galton's main argument, it still seems to render the column $C$ of his Tables unsatisfactory.

In the case of 'Commanders' Galton points out the early age at which they generally achieved greatness. He also points out that they have not a long life, and that as their relative chance of being shot varies as the square root of the product of their height and weight, the man who lives to be a great commander will probably be small.

"Had Nelson been a large man, instead of a mere feather-weight, the probability is that he would not have survived so long." (p. 145.)

Galton does not draw the obvious moral that it is good policy not to face your foe, but to approach him edgewise!

"The enemy's bullets are least dangerous to the smallest men, and therefore small men are more likely to achieve fame as commanders than their equally gifted contemporaries whose physical frames are larger." (p. 146).

Under 'Men of Science,' a subject which Galton was to take up again later, there are many topics of interest raised. Galton notes that in the case of science the mother appears to play a greater part than the father. There is a long passage here—one of the finest Galton ever wrote—and one, notwithstanding its length, I feel bound to cite. It runs:

"It therefore appears to be very important to success in science that a man should have an able mother. I believe the reason to be, that a child so circumstanced has the good fortune to be delivered from the ordinary narrowing, partisan influences of home education. Our race is essentially slavish; it is the nature of all of us to believe blindly in what we love, rather than in that which we think most wise. We are inclined to look upon an honest, unshrinking pursuit of truth as something irreverent. We are indignant when others pry into our idols, and criticise them with impunity, just as a savage flies to arms when a missionary picks his fetish to pieces. Women are far more strongly influenced by these feelings than men; they are blinder partisans and more servile followers of custom. Happy are they whose mothers did not intensify their naturally slavish dispositions in childhood, by the frequent use of phrases such as, "Do not ask questions about this or that, for it is wrong to doubt"; but who showed them, by practice and teaching, that inquiry may be absolutely free without being irreverent, that reverence for truth is the parent of free inquiry, and that indifference or insincerity in the search after truth is one of the most degrading of sins. It is clear that a child brought up under the influences I have described is far more likely to succeed as a scientific man than one who was reared under the curb of dogmatic authority. Of two men of equal abilities, the one who has a truth-loving mother would be the more likely to follow the career of science; while the other, if bred up under extremely narrowing circumstances, would become as the gifted children in China, nothing better than a student and professor of some dead literature.

It is, I believe, owing to the favourable conditions of their early training that an unusually large proportion of the sons of the most gifted men of science become distinguished in the same career. They have been nurtured in an atmosphere of free inquiry, and observing as they grow older that myriads of problems lie on every side of them, simply waiting for some moderately capable person to take the trouble of engaging in their solution, they throw themselves with ardour into a field of labour so peculiarly tempting. It is and has been, in truth, strangely neglected. There are hundreds of students of books for one student of nature; hundreds of commentators for one original inquirer. The field of real science is in sore want of labourers. The mass of mankind plods on, with eyes fixed on the footsteps of the generations that went before, too indifferent or too fearful to raise their glances to judge for themselves whether the path on which they are travelling is the best, or to learn the conditions by which they are surrounded and affected." (pp. 196–7.)

Such was Galton's view of the relation of the higher freethought to science, and those who know his writings closely and knew the man himself will recognise how much of his own course in life these sentences depict; he threw himself with ardour into almost every field of inquiry, absolutely free of customary opinion, regardless, perhaps occasionally too regardless, of the footsteps of the generations that went before; he was essentially a student of phenomena and not of books.

In mathematicians among the men of science he seems somewhat disappointed, for

"when we consider how many among them have been possessed of enormous natural gifts it might have been expected that the lists of their eminent kinsmen would have been richer than they are." (p. 198.)

Galton realised fully the Bernoullis and the Gregorys, but wrote too early to have realised the Darwins, who combine mathematics and natural science. For his criticism is that while he knows senior wranglers related to other mathematicians, he does not think they have enough kinsmen eminent in *other* ways.

"I account for the rarity of such relationship in the following manner. A man given to abstract ideas is not likely to succeed in the world, unless he is particularly eminent in his peculiar line of intellectual effort. If the more moderately gifted relative of a great mathematician can discover laws, well and good; but if he spends his days in puzzling over problems too insignificant to be of practical or theoretical import, or else too hard for him to solve, or if he simply reads what other people have written, he makes no way at all, and leaves no name behind him. There are fewer of the numerous intermediate stages between eminence and mediocrity adapted for the occupation of men, who are devoted to pure abstractions, than for those whose interest is of a social kind." (p. 198.)

I think there is also another point which applies to all men of science but particularly to the mathematician. Two factors or qualities are needful for a great man of science, namely accurate power of analysis, and an intense power of imagination which equals, if it does not transcend, that of poet or painter. Imagination and analytical power do not seem correlated characters; they may be most fortunately combined in some individual but separated in his kinsmen. There are many mathematicians who are brilliant algebraists, but lack the imagination which finds problems worth solving and suggests the solution to be attempted analytically. That is why so many mathematicians are dull socially, they are inclined to spend their leisure playing chess or solving mathematical puzzles propounded by their fellows in educational journals—a pursuit akin to solving conundrums.

The really eminent man of science does, however, possess imagination; in fact, it is probably the most marked characteristic of all forms of genius, and with it comes the width that counteracts the dangers of specialisation. Galton saw this as fully as Huxley, and would smile in his quiet way when a committee of mediocrities turned down the proposal of a man of great imagination on the ground that it was not 'practical politics.'

"People lay," he writes, "too much stress on apparent specialities, thinking overrashly that, because a man is devoted to some particular pursuit, he could not possibly have succeeded in anything else. They might just as well say that, because a youth had fallen desperately in love with a brunette, he could not possibly have fallen in love with a blonde. He may or may not have more natural liking for the former type of beauty than the latter, but it is as probable as not that the affair was mainly or wholly due to a general amorousness of disposition. It is just the same with special pursuits. A gifted man is often capricious or fickle before he selects his occupation, but when it has been chosen he devotes himself to it with a truly passionate ardour. After a man of genius has selected his hobby, and so adapted himself to it as to seem unfitted for any other occupation in life, and to be possessed of but one special aptitude, I often notice with admiration how well he bears himself when circumstances suddenly thrust him into a strange position. He will display an insight into new conditions, and a power of dealing with them, with which even his most intimate friends were unprepared to accredit him. Many a presumptuous fool has mistaken indifference and neglect for incapacity; and in trying to throw a man of genius on ground where he was unprepared for attack, has himself received a most severe and unexpected fall. I am sure that no one who has had the privilege of mixing in the society of the abler men of any great capital, or who is acquainted with the biographies of the heroes of history, can doubt the existence of grand human animals, of natures pre-eminently noble, of individuals born to be kings of men. I have been conscious of no slight misgiving that I was committing a kind of sacrilege whenever, in the preparation of the materials of this book, I had occasion to take the measurement of modern intellects vastly superior to my own, or to criticise the genius of the most magnificent historical specimens of our race. It was a process that constantly recalled to me a once familiar sentiment in bygone days of African travel, when I used to take altitudes of the huge cliffs that domineered above me as I travelled along their bases, or to map the mountainous landmarks of unvisited tribes, that loomed in faint grandeur beyond my actual horizon." (pp. 24–5.)

As I have tried to impress on the reader, if Galton was a freethinker, he still worshipped as one of simpler faith at his own peculiar shrine—a shrine dedicated to the genius of his race. Those, who have not recognised that the inheritance of mental ability is the essential doctrine of Galton's faith—the

PLATE XI

Francis Galton, aged 60.
From the painting made in 1882 by Professor Graef, now in the possession of Mr Cameron Galton.
It was during the painting of this picture that Galton counted the strokes of the artist's brush.

keystone of the arch by aid of which mankind shall pass to a higher future—those, I say, have never properly understood his message to his generation; it is hard to believe that they have read, even superficially, his writings.

With 'Poets,' 'Musicians' and 'Painters' Galton is more brief than he has been in the case of men of action and of reason.

"The Poets and Artists generally are men of high aspirations, but, for all that, they are a sensuous, erotic race exceedingly irregular in their way of life. Even the stern and virtue-preaching Dante is spoken of by Boccaccio in most severe terms[1]. Their talents are usually displayed early in youth, when they are first shaken by the tempestuous passion of love." (p. 225.)

Almost all the able kindred of poets are in the first degree.

"Poets are not the founders of families. The reason is, I think, simple and it applies to artists generally. To be a great artist requires a rare and so to speak unnatural correlation of qualities. A poet, besides his genius, must have the severity, and stedfast earnestness of those whose dispositions afford few temptations to pleasure, and he must, at the same time, have the utmost delight in the exercise of his senses and affections. This is a rare character, only to be formed by some happy accident, and is therefore unstable in inheritance. Usually people who have strong sensuous tastes go utterly astray and fail in life, and this tendency is clearly shown by numerous instances mentioned in the following appendix, who have inherited the dangerous part of a poet's character and not his other qualities that redeem and control it." (p. 227.)

Dante having been put aside, and recent revelations having rather discredited Wordsworth, must we look on Goethe alone as the one poet who with stedfast earnestness took the utmost delight in the exercise of his senses? Or, shall we pin our faith solely to Milton, with perhaps Samuel Rogers lurking in the background? Galton was writing in his 47th year, and as we all know 20 and 60 are the dangerous ages for men. Would he have written forty years later with less condemnation, with a greater sense of the impenetrable mist which screens from our gaze the links between creative genius and sex? One of the most brilliant women of our day once said to the present writer: "I am most creative, when my senses dominate me most." The wise man strives to hold the bridle firmly, but in his old age will echo Galton's Boccaccio-like cry: "But who am I to pass judgment?"[2]

Galton could sit to an artist and count the number of his brush- or mallet-strokes[3], while to another the fascination is to watch the features of the man oscillating between gratification and despair as he strives to impart his ideal version of a crude reality to a refractory medium. I do not think Galton had a very clear appreciation of the artistic nature—I do not say of art.

But if he had less feeling for men of poetry than for men of science, he had almost no sympathy at all for 'Divines.' This is not to say that his accounts in the "Appendix to Divines" (pp. 283–98) are not fair, I believe they are, and, I think, he even exaggerates the importance of some of his selected men of piety. In order to be impartial he selects the 196 names

[1] "Amid so much virtue, amid so much knowledge as have been above shown existing in this wondrous poet, lustfulness found most ample place, and not only in his youthful, but in his mature years; which vice, although it is natural, and common and almost of necessity, of a truth cannot be rightly excused, much less commended. But who among mortals shall be the just judge to condemn him? Not I." Thus Boccaccio, and these the words Galton had in mind.

[2] Letter of February 6, 1907.

[3] See *Nature*, June 29, 1905.

contained in Middleton's *Biographia Evangelica*, 1786, and confines his attention to these. In this way he excludes in the first place practically all the founders of great religious movements. In the next place he excludes all Roman Catholic divines, and this on the ground of their celibacy; but still we might question whether research into their ancestry and collaterals would not have been of great interest, even if no descendants are available. But again, before the Reformation and for two centuries after, to a lesser extent however, churchmen of genius reached eminence largely because they were politicians, or the more illustrious because they were statesmen. On the other hand some information would surely be available as to the stirps of St Augustine, St Thomas Aquinas, St Francis, Meister Eckhardt, Johannes Tauler, George Fox, Swedenborg, John Wesley and many others who have left their impress on religious thought, and are excluded from the selection of Middleton. If I may say so, that selection seems to me to contain names of persons who in their day may have been eminent for piety, but did not possess intellectual ability amounting to genius. Yet Galton himself says that after reading Middleton's work he gained a much greater respect for the body of Divines than he had before:

"One is so frequently scandalised by the pettiness, acrimony and fanaticism shown in theological disputes, that an inclination to these failings may reasonably be suspected in men of large religious profession. But I can assure my readers, that Middleton's biographies appear, to the best of my judgment, to refer, in by the far greater part, to exceedingly noble characters. There are certainly a few personages of very doubtful reputation, especially in the earlier part of the work, which covers the turbid period of the Reformation......Nevertheless, I am sure that Middleton's collection, on the whole, is eminently fair and trustworthy." (p. 259.)

Now you may think a bishop an excellent fellow and find a boon companion in a dean; but even friend 'Punch' will smile at the gravity of the one, and venture to laugh out-loud at the seriousness with which the other takes himself. To make merry over the Divines is not to mock the Divinity—but I fear many of Galton's readers thought so in 1869, and his book for this and other reasons was by some strongly condemned. Yet, perhaps, Galton's chapter on the 'Divines' with all the irony and genial ridicule which make it such good reading is perhaps the subtlest in its analysis.

"I am now about to push my statistical survey into regions where precise inquiries seldom penetrate, and are not very generally welcomed. There is commonly so much vagueness of expression on the part of religious writers that I am unable to determine what they really mean when they speak of topics that directly bear on my present inquiry. I cannot guess how far their expressions are intended to be understood metaphorically, or in some other way to be clothed with a different meaning to what is imposed by the grammatical rules and plain meaning of language. The expressions to which I refer are those which assert the fertility of marriages and the establishment of families to be largely dependent on godliness. I may even take a much wider range and include those other expressions which assert that material well-being generally is influenced by the same cause......What I undertake is simply to investigate whether or no the assertions they contain, according to their *primâ facie* interpretation, are or are not in accordance with statistical deductions[1]. If an exceptional providence protects the families of godly men, it is a fact that we must take into account. Natural gifts would then have to be conceived as due, in a high and probably measurable degree to ancestral piety, and in a much

[1] We trace here the germ of Galton's later work on "The Efficacy of Prayer."

lower degree than I might otherwise have been inclined to suppose to ancestral natural peculiarities. All of us are familiar with another and an exactly opposite opinion. It is popularly said that the children of religious parents frequently turn out badly, and numerous instances are quoted to support this assertion. If a wider induction and a careful analysis should prove the correctness of this view, it might appear to strongly oppose the theory of heredity.

On both these accounts, it is absolutely necessary, to the just treatment of my subject, to inquire into the history of religious people, and learn the extent of their hereditary peculiarities, and whether or no their lives are attended by exceptionally good fortune." (pp. 257–8.)

Galton then starts on his analysis and finds that:

"As a general rule, the men in Middleton's collection had considerable intellectual capacity and natural eagerness for study, both of which qualities were commonly manifest in boyhood. Most of them wrote voluminously, and were continually engaged in preachings and religious services. They had evidently a strong need of utterance. They were generally, but by no means universally, of religious parentage......There is no case in which either or both parents are distinctly described as having been sinful, though there are two cases of meanness and one of over-spending......The Divines, as a whole, have had hardly any appreciable influence in founding the governing families of England, or in producing our judges, statesmen, commanders, men of literature and science, poets or artists. The Divines are but moderately prolific." (pp. 260–2.)

Those who marry often marry several times; thus out of Galton's 100, three married four times, two three times, and twelve had two wives apiece. Galton accounts for the early deaths of the wives of Divines by the hypothesis that their constitutions were on the whole weak. They were usually women of great piety, and

"there is a frequent correlation between an unusually devout disposition and a weak constitution." (p. 264.)

Galton finds the median age at death of Divines to be 62 to 63, which is rather *less* than that of eminent men dealt with in other parts of his volume.

"As regards health, the constitutions of most of the divines were remarkably bad." (p. 265.)

Studying young scholars or students he finds that they either die young, or strengthening as they grow retain their scholarly tastes and indulge them with sustained energy, or lastly live on in a sickly way. The Divines are largely recruited from the last class.

"There is an air of invalidism about most religious biographies......It is curious how large a part of religious biographies is commonly given up to the occurrences of the sick room[1]....... I can add other reasons to corroborate my very strong impression that the Divines are, on the whole, an ailing body of men." (pp. 265–74.)

Those who were of vigorous constitution had too frequently been wild in their youth. Galton generally concludes that a pious disposition is decidedly hereditary, but there are also frequent cases of the sons of pious parents turning out badly.

[1] Thus Rivet's biography is cited. He died after twelve days' suffering of strangulation of the intestines, the remedies attempted, each successive pang and each corresponding religious ejaculation is recorded; the history of his bowel attack being protracted through forty-five pages or as much space as is allotted to the entire biographies of four average divines. Where the piety of the divine is not witnessed by his martyrdom by men, it must be illustrated by his martyrdom by disease.

"I therefore see no reason to believe that the Divines are an exceptionally favoured race in any respect; but rather, that they are less fortunate than other men." (p. 274.)

Galton's statistical table indicates that the influence of the female line has an unusually large effect in qualifying a man for eminence in the religious world.

"The only other group in which the influence of the female line is even comparable in its magnitude is that of scientific men; and I believe the reasons laid down when speaking of them will apply, *mutatis mutandis*, to the Divines. It requires unusual qualifications and some of them of a feminine cast, to become a leading theologian. A man must not only have appropriate abilities, and zeal, and power of work, but the postulates of the creed that he professes must be so firmly ingrained into his mind, as to be the equivalents of axioms. The diversities of creeds held by earnest, good and conscientious men, show to a candid looker-on, that there can be no certainty as to any point on which many of such men think differently[1]. But a divine must not accept this view; he must be convinced of the absolute security of the groundwork of his peculiar faith,—a blind conviction which can best be obtained through maternal teachings in the years of childhood." (p. 276.)

The chapter concludes with a discussion which, whether it be correct or incorrect, is certainly subtle, of the relative views of the pious man and the sceptic. The contented sceptic having no faith in an external power tends to have confidence in himself and is therefore more stable.

"The sceptic, equally with the religious man, would feel disgust and shame at his miserable weakness in having done yesterday, in the heat of some impulse, things which to-day, in his calm moments, he disapproves. He is sensible that if another person had done the same thing, he would have shunned him; so he similarly shuns the contemplation of his own self. He feels he has done that which makes him unworthy of the society of pure-minded men; that he is a disguised pariah, who would deserve to be driven out with indignation, if his recent acts and real character were suddenly disclosed. The Christian feels all this and something more. He feels he has committed his faults in the full sight of a pure God; that he acts ungratefully and cruelly to a Being full of love and compassion, who died for sins like those he has just committed. These considerations add great poignancy to the sense of sin......

The result of all these considerations is to show that the chief peculiarity in the moral nature of the pious man is its conscious instability. He is liable to extremes—now swinging forward into regions of enthusiasm, adoration and self-sacrifice; now backwards into those of sensuality and selfishness. Very devout people are apt to call themselves the most miserable of sinners, and I think they must be taken to a considerable extent at their word. It would appear that their disposition is to sin more frequently and to repent more fervently than those whose constitutions are stoical, and therefore of a more symmetrical and orderly character. The *amplitude* of the moral oscillations of religious men is greater than that of others whose *average* moral position is the same." (pp. 280–2.)

On this hypothesis Galton explains the apparent anomaly why children of extremely pious parents occasionally turn out very badly.

[1] The agnostic standpoint has rarely been better put; yet while Spencer, Huxley and Clifford have been acknowledged as protagonists in the mid-Victorian contest of science and theology, Galton's attitude, in many respects more logical, and which caused much opprobrium at the time, has been largely forgotten or overlooked. And yet Galton's view of religion in 1869 was that of the Galton of 1894 and of 1907. He thought that a religion required no ultra-rational sanctions for conduct, and that a passionate aspiration to improve the heritable powers of man would suffice as the basis of a national religion, when the old religious notions and social practices had avowedly failed. See "The Part of Religion in Human Evolution," *National Review*, 1894, pp. 755 *et seq.*

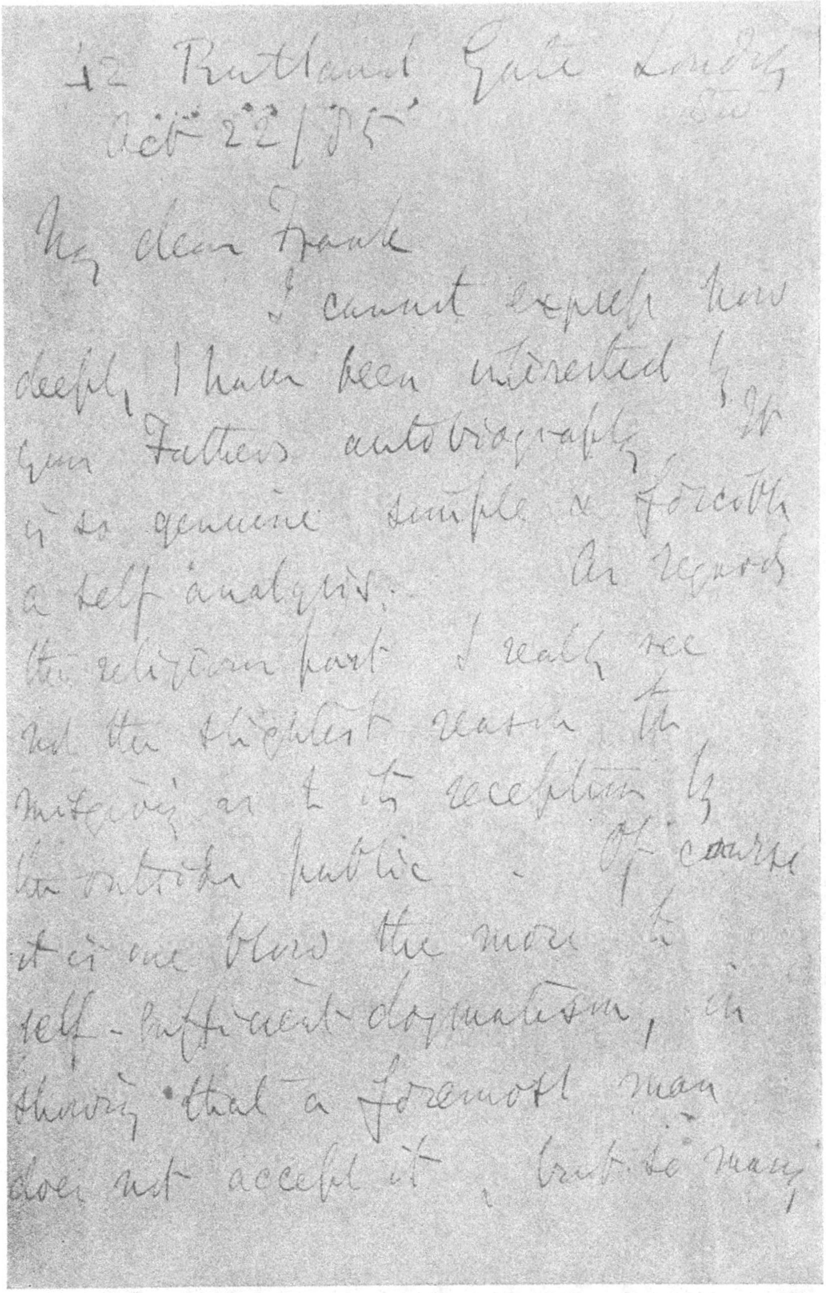

PLATE XII

42 Rutland Gate London SW
Oct 22/85

My dear Frank

I cannot express how deeply I have been interested by your Father's autobiography. It is so genuine simple & forcible a self analysis. As regards the religious part I really see not the slightest reason for misgiving as to its reception by the outside public. Of course it is one blow the more to self-sufficient dogmatism, in showing that a foremost man does not accept it, but so many

Letter of Francis Galton to Charles Darwin's son Francis, indicating the religious views of both Galton and Darwin.

blows of the sort have been given to it, that this can hardly be taken as a serious new assault.

Still, I quite see the difficulty about the Westminster Abbey question, that it would not be in good taste to ask for it just about the time that so firm a disavowal of Xianity was being prepared for publication.

I have kept the papers these three or four days, to read them out in the evening to my wife & to reread them. She is as interested as I am.

Very sincerely yours

Francis Galton

Have you happened to see
Herbert Spencer's just published
book on Ecclesiastical
Institutions !!!

If there still exists a service of
bell, book & candle, I can
fancy all the Bishops in
procession performing it over
the book & burning it and
its author together amidst howls
of anathemas

"The parents are naturally gifted with high moral characters combined with instability of disposition, but these peculiarities are in no way correlated. It must, therefore, often happen that the child will inherit the one and not the other. If his heritage consist of the moral gifts without the great instability, he will not feel the need of extreme piety; if he inherits great instability without morality, he will be very likely to disgrace his name." (p. 282.)

As I have said it is a very subtle hypothesis and to be convinced of its adequacy one would need to examine the facts of the instability with statistical categories. Galton had read more than 200 lives of Divines, which is immensely more than his biographer can lay claim to, and Galton had a very shrewd appreciation of character. Still he has not graded his Divines by instability of disposition and compared his graduation with that of other groups in the community, and until that is done his suggestion must remain hypothetical.

But there is a far more valuable idea at the root of the matter than its application to Divines, and that is where the subtlety arises. We are accustomed to speak of the quality or faculty of an individual for a given characteristic and measure it quantitatively if we can on a single occasion, or by a given test. We speak of a man's intellectual power and consider it as exhibited in his actions. But in all his actions he does not necessarily exhibit the same degree of wisdom; his intelligence fluctuates about a mean, and if we examine a man's life as a whole, it is this mean intelligence that we roughly appreciate. The same applies to all psychical characters, and indeed to many physical. Now Galton asserts that two individuals who have the same mean character will or may have widely different fluctuations from the mean. I think no man who has to deal with students or measure them anthropometrically would dispute this view. Personal equations fluctuate round an average and the intensity of the fluctuation or the stability of judgment varies from individual to individual. So far so good, but now comes Galton's subtle suggestion. It is that the magnitude of a character and its stability are independent units and may be inherited independently. As far as I am aware no attempt has been made to correlate the magnitude and the stability of any characters, psychical or physical, still less to test their independence in heredity. It should not be a hard piece of investigation and might lead to very valuable results, especially in economic breeding. It is peculiar to Galton's suggestions that they lead one so far afield. One passes almost unconsciously from the moral character of Divines to problems of root-growing and cattle-breeding[1]!

Of the chapter on 'Senior Classics' there is little to be said; it marks the grip of our *Alma Mater*, no less powerful on Galton, than on less considerable sons. The final chapters on 'Oarsmen' and 'Wrestlers' show that Galton gave rather a wide meaning to the term 'genius.' The material is interesting for two reasons. The inquiry for it brought Galton, the descendant of Quakers, into touch with that fine old Friend, Dr Robert Spence Watson,

[1] The seed from two turnip plants gives daughter-plants, say, of the same *average* weight, but in one case the fluctuations from the mean are large and in the other small. The stable crop would probably be more valuable. Is this stability an independent unit?

and I know from the personal accounts of both, how these two men, in many respects of kindred mind, appreciated each other. And secondly, because Galton endeavoured to destroy dogmas about muscle, so similar to those held by many about brain.

"No one doubts that muscle is hereditary in horses and dogs, but humankind are so blind to facts and so governed by preconceptions, that I have heard it frequently asserted that muscle is not hereditary in men. Oarsmen and wrestlers have maintained that their heroes spring up capriciously, so I have thought it advisable to make inquiries into the matter. The results I have obtained will beat down another place of refuge for those who insist that each man is an independent creation, and not a mere function, physically, morally and intellectually, of ancestral qualities, and external influences." (p. 305.)

We must now turn to Galton's final summarising chapters. I must confess frankly that while I consider that Galton has demonstrated the hereditary character of ability as judged by eminence, I find it very hard to fit in his statistical results with our present knowledge of the inheritance of ability.

One thing of course follows certainly and conclusively from the data, namely the farther removed, either directly or collaterally, a kinsman is from his eminent relative the smaller is his chance of being eminent. A son has the best chance of all and then comes the brother, and the probability tails away as we come to more distant relatives. This is reasonable because the ability has been usually diluted by what Galton would term 'mongrel' marriages, i.e. marriages with the intellectually mediocre. But there are two great difficulties in my mind about the analysis. The first is that of his grade of ability. On pp. 33–34, he defines his conception of eminence to be 250 men per million or one man in 4000. He also assumes a normal distribution for intelligence. Now, I think, that the student of *Hereditary Genius*, who considers the men, whether Judges, or Statesmen, or Men of Science, and still more the Divines, in Galton's lists, will hardly credit them with this degree of rarity. I confess that limiting the selection to the class of men educated professionally or by class tradition to aim at distinctions of this kind, I felt in my recent re-perusal that 1 in 500 was an adequate measure of the eminence, and before I came to the end of the book, I doubted whether it was more than 1 in 100. That is to say that while some of Galton's lists indicated men with a grade of 1 in 10,000 or even more, there was a very considerable tail, some of whom had not a greater ability than you would find in one in a hundred *or even fewer*.

I now started to test this on Galton's hypothesis that the distribution of capacity is normal, and on the result of much recent work that in a stable population the son will on the average inherit half his father's deviation from mediocrity[1], the mothers *not being selected*. In this manner I was able to form the following tables which indicate in a population of a million the probable number of eminent sons of eminent fathers for each standard of eminence.

[1] In technical language, if the standard deviations of the population in the two generations are equal, the coefficient of correlation will be 0·5.

But Galton found 48 sons per 100 fathers! Now I have already referred (p. 96) to my doubts as to Galton's estimate of the number of relatives to be attributed in each grade to an eminent man. He was perhaps biased by the wickedness of Judges and the misogyny of Statesmen! Anyhow I feel certain that the columns *C* of his tables and consequently the columns *D* are incorrect[1]. Had he attributed 200 or 250 sons to 100 eminent fathers or families, say, of 4 to 5, he would have found 19 to 24 eminent sons to 100 eminent fathers—still far too many—but approaching nearer our 13 with a

| | Eminence 1 in 1000 Father | | | Eminence 1 in 500 Father | | | Eminence 1 in 100 Father | | |
|---|---|---|---|---|---|---|---|---|---|
| | Non-eminent | Eminent | Totals | Non-eminent | Eminent | Totals | Non-eminent | Eminent | Totals |
| Son: Non-eminent | 998,054 | 946 | 999,000 | 996,139 | 1,861 | 998,000 | 981,298 | 8,702 | 990,000 |
| Son: Eminent | 946 | 54 | 1,000 | 1,861 | 139 | 2,000 | 8,702 | 1,298 | 10,000 |
| Totals ... ... | 999,000 | 1000 | 1,000,000 | 998,000 | 2,000 | 1,000,000 | 990,000 | 10,000 | 1,000,000 |
| No. of eminent sons per 100 eminent fathers | | 5·5 | | | 7 | | | 13[2] | |

much lower degree, however, of eminence. An explanation of the remaining discrepancy may, however, be found in the hint[3] thrown out by Galton in this chapter, that "a large number of eminent men marry eminent women[4]." He had already emphasised this point of view when discussing Men of Science and Divines. But such a mating of 'like with like' raises the correlation between offspring and parents slightly under 50 °/。[5]. Forming a table under these conditions we find for 1 in 100 degree of eminence:

[1] *Loc. cit.* p. 317 for general table, and compare tables at end of each section.

[2] This is much of the order one finds for number of insane sons of insane fathers.

[3] *Loc. cit.* p. 325.

[4] "The large number of eminent descendants from illustrious men must not be looked upon as expressing the results of their marriage with mediocre women, for the average ability of the wives of such men is above mediocrity. This is my strong conviction, after reading very many biographies, although it clashes with a commonly expressed opinion that clever men marry silly women. It is not easy to prove my point without a considerable mass of quotations to show the estimation in which the wives of a large body of illustrious men were held by their intimate friends, but the following two arguments are not without weight. First, the lady whom a man marries is very commonly one whom he has often met in the society of his own friends, and therefore not likely to be a silly woman. She is also usually related to some of them, and therefore has a probability of being hereditarily gifted." (p. 324.)

[5] The multiple correlation coefficient between parentage and offspring is now ·7071.

*Parentage*

| | | Non-eminent | Eminent | Totals |
|---|---|---|---|---|
| Son | Non-eminent ... | 982,675 | 7,325 | 990,000 |
| | Eminent ... ... | 7,325 | 2,675 | 10,000 |
| Totals ... ... ... | | 990,000 | 10,000 | 1,000,000 |
| No. of eminent sons per 100 eminent parentages | | | 27 | |

This result is well within Galton's limits if we suppose the number of sons born to an eminent father to average 1·8. With more sons, we could afford to give somewhat less credit to the mothers or increase the standard of eminence in the fathers, or take partly both these steps. I believe that the points just referred to: (1) under-estimate of the number of sons of men of eminence; (2) over-estimate of the rarity of the ability of many men, which flowed from Galton's intense admiration for all forms of intellectual power and all grades of originality; and (3) due appreciation of the extent to which assortative mating, or marriage within the caste, already exists among men of ability, suffice to bring the statistical results of Galton's book into line with more recent work. Let it be remembered that what Galton's researches really show is not that talent is under-inherited, but, if his treatment were correct, that it is *markedly over-inherited* as compared with other characters we know of. Genius must in that case be a sport and inherited in a peculiar and intense fashion. Galton felt considerable doubts as to his data of the number of relatives of a man (pp. 318–19), and many years later the Biometric Laboratory provided him with more ample material. He recognised (3), but at that time was not able adequately to measure its effect. While as to (2) I know from personal experience that he was apt to exaggerate the intellectual ability of men, who by a certain brilliancy of expression and adequate self-assertion, obtained temporary notoriety[1].

Galton's next chapter on "The Comparative Worth of Different Races" is, at the same time, more sketchy and more suggestive than the last. He compares the intellectual ability of various races; he places a difference of three grades between Anglo-Saxon and Negro, one being due to relative demerits of native education and two to natural ability. He puts the Australian native one grade below the African negro, and North of England and Lowland Scottish a fraction of a grade above the ordinary English.

[1] I do not mean that Galton could not discriminate a charlatan, but where there was *some* originality, he was apt to exaggerate it into *all* originality; it seemed a natural consequence of his own modesty, of his geniality and above all of the weight he laid on *any* originality.

"The peasant women of Northumberland work all day in the fields and are not broken down by the work; on the contrary, they take a pride in their effective labour as girls, and when married, they attend well to the comfort of their homes. It is perfectly distressing to me to witness the draggled, drudged, mean look of the mass of individuals, especially of the women, that one meets in the streets of London, and other purely English towns. The conditions of their life seem too hard for their constitutions, and to be crushing them into degeneracy." (p. 340.)

Galton then turns to Greece, and having equated Plato and Bacon, by what must largely be the impressionism of individual personal judgment, concludes that the Athenian race from 530 B.C. to 430 B.C. was very nearly two grades above our own,

"that is, about as much as our race is above that of the African negro. This estimate, which may seem prodigious to some, is confirmed by the quick intelligence and high culture of the Athenian commonalty, before whom literary works were recited, and works of art exhibited, of a far more severe character than could possibly be appreciated by the average of our race, the calibre of whose intellect is easily gauged by a glance at the contents of a railway bookstall." (p. 342.)

I do not think Galton's comparison is justified, for he leaves out in the one case the labouring and artizan population of 400,000 slaves, and includes such a population in the other when reckoning his percentages of extreme ability. Again he writes:

"We have no men to put by the side of Socrates and Phidias, because the millions of all Europe, breeding as they have done for the subsequent 2000 years, have never produced their equals." (p. 342.)

Without belittling Phidias we may reasonably question whether his genius was really greater than that of the designer of any one of the great medieval Gothic cathedrals. Who can determine whether Raphael or Phidias was the greater artist? As for Socrates we see him through the mists; we do not know the man himself, but still only perceive him amid the glamour of his contemporaries and the veneration of renascent humanists[1]. If we judge him by the Socrates of the Platonic dialogues, his subtlety is not always deep and his wisdom does not invariably appear very fundamental to the modern cultured mind. If we require a fair test of relative fineness of intellect, in two ages, surely we may ask this: Would the ablest minds of Age $A$ have grasped the subtlest thought of Age $B$, and would the genius of $B$ have failed to appreciate the intellectual product of $A$'s most eminent minds? Judged by this test, I think both Kant and Einstein could fully grasp and duly appreciate what the Platonic Socrates had to say, but I gravely doubt whether the ideas of both Kant and Einstein would not have transcended Socrates' mental capacity, even as the modern geometrician himself fully understands Euclid, but Euclid would have failed to understand him. And this is not a matter of the accumulated *knowledge* of the intervening centuries, it is a result of the ablest intellects being more subtle, more capable of forming generalised conceptions than the most capable of ancient Greeks.

Again, it is true that 9 °/₀ of the Athenian population[2] did enjoy the

[1] 'Sancte Socrates, ora pro nobis.'

[2] Rather 2 to 3 °/₀, if we take no account of the women and children, who did not of course witness the plays.

tragedies of Aeschylus, Sophocles and Euripides; but it is certain that those tragedies appealed to the primal passions of mankind, stronger and less bridled the closer civilised man is to the primitive savage; and it is not certain that nine-tenths of that audience did not prefer the buffoonery scenes from "The Birds," just as the 5 °/₀ more highly educated class of to-day professes interest in problem plays but attends the 'revues.'

Galton uses the Greeks as an illustration of a race two to three 'grades' higher in intelligence than our own, and hence as an argument that what man has been man can be. Its failure was due, he holds, to lax morality. But surely that want of moral stability indicated an inferiority in certain aspects of the psychical side, a lack which permitted the shadow of the doctrines of Paul to dim the brilliance of Greek culture. Galton traces with emphasis how the more intellectual rather than the physically stronger nations have dominated the world, the survival of the fitter meaning rather the mentally than the physically fitter. In this evolution of fairly consistent trend, the collapse of the Attic race appears as a most disturbing factor. Galton distinctly felt this, but I believe he took too much on faith. Our confidence in the superiority of the Greek intellect has been too largely based on the judgment of men, the classical scholars, who have devoted a disproportionately large period of their lives to the study of a single, if undoubtedly important, phase of human culture. You cannot judge the relative value of a human culture—especially if you approach it from a literary side only—unless you have a fairly comprehensive knowledge of the achievements of other cultures, and it is needful to study them not from one but from many sides. Our judgment of Greek culture has, I venture to think, not been made with a due appreciation of other cultures even up to our own; it has not been in the highest sense an anthropological judgment—we have taken at second-hand the opinion of men whose lives have been devoted to the study of an isolated, if brilliant incident in the hundreds of thousands of years of human evolution, and we have accepted their justifiable enthusiasm, as if it must be the whole truth as seen from a more distant but wider point of view.

It is a strange illustration of human love of dogmas that Galton's appraisal of the Greek intellect has, perhaps, been the most frequently remembered and cited passage of a book remarkable for its novel and reasoned opinions. Of course its citation is generally associated with the suggestion that the history of man is not one of advancing mental development; whereas Galton used it to point out that races could by judicious organisation raise their intellectual grade.

"And we too, the foremost labourers in creating this civilisation, are beginning to show ourselves incapable of keeping pace with our own work. The needs of centralisation, communication, and culture, call for more brains and mental stamina than the average of our race possess. We are in crying want of a greater fund of ability in all stations of life; for neither the classes of statesmen, philosophers, artizans, nor labourers are up to the modern complexity of their several professions. An extended civilisation like ours comprises more interests than the ordinary statesmen or philosophers of our present race are capable of dealing with, and it exacts more intelligent work than our ordinary artizans and labourers are capable of performing. Our race is overweighted, and appears likely to be drudged into degeneracy by demands that exceed its powers. If its average ability were raised a grade or two, our new classes *F* and *G* would conduct the complex affairs of the state at home and abroad as easily as our present *F* and *G*, when in the position of country squires, are able to manage the affairs of their establishments and tenantry. All other classes of the community would be similarly promoted to the level of the work required by the nineteenth century, if the average standard of the race were raised." (p. 345.)

The Greek statesman or commander had to deal with hundreds or thousands of men, where ours have to deal with millions in a society where the relations are of immensely increased complexity; that must be borne in mind when we compare the intellectual ability of the two cultures. Foch could have done the work of Themistocles, but the latter would have broken down under the complexity of the work of a modern commander. He would, as Galton does, have certainly called for a superman. One wonders if the ancestry of Mr Bernard Shaw's 'superman' cannot be traced to Galton; for Mr Shaw took him from Nietzsche, and the latter knew of Galton's work[1].

Galton's penultimate chapter "Influences that affect the Natural Ability of Nations" contains results almost commonplaces now, but in 1869 they were original suggestions of the highest value, because he was practically the first to apply the Darwinian doctrines to man and his communities. He notes how careless Nature is of the lives of individuals, she is equally careless of the lives of eminent families, they arise, flourish and decay; and the same may be said of races and of nations, they have arisen in the past, reached grandeur and then perished, often leaving but the slenderest shreds of their culture to be preserved among the mental traditions of humanity as a whole. Nay it is possible that such may be the story of our earth itself relative to other possible scenes of existence in the cosmos around us.

"We are exceedingly ignorant of the reasons why we exist, confident only that individual life is a portion of some vaster system that struggles arduously onwards towards ends that are dimly seen or wholly unknown to us, by means of the various affinities—the sentiments, the tastes, the appetites of innumerable personalities who ceaselessly succeed one another on the stage of existence." (p. 351.)

But such an outlook produced by the growing physical and historical knowledge of man, while it depressed many of lesser mental calibre, who found themselves torn from their old supernatural moorings and carried helplessly along on the overwhelming tide of new thought, such an outlook only led Galton to proclaim that Man—if at last he would stand on his own feet, and discard his ages-old crutches—could to a large extent make his own future. Confidence in himself, and in his own knowledge—faith in his own intellectual leaders and not in any supernatural kindliness of cosmic purpose—were for Galton the thorny but certain path towards man's salvation.

"Our world appears hitherto to have developed itself, mainly under the influence of unreasoning affinities; but of late, man slowly growing to be intelligent, humane, and capable, has appeared on the scene of life and profoundly modified its conditions. He has already

[1] Frau Förster Nietzsche in *The Lonely Nietzsche* gives (p. 191) a letter of Dr Panneth (15/12/1883) and the latter reports a talk with Nietzsche at Nice, when "the conversation turned on Galton," but there are unfortunately no particulars.

become able to look after his own interests in an incomparably more far-sighted manner, than in the old prehistoric days of barbarism and flint knives; he is already able to act on the experiences of the past, to combine closely with distant allies, and to prepare for future wants, known only through the intelligence, long before their pressure has become felt. He has introduced a vast deal of civilisation and hygiene which influence, in an immense degree, his own well-being and that of his children; it remains for him to bring other policies into action that shall tell on the natural gifts of his race." (p. 352.)

"How consonant it is to all analogy and experience to expect that the control of the nature of future generations should be as much within the power of the living, as the health and well-being of the individual is in the power of the guardians of his youth." (p. 351.)

Galton puts on one side such social arrangements as existed in Sparta "as alien and repulsive to modern feelings"[1] and confines his discussion to "agencies that are actually at work, and upon which there can be no hesitation in speaking." (p. 352.)

He now takes in succession a series of factors which affect the natural ability of nations. He first stresses differential fertility and says that the wisest policy is that which retards marriage among the weak and hastens it among the vigorous classes. He was the first, I believe, to draw attention to the fact that many social agencies have been "strongly and banefully exerted in the precisely opposite direction." He points out how a very slight difference in fertility of two classes of the community will in one or two centuries enormously change the constituents of a population. He indicates that early marriage not only increases fertility, but by causing more overlapping of generations largely increases population apart from increased fertility. After referring to the rapidly waning influence of any subsection of a race which postpones marriage, Galton continues:

"It is a maxim of Malthus that the period of marriage ought to be delayed in order that the earth may not be overcrowded by a population for whom there is no place at the great table of nature. If this decline influenced all classes alike, I should have nothing to say about it here, one way or another, for it would hardly affect the discussions in this book; but as it is put forward as a rule of conduct for the prudent part of mankind to follow, whilst the imprudent are necessarily left free to disregard it, I have no hesitation in saying that it is a most pernicious rule of conduct in its bearing upon race. Its effect would be such as to cause the race of the prudent to fall, after a few centuries, into an almost incredible inferiority of numbers to that of the imprudent, and it is therefore calculated to bring utter ruin on the breed of any country where the doctrine prevailed. I protest against the abler races being encouraged to withdraw in this way from the struggle for existence. It may seem monstrous that the weak should be crowded out by the strong, but it is still more monstrous that the races best fitted to play their part on the stage of life should be crowded out by the incompetent, the ailing, and the desponding.

The time may hereafter arrive, in far distant years, when the population of the earth shall be kept as strictly within the bounds of number and suitability of race, as the sheep on a well-ordered moor or the plants in an orchard-house; in the meantime, let us do what we can to encourage the multiplication of the races best fitted to invent and conform to a high and generous civilisation, and not, out of a mistaken instinct of giving support to the weak, prevent the incoming of strong and hearty individuals." (pp. 356–7.)

[1] This point is very important, for superficial critics of eugenics have asserted that Galton advocated 'Spartan' methods of mating. The creation of a superior intellectual caste, with a religious feeling against mating outside it, and a national encouragement of its early marriage and its fertility formed Galton's policy. The adequate endowment of superior motherhood so that women of marked intelligence shall have greater freedom in the choice of the father of their children is possibly the only considerable addition which has been made since by cautious eugenists to Galton's positive policy.

The above is one of the most noteworthy passages in which Galton condemns the teaching of unrestricted birth-control[1]. And its wisdom appears convincing to any but Neo-Malthusian fanatics! The imprudent, the feckless, and the feebleminded by their very nature will not control their births, and the higher intelligences will and do. Birth-control is poison to a race which has not legislatively organised a differential fertility of its castes; it is death to a race which has not regarded its own fertility in relation to that of its neighbours and possible enemies. The fear France exhibits before Germany to-day, even after a successful war, is largely the outcome of her neglect of these Galtonian axioms[2].

Galton's condemnation of unthinking birth-control is followed by a still stronger condemnation of the teaching of the Catholic Church, which chose to preach and exact celibacy from its most earnest devotees.

"The long period of the dark ages under which Europe has lain is due, I believe to a very considerable degree, to the celibacy enjoined by the religious orders on their votaries. Whenever a man or woman was possessed of a gentle nature that fitted him or her to deeds of charity, to meditation, to literature or to art, the social condition of the time was such that they had no refuge elsewhere than in the bosom of the Church. But the Church chose to preach and exact celibacy. The consequence was that these gentle natures had no continuance, and thus by a policy so singularly unwise and suicidal that I am hardly able to speak of it without impatience, the Church brutalised the breed of our forefathers. She acted precisely as if she aimed at selecting the rudest portion of the community to be, alone, the parents of future generations. She practised the arts which breeders would use, who aimed at creating ferocious, currish and stupid natures. No wonder that club law prevailed for centuries over Europe; the wonder rather is that enough good remained in the veins of Europeans to enable their race to rise to its present, very moderate level of natural morality." (p. 358[3].)

But the destruction of moral gentleness was not the only or perhaps the most culpable result of Catholic policy. Galton cites the effects of the Inquisition, of the martyrdom and imprisonment of original thinkers in Spain, Italy and France.

"The Spanish nation was drained of free-thinkers at the rate of 1000 persons annually for the three centuries between 1471 and 1781."

In Italy

"in the diocese of Como alone more than 1000 were tried annually by the inquisitors for many years, and 300 burnt in the single year of 1416."

In France during the seventeenth century three to four hundred thousand

[1] Darwin strongly supported Galton's opinion; see *More Letters of Charles Darwin*, Vol. II, p. 50.

[2] It would be amusing, were it not sad, to note how large and influential a section of Galton's creation, the English Eugenics Education Society, has recently been satisfying its thirst for education at Neo-Malthusian rather than Galtonian springs.

[3] Galton refers to a relic of this monastic spirit which in his day gave an able young man a fellowship at the University, not in order that he might marry, but on condition that he did *not*. That is now abolished, but the lay councils of academic institutions are still imbued with the ignorance of the Catholic Church, they just as effectually check the fertility of the able by allowing only celibate pittances to the young men and women whose instinct impels them to make research and learning their calling in life. Such a layman recently pointing to the list of the ill-paid staff of the Galton Laboratory wanted to know why it should be maintained, if there were so few students to be taught!

Huguenots perished in prison, at the galleys, on the scaffold or in attempting to escape, and an equal number emigrated. The Huguenots

"were able men, and profoundly influenced for good both our breed and our history."

This cruel policy degraded future generations, for it brought

"thousands of the foremost thinkers and men of political aptitudes to the scaffold, or imprisoned them during a large part of their manhood or drove them as emigrants to other lands."

Thus it came about that the Church,

"having first captured all the gentle natures and condemned them to celibacy, made another sweep of her huge nets, this time fishing in stirring waters, to catch those who were the most fearless, truth-seeking and intelligent in their modes of thought, and therefore the most suitable parents of a high civilisation, and put a strong check, if not a direct stop, to their progeny. Those she reserved on these occasions, to breed the generations of the future, were the servile, the indifferent and again the stupid. Thus as she—to repeat my expression—brutalised human nature by her system of celibacy applied to the gentle, she demoralised it by her system of persecution of the intelligent, the sincere, and the free. It is enough to make the blood boil to think of the blind folly that has caused the foremost nations of struggling humanity to be the heirs of such hateful ancestry, and that has so bred our instincts as to keep them in an unnecessarily long-continued antagonism with the essential requirements of a steadily advancing civilisation." (pp. 358–9.)

Such is Galton's terrible indictment of the effect of the Roman ecclesiastical policy. It has not been refuted, and it cannot be, except either by denying the value of original thinking to mankind, or demonstrating that originality of mind is not an hereditary characteristic. It is little wonder that eugenics has met with small appreciation from Catholic writers. Yet the charge has no longer other than historic value; the will to persecute may still exist in the ecclesiastically minded, but there is little force behind it; the old religions, except in savage races, have lost their hold on tribal imagination; we are seeking new religious ideals. And, as for the Roman Catholic celibacy, it may now, with a few if notable exceptions, be looked upon as a eugenic rather than a dysgenic factor.

Galton finally points out how in a young colony

"the strong arm and enterprising brain are the most appropriate fortunes for a marrying man,"

but in an old civilisation the factors at work are far more complex.

"Among the active ambitious classes, none but the inheritors of fortune are likely to marry young."

Men of moderate but more than average ability will not do so because

"their future is not assured except through a good deal of self-denial and effort."

Men of great ability, even if they marry young, think of social position and desire to found families and are attracted by wealth in the first place. Thus Galton holds that in an old civilisation there is a steady check on the fertility of the abler classes, so that the race gradually deteriorates, until

"the whole political and social fabric caves in, and a greater or less relapse to barbarism takes place, during the reign of which the race is perhaps able to recover its tone." (p. 362.)

"The best form of civilisation in respect to the improvement of the race would be one in which society was not costly; where incomes were chiefly derived from professional sources, and not much through inheritance; where every lad had a chance of showing his abilities, and, if highly-gifted, was enabled to achieve a first-class education and entrance into professional life by the liberal help of the exhibitions and scholarships which he had gained in his early youth; where marriage was held in as high honour as in ancient Jewish times; where the pride of race was encouraged (of course I do not refer to the nonsensical sentiment of the present day that goes under that name); where the weak could find a welcome and a refuge in celibate monasteries or sisterhoods, and lastly, where the better sort of emigrants and refugees from other lands were invited and welcomed, and their descendants naturalised." (p. 362.)

Such was the gospel of national welfare that Galton taught in 1869—more than half a century ago—then as now indicating a fundamental truth. What progress have we made towards his ideal in the fifty intervening years? Well, we have the educational ladder, and men can and actually do to some extent mount it. But wealth, especially in these last years, has grown a still more destructive factor of social stability, and the relatively low fertility of the abler families has been emphasised, not reduced. Yet, if the need for race-betterment has become greater, the recognition of that need has undoubtedly become more widespread; and that recognition we owe, not a little, to the labours of Galton in the last ten years of his life.

The last chapter of Galton's work reads somewhat quaintly now, partly in the light of later researches by others, and partly because of the progress Galton himself made later in hereditary theory. He adopts Darwin's theory of Pangenesis which was clearly much exercising his mind at this time. He speaks twice of the "gemmules"—i.e. the germs thrown off by each cell and carrying its hereditary qualities—as "circulating in the blood" (pp. 363 and 367) and even propagating there[1]. Darwin did not at this time correct the error, if error it was, in Galton's interpretation, although he wrote very enthusiastically about the book. Galton illustrates what he considers would be the results of the theory of Pangenesis by a series of rather quaint analogies in the midst of which we find his theory of stability—later more fully developed—illustrated by the oscillations of a rocking-stone stable until violent movement throws it over into a new position of equilibrium. In a footnote, pp. 371–2, we find Galton, on the basis of Pangenesis, feeling his way towards the Law of Ancestral Heredity—namely that the influence of an individual ancestor in the $n$th generation diminishes in geometrical progression. He states that the treatment of heredity on the basis of Pangenesis

"seems well within the grasp of analysis, but we want a collection of facts, such as the breeders of animals could well supply, to guide us for a few steps out of the region of pure hypothesis." (ftn. p. 372.)

Herein lies the germ of the quantitative or statistical theory of heredity. Again Galton points out that the artificial breeder of fish by taking milt from the male and allowing it to fall on the ovum deposited by the female can produce a new individual life, and that the characteristics of this individual are largely under his control, if he has studied the parents. But

[1] "Mr Darwin maintains, in the theory of Pangenesis, that the gemmules of innumerable qualities, derived from ancestral sources, circulate in the blood, and propagate themselves, generation after generation, still in the state of gemmules, but fail in developing themselves into cells; because other antagonistic gemmules are prepotent and overmaster them, in the struggle for points of attachment, etc." (p. 367.)

"all generation is physiologically the same, and therefore the reflections raised by what has been stated of fish are equally applicable to the life of man." (p. 375.)

"Nature teems with latent life, which man has large powers of evoking under the forms and to the extent which he desires. We must not permit ourselves to consider each human or other personality as something supernaturally added to the stock of nature, but rather as a segregation of what already existed, under a new shape, and as a regular consequence of previous conditions. Neither must we be misled by the word 'individuality,' because it appears from many facts and arguments in this book that our personalities are not so independent as our self-consciousness leads us to believe. We may look upon each individual as something not wholly detached from its parent source—as a wave that has been lifted and shaped by the normal conditions of an unknown, illimitable ocean. There is decidedly a solidarity as well as a separateness in all human, and probably in all lives whatsoever,.......

It points to the conclusion that all life is single in its essence, but various, ever varying and interactive in its manifestations, and that men and all other animals are active workers and sharers in a vastly more extended system of cosmic action than any of ourselves, much less of them, can possibly comprehend. It also suggests that they may contribute, more or less consciously, to the manifestation of a far higher life than our own, somewhat as—I do not propose to push the metaphor too far—the individual cells of one of the more complex animals contribute to the manifestation of its higher order of personality." (p. 376.)

This wonderful final passage of Galton's work foreshadows the doctrine of the continuity of the germ-plasm, the one eternal, amid transitory individual bearers of the life-giver. But it goes further, it reminds us that on the theory of evolution all present forms of germ-plasm are ancestrally related, are all descended from a single primary form of plasma to which we can give the name Life. This is the solidarity of all living forms which Galton refers to, and which leads him to look upon the living Universe as a pure theism—if indeed he did not mean a pure pantheism. It needed the imagination of a great scientist to give such a turn to the inspiration of the poet:

Was wär' ein Gott, der nur von aussen stiesse,
Im Kreis das All am Finger laufen liesse,
Ihm ziemt's, die Welt im Innern zu bewegen,
Natur in sich, sich in Natur zu hegen,
So dass, was in Ihm lebt und webt und ist,
Nie Seine Kraft, nie Seinen Geist vermisst[1].

How many men have talked glibly of the continuity of the germ-plasm without realising the solidarity of all life which flows from it! How few reading Galton's views on the *religious* nature of eugenic belief have grasped how closely his doctrine of race-betterment was associated with the pantheism, to which his view of the germ-plasm had led him. For Galton the Deity was synonymous with Life in its entirety and he asked us to aid Life struggling to fuller expression by elevating the race of man. Georgians may term this idea the sentimentality of a mid-Victorian: but after all it is less of a dogma than that of some of their number, who tell us that the Deity is limited in his powers and ask us to come—in some unexplained manner—to his assistance.

*Hereditary Genius* is one of the great books of the world, not so much by what it proves, as by what it suggests. Detailed proof was to come afterwards, step by step. Its publication formed the central epoch of Galton's life and nearly all his later work may be seen therein to take its origin. If

[1] Goethe, *Gott und Welt. Proemion.*

it met with a cool reception, it was because the world was not ripe for it. Two men, however, perceived its value; Darwin wrote: "I congratulate you on producing what I am convinced will prove a memorable work" (see our Plate I, Vol. I); and Alfred R. Wallace said of it in the just-born *Nature*[1]:

"Turning now to the concluding chapters of the book, we meet with some of the most startling and suggestive ideas to be found in any modern work....These concluding chapters stamp Mr Galton as an original thinker, as well as a forcible and eloquent writer, and his book will take rank as an important and valuable addition to the science of human nature."

Those judgments, not contemporary newspaper opinions, have stood the test of time.

### C. PAPERS CLOSELY ASSOCIATED WITH THE THEME OF *HEREDITARY GENIUS*

Two further popular papers are closely related to Galton's *Hereditary Genius* and may be considered here. The first is entitled: "Statistical Inquiries into the Efficacy of Prayer"; it appeared in the *Fortnightly* for August 1872[2]. It led to a certain amount of controversy in the pages of *The Spectator* in which Galton took part, but it also so pained—I think unreasonably—many worthy folk that Galton was treated for a time as a very flippant freethinker. His opponents asserted first that the desire to pray is intuitive in man, and secondly that the cogency of intuition is greater than that of observation. If the word 'intuitive' be interpreted to mean 'instinctive' and the words 'to pray' be interpreted 'to cry out in despair or in agony,' then the terrible cry of the young rabbit when the stoat springs upon him—a cry which is made to no one in particular—is an intuitive prayer. But if prayer means an appeal for temporal aid to a supernatural power, then the savage does not pray until the missionary teaches him. As Galton points out, obedience to dreams, belief in incantations, fear of witchcraft, fetish worship and tabu are intuitive, for they occur in uncivilised peoples all the world over. In modern civilisation the mother replaces the missionary and the child is taught with caressing earnestness to pray for temporal blessings, and in distress to appeal for aid to an all-seeing and all-loving deity. What wonder that this nursery-theology pervades human life, and being associated with a child's earliest and deepest feelings, should come to be looked on as intuitive! The habit of prayer, until its source has been analysed, is held to be of primeval origin. The theologians who accept the *objective* efficacy of prayer—i.e. not merely its subjective value to certain natures, but its power to produce temporal blessings—are the descendants of those who only a few centuries ago believed in the efficacy of auguries, of ordeals, of ecclesiastical blessings and cursings, in the existence of demoniacal possession and the value of exorcisms, in the possibility of witchcraft and of miraculous cures. All these the English Church has now suppressed as of 'superstitious' origin. But it was the more or less unconscious use of statistics which demonstrated the idle character of these 'intuitive' beliefs. Observation proved greater than

[1] March 17, 1870, Vol. I, p. 501. [2] Vol. XII, N.S. pp. 125–35.

the cogency of intuition in these cases; then why should the theologians of to-day, if summoned on the grounds of observation or statistics to give up a belief which has far less claim to be considered an intuition, start with naïve indignation, as at a previously unheard-of and most unreasonable interference[1]?

I do not think Galton propounded his thesis of the statistical inefficacy of prayer—as Clifford in other like matters stated he did—with the view of "drawing" *The Spectator*. He came to his topic naturally and unexpectedly. In his study of the 'Divines' for his *Hereditary Genius*, he had been struck by "their wretched constitutions" (see our p. 101). To obtain a measure of this Galton investigated their age at death, and compared it with that of other classes. Using *Chalmers's Biography* and *The Annual Register*, Galton found

| | | |
|---|---|---|
| Artists | 64·74 | Mean age at death. |
| Men of Literature and Science | 65·22 | |
| Clergy | 66·42 | |
| Lawyers | 66·51 | |
| Medical Men | 67·07 | |

Galton holds that the clergy are a far more 'prayerful class' than lawyers or doctors, and yet, although the numerous published collections of family prayers are full of petitions for temporal benefits, and the prayers of the clergy are for protection against the perils and dangers of the night and of the day and for recovery from sickness, such prayers appear to be futile in result[2]. The above statistics are for eminent men, and therefore may be supposed to be in the case of 'Divines' for those of marked piety. Galton also cites Guy's data[3] which provide the following figures:

| | | |
|---|---|---|
| Members of Royal Houses | 64·04 | Mean age at death. |
| Artists | 65·96 | |
| Medical Men | 67·31 | |
| Men of Literature and Science | 67·55 | |
| Lawyers | 68·14 | |
| Clergy | 69·49 | |
| Gentry | 70·22 | |

The members of the Royal Houses are the persons whose longevity is most widely and continuously prayed for, and they have the least average length of life! But the mass of clergy—as distinct from eminent divines—have a longer life than the mass of lawyers or medical men. Galton attributes this to the easy country life, family repose, and sanitary conditions, but his critics might well have attributed the result to the greater prayerfulness of the lesser clergy. The greater length of life of the clergy as a whole is now a well-established actuarial fact, but probably to-day no one associates it with prayerfulness. Galton gives a good many illustrations of the want of efficacy in prayer, e.g. the relatively short lives of missionaries, the distribution of still-births as between clergy and laymen being wholly unaffected by

[1] Letter of Galton to *The Spectator*, 1872, August 24. In editorials and correspondence the discussion lasted from the issue of August 3 until that of September 7.

[2] *Fortnightly* (*loc. cit.*), p. 129.

[3] *Journal of R. Statist. Society*, Vol. XXII, p. 355.

piety, the fact that the nobility are peculiarly subject to insanity notwithstanding that our Liturgy prays that they may be endued "with grace, wisdom and understanding," the fact that insurance offices make no differences in the insurances of pious and profane persons, or of ships fitted out for pious or profane purposes, although they are for ever measuring slight differences of risk, etc.

In his last paragraphs Galton turns to the *subjective* value of prayer:

"Nothing that I have said negatives the fact that the mind may be relieved by the utterance of prayer. The impulse to pour out the feelings in sound is not peculiar to man. Any mother that has lost her young, and wanders about moaning and looking piteously for sympathy, possesses much of that which prompts men to pray in articulate words. There is a yearning of the heart, a craving for help, it knows not where, certainly from no source it sees." (p. 135.)

The paper concludes with a fine statement which at least emphasises the religious comfort Galton found in his own pantheistic views and from which freethinkers without those views may still draw consolation:

"A confident sense of communion with God must necessarily rejoice and strengthen the heart, and divert it from petty cares; and it is equally certain that similar benefits are not excluded from those who on conscientious grounds are sceptical as to the reality of a power of communion. These can dwell on the undoubted fact, that there exists a solidarity between themselves and what surrounds them, through the endless reactions of physical laws, among which the hereditary influences are to be included. They know that they are descended from an endless past, that they have a brotherhood with all that is, and have each his own share of responsibility in the parentage of an endless future. The effort to familiarise the imagination with this great idea has much in common with the effort of communing with a God, and its reaction on the mind of the thinker is in many respects the same. It may not equally rejoice the heart, but it is quite as powerful in ennobling the resolves, and it is found to give serenity during the trials of life and in the shadow of approaching death." (p. 135.)

I now turn to the last popular appeal which Galton made for conscious race-betterment for more than 30 years. As he himself has said, the time was not ripe for such a programme as he had in mind, and he did not recur to the topic until 1901. The paper appeared in *Fraser's Magazine* in January 1873[1], under the title: "Hereditary Improvement." It opens as follows:

"It is freely allowed by most authorities on heredity, that men are just as subject to its laws, both in body and mind, as are any other animals, but it is almost universally doubted, if not denied, that an establishment of this fact could ever be of large practical benefit to humanity. It is objected that, philosophise as you will, men and women will continue to marry, as they have hitherto done, according to their personal likings; that any prospect of improving the race of man is absurd and chimerical, and that though inquiries into the laws of human heredity may be pursued for the satisfaction of a curious disposition, they can be of no real importance. In opposition to these objections, I maintain, in the present essay, that it is feasible to improve the race of man by a system which shall be perfectly in accordance with the moral sense of the present time." (p. 116.)

Galton holds that conscious race-betterment must arise as soon as the

[1] Vol. VII, New Series, pp. 116–30. I may, perhaps, be permitted to interpolate here a remark, which is true if pessimistic. The impression which has remained to me from younger days of the relatively high intellectual standard of mid-Victorian magazines has been confirmed by my re-examination of them for the purposes of this biography. These old magazines—many of them now dead—are full of good work by the best minds of that age, both literary and scientific; the magazines of to-day—from big to small—are almost entirely written by professional journalists to amuse an uncultured public. The writers bear as a rule names which have made no permanent mark on literature, science or politics, and their readers leave these productions to litter the railway carriage or the sea-beach.

mass of educated men shall have learnt to appreciate the truth of the ordinary doctrines of heredity. In this paper he begins his long inquiry as to the relative influence of nature, or as he here calls it 'race,' and nurture.

"There is nothing in what I am about to say that shall underrate the sterling value of nurture, including all kinds of sanitary improvements; nay I wish to claim them as powerful auxiliaries to my cause; nevertheless, I look upon race as far more important than nurture. Race has a double effect, it creates better and more intelligent individuals, and these become more competent than their predecessors to make laws and customs, whose effects shall favourably react on their own health and on the nurture of their children[1]......Constitutional stamina, strength, intelligence, and moral qualities cling to a breed, say of dogs, notwithstanding many generations of careless nurture, while careful nurture, unaided by selection, can do little more to an inferior breed than eradicate disease and make it good of its kind." (pp. 116–7.)

Galton points out that the mass of the population is never likely to enjoy sanitary conditions as good as are now enjoyed by the wealthy classes, but in these classes we frequently find narrow-chested men, delicate women and sickly children; they are very far from possessing those high physical and mental qualities which are the birthright of a good race. Their physical and mental failures are much more frequent than the sickly and misshapen contingent which is found in the stock of any of our breeds of domestic animals. The best environment will not free mankind from weaklings, they can only be 'bred out.' Galton considers that the forms of civilisation at present prevailing tend to spoil a race, and two of their chief factors are the following:

"The first is, the free power of bequeathing wealth, which interferes with the salutary action of natural selection, by preserving the wealthy, and by encouraging marriage on grounds quite independent of personal qualities; and the second is the centralising tendency of our civilisation, which attracts the abler men to towns, where the discouragement to marry is great, and where marriage is comparatively unproductive of descendants who reach adult life." (p. 117.)

Galton at this time believed strongly in the evil influences of town-life, and we shall shortly discuss a paper by him on the subject. He thought town-life selected those who are able to withstand best zymotic diseases and impure and insufficient food, but such a population is not necessarily foremost in the qualities which make a nation great. But to this it may be replied that if town-life does attract some abler men, it also attracts the men who can stand behind a counter, or operate a machine, but whose physique is quite inadequate to plough a straight furrow or to collect the sheep from the high moor. The problem cannot really be answered by a comparison of existing factory operatives and rural labourers, unless we know the nature of the town-immigrants at the time of their migration. Galton appears to me on safer ground when he turns to the mental characters and emphasises his earlier conclusions that the intelligence of men to-day has not kept pace with the growing complexity in trade and profession, nor with the increasingly difficult duties of the citizens of modern large nations.

"Great nations, instead of being highly organised bodies, are little more than aggregations of men severally intent on self-advancement, who must be cemented into a mass by blind feelings of gregariousness and reverence to mere rank, mere authority, and mere tradition, or they will assuredly fall asunder......But the case would be very different in those higher forms of

[1] The 'hard cradle' may be a distinct advantage, as the biographer has observed in visiting the kennels of wealthy breeders of pet dogs.

civilisation, vainly tried as yet, of which the notion of personal property is not the foundation, but which are, in honest truth, republican and cooperative, the good of the community being literally a more vivid desire than that of self-aggrandisement or any other motive whatever. This is a stage which the human race is undoubtedly destined sooner or later to reach, but which the deficient moral gifts of existing races render them incapable of attaining. It is the obvious course of intelligent men—and I venture to say it should be their religious duty—to advance in the direction whither Nature is determined they shall go; that is towards the improvement of their race. Thither she will assuredly goad them with a ruthless arm if they hang back, and it is of no avail to kick against the pricks......For my part, I cling to the idea of a conscious solidarity in Nature, and of its laborious advance under many restrictions, the Whole being conscious of us temporarily detached individuals, but we being very imperfectly and darkly conscious of the Whole. Be this as it may, it becomes our bounden duty to conform our steps to the paths which we recognise to be defined, as those in which sooner or later we have to go. We must, therefore, try to render our individual aims subordinate to those which lead to the improvement of the race. The enthusiasm of humanity, strange as the doctrine may sound, has to be directed primarily to the future of our race, and only secondarily to the well-being of our contemporaries. The ants who, when their nest is disturbed, hurry away each with an uninteresting looking egg, picked up at hazard, not even its own, but none the less precious to it, have their instincts curiously in accordance with the real requirements of Nature. So far as we can interpret her, we read in the clearest letters that our desire for improvement of our race ought to rise to the force of a passion; and if others interpret Nature in the same way, we may expect that at some future time, perhaps not very remote, it may come to be looked upon as one of the chief religious obligations. It is no absurdity to expect, that it may hereafter be preached, that while helpfulness to the weak, and sympathy with the suffering, is the natural form of outpouring of a merciful and kindly heart, yet that the highest action of all is to provide a vigorous, national life[1], and that one practical and effective way in which individuals of feeble constitution can show mercy to their kind is by celibacy, lest they should bring beings into existence whose race is predoomed to destruction by the laws of Nature. It may come to be avowed as a paramount duty, to anticipate the slow and stubborn processes of natural selection, by endeavouring to breed out feeble constitutions, and petty and ignoble instincts, and to breed in those which are vigorous and noble and social." (pp. 118–20.)

I have given nearly the whole of this lengthy passage because it contains the whole gospel of eugenics—then termed by Galton (p. 119) 'viriculture.' The world, in a crude sort of way, looks upon Galton as a 'eugenist,' but hardly knows what the word means, still less recognises why it was a religious faith to him. Of his pantheism, based upon the solidarity of nature as evidenced by the continuity of the germ-plasm, it realises nothing; that he wanted race-improvement in order that men might be good socialists in the highest sense—which their lack of intelligence at present denies them—has scarcely been whispered. Even now if I characterised Galton as a freethinking pantheist, who desired to reach a socialistic state by breeding supermen for intellect—and every word of this characteristic is a literal fact—I should be accused by the bulk of his acquaintances of misrepresentation, and by many of his friends of sensationalism, the explanation being, that Galton wrote far more than he spoke of his philosophy of life, while the majority of men talk more of their heroes than they read of their written thoughts;

[1] And again p. 123: "We shall come to think it no hardheartedness to favour the perpetuation of the stronger, wiser and more moral races, but shall conceive ourselves to be carrying out the obvious intentions of Nature by making our social arrangements conducive to the improvement of their race." I cannot but believe that Nietzsche took his doctrine of scorn and contempt for the feeble—with the cynicism (I should like to write the 'sardony') of a social invert—from Galton.

whence we can easily explain why the popular conception of a great leader of thought is nearly always vacuous, for it lacks that terse characterisation of individuality which can spring only from first-hand study of a man's mind.

Galton's thesis is that artificial selection in our race is lowering its type; the 'typical condition' of a race is that in which there is a moderate amount of healthy natural selection and fair conditions of nurture. He illustrates the lowest quarter of a race by statistics of French conscripts in which 30 °/。 were rejected in 1859. He estimates that some 5 °/。 of these may have been rejected owing to injury or accident, but holds that one-quarter of the French youths are naturally and hereditarily unfitted for active life. To illustrate the uppermost quarter he cites the lads of the *St Vincent* training-ship for seamen of the Royal Navy, where about one boy in four applicants is admitted, and he cites the conditions physical, mental and moral for admission.

"When I stood among the 750 boys who composed the crew, it was clear to me that they were decidedly superior to the mass of their countrymen. They showed their inborn superiority by the heartiness of their manner, their self-respect, their healthy looks, their muscular build, the interest they took in what was taught them, and the ease with which they learnt it......If the average English youth of the future could be raised by an improvement in our race to the average of those on board the *St Vincent*, which is no preposterous hope, England would become far more noble and powerful than she now is......The present army of ineffectives which clog progress would disappear, and the deviations of individual gifts towards genius would be no less wide or numerous than they now are; but by starting from a higher vantage ground they would reach proportionately farther." (p. 123.)

I think many of us would now admit not only the advantage but the possibility of such a degree of betterment of our race. But it may be permissible to doubt whether Galton's solution of 1873 was a feasible one.

"My object is to build up, by the mere process of extensive inquiry and publication of results, a sentiment of caste among those who are naturally gifted, and to procure for them, before the system has fairly taken root, such moderate social favour and preference, no more and no less, as would seem reasonable to those who were justly informed of the precise measure of their importance to the nation." (p. 123.)

The "extensive inquiry and publication of results" were to be undertaken by the organisation of a widely extended "Eugenics Records Society," with branches all over the country, which was to collect and digest information as to the physique, mentality and ancestry of individuals.

"My proposition certainly is not to begin by breaking up old feelings of social status, but to build up a caste *within* each of the groups into which rank, wealth and pursuits already divide society, mankind being quite numerous enough to admit of this sub-classification." (p. 123.)

It is abundantly clear that in 1873 Galton had not fully realised how unprepared the nation was for such a scheme. In the first place had such a society or institution been created, it would have met with an impenetrable barrier of real, if mistaken, opposition to what would have been looked upon as prying inquiries. But still more important factors of failure would have been the absence of any properly trained mental, medical and physical anthropometricians who could have carried out adequate researches of this kind. The work cannot be done by untrained, however enthusiastic, volunteers, but only under the direction of first-class scientists, and the very sciences these men were to be adepts in—experimental psychology, medical histography and physical anthropometry—were either unborn or in the most infantile stages at that date. Had the sciences existed, and the scientists been forthcoming in adequate numbers, the cost of Galton's network of Eugenics Record Offices would have been prohibitive! All this Galton very soon realised, and it forms the key to his later labours—he set about creating, or at any rate building up from feeble beginnings, the requisite branches of science. He later differentiated the science of eugenics from eugenics propagandism, and realised how the latter, if not adequately based on the former, might easily, if not discreet, delay rather than accelerate the spread of fundamental truths.

We are, fifty years later, scarcely yet ripe for the registration of the fitter and abler members of our society. Indeed, while the youth of our professional classes is now far more open to an appreciation of the fundamental importance of sex-questions on the future of the race, the main conception of eugenics has scarcely reached the artizan classes, and many of the fundamental ideas of trades-unionism are retrogressive from the racial standpoint[1]. Galton's belief that racial improvement must depend on the creation of a caste in each social class, a caste which will seek intermarriage, and to which social recognition will give differential opportunities for starting a home and founding families, will long outlive the scheme by which he proposed in 1873 to attain it. We may, however, learn something still from Galton's proposed national register. For example, that he did not think it would be immediately adequate and successful:

"A vast deal of work would be, no doubt, thrown away in collecting materials about persons who afterwards proved not to be the parents of gifted children. Also many would be registered on grounds which our future knowledge will pronounce inadequate. But gradually, notwithstanding many mistakes at first, much ridicule and misunderstanding, and not a little blind hostility, people will confess that the scheme is very reasonable, and works well of its own accord. An immense deal of investigation and criticism will bear its proper fruit, and the cardinal rules for its successful procedure will become understood and laid down. Such, for example, as the physical, moral and intellectual qualifications for entry on the register, and especially as to the increased importance of those which are not isolated, but common to many members of the same family. It will be necessary also to have a clear idea of the average order of gifts to aim for, in the race of the immediate future, bearing in mind that sudden and ambitious attempts are sure to lead to disappointment. And again, the degree of rigour of selection necessary among the parents to insure that their children should, on the average, inherit gifts of the order aimed at. Lastly, we should learn particulars concerning specific types, how far they clash together or are mutually helpful." (p. 126.)

And again, referring to voluntary marriage within the caste:

"So a man of good race would feel that marriage out of his caste would tarnish his blood, and his sentiments would be sympathised with by all. As regards the democratic feeling, its assertion of equality is deserving of the highest admiration so far as it demands equal consideration for the feelings of all, just in the same way as their rights are equally maintained by the law. But it goes further than this, for it asserts that men are of equal value as social

[1] Wages as a standard of craft-ability, and as a rough measure of capacity for founding a home and family, have been not entirely, but very largely interfered with by Trade Union action.

units, equally capable of voting, and the rest. This feeling is undeniably wrong and cannot last. I therefore do not hesitate in believing that if the persons on the register were obviously better and finer pieces of manhood in every respect than other men, democracy notwithstanding, their superiority would be recognised as just what it amounted to, without envy, but very possibly with some feeling of hostility on the part of beaten competitors." (p. 127.)

Thus Galton ploughed his lonely furrow, or rather Odysseus-like ploughed that track of the restless sea which carried him between the Charybdis of Democracy and the Scylla of an Hereditary Peerage; and the unwonted path gave him few social or political comrades. Men do not trouble when a brand-new halo is affixed to their old sanctities, but to suggest even that the old halo would befit better a new sanctity than a worn-out fetish is sacrilege indeed.

There is still a law in France, due to Colbert in 1669, touching the wide-spread oak forests; it runs:

"in none of the forests of the State shall oaks be felled until they are ripe, that is, are unable to prosper for more than thirty years longer."

Thus France legislated for the parquet-floors and wine-casks of to-day, for she did not repeal the ordinance when oak became of little service for shipping.

"Is not man," asks Galton, "worthy of more consideration than timber? If a nation readily consents to lay costly plans for results not to be attained until five generations of men shall have passed away, for a good supply of oak, could it not be persuaded to do at least as much for a good supply of Man? Marvellous effects might be produced in five generations (or in 166 years; allowing three generations to a century). I believe when the truth of heredity as respects man shall have become firmly established and be clearly understood, that instead of a sluggish regard being shown towards a practical application of this knowledge, it is much more likely that a perfect enthusiasm for improving the race might develop itself among the educated classes[1]." (p. 120.)

Thus testifies Francis Galton, much as George Fox had testified two centuries earlier. Both were men with a strong religious enthusiasm behind their doctrines. But the former, who combined knowledge with intense conviction, has so far failed relatively to the latter, who combined forcible ignorance with intense conviction. Are we to attribute the difference to the 'atmosphere' peculiar to their ages, to market-place methods, or to the fact that while Fox appealed to the individual's dubiety as to his own future, Galton asked the individual to consider the welfare, not of himself, but of his offspring? The sanctions behind past religious beliefs will be found in the great mass of men to have a selfish origin, welfare now or in the future.

It is conceivable that Galton's teaching, depending largely on the intensification of the herd-instinct, may need another ice-age, or defeat in a war greater even than the 'Great War,' to impress itself upon our race. The immediate future social history of Germany may not be uninstructive from this standpoint to the philosophical onlooker.

We have seen that Galton in his last paper laid stress on the ill-effects on race of town-life, and that in an earlier paper he had fully recognised the importance of measuring the relative influence of nurture and of nature.

[1] Let us remember that the passion to breed animals and to breed them for points is almost instinctive in the Briton, and we see why our race ought to be foremost in the study and practice of eugenics.

We now turn to certain papers dealing with these points. The first paper we have to notice is entitled: "The Relative Supplies from Town and Country Families to the Population of Future Generations"; it was the first paper Galton read before the Statistical Society of London[1].

"This is an inquiry into the relative fertility of the labouring classes of urban and rural populations, not as regards the number of children brought into the world, but as regards that portion of them who are destined to live and become the parents of the next generation. It is well known that the population of towns decays, and has to be recruited by immigrants from the country, but I am not aware that statistical measurement has yet been attempted of its rate of decay. This inquiry is part of a larger one, on the proportionate supply to the population from the various social classes, and which has an obvious bearing on investigations into the influences that tend to deteriorate or to improve our race. If the poorer classes, that is to say, those who contain an undue proportion of the weak, the idle, and the improvident, contribute an undue supply of population to the next generation, we are justified in expecting that our race will steadily deteriorate, so far as that influence is concerned. The particular branch of the question to which I address myself in this memoir is very important, because the more energetic of our race, and therefore those whose breed is the most valuable to our nation, are attracted from the country to our towns. If, then, residence in towns seriously interferes with the maintenance of their race, we should expect the breed of Englishmen, so far as that influence is concerned, to steadily deteriorate." (p. 19.)

It will be seen that Galton makes two great assumptions: (*a*) that the population of the town decays, and (*b*) that the most energetic of our race are attracted to the towns. Now there is no doubt that a considerable number of energetic men do come from the country into the towns, but also many weaklings and the general human refuse also migrate, and it is not at all clear where the balance of gain may lie. If there be a large contingent of the loafing, pauper and even criminal sections of the community who have come from country to town, the want of fertility in the town may be in part due to this selection.

Captain John Graunt in his "Observations on the Bills of Mortality," 1662, was perhaps the first to assert that the town was recruited from the country, but he had the marked experience of London being rapidly refilled after great plagues. Galton got Dr Farr to provide him with the size of family of 1000 mothers between ages 23 and 40 from Coventry, and the same series of mothers from the rural districts of Warwickshire; the former were the wives of factory hands, and the latter of agricultural labourers. He does not say, however, whether the wives of the factory hands were employed or not, and he does not know whether the ages at marriage of the town wives were differentiated from those of the rural wives. Now the town returns show 510 wives under 32 and the rural returns only 466. It follows therefore that the town wives were *younger* in the selection made than the rural wives; or quite apart from the possibility of an evil influence of town-life on fertility, we might well anticipate that the 1000 town wives would show fewer children. Accordingly, I reconstructed Galton's table, by considering the ages of the wives and reducing the town and country

[1] *Journal*, March 1873, Vol. XXXVI, pp. 19–26. Galton was elected to the Society in 1860 and served on the Council from 1869 to 1879.

populations to a standard population (the average of the two) of wives. The following resulted:

| Mother's age | Number of wives | | Number of children | | Average number of children per wife | | Standard Population | Standard population. Total children | |
|---|---|---|---|---|---|---|---|---|---|
| | Factory | Rural | Factory | Rural | Factory | Rural | | Factory | Rural |
| 24 and 25 | 107 | 91 | 137 | 130 | 1·28 | 1·43 | 99 | 127 | 142 |
| 26 and 27 | 122 | 104 | 209 | 190 | 1·71 | 1·83 | 113 | 193 | 207 |
| 28 and 29 | 132 | 119 | 304 | 279 | 2·30 | 2 34 | 125·5 | 289 | 294 |
| 30 and 31 | 120 | 119 | 299 | 368 | 2·49 | 3·09 | 119·5 | 298 | 369 |
| 32 and 33 | 128 | 124 | 371 | 403 | 2·90 | 3·25 | 126 | 365 | 410 |
| 34 and 35 | 132 | 120 | 430 | 401 | 3·26 | 3·34 | 126 | 411 | 421 |
| 36 and 37 | 100 | 117 | 369 | 395 | 3·69 | 3·38 | 108·5 | 400 | 367 |
| 38 and 39 | 100 | 128 | 354 | 469 | 3·54 | 3·66 | 114 | 404 | 417 |
| 40 | 59 | 78 | 208 | 276 | 3·52 | 3·54 | 68·5 | 241 | 242 |
| Totals | 1000 | 1000 | 2681 | 2911 | 2·68 | 2·91 | 1000 | 2728 | 2869 |

From this table we see that Galton overlooked the fact that his Coventry sample consisted of *younger* women than his rural Warwickshire mothers, and therefore would naturally have fewer children. The average difference on the standard population is only ·14 of a child, or if we take the *average* interval between births to be 2·5 years, it follows that a postponement of marriage on the average for four months would explain this difference[1]. Thus far Galton's paper would not justify any statement as to deterioration arising from town-life. The lesser apparent fertility would be fully accounted for by emigration of the younger women into the towns and a slight postponement of marriage. Galton next proceeds to take the influence of mortality on the town and rural populations. Failing other data he applies to Coventry the mortality table of Manchester and to the rural districts of Warwickshire that of the 'Healthy Districts.' I do not think either of these steps is justifiable, nor again the method by which he applies these life-tables. He draws the conclusion that:

"the rate of supply in towns to the next adult generation is only 77 per cent., or say, three-quarters of that in the country. In two generations the proportion falls to 59 per cent., that is the adult grandchildren of artisan townsfolk are little more than half as numerous as those of labouring people who live in healthy country districts." (p. 23.)

This conclusion has been often cited as if it were rigorous, whereas it is rather an illustration of the grave difficulty of inquiries of this nature, even

[1] This suggestion is supported by the fact that as the women get older there is a less difference between their number of children, e.g. Mothers 36–40, Factory 3·59 and Rural 3·52, or the town is in excess.

when undertaken by a man of keen insight[1]. If the social conditions be such that young mothers with a slightly postponed age at marriage are collected in one district and old mothers with no such postponement in a second, it is clear that the fertility of the first would appear very much larger than the second, and need not in any way be due to differences in environment. The question of whether a town population is decadent is really not touched by this paper; that the gross fertility was less in Galton's instance he demonstrated, but gross fertility means very little without a knowledge of the distribution of the ages and durations of marriage in the population. The main value of the paper lies in Galton's recognition of the problem as of first-class importance.

The idea of the above paper—comparison of the effect of town and country life—was the basis of another paper by Galton at a somewhat later date. It is entitled "On the Height and Weight of Boys aged 14 in Town and Country Public Schools"[2]. Galton's conclusion in this paper is as follows:

"It appears that the boys of the above-mentioned ages in the country group are $1\frac{1}{4}$ inch taller than those in the town group, and 7 lbs. heavier; also that this difference of height is due, in about equal degrees, to retardation and to total suppression of growth; and lastly, that the distribution of heights in both cases conforms well to the results of the 'Law of Error'." (p. 174[3].)

Galton appears to have been inclined to ascribe the differentiation as not

"altogether due to bad effects of nurture on the individual boys" but "much of it to the town-life of their parents, and probably of other ancestors." (p. 181.)

Galton's data for country schools consist in the measurement of 296 boys aged 14 at Clifton, Eton, Haileybury, Marlborough and Wellington, i.e. essentially of boys of the Upper Middle, professional and administrative classes. His town boys are from Christ's Hospital, City of London School, King Edward's School at Birmingham and Liverpool College; these schools in 1871 drew boys largely from the Lower Middle, shopkeeping, clerking and similar classes. Galton's conclusion that the boys of the former series of schools were of better physique would, I feel sure, be confirmed to-day, but, I think, it is a *class* and not an environmental distinction. He did not record the categories of town and country origin in the boys at these schools, but only says that it is fair to assume their origins in bulk to be country and town respectively. Again, I think, although he made some, he did not make adequate allowance for the fact, that while in the first grade public schools the most numerous boys were those aged 15, in the lower grade public schools, the town schools, the boys of 14 and in the case of Birmingham the boys of 13 are most numerous. In other words, at 14 the boys are

[1] The President of the Royal Statistical Society cited it on February 16, 1922 at the Galton Centenary Celebration of the London Eugenics Education Society, without a word of comment or of caution in his speech. *Eugenics Review*, Vol. XIV, p. 4.

[2] *Journal of the Royal Anthropological Institute*, Vol. V, 1876, pp. 174–80.

[3] We have in this paper an early practical application of Galton's method of percentiles; the 'median' and the 'quartiles' are provided.

already leaving the 'town' schools, while they appear to be still coming to the 'country' schools. The former are doubtless going out into business and other occupations, and the boys that leave earliest will, as a rule, be those most fully developed. Galton has clubbed together his schools so that we cannot separate them out, but one would anticipate that Liverpool College, if it draws from Lancashire, would racially have a low stature. Here again Galton reached a right conclusion, if the data on which he bases it are open to criticism. If we take, as the Galton Laboratory has done, adjacent rural and urban districts in Worcestershire or Staffordshire and compare the children of like ages in the *primary* schools, then the balance of physique is in favour of the rural. In doing this we work within the same social class, we attempt to get the same local race, but we cannot be certain that the town occupations have not attracted the less physically fit parents.

### D. HEREDITY IN TWINS

It seems best to consider here two papers on the subject of twins, because although they to some extent were associated with Galton's ideas on heredity, yet they sprung, I think, from his work on the influence of environment. The first paper is entitled "The History of Twins, as a Criterion of the Relative Powers of Nature and Nurture[1]." Among the claims which twins have to attention, Galton tells us, is the fact that:

"their history affords means of distinguishing between the effects of tendencies received at birth, and of those that were imposed by the circumstances of their after lives; in other words, between the effects of nature and of nurture. This is a subject of especial importance in its bearings on investigations into mental heredity, and I, for my part, have keenly felt the difficulty of drawing the necessary distinction whenever I attempted to estimate the degree in which mental ability was, on the average, inherited. The objection to statistical evidence in proof of its inheritance has always been: 'The persons whom you compare may have lived under similar social conditions and have had similar advantages of education, but such prominent conditions are only a small part of those that determine the future of each man's life. It is to trifling accidental circumstances that the bent of his disposition and his success are mainly due, and these you leave wholly out of account—in fact they do not admit of being tabulated, and therefore your statistics, however plausible at first sight, are really of very little use.' No method of inquiry which I have been able to carry out—and I have tried very many methods—is wholly free from this objection." (p. 391[2].)

Accordingly Galton turns to try and appreciate what *relative* effect nature and nurture have. Galton's scheme was to consider twins who were closely alike in youth and whether after being separated they grew unlike, and again whether twins who being unlike in childhood were subjected to the same nurture grew more alike. Galton collected his material by circu-

[1] *Fraser's Magazine*, Nov. 1875, issued with revision and additions in *Journal of the Royal Anthropological Institute* (1875), Vol. v, pp. 391–406, 1876.

[2] The passage is clearly written before the idea of correlation had reached Galton's mind. The true test is whether the degree of association in mental characters between relatives is the same as that between physical characters. If it be, then it is exceedingly improbable that nurture on the one hand should have produced exactly the same quantitative association as nature on the other, for we can always select physical characters which are not materially influenced by nurture.

larising twins or relatives of twins, and 'snowballed' by asking them for the addresses of other twins, which he remarks led to a continually widening circle of correspondence. Finally Galton obtained information concerning 94 sets of twins (see his statement in second paper). He considers their resemblances in the case of 35 twin-sets—each of like sex—in which there was detailed evidence of close similarity. He finds that this likeness—mental, physical and pathological—is maintained even when life has carried the twins into different environments; they appear to have the same illnesses at the same times. The answers showed that in the bulk of cases the resemblance of body and mind continued unaltered up to old age and under very different conditions of life; in other cases dissimilarity was attributed wholly to some form of illness or accident which had befallen one twin and not the other. Galton then turns to the 20 sets of twins he had detailed accounts of unlikeness between. He has not a single case in which his correspondents speak of originally dissimilar characters having become assimilated through identity of nurture.

"The impression that all this evidence leaves on the mind is one of some wonder whether nurture can do anything at all beyond giving instruction and professional training. It emphatically corroborates and goes far beyond the conclusions to which we had already been driven by the cases of similarity. In these the causes of divergence began to act about the period of adult life, when the characters had become somewhat fixed; but here the causes conducive to assimilation began to act from the earliest moment of the existence of the twins, when the disposition was most pliant, and they were continuous until the period of adult life." (p. 404.)

And then follows the passage cited on p. 8 of our first volume.

There is another passage also which is of great suggestiveness and may be cited here:

"Much stress is laid on the persistence of moral impressions made in childhood, and the conclusion is drawn, that the effects of early teaching generally must be important in a corresponding degree. I acknowledge the fact, but doubt the deduction. The child is usually taught by its parents, and their teachings are of an exceptional character for the following reason. There is commonly a strong resemblance, owing to inheritance, between the dispositions of the child and its parents. They are able to understand the ways of one another more intimately than is possible to persons not of the same blood, and the child instinctively assimilates the habits and ways of thought of its parents. Its disposition is educated by them, in the true sense of the word; that is to say, it is evoked earlier than it would otherwise have been. On these grounds I ascribe the persistence of habits that date from the early periods of home education, to the peculiarities of the instructors, rather than to the period when the instruction was given. The marks left on the memory by the instructions of a foster-mother are soon spunged clean away." (p. 405.)

Consider, says Galton, the history of the cuckoo, which is reared exclusively by foster-parents!—Neither its note, nor its habits, nor its sympathies are influenced by those of its foster-parents. Galton concludes generally that with reasonable care in the collection of our data, we may ignore the many small differences in nurture which characterise individual cases.

The reader of Galton's first paper may possibly hold that he ought to have given more of his data, but it must be remembered that his paper was written originally as a popular article for *Fraser's Magazine*, and Galton was not yet a practised statistician. His material is worthy of a fresh analysis, and we hope that this may be forthcoming. At the same time his general conclusion that the differential effects of nurture within the range of existing social or possible political conditions are extremely small has been amply confirmed by other methods of approaching the problem.

Galton felt that this conclusion was very essential in its bearing on his theory of the heredity of the mental characters. The two principles together—i.e. inheritance of ability, and the relatively small influence of nurture as compared to nature—form the basis of his scheme for racial betterment. While Galton always acknowledged the importance of a good physique, and suggested that marks should be allotted to it in competitive examinations, he was deeply impressed with the fact that the future lies rather with the man of brains than with the man of muscles; he repeatedly asserted, however, that the man of ability was very often of good physique, and the stock wherein the two were combined was the ideal stock for race-betterment.

In his second paper on twins, entitled "Short Notes on Heredity, etc., in Twins[1]," Galton turns somewhat more to the numerical and physiological sides of the subject.

"The word 'twin' covers different classes of events—those in which each twin is derived from a separate ovum, and those in which they come from two germinal spots in the same ovum. In the former case they are enveloped, previously to their birth, in separate membranes; and in the latter in the same membrane. Now it appears that twins enveloped in the same membrane are invariably of the same sex, and these according to the cases of Späeth, who has evidently taken great pains to secure reliable data[2], are 24 per cent. of the whole number." (p. 329.)

Galton finds great variation in the statistics as to twins of various types, but concludes that about twice as many twins are born of the same sex as of opposite sex[3]. This would mean that Späeth's percentage should have been 33 °/₀ instead of 24 °/₀. It seems probable that the divergence arises from Galton's "twice" being a rough approximation. He considers that twins do not marry as often as other people and are less fertile. He contradicts, however, the popular belief that both twins whether of the same or opposite sexes *never* have children, for he had many instances to the contrary. He says that as far as he is aware nothing corresponding to the 'free-martin' of cattle occurs with the human twin. Still the infertility of twins Galton considers to be so great that it renders the question of direct inheritance of twinning difficult. He accordingly asks whether the number of twins among the uncles and aunts of twins is in excess of the probable on the basis of the non-inheritance of the twinning tendency. Galton shows from Ansell's *Statistics of Families* that there is 1 twin born to every 100 births,

[1] *Journal of the Royal Anthropological Institute* (1875), Vol. v, pp. 324–29, 1876.

[2] "Studien über Zwillingen," *Zeitschrift der Wiener Gesellschaft der Aerzte*, 1860, Nos. 15 and 16.

[3] Let $x$ = number of single membrane or 'like' twins, $y$ = number of double membrane or 'unlike' (i.e. only as like as normal siblings). Then according to Späeth, $x = \frac{1}{4}(x+y)$, nearly. Now in $x$ all twins are of the same sex, and in $y$ only half will be of the same sex; hence total number of all twins of same sex $= x + \frac{1}{2}y$ and of different sexes $= \frac{1}{2}y$, but $x = \frac{1}{3}y$, and accordingly $\frac{x+\frac{1}{2}y}{1\,y} = \frac{5}{3}$, or there are 1·7 times as many twins of same sex as of unlike sex.

or say 2 persons twins in 101 persons. Now in his own data for 94 sets of twins he found that they had 1065 uncles and aunts, and among these were 27 sets of twins or 54 twins in 1065 persons, or, say, 1 person a twin in 20. Galton accordingly concludes that twins are far more frequent among the relatives of twins than in the general population or twinning must be an hereditary character. Galton further noted that on the father's side there were 538 uncles and aunts with 28 twins among them, and on the mother's side 527 uncles and aunts with 26 twins among them. Hence he concluded that the hereditary tendency to twinning was the same in the male and female lines[1].

Two other interesting points are given in this paper: (*a*) the case of a woman in a family remarkable for twins, whose single children were polydactyle, but not the twins, and (*b*) a pedigree—unfortunately omitting single births—in which the intermarriages of three twinning stocks show eight sets of twins, one triplet, and one quadruplet[2].

I have endeavoured in this chapter to trace one strand of Galton's labours, that which shows him passing from the human side of geography to anthropology, and so to heredity and race-betterment. The reader, who has had the patience to follow my analysis of Galton's papers, will have observed that Galton was becoming more and more conscious that a statistical treatment of both anthropology and heredity was necessary, and he steadily set himself to understand the then existing statistical processes and to develop them where necessary. I have not hesitated to indicate the statistical weakness displayed in some of the papers discussed in this chapter, because it was Galton's growing consciousness in this matter that largely led to his later contributions to statistical theory.

In 1872 Mrs Galton's mother had died and in 1874 Galton's mother

[1] Assuming one person in fifty to be a twin, Galton's 94 sets of twins lead us to a table of the following character:

| | | Uncle or Aunt | | |
|---|---|---|---|---|
| | | Twin | Not-twin | Totals |
| Nephew or Niece | Twin | 108 | 2,022 | 2,130 |
| | Not-twin | 2,022 | 102,348 | 104,370 |
| | Totals | 2,130 | 104,370 | 106,500 |

If we assume for a moment that twinning is the result of some practically continuous variable exceeding a certain value, the correlation for uncle and nephew would be ·1817, rather low for this relationship (·2 to ·3), although such a value has occurred for physical characters occasionally. It is possible that Ansell's value for the frequency of twin births is somewhat exaggerated.

[2] A number of pedigrees of twinning stocks were presented some years ago to the Galton Laboratory, and will ultimately be published, but having been collected because of the twinning frequency, it is not easy to use them for measuring the intensity of heredity in twinning.

(Violetta Darwin) also passed away. We may conclude this chapter with an extract from *L. G.'s Record*, for it indicates that the large amount of work which Galton published in 1874 must have been done amid much stress.

"1874. Uneasy from the very beginning about dear Mrs Galton. Frank went to see her early in February, and she died February 12th aged 90. This coming so soon after my dear mother made a sad blank; both homes gone. We went to dear Emma ['Sister Emmy'] at Easter for a week and then made a few days visit to Winchester, Salisbury and Lyndhurst, the weather not good. At Whitsuntide we visited the Jenkinsons and greatly enjoyed their lovely country. George and Josephine [Canon Butler and his still well-remembered wife, Mrs Josephine Butler[1]] came to us at the end of June. We were prevented going away by domestic bothers, which have made the whole of this year sadly trying......We paid many visits during July and at the end went down to Cornwall spending a fortnight with Adèle and Milly ['Sister Delly' Mrs Bunbury and her daughter, afterwards Mrs Lethbridge] and visited Boscastle and Tintagel. After this we went to N. Devon till September, when we visited the Groves at Cheetle near Blandford. On the 14th I broke a blood vessel and was very near dying, but thro' God's mercy I came back to life and felt so peaceful and happy in my quiet sickroom, that it was not a time of misery. And all were so kind and good to me, and Frank especially, that I felt sustained by love. We moved to Bournemouth as soon as I was able and then in November to dear Emma, and found her well in her newly arranged house. We came home November 18th. Very severe weather soon set in and lasted to the end of the year. Frank was ill in December and had Dr A. Clarke[2]. We had a quiet dull Xmas, no going out, and Frank had to give up his promised lectures at Newcastle[3]. His book on the Nature and Nurture of Scientific Men came out in December; occupied on inquiry about Twins. On the whole a year of sad memories."

And yet Galton published one fundamental book, *English Men of Science*, and three memoirs in this year and wrote at least four others! He depended singularly little upon a stable environment; yet it must be remembered that years were not needed then for the collection of data and its numerical reduction, as in the case of modern biometric studies.

[1] One of the protagonists for social purity. I remember many years ago, one evening in Grindelwald, being struck by a very commanding personality, one of the most 'regal' women I had yet met; it was Mrs Josephine Butler, Mrs Galton's sister-in-law.

[2] "What a pleasant man Dr Andrew Clarke is. He examined me most thoroughly, pronounced it a concurrence of irregular gout with influenza and that my heart was weak. I mend, but not overfast." Letter of Galton to George Darwin, Xmas Day 1874. Strange to say Sir Andrew Clarke's directions for treatment, principally diet, have survived almost the half-century. Perhaps the wisest was: "Walk at least half-an-hour twice a day, and do the most important headwork after breakfast, *not* after dinner."

[3] The manuscript draft of these lectures has survived.

PLATE XIII

Francis Galton when about fifty years of age.

# CHAPTER X

## THE EARLY STUYD OF HEREDITY: CORRESPONDENCE WITH ALPHONSE DE CANDOLLE AND CHARLES DARWIN

"There is a vast difference between an intellectual belief in any subject and a living belief which becomes ingrained, sometimes quite suddenly, into the character."

FRANCIS GALTON, *Hereditary Improvement*, p. 123.

### A. DE CANDOLLE AND *ENGLISH MEN OF SCIENCE*

As I have already indicated, Galton's writings of 1865–1875 on social topics met with a very mixed reception. The paper on the "Efficacy of Prayer" had been refused by Grove and Knowles for their respective journals before it found a place in the *Fortnightly*[1]. His views on race-betterment met with a variety of remonstrances from the mediocrity which Galton would efface, and which is still the blindest opponent of his ideas. Mrs Grundy—whom Galton desired to raise two 'grades' in intelligence—was naturally outrageously shocked. She is still shocked but, while dumb herself, has a capacious petticoat pocket whence she extracts ample sweetmeats for her expostulatory and filial scribes. Galton throughout his life rarely permitted himself to be drawn into controversy, and, as Mrs Galton records, his work was approved by men of note. We may add by some women of note too, although they took a line of their own, with which I am far from certain Francis Galton fully sympathised. While advanced in many ways far beyond his contemporaries, he never struck me as fully recognising the need for the oncoming change in the status of women. He invariably treated them with an old-fashioned courtesy, which had an irresistible and undefinable charm; but the gradual entry of highly trained women, even into his own statistical studies, seemed to some extent to find him unprepared and puzzle him. He would accept the ability, but hardly appreciated it fully, if it were not accompanied by personal presence and a reciprocating charm. He could rejoice in the able woman of the salon, but I am less certain that he sympathised as fully with the equally able, but more highly trained modern academic woman. But the men who did in 1870 could be almost counted on the fingers of one hand! I cannot refrain from citing (by permission) the following fine letters from Miss Emily Shirreff, a woman of much

[1] "I am afraid," wrote Knowles, "that after all my courage is not greater than Grove's. You will think that editors are a 'feeble folk,' and so perhaps they are, but it is certain that our constituents (who are largely clergymen) must not be tried much further just now by proposals following Tyndall's friend's on prayer—and of a similar bold,—or as you yourself say 'audacious character'."

distinction[1], because they indicate the impression Galton's work made on the best minds of that moiety of humanity, to which I do not think even later he ever made the great appeal—the utterance of the few words which would have shown he realised their problems[2].

The letters referred to run as follows:

THE COLLEGE, HITCHIN. *March* 7th '70

SIR, Having just read your remarkable book on *Hereditary Genius*, I hope I may be excused for addressing these few lines to say with what extreme interest I have followed your speculations and how valuable seem to me many of the suggestions you throw out. But what practical result can we hope from any such theories addressed to a society not yet civilised enough to refrain from marriages and intermarriages in families known to bear the taint of consumption and of madness? Money and position are the only influences that we have yet seen powerful enough to counteract the passion of the hour.

Your very able and original remarks on some of the causes which tended under the sway of the church to deteriorate the race of men, especially in Spain and Italy, seem to me to bear with great force upon some things going on around us now in this country, and still more I believe in America. I mean the various causes which are combining to turn a large number of the ablest and most active women away from marriage. The luxury of modern life pernicious in so many ways has made marriage more difficult, and at the same time made the condition of single women more precarious by making it harder for parents to provide for them. Wider freedom of action and opinion make the conditions of married life with its fearful possibilities of legal oppression more revolting to many women, while the just claim they put forward for freedom to work their own way, and the resistance that claim meets with from the majority create a painful and mischievous sense of antagonism against men. No one would hail more gladly than I should a rebellion against the miserable system which has driven women to marry for subsistence or position, but unfortunately the feeble will still be content to do so. Fathers may still reckon on the larger number of their daughters willingly enough living in idleness till a husband takes them off their hands, and saves them the necessity of providing for their respectable independence—while the abler, the more energetic, the most fit to be the mothers of a better generation will revolt against the injustice of our social arrangements, and struggle singly for an independent position; thereby sacrificing at once the interests of society and some of the highest cravings of their own nature. I hold the individuals to be blameless, but it seems strange that those who watch the workings of society see apparently without an effort at resistance that fatal tide of luxury rise, and rise aided by much that is really refined and by all that is base and coarse in our present civilisation; and take no heed of the probable effects upon another generation of this 'woman movement,' which judiciously met might be made productive of almost unmixed good.

I have written at greater length than I at all intended and can only hope you will excuse my following the train of thought roused by your speculations.

Believe me, Yours sincerely, EMILY SHIRREFF.

In a postscript the writer gives a number of corrections for a later edition of the *Hereditary Genius*. It is a misfortune that we have not Francis Galton's reply, but clearly he asked the writer for information on what was to him a novel point. If he hoped for statistical information, it certainly was not forthcoming in 1870.

[1] She was on the first Girton College Committee and for a short time took charge of the original college at Hitchin.

[2] The modern 'master' is bound to disappoint the critical mind of the modern disciple in some one or other aspect, and such an occasion came to the present writer on hearing a lecture by Francis Galton many years later on eugenic propagandism. The only special appeal he made to the modern woman—to whom eugenics is a vital interest—was to throw open her house for drawing-room meetings. The suggestion seemed to be out of touch with the modern centre of gravity of women's activities and opinions.

THE COLLEGE, HITCHIN. *March* 16th '70

DEAR SIR, I have been thinking very much about the opinion I expressed to you concerning the disinclination to marriage among women of a certain stamp in England and America, and wishing very much that I could give you evidence of what to me is certainly a fact, but which has grown into that through a multitude of channels, minute observation, scattered opinion in books and some aspects of opinion upon social questions. As regards America I know I gathered much from Mr H. Dixon's two works, also I remember some papers of Mr F. Newman which left me the same impression. Generally, the tone of what some call the most advanced (and which appear to me the most exaggerated) views of the 'woman question' treats marriage and the home position of the wife and mother with a sort of half contempt which indicates the feeling I speak of. Mr Mill has expressed what women have been feeling more and more for years past concerning the injustice of the irresponsible despotism under which they live; it may be and often is a benevolent despotism, but absolute governments are not in fashion and the reaction of liberal politics has doubtless affected the views of home life. Women have little hope of any change that law can make in their destiny,—better therefore they say to be independent of men. All that foolish talk about *equality* (foolish because it never can be proved one way or the other and has very little bearing upon the practical question) has stirred up feelings of antagonism and these are most unfavourable to marriage. I believe men do not realise to what a miserable extent women have married for position or independence, and degrading as the system has been it must be owned that to the larger number there was no other resource,—they had no openings for employment and their families did not provide for them. Now therefore that there is so much stir about occupation for women and that they see for the first time a vision of independence to be earned by their own work, it is, perhaps, a necessary—at any rate a natural—result that they should look with some dislike to what seems the refuge for weaker minds. I believe that the feeble influence of passion over women—educated women at any rate—as compared with affection aids this state of things. If they do not meet with the individual who calls forth the strong affection, and for whom any sacrifice is light, a single life has nothing from which they shrink. If they can live without hardship they feel that they have much in the sense of freedom to compensate for any advantages marriage might have given. In France they will not bear the social discredit of old maidism, but the same struggle for independence is going on and is more easy to carry on within the limits of married life than in England. I think that the frequency with which we see women of property remain single is a corroboration of what I have said. But I feel that all I have written is most vague and can only show you the nature and direction of the evidence that has produced this conviction on my mind—one to which I give no welcome—for with all its ignorant shortcomings in practice, I am sure the old theory of life is that which is true to Nature, and that the existence of man or woman is abortive without the other. I say this only to let you know that my belief is not born of my own prepossession. Should I come across any facts or views I think might be interesting to you on this point, I shall have great pleasure in sending them to you. Yours truly, EMILY SHIRREFF.

I do not think Galton fully heard this knocking of the younger generation on the door. He had assumed that the abler men could have the abler women for the asking, but what if the social evolution were to be such that the latter tended to stand aloof from marriage altogether? The writer of these letters may have put the matter vaguely in 1870—she was only at the beginning of the great movement of the last fifty years, and must fail to foresee all its phases. Yet she grasped a danger to the race, that the author of *Hereditary Genius* had not hinted at—the danger that the abler woman, when she has once realised her powers, may prefer to remain unmarried[1]. You may press upon her the religious duty of race-betterment,

[1] I think this 'fear' of the 'seventies has largely materialised into a statistically demonstrable fact in the last thirty years. The abler men, either by choice or necessity, do not mate with the abler women, and the latter either by choice or necessity remain to a very large extent unmarried.

and she may put you to confusion by distinguishing between the social and honourable duty of parentage, and the individualistic and onerous duties of marriage. We may much desire, I fear now in vain, to see Galton's replies to these letters; but I do not think he saw how the emancipation of women must change the aspect of his proposals for race-betterment[1]. Had he done so, he must surely have given us some light on one of the most difficult problems of our present day civilisation. I cannot find trace of it in Galton's later writings on eugenics, and the present legislative pittance to lying-in mothers of the artizan classes is only a travesty of what the eugenist understands by the endowment of the physically and psychically abler mothers in each caste of our modern highly differentiated social organisation.

We have further evidence, however, of the manner in which *Hereditary Genius* stirred up men by its suggestions, if not to new investigations, at any rate to the publication of work they had in mind. Noteworthy in this respect is Alphonse de Candolle's book: *Histoire des Sciences et des Savants depuis deux Siècles*, published in 1872. It shows undoubted signs of Galton's influence, if only as an irritant, and one may reasonably question how and when it would have been issued, but for the *Hereditary Genius* of 1869. De Candolle, son of the still more distinguished botanist, was, as befits a botanist, more impressed than Galton with the influence of environment, and he quite possibly gave too little weight to the selective immigration into Switzerland, which has gone on at various periods to the great advantage of the intellectual life of that little country. Certainly some of his criticisms of Galton appear to me hasty and unjustified, and counter-criticisms against his own work could be easily raised, were it worth doing. The main point is that after a certain amount of friction these two able men became friends and mutually inspired each other's ideas and work. Just as we may consider De Candolle's work as brought to birth—perhaps prematurely—by Galton's *Hereditary Genius*, so Galton's *English Men of Science, their Nature and Nurture* of 1874, while an extension of the corresponding chapter in the *Hereditary Genius*, was prompted by De Candolle's *Histoire des Sciences*[2].

The correspondence between these two men gives a greater insight into Galton's manner of thought at this period than any other record—except his writings—which has been preserved from that date. Communications with English scientific friends—Darwin excepted—were largely verbal. They are lost in the after-dinner talks of the Athenaeum and of the Royal Society clubs.

[1] He was not opposed to the academic education of women—witness the following letter of Charles Darwin to his son George of Feb. 27 [1881]:

"You will have heard of the triumph of the Ladies at Cambridge [Grace of Senate admitting women to University examinations carried by 398 to 32]. The majority was so enormous that many men on both sides did not think it worth voting. The minority was received with jeers. Horace was sent to the Ladies College to communicate the success and was received with enthusiasm. Frank and F. Galton went up to vote. We had F. Galton to Down last Sunday. He was splendid fun and told us no end of odd things." *A Century of Family Letters*, 1904, Vol. II, p. 315.

But Galton naturally did not realise all that would flow from the movement.

[2] "I thought that a somewhat similar investigation might be made with advantage into the history of English men of science." *Memories*, p. 291.

PLATE XIV

Down (Beckenham, Kent), the home of Charles Darwin from 1842 to his death in 1882.
From a photograph by Dr David Heron.

42, RUTLAND GATE, LONDON. *Dec.* 27/72.

DEAR SIR, I thank you much for your volume which I received about a fortnight since and which I have read and re-read with care and with great instruction to myself. Allow me to congratulate you on the happy idea of accepting the nominations of the French Academy and similar bodies as reliable diplomas of scientific eminence, and on thus obtaining a solid basis for your reasoning. I must however express no small surprise at the contrast between your judgement on my theories and your own conclusions. You say and imply that my views on hereditary genius are wrong and that you are going to correct them; well, I read on, and find to my astonishment that so far from correcting them you re-enunciate them. I am perfectly unable to discover on what particulars, speaking broadly, your conclusions have invalidated mine. They have largely supplemented them, by thoroughly working out a branch of the inquiry into which I never professed to enter, but I literally cannot see that your conclusions, so far as heredity is concerned, differ in any marked way from mine. You say that race is all-important (p. 253 etc.)—that families of the same race differ from each other more widely than the races themselves (p. 268)—that physical form is certainly hereditary and that intellect is dependent on structure and must therefore be inherited (p. 326)—that for success, an individual must both "vouloir et *pouvoir*" (p. 92)—that the natural faculties must be above mediocrity (p. 106) and very many other similar remarks. I never said, nor thought, that special aptitudes were inherited so strongly as to be irresistible, which seems to be a dogma you are pleased to ascribe to me and then to repudiate. My whole book, including the genealogical tables, shows that ability—the "*pouvoir*"—may manifest itself in many ways. I feel the injustice you have done to me strongly, and one reason that I did not write earlier was that I might first hear the independent verdict of some scientific man who had read both books. This I have now done, having seen Mr Darwin whose opinion confirms mine in every particular. Let me, before proceeding to more agreeable subjects, complain of yet another misrepresentation. You say (p. 380) that I deny or doubt (contester) the good tendencies of children reared in the families of clergymen—I never said anything of the sort. What I did say was against the "pious," that is the *over-religious*. My genealogies are full of clergymen:—in the list you give, p. 381, I doubt if any of the parents are known to have been "pious"—though you might have quoted *Haller* in your favour. Let me en passant remark about the last paragraph of your footnote to p. 383, the sons of English clergy are or were hardly ever sent into the army, because their parents could not afford it, and therefore their sons could not become Generals. Sometimes, but very rarely, they *were* put into the *Navy*, which is a less costly profession, and Admiral Lord Nelson was one of such.

I regret very much that you did not succeed in working out the genealogies of the scientific discoverers, on whom you rely, and on both sides. However there is no denying the fact, that as a whole they are specialists, rather than illustrious men, and are therefore somewhat obscure to fame. Man against man, they would be nowhere in competition with a great statesman—but they have owed more to concentration and the narrowing of their faculties than to a general prodigality of their nature. Such men are more easily affected by circumstances than the born geniuses about whom I chiefly busied myself, and are therefore all the more suitable subjects for an inquiry like yours, into the effects of different circumstances.

One of the most striking things to me in your book is the chilling influence on scientific curiosity you prove to result from religious authority. The figures you give seem to me of the highest importance. I am also greatly impressed with the conditions of fortune (funds not land) and the desire for an European rather than a local reputation which you ascribe to religious and other refugees.—Switzerland's reputation seems made by the Huguenots, Euler and Haller being the only two in your list of purely native birth. I wish you had given the genealogies of the rest in full.—Have you not made some slip of the pen in p. 125—at the bottom? If you cut off the sons of pasteurs I do not find that equality *is* re-established, nearly.—Then see p. 40 where out of 20 fathers of associates only 4 were pasteurs and of all these associates only one was Catholic.—There remain 14 non-clergy and 2 Quakers as parents of 16 Protestants to that 1 Catholic. Is not 'Protestant' a deceptive word? I fear most of the scientific men would be more truly described as 'infidel' or 'agnostic.'

How remarkable are your conclusions about teaching. I suppose severe teaching sacrifices many original minds but raises the level. We in England are in the throes of educational reform, wanting to know how best to teach "How to observe."

In your table XI of the scientific value of a million of different races, I note, what appears to me, a serious statistical error. You disregard the fact that some populations increase faster than others and have therefore always a plethora of children and of persons too young to be academicians. Take as sample and not very incorrect figures, that America U.S. doubles in 25 years, England in 50 and that France remains stationary. Then your calculation would do about a *four*-fold injustice to America, and a double injustice to England as compared to France, because it is at the age of 50 or thereabouts that people become academicians. The true comparison would be with the number of persons in the nations *above the age of* 50. This would avoid another great source of error arising from the very different chances of life of a child in different countries.

I fear the English physiologists will exclaim at your "état momentané lors de la conception." Am I doing you an injustice in supposing that you argue on the hypothesis that conception and copulation are simultaneous? I certainly understand you do (pp. 311–2) but how can the argument stand? The spermatozoa do not get at the ovum for hours, perhaps many days, after copulation, and the ovum itself, when fecundated, has long been detached from the ovary.

I feel, now that I have come to the end of this letter, that I have done little else than find fault, but I beg you to be assured that my general impression of the book is of another kind. I feel the great service you have done in writing it, and I shall do what I can to make it known, as it ought to be, in England. Can you get any facts out of Foundling Hospitals about heredity? The people here who administer ours are not scientific.

I have written an audacious article for Fraser's Magazine in Jan./73 of which I will send you a copy. Believe me, faithfully yours, FRANCIS GALTON.

To M. ALPHONSE DE CANDOLLE.

GENÈVE. 2. *Janvier* 1873.

MONSIEUR ET HONORÉ COLLÈGUE, Le volume que j'ai publié vous a causé un mélange d'impressions agréables et désagréables. Je puis en dire autant de votre lettre du 27 Décembre mais avant de discuter certains points, je désire vous faire une déclaration générale. S'il m'échappe, dans les 482 pages de mon livre, une phrase, un mot pouvant faire douter de mon respect pour votre impartialité, votre caractère et votre talent d'investigation, ce ne peut être absolument que par erreur et contrairement à mes intentions. Vous avez toujours cherché la vérité. J'ai apprécié beaucoup votre travail et s'il n'était pas inusité de transcrire de nombreux articles d'un auteur je vous aurais cité encore plus souvent.

L'idée de consulter les nominations par les Académies m'est venue il y a 40 ans! J'avais prié un de mes amis de prendre au sécretariat de l'Institut les listes des Associés étrangers et Correspondants de 1750 en 1789. Les noms modernes sont aisés à trouver ailleurs. J'avais redigé en 1833 un mémoire sur ces listes de Paris et sur celles de la Société Royale. Si je ne l'ai pas publié alors c'est qu'il me semblait un peu présomptueux chez un jeune homme de mesurer ainsi la valeur de savants illustres, parmi lesquels se trouvait son père et quelques hommes distingués à côté de lui. Une fois moi-même sur certaines listes, il me répugnait d'en parler. Enfin, à 66 ans, après une série de travaux spéciaux propres à justifier ma position, le courage m'est venu et j'ai pensé pouvoir m'élever au dessus des considérations personnelles de toute nature.

Ma rédaction était fort avancée quand j'ai connu votre ouvrage. Je l'ai lu avec infiniment de plaisir, comme je viens d'en relire les chapitres les plus importants.

Nous sommes admirablement d'accord sur les faits. Nous avons les mêmes idées sur les races. Vous avez envisagé un plus grand nombre de catégories d'hommes, mais celle des savants que j'ai étudiée d'une manière plus spéciale, avec une méthode différente, m'a donné de résultats extrêmement semblables aux vôtres quant aux faits.

Je persiste à croire qu'il y a, non pas une opposition mais une *différence* assez sensible dans l'appréciation des causes qui ont influé sur les faits.

Vous faites habituellement ressortir, comme cause principale, l'hérédité. Quand vous parlez des autres causes elles sont indiquées accessoirement et sans chercher à démêler ce qui tient particulièrement à elles ou à chacune d'entre elles. De loin en loin vous mentionnez ces autres causes. Ainsi on peut lire bien des pages où vous démontrez l'influence de l'hérédité avant de rencontrer une ligne comme au haut de la page 88 sur les *social influences*. Le titre même de l'ouvrage implique l'idée de rechercher uniquement sur l'hérédité, ses lois et ses conséquences, autrement vous auriez dit: *On the effect of heredity and other circumstances as to genus.* Assurément vous avez rendu un vrai service à la science, mais votre point de vue était essentiellement l'hérédité.

Quant à moi j'ai eu l'avantage de venir après vous. Il ne m'a pas été difficile de confirmer par de nouveaux faits l'influence de l'hérédité, mais je n'ai jamais perdu de vue les autres causes, et la suite de mes recherches m'a convaincu qu'elles ont en général plus d'importance que l'hérédité, du moins parmi les hommes de même race. Si l'on compare des nègres avec des blancs, ou même des blancs asiatiques avec des blancs européens, l'effet de la race est prédominant, mais parmi les hommes de nos pays civilisés l'effet des traditions, exemples et conseils dans l'intérieur des familles m'a paru exercer plus d'influence que l'hérédité proprement dite. Vient ensuite l'éducation extérieure, l'opinion publique, les institutions etc. Je me suis appliqué à distinguer la part d'influence de toutes ces causes, part qui varie suivant les pays et les époques, et qui favorise ou contrarie les effets de l'hérédité. Le but de mes recherches était donc différent du vôtre et les résultats en ont été *différents* sans être opposés. C'est ce que j'ai dit à la p. 93.

J'en viens aux observations de détail.

Au bas de la p. 273 vous parlez de faits et d'une certaine opinion courante défavorables aux enfants de personnes religieuses (religious). Dans mon idée l'immense majorité des clergymen est religieuse, et comme j'ai remarqué cependant moi-même des cas dans lesquels leurs enfants avaient mal tourné, j'ai examiné les faits, et j'ai trouvé (à ma grande surprise) qu'une forte proportion de savants célèbres avaient été fils de pasteurs ou ministres. J'en ai tiré un argument en faveur de l'importance d'éducation simple et morale.

Je n'ai pas compris votre observation sur une erreur de chiffre à la p. 40. En comptant de nouveau sur la dernière colonne des pp. 36–40, je trouve 13 fils de pasteurs ou ministres comme je l'avais dit. Par parenthèse, j'ai soupçonné quelquefois que Sir David Brewster était un 14[ième]. Il était fils d'un *Rector* of the Grammar school of Jedburgh, ce qui d'après le pays peut faire croire qu'il était peut-être ministre. Une notice dans un journal religieux m'a appris que Sir David avait été élevé dans une atmosphère très pieuse.

Je n'ai pas pu savoir l'origine de famille de Mr Owen, de même que j'ignore celle de Mr Airy, que l'Académie de Paris a nommé depuis 1869, associé étranger. Agassiz, nommé également depuis mon tableau, est fils d'un pasteur suisse (de famille indigène).

Si l'on retranchait du nombre des savants suisses ceux qui descendent de familles étrangères, il resterait encore un nombre assez respectable qui placerait notre petit pays à côté des Etats scandinaves et de la Hollande, selon les époques. Ce ne serait pas juste en soi, parce que nos savants d'origine étrangère étaient tous nés en Suisse, et même petit-fils ou arrière-petit-fils de réfugiés nés en Suisse. Pour eux l'influence d'hérédité avait déjà été attenuée énormément par la loi géométrique des degrés.

Assurément l'arbre que Calvin et ses amis avaient planté à Genève avec ses rameaux de Hollande, d'Ecosse, des puritains anglais et d'Amérique, était doué d'une grande vigueur. Notre souche à Genève s'est modifiée dans un sens libéral en 1720 (pp. 127 et 205), comme plus tard à Boston et même un peu en Hollande et en Ecosse, mais il est resté dans tous ces pays un esprit d'indépendance et une persistance de volonté qui ont été favorables aux sciences lorsqu'il a convenue aux individus de s'en occuper. Je n'ai pas voulu m'étendre davantage sur un aussi petit pays que Genève, mais voici quelques faits qui peuvent vous intéresser. A l'époque de la Réformation beaucoup de familles nobles quittèrent Genève pour demeurer catholiques. Il vint à la place une foule de gentilhommes et bourgeois instruits de France et d'Italie, qui étaient zélés pour la nouvelle religion. Grâce à leurs antécédents et leur éducation, ils entrèrent dans la classe des familles notables du pays, qu'ils dominèrent et ils devinrent le fond d'une aristocratie locale qui a subsisté de fait, sans titre ni privilège légal, jusqu'en 1841. Cette sorte de patriciat ne visait pas seulement aux places du gouvernement et des Conseils: elle occupait à l'origine les charges de pasteurs et dans le XVIIIe siècle et jusqu'en 1841 celles de professeurs de l'Académie (avec ou sans enseignement) qui donnait la surveillance de l'instruction publique et un rang honorable dans l'opinion. Grâce à ces mœurs, un jeune homme studieux, d'une famille notable, pouvait se contenter d'une fortune médiocre. Il avait d'ailleurs une bonne chance pour se marier richement. Vous voyez quels encouragements existaient en faveur des sciences, précisément dans les familles d'anciens réfugiés.

J'aurais pu donner plus d'informations sur les mères et autres ascendants des savants suisses, surtout genévois, mais vous avez démontré que passé les premiers degrés l'hérédité influe fort peu. A vrai dire toutes nos familles un peu anciennes à Genève ont du sang de huguenot (si l'on ose employer cette vieille locution après votre curieux travail sur la transfusion), seulement ce sang est tout à fait dilué. Ce sont les institutions et les mœurs implantées par nos ancêtres réfugiés qui ont passé bien plus longtemps que l'hérédité. Maintenant une affluence d'ouvriers catholiques des pays voisins et une série de révolutions nous ont donné un nouvel état social. Nous devenons américains. Désormais la distinction presque unique entre les familles sera la fortune. Ce ne sera pas au profit de la science. D'autres cantons de la Suisse (Zurich, Vaud, Neuchâtel) se préparent à nous succéder, les bonnes conditions étant mieux réunies chez eux.

Dans les proportions de savants à l'égard des populations de pays il aurait mieux valu pouvoir calculer sur les hommes d'un certain âge, soit celui auquel on est ordinairement élu aux Académies, soit celui auquel on commence à travailler utilement. Malheureusement c'était impossible pour les années 1750 et 1789, dont je me suis occupé, et fort difficile pour 1829 et même 1869. La Suède est le seul pays qui ait eu un dénombrement par âges dans le siècle dernier. Pour le reste de l'Europe j'ai été forcé de recourir à des estimations même pour la population totale. On ne pouvait pas avoir la division par âges, en 1829 et 1869, dans les pays comme l'Allemagne et l'Italie où chaque Etat faisait ses recensements sur la base qu'il imaginait, à des époques différentes et souvent n'en faisait pas ou ne publiait pas les détails. Remarquez d'ailleurs que la proportion des décès d'enfants est d'autant plus forte qu'il y a plus d'enfants à soigner, d'où il resulte un nombre d'adultes moins différent qu'on ne croirait d'un pays à l'autre.

Il y aurait une correction importante à faire—celle de défalquer de chaque pays les individus nés à l'étranger et d'ajouter les nationaux qui se sont établis ailleurs. On ôterait de cette manière aux Etats Unis, en 1869, environ 5½ millions, qu'il faudrait repartir sur les Royaumes britanniques et l'Allemagne principalement. La correction serait équitable, car s'il y avait eu dans les étrangers établis en Amérique des titulaires d'Académie je les aurais imputés à leurs pays de naissance. Cela ne sortirait pourtant pas les Etats Unis de la région inférieure de mes tableaux. Et comment savoir la quantité de sujets britanniques établis sur le Continent ou ailleurs qu'en Amérique? Celle des Allemands établis en Russie, en France etc.? Le sujet heureusement n'exige pas une si grande précision. Je l'ai dit plusieurs fois, les chiffres de population ne sont pas en corrélation avec les groupes exceptionnels d'hommes s'occupant de science. On est prié d'englober dans chaque pays des parties considérables de population qui jouent un rôle scientifique insignifiant, comme l'Autriche en Allemagne, le royaume des Deux Siciles en Italie, l'Irlande dans le Royaume Uni, les cantons catholiques en Suisse. Les calculs sur les populations ne peuvent donc pas avoir une véritable valeur statistique, mais ils sont utiles pour pouvoir apprécier les causes qui ont influé en divers pays, à divers époques, en tenant compte des détails accessoires propres à modifier l'impression déterminée par les chiffres.

Sur la fécondation, distincte de la copulation, je ne vois pas bien la valeur des objections. Il y a dans ce que j'ai dit 1° des faits, 2° des conjectures. Les faits sont que des hommes en état d'ivresse (affection temporaire du cerveau) ont engendré souvent des idiots, épileptiques, etc., et qu'une chienne fortement blessée sur l'arrière partie du dos pendant l'accouplement a donné naissance à des petits défectueux du train du derrière (Lucas 2, p. 250). Donc chez l'homme les spermatozoaires peuvent être modifiés par l'état momentané maladif du système nerveux, et chez l'espèce canine les ovules peuvent être modifiés par un accident survenu au moment même de la copulation. Voilà des faits. Maintenant il ne me parait pas hasardé de croire que d'autres affections temporaires pourraient aussi influer comme l'alcoolisme ou comme une lésion. Dans l'espèce humaine une terreur, une idée dominante ou exclusive (espèce de monomanie) peuvent durer plusieurs jours et influeraient non seulement sur les spermatozoaires mais aussi sur les ovules au moment où ils vont se détacher. Je me souviens qu'à l'époque du siège de Sébastopol il y avait des personnes qui avaient perdu le sommeil de l'effroi et de la commisération des souffrances racontées par les journaux. Les évènements de 1870–71 ont causé beaucoup d'aliénations mentales et sûrement ont troublé à un moindre degré beaucoup d'esprits. Je ne serais pas étonné que ce n'ait été une cause d'augmentation de folie ou d'idiotisme chez les enfants nés en 1871 dans une partie de l'Europe. Dans les pays où les fortunes doivent être partagées également entre les enfants vous ne pouvez pas vous figurer la terreur plus ou moins secrète de plusieurs femmes à l'idée d'une nouvelle grossesse. Représentez vous aussi l'état nerveux de certaines femmes quand elles ont été infidèles à leurs maris et celui d'un mari qui déteste sa femme sans vouloir le lui montrer, du Duc de Praslin, par exemple, qui maintenait ses rapports conjugaux avec l'intention arrêtée d'assassiner la duchesse.

Je ne connais rien de fait su à faire ici sur les enfants trouvés. Ces enfants proviennent de parents inconnus et très variés. Je ne sais ce qu'on pourrait conclure de leurs aptitudes. Rousseau avait mis ses enfants à l'hôpital. On s'est demandé s'ils étaient devenus quelquechose. Je crois qu'ils sont morts jeune vu les conditions détestables des anciens hôpitaux de Paris.

Je vous serai fort obligé de m'envoyer l'article de Fraser's Magazine dont vous me parlez, de même que de toute rédaction avec ou sans critique de mon travail que vous auriez la bonté de publier. En attendant je vous prie de me croire, Monsieur, votre très dévoué collègue,

ALPH. DE CANDOLLE.

P.S. J'ai fait achever par le libraire un exemplaire de mon livre à la Société Royale. J'espère qu'il est parvenu.

42, RUTLAND GATE, LONDON. *May* 7/73

MY DEAR SIR, It gave me much pleasure to receive your letter. I assure you I feel like yourself, that the subjects on which we differ are altogether subordinate to the common interest we have in arriving at the truth on the same line of inquiry. My article in the *Fortnightly* was much shorter than I should have liked to have made it, but there was a difficulty about space and I crammed all I could in what was given to my disposal.

Of the many topics in your work left unnoticed I regretted much not being able to speak of your most just criticism of the misuse of the word 'Nature.' For my part, I will never offend again unless through a slip of the pen. Your work has been read by many of my scientific friends here, and a passage in it prompted one of the most effective parts of by far the most effective speech,—that of Dr Lyon Playfair,—in the recent Parliamentary debate upon Irish University Education. The debate, as perhaps you may have seen, was one of extreme importance to the future of science in Ireland, and the question was how far it should be submitted to or emancipated from Catholic control. Lyon Playfair quoted the effect of Calvinism in Geneva on science, during the time of its ascendancy in wholly suppressing it, which was shown by the immediate start made by science as soon as the strict dogmatic influence began to wane. He spoke with excellent effect and success, and I know that he derived at least that part of his argument from you, because I had myself directed his attention to your work previously as having a direct bearing on his then proposed speech.

Thank you for your interesting fact about impregnation under the effect of alcoholism. One of course needs many such facts and it occurs to me that perhaps some direct experiment might be made, say with white mice, which breed very frequently and largely, are easily reared and cheap to keep. The he-mouse might be fed on some suitable narcotic stimulant before being put in with the female. I have no idea what stimulant would be suitable, one would have to try cannabis sativa, belladonna, opium, etc. A strong instance (if accurately recorded) of alcoholism combined both with the evil influences of close interbreeding and of old age on the part of one of the parents, in producing no *bad* effect on the offspring, is that of Lot and his daughters (Genesis xix. 31).

You are good enough to remark on my views about improving the human breed, showing the difficulty of detecting and of discovering defects which families scrupulously conceal. But then on the other hand, it must be borne in mind that my primary object is not to deter the bad from, but to encourage the good breeds in, making early marriages. Those who are conscious of being of a good stock would court inquiry, for by having a warranty they would be advantaged. People take such extraordinary pains to found families that they could easily be taught the importance of marrying their sons and daughters to persons likely to cooperate in begetting children capable of supporting the dignity of the family. Hence youths having warranties would be sought after far more than the same persons are sought after now. After many generations, the absence of a warranty would look suspicious.—Encouragement of the best is the surest and safest way of discouraging the inferior. We are such a set of mongrels that except in extreme cases we should not be justified in 'banning' any marriage. All we can say is, that some marriages are more hopeful than others. I therefore go no further at present, than urging that hopeful marriage should be encouraged.

If an autumn's tour should take me to Geneva, I trust you will not think it a liberty if I do myself the pleasure of seeking your personal acquaintance, with a view to some conversation on the many subjects in which we have a strong common interest.

Believe me very faithfully yours, FRANCIS GALTON.

To M. ALPHONSE DE CANDOLLE.

GENÈVE. 16 *Juin* 1873.

MON CHER MONSIEUR, Vous me faites espérer une visite dans le courant de l'automne et comme je serais bien fâché de vous manquer, permettez-moi de vous dire que je serai sans doute à Genève depuis le milieu ou au plus tard la fin de septembre. J'ai l'intention de m'absenter pendant le mois d'août, mais il est bien douteux que mon excursion se prolonge au delà de 5 ou 6 semaines.

Si vous passez dans notre ville vous n'avez qu'à me demander *Cour St Pierre* 3 (au 2[ième] étage). C'est là qu'est mon herbier et aussi ma bibliothèque. J'y viens tous les jours de 11 à 2 h., excepté le dimanche.

Dans la soirée et les dimanches vous me trouveriez *au Vallon*, près Chêne et de Chêne à ma maison de campagne il y a 8 minutes.

Votre idée d'expérimenter sur les souris serait excellente à suivre. Il faudrait voir d'abord quelle substance toxique ces animaux mangent volontiers et quelle dose on peut leur en donner sans leur faire trop de mal et en produisant cependant des effets sensibles. Le Cannabis a moins de matière inébriante dans la graine que dans les feuilles, mais il doit pourtant y en avoir. Une pâtée alcoolisée serait peut-être plus commode, parce qu'on saurait bien la dose d'alcool administrée.

Des pigeons qui s'accoupleraient sous l'empire de l'alcool donneraient peut-être des petits tumblers? Mr Darwin pourrait l'essayer mieux que personne. Mais il dirait peut-être que la fécondation des œufs dans les oiseaux est entourée de trop d'incertitude. On ne saurait peut-être pas exactement quels œufs auraient été fécondés à tel moment par le mâle alcoolisé.

Recevez, mon cher Monsieur, l'assurance de mes salutations empressées.

ALPH. DE CANDOLLE.

42, RUTLAND GATE, LONDON. *Oct.* 18/73

MY DEAR SIR, Thank you very much for your kind letter this morning telling me of your whereabouts this autumn. I heartily wish I could have managed to meet you in Paris or elsewhere, for I have much in hand at this moment concerning the topic which interests both of us so much, and which I hope to publish in 3 or 4 months.

My wife and myself passed our summer in the heart of Germany, in the Thüringer Wald, and I there continually consoled myself with your prophecy that that tiresome German language is doomed to extinction, as one of the dominant tongues. That, and our atrocious English spelling! for which I, not being a classical scholar, entertain no respect whatever!

Believe me very faithfully yours, FRANCIS GALTON.

I have just left the house of a friend, where I had paid a short visit at the same time as our mutual friend Dr Hooker,—who very shortly will occupy the most distinguished of English scientific posts, namely the Presidency of the Royal Society. He will be a most acceptable President over us.

To M. ALPHONSE DE CANDOLLE.

42, RUTLAND GATE, LONDON. *May* 5/74

MY DEAR SIR, A few weeks back I gave a lecture at our Royal Institution on the subject that interests both of us, and only delayed sending you a printed account of it in the hopes of sending at the same time another and different memoir, which however will not be in print for some time.

My lecture will reach you by book post, and I hasten to send it because I see that the *Revue Scientifique* has been so good as to publish a translation in French, which however does not render some phrases quite exactly and gives a small but decided modification to their meaning. When the book will be complete, to which this lecture is a prelude, I cannot say;—but as soon as it is out, I will send you a copy for acceptance.

It seemed to be well worth while to select, as I have done, a group of men of the same nationality, similar education, race, religion, and period, in order to eliminate the disturbing influences of as many large variable causes as possible, and to bring out into stronger relief the effect of the residuum. I think you will be interested, when all the results are before you, in tracing the differences between them and your conclusions derived from the study of a selection of much more able men but under more varied circumstances.

My scientific men, for example, are mostly born in *towns*. Out of every 5 of these men, 1 is born in London, 1 in other very large towns, 1 in moderately sized towns, and 2 in villages or in country houses. The Heredity comes out very markedly and follows unexpected and peculiar "laws" (if I may be allowed, for want of a better, to use so grand an expression). I have not touched on these in the lecture, there was no time.

The other memoir alluded to at the opening of this letter, is a partially successful attempt to solve that *very* difficult mathematical question to which you drew attention in your book and about which I had plagued my not very brilliant mathematical head, at intervals, for many years, namely the extinction of families by the ordinary laws of chance. I contrived to *state* the problem, in a not unreasonable form that was at the same time fitted for mathematical investigation and did my best to persuade friends to work it out. At length one friend has got the thing into a shape that admits of some general conclusions being drawn but it is by no means a 'solution' of the problem in the ordinary sense of that word. When it is printed I will send you copies. Perhaps you may persuade Swiss mathematicians to investigate further. It really ought to be solved *if possible*, but it is only too probable that a direct solution is an improbability.

Our mutual friend Mr Bentham is, I am glad to say, abroad on a holiday with the Hookers. His domestic griefs due to the long-continued mental ill-health of his wife seem to have preyed much upon his spirits.

Fearing lest I have wearied you by this long and somewhat egotistic letter, I will now conclude and beg you to believe me, Faithfully yours, FRANCIS GALTON.

To M. ALPHONSE DE CANDOLLE.

GENÈVE. 11 *Sept.* 1874.

MON CHER MONSIEUR, Plusieurs absences et une indisposition de quelques semaines m'ont empêché de vous écrire au sujet de votre séance de la Royal Institution: *On men of science* etc. J'ai pourtant lu deux fois votre opuscule avec beaucoup de plaisir. La base sur laquelle vous vous appuyez est originale. Je doute qu'on pût obtenir ailleurs qu'en Angleterre un aussi grand nombre de réponses faites franchement et consciencieusement. En partant de la liste de la Société royale je crains un peu que le nombre des Ecossais n'ait pas été assez élevé. Comme leur éducation était naguère très différente de celle des Anglais ce serait regrettable. Sans doute vous parlerez de cette différence dans votre ouvrage que je me réjouis beaucoup de lire.

Plusieurs des conditions réputées favorables par vos correspondants se retrouveraient certainement hors des pays anglo-saxons, mais à des degrés autres, et quelques unes se retrouveraient probablement pas du tout.

*Energie*—Ce doit être général, seulement la persévérance en tient bien quelquefois.

*Santé*—J'ai connu bien des savants d'une mauvaise santé. Lorsqu'il en resulte, chez nous, la dispense du service auxiliaire, et que néanmoins la tête est bonne, la mauvaise santé est une chance de succès. Quand j'étudiais le droit mon meilleur professeur, un très habile juriconsulte, était estropié des jambes et des bras. Dans ce moment, à Genève, un des hommes les plus habiles, comme horticulteur et naturaliste, est un ancien pasteur chez lequel la vie s'est retirée peu à peu des extrémités à la tête depuis 25 ans. Il ne peut plus porter sa nourriture à la bouche, mais il continue à lire des ouvrages scientifiques en 3 langues et dicte des traductions. Il fait porter son tronçon supérieur (le seul vivant) dans son jardin pour vérifier la suite d'expériences qu'il a ordonnées. Le premier de ces deux estropiés remarquables avait une grosse tête et ressemblait à Napoléon I^er^: le second, également énergique, en a une petite. J'ai vu souvent les exercices de corps détourner des études ou détruire l'habitude d'observer soigneusement. Sans doute ils doivent être proportionnés, au degré de force de chacun, mais pas au delà.

*Practical business*—Ceci est bien anglais! Sur le Continent vous trouveriez une foule de savants qui n'entendent rien aux affaires, qui s'en moquent, et dont la fortune diminue plutôt que d'augmenter. La négligence des affaires est plus commune encore chez les hommes de lettres. Je crois bien qu'un savant réussit rarement s'il n'a pas de l'ordre dans ses papiers, ses travaux et même dans les habitudes ordinaires de la vie, mais ce n'est pas ce que vous appelez practical business (aptitude aux affaires commerciales, industrielles etc.).

Vos Réflexions p. 6....*All tends to...divine*, sont extrêmement justes, surtout le caractère anti-féminin. Je ne dirai pas pourquoi ce serait offenser la plus belle moitié de notre espèce.

J'en dirai autant de l'avantage d'une instruction *variée*. Contrairement à l'opinion des pedants de collèges, j'estime avec vous que c'est une grande source de curiosité et d'autres avantages, qui se retrouvent ensuite dans la spécialité d'une carrière. Seulement il ne peut pas continuer son éducation variée indéfiniment. Il faut savoir devenir spécial.

Dans l'espoir de lire bientôt le volume projeté de vos recherches intéressantes, je suis toujours, mon cher monsieur, votre très dévoué collègue ALPH. DE CANDOLLE.

GENÈVE. 11 *Janvier* 1875.

MON CHER MONSIEUR, Si j'ai tardé à vous remercier de l'envoi de votre volume *English Men of Science*, ce n'est pas que je l'aie négligé. Au contraire je l'ai lu deux fois avec beaucoup d'intérêt et me propose de le citer souvent si je fais une seconde édition de mon ouvrage sur *l'Histoire des Sciences*. Pour le moment je prépare une autre seconde édition, celle de ma *Géographie botanique raisonnée*, mais j'espère revenir ensuite à l'objet qui nous a tous deux occupés.

Nous avons employé deux méthodes bien différentes, qui nous ont souvent conduits aux mêmes déductions. C'est une preuve en faveur de toutes les deux. Lorsque les résultats diffèrent c'est probablement que les conditions spéciales aux Anglais ne sont pas celles de la plupart des pays, et je crois bien, en effet, que si l'on posait ailleurs les mêmes questions, on aurait de réponses assez souvent différentes. Il ne serait pas faute d'obtenir ces réponses. Peut-être les individus seraient-ils moins vrais dans les réponses, tantôt le voulant, et tantôt sans le vouloir. En Angleterre même, où l'on est plus véridique, certaines personnes se font des illusions. Je n'ai pas pu m'empêcher de sourire en lisant que quelques uns de vos ecclésiastiques prétendent n'avoir *nullement* été contrariés par leurs opinions dans les recherches scientifiques. Je l'ai entendu dire, avec la même bonne foi, à quelques uns des nôtres, mais quand on leur parle de certains faits, ils les nient ou les éludent, et souvent ils évitent certaines sciences ou certaines recherches. Comme l'expose très bien Herbert Spencer, dans sa *Social Science* l'homme est naturellement inconséquent. Vos ecclésiastiques paraissent plus éloignés que les nôtres des tendances scientifiques, exceptés les unitairiens. De même, chez nous les trinitairiens sont souvent ennemis de la science et non les unitairiens (protestants libéraux).

L'origine géographique de vos hommes de science (p. 20) tient peut-être à la prépondérance des villes dans la partie ombrée. La proportion favorable à l'Ecosse et défavorable à l'Irlande, dont vous parlez ailleurs, résultait aussi très clairement de mes listes, et vous expliquez, je crois, avec raison que les écossais reçoivent une éducation meilleure. Genève, la Hollande, l'Ecosse et la Nouvelle Angleterre sont ou étaient des branches d'un arbre intellectuel vigoureux planté par Calvin.

La quantité de savants fils uniques ou 1ers nés (p. 33) m'a étonné, mais avec un peu de réflexion cela se comprend.

Vous aimez les recherches statistiques, entre autres sur la fécondité. Elles peuvent se faire en Angleterre mieux que sur le Continent, mieux surtout qu'en la France, parce que l'absence d'enfants y est moins souvent volontaire. Je n'avais pas d'idée de l'étendue de cette cause dans les provinces françaises avant d'avoir lu un ouvrage sérieux, mais peu décent, d'un vieux médecin qui pratique dans la petite ville d'Arbois: *Bergeret—Des fraudes dans l'accomplissement des fonctions génératrices, leurs Causes, Dangers* etc. 1 vol. 8°, Paris 1873, chez J. B. Baillières, prix 2 fr. 50 c. L'auteur exagère je crois beaucoup les inconvénients des abus qu'il signale. Comme tous les médecins il ne voit que ceux qui souffrent de certaines pratiques et ne peut pas à tous ceux qui n'en ont pas éprouvé d'inconvénient. Mais le libertinage qu'il signale dans de très petites villes et même dans les communes rurales de la France explique bien le nombre très faible des naissances et fait naître de singulières réflexions. Le partage égal entre les enfants est évidemment une cause d'immoralité chez les parents et de paresse chez les enfants. C'est aussi, il est vrai, une cause d'économie dans les familles et de richesse totale. En Angleterre les accroissements de capitaux se divisent entre des individus toujours plus nombreux; en France ils restent accumulés dans une population stationnaire.

Je ne juge pas la condition de la *santé* tout-à-fait comme vous. Il est possible que les fils de parents robustes le nient moins[1] lorsqu'ils deviennent des hommes de science, vivant dans les

[1] Wording appears wrong, but I cannot correct it.

villes, mais ils n'en sont pas moins adaptés à leur milieu et c'est l'essentiel. A Paris j'ai remontré souvent des fils et petit-fils de Parisiens qui n'avaient pas l'air robuste, mais qui supportaient mieux que les campagnards les fatigues de la vie urbaine. Plusieurs étaient remarquables par leur intelligence et leur activité. En se tenant à l'abri du froid et se soutenant par une bonne nourriture ces citadins, sans avoir de bons muscles, vivent longtemps et font un travail excellent. Ils sont naturalisés—mais pour cela il faut deux générations au moins. Les fils de campagnards m'ont paru souffrir fréquemment de l'éducation à la ville de même que les fils d'ouvriers dans la campagne.

Vos idées sur l'éducation résultent des faits. Elles sont excellentes et je leur souhaite du succès. Le journal *Nature* dit souvent les mêmes choses, mais il les mélange d'assertions en mêmes choses, et de préventions que je ne puis pas toujours approuver. Il a l'air de croire que les gentlemen sont naturellement inférieurs aux ouvriers. Nous avons prouvé l'un et l'autre que les classes supérieures donnent partout une proportion considérable d'hommes éminents. Il serait donc utile de favoriser ceux de cette classe qui montrent quelque goût pour les siennes et ce serait plus profitable que d'élever artificiellement quelques individus de la classe inférieure, au moyen de subventions pécuniaires difficiles à bien placer.

Votre dernier volume complète admirablement le premier. Je vous en félicite et vous prie d'agreer, mon cher Monsieur, l'assurance de tout mon dévouement. ALPH. DE CANDOLLE.

42, RUTLAND GATE, LONDON. *March* 5/75

MY DEAR SIR I have left your welcome letter unanswered for two months, being desirous of sending, which I now do, a daily expected copy of a paper read at the Anthropological Society, on a subject to which you directed attention in the "Histoire des Sciences et des Savants." It is on the probability of the decay of families by purely accidental causes. If you could persuade any of your Swiss mathematicians to pursue the subject, it would be very advantageous. What seems to be wanted now, is some simple function which approximately represents the distribution of children in families, and to work this with Watson's general formulae. You will see how complicated the problem is. I send the memoir by book post.

Thanks greatly for your helpful criticism on my 'Men of Science.' I greatly value your hints. I got Bergeret's strange book and read it with no little alarm; but after all, even supposing he does not exaggerate, it seems to me that his own observations go to prove that strict Malthusian restraint may generally coexist with a pure life, because he states that he never found those uterine maladies in nunneries, which he seems to think so frequent in ordinary social life, in France. But he certainly reveals a strange state of things, unknown in England generally.

Your remarks on the quality of health of townsfolk,—of their being acclimatised to town conditions and of being able under these conditions to do good work,—are very instructive. Still, if their race dies out rapidly, it shows, does it not? that their health has suffered. It would be instructive to learn the social statistics of the numerous small Italian towns, where the same families have resided for centuries and whose population appears to vary but little.

Trusting that your labours in the 2nd Edition of the 'Géographie botanique' are happily concluded, Believe me very faithfully yours, FRANCIS GALTON.

To M. ALPHONSE DE CANDOLLE.

(English address) 42, RUTLAND GATE, LONDON.
GRAND HOTEL, THUN. *July* 22/75

DEAR SIR, Thank you much for the pamphlets on "effets différents d'une même temp. etc." in which the very interesting remarks about the struggle for existence among the buds, and the persistence of character in the produce of different boughs, are most instructive.

I am not acquainted with the memoir of Carl Linseer, who very probably has anticipated much of what I am about to say, namely that one might, perhaps with profit, compare "les sommes de température" not only "au-dessus de zéro" but above other fixed points of departure. Thus if the broken line represent the well-known "thermogram" made by a self-recording instrument, or protracted from eye observations, the ratios of the areas above the lines $B$ and $C$ would have *no relation to the ratios of AB to AC.* Therefore some general law might exist for plants, which would be clear enough when a correct base line was taken, but which would be wholly obscured when any other base line was employed.

No doubt this has been thought of, but what I would point out, is the *great facility* of obtaining these sums of temperature, from different base lines, by the use of that most ingenious little instrument of Swiss invention and manufacture, "Amsler's Planimeter." I have had it largely tested and employed at the Meteorological Office of England (of which I am one of the managing Committee) with perfect success. A full description of its employment will be found in our Meteor. Office 'Quarterly Weather Reports' of last year which are in the Geneva Observatory.

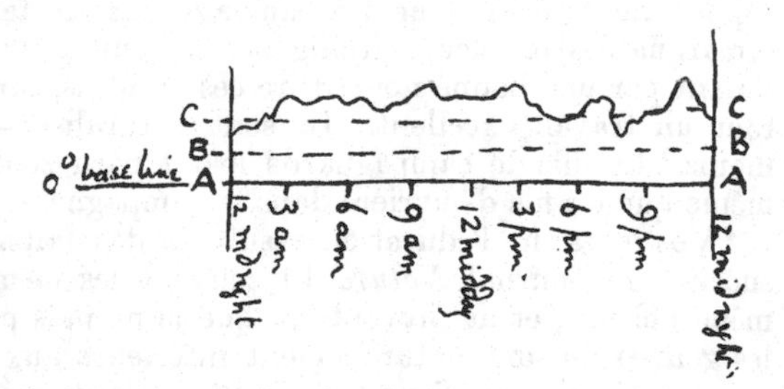


If the desire be, to try sums of the *squares* of excess of temperature, or of any other function, the same method of summation is of course equally applicable.

Pray excuse my prolixity; I write on the *chance* that our meteorological experience of rapid methods for avoiding tedious computation may prove of service in your further inquiries.

I am writing from Switzerland, from Thun, whither my wife and I are shortly going towards the Lake of Geneva; I had your pamphlet sent to me here. Should I be in the neighbourhood of Geneva, I will certainly do myself the pleasure of calling at your house in the hope (I fear a faint one at this season) of finding you at home.

Yours very faithfully, FRANCIS GALTON.

To M. ALPHONSE DE CANDOLLE.

GENÈVE. 24 *mai* 1876.

MON CHER MONSIEUR Je vous envoie (sous bandes) un article sur votre intéressant opuscule relatif aux jumeaux. Il vient de paraître dans les *Archives des Sciences physiques et Naturelles* de mai 1876 qui se publie ici. J'ai ajouté çà et là quelques réflexions pour montrer mieux l'intérêt de vos recherches. Je regrette que votre départ de la Suisse ait coincidé avec mon excursion dans l'Engadine, où ma santé m'oblige à aller pendant les grandes chaleurs. Agréez, je vous prie, l'assurance de mes salutations empressées. ALPH. DE CANDOLLE.

It is not necessary now to settle whether the emphasis placed by Galton on heredity or by De Candolle on environment was the more scientific; they both in fact cited individual cases, and discussed, where at present we should set about measuring and statistically analysing. Nor ought we to judge a man by hastily written letters, but there is much in De Candolle's book and letters, such as his belief—for it amounts to little more—that it was the conditions of Geneva and not the hereditary ability of the protestant immigrants, which produced a scientific revival, or his faith in tales of mental or physical state during sexual union producing marked results on offspring, that will not be accepted as proven by the calmer judgment of modern science. That science would sympathise far more with Galton's suggestion that one needs "many such facts[1]," and that it would be better to experiment with white mice, giving the male an intoxicant before admitting him to the female. This reference to appeals to experiment is of interest, for it indicates the change that had been taking place in Galton's own procedure.

[1] The assertion of De Candolle that men intoxicated at the time of coition have often produced idiots. Galton's humorous citation of Lot as evidence to the contrary was equally valid. For a modern discussion see *Eugenics Laboratory Memoirs*, No. XIII, pp. 19–25, Cambridge University Press.

It is not possible to say now what form De Candolle's *Histoire des Sciences et des Savants* would have taken, had he not seen Galton's chapter on men of science in the *Hereditary Genius*, but I doubt whether De Candolle would have emphasised, even as much as he did, the hereditary factor as a "pre-efficient of eminent men[1]." At any rate only crass ignorance could allow a man to do what a German does, namely speak of De Candolle as the pioneer and Galton as a later, but independent, worker[2]! When Galton started his investigation of *English Men of Science*, it really was on very different lines to De Candolle; the latter took as his field the foreign members of scientific academies and sought what facts he could ascertain about them. Galton issued a *questionnaire* to the Fellows of the Royal Society, and he did this because he had studied De Candolle's work,

"finding in it many new ideas and much confirmation of my own opinions; also not a little criticism (supported, as I conceive, by very imperfect biographical evidence) of my published views on heredity. I thought it best to test the value of this dissent at once, by limiting my first publication to the same field as that on which M. de Candolle had worked—namely to the history of men of science, and to investigate their sociology from wholly new, ample, and trustworthy materials." (Preface, p. vi.)

Galton had been leisurely working on an extended investigation as regards men of ability of all descriptions, to supplement his *Hereditary Genius*, when De Candolle's work appeared, and it was the latter's criticism of Galton's *prior* work that produced *English Men of Science*[3], a book which Galton held justified the utmost claims he had ever made for the recognition of the importance of hereditary influence. Galton's work was thus a defence of his own pioneer writings, and cannot possibly be regarded by any careful historian as a later and independent product in De Candolle's original field.

As a matter of fact the influence was essentially the other way round. In the 1873 first edition the title of De Candolle's book is: *Histoire des Sciences et des Savants depuis deux siècles*, with the additional words: "Suivie d'autres Études sur des sujets scientifiques en particulier SUR LA SÉLECTION DANS L'ESPÈCE HUMAINE." But in the edition of 1885 the additional words are changed to "Précédée et suivie d'autres Études sur des sujets scientifiques, en particulier sur *L'HÉRÉDITÉ ET LA SÉLECTION*, DANS L'ESPÈCE HUMAINE," the words "L'Hérédité et la Sélection" now standing out in the manner of a sub-title. In fact, while De Candolle minimised the effect of heredity in his first edition, he had not, as Galton very properly observes,

[1] A term introduced by Galton to denote anything "which has gone to the making of": see *English Men of Science*, Preface, p. vi

[2] Ostwald in the preface to his German edition of De Candolle's book writes: "Um das was Alphonse de Candolle mit so grossem Erfolge begonnen hatte und was seitdem namentlich von Francis Galton in England unabhängig geleistet worden ist u.s.w." Preface, S. vi, Edition 1911. If Ostwald had known the relative dates of *Hereditary Genius* and the *Histoire des Sciences*, he could hardly have said 'seitdem'; if he had opened *English Men of Science* he surely would have avoided the word 'unabhängig.' If "the Germans in Greek, are sadly to seek," then in history their searchings are deserving of birchings!

[3] As early as Feb. 27, 1874 Galton gave a Friday evening lecture at the Royal Institution "On Men of Science, their Nature and their Nurture."

collected the data on which an answer as to the value of the hereditary factor could really be given.

That Galton was somewhat moved by the attitude of De Candolle to his own work is proved not only by the letters cited above, but by an interesting paper he contributed to the *Fortnightly Review* entitled: "On the causes which operate to create Scientific Men[1]." Galton begins by referring to his difficult paper "Blood Relationship" of the previous year which we shall consider later. In that paper he distinguishes clearly, if in now unusual language, between somatic and genetic characters, and in the *Fortnightly* he refers to the "paradoxical conclusion"—as it certainly was in 1872—that the child must not be looked upon as directly descended from his own parents. The bearing of this statement lies in the explanation it affords of the reason why children differ frequently in mental character from their parents[2]—an observation which had been raised as an argument against Galton's theory of the inheritance of ability. Ability in an individual marks as a rule ability in the ancestry, not necessarily in the immediate parents. Galton then turns to

"a volume written by M. de Candolle......in which my name is frequently referred to and used as a foil to set off his own conclusions. The author maintains that minute intellectual peculiarities do not go by descent, and that I have overstated the influence of heredity, since social causes, which he analyses in a most instructive manner, are much more important. This may or may not be the case but I am anxious to point out that the author contradicts himself, and that expressions continually escape from his pen at variance with his general conclusions. Thus he allows (p. 195) that in the production of men of the highest scientific rank the influence of race is superior to all others; that (p. 268) there is a yet greater difference between families of the same race than between the races themselves; and that (p. 326) since most, and probably all, mental qualities are connected with structure, and as the latter is certainly inherited, the former must be so as well. Consequently I propose to consider M. de Candolle as having been my ally against his will, notwithstanding all he may have said to the contrary. The most valuable part of his investigation is this: What are the social conditions most likely to produce scientific investigations, irrespective of natural ability, and *a fortiori* irrespective of theories of heredity? This is necessarily a one sided inquiry......But......it admits of being complete in itself, because it is based on statistics which afford well-known means of disentangling the effect of one out of many groups of contemporaneous influences. The author, however, continually trespasses on hereditary questions, without, as it appears to me, any adequate basis of fact, since he has collected next to nothing about the relatives of the people upon whom all his statistics are founded. The book is also......deficient in method......but it is full of original and suggestive ideas." (pp. 346–7.)

As an example of deficiency of method, Galton cites De Candolle's statistical treatment:

"The author's tables of the scientific productiveness per million of different nations at

[1] *Fortnightly Review*, Vol. XIII, New Series, pp. 345–510. No copy of this exists among Galton's collected papers and no reference to it in any of the manuscript or printed bibliographies.

[2] Even De Candolle (p. 282) admits that "Une aptitude naturelle est toujours probablement héritée, puisque les parents sont la cause qui à précédé et déterminé l'existence de l'individu. Les exceptions s'expliquent par la diversité des parents, leur état momentané lors de la conception......" to which latter source of divergence Galton in his much annotated copy puts the note "Stuff!!" I cite this to indicate how strongly he held that the environment could not immediately affect the gametic characters, and how fully Galton rejected factors like 'maternal impressions' and 'momentary states at conception' as influencing the mental character of offspring.

different times are affected by a serious statistical error. He should have reckoned per million of men above fifty instead of the population generally. In a rapidly increasing country like England, the proportion of the youthful population to those of an age sufficient to enable them to become distinguished is double what it is in France, where population is stationary, and an injustice may be done by these tables to England in something like that proportion. They require entire reconstruction." (p. 347.)

In this essay Galton discusses the cases of apparent inheritance of acquired characters, and he gives expression to views which I doubt whether he held at a later date. In the first case he refers to De Candolle's instance of tame birds on a desolate island acquiring a fear of man, and *transmitting that fear as an instinctive habit to their descendants.* There is no evidence of a congenital transmission in such cases; the flock may acquire a habit which may be handed down from living member to living member, or the tamer members may be killed by their tameness, and so the species grow wilder. In the same way traditional habit of mind or emigration can adequately explain why "a population reared for many generations under a dogmatic creed" becomes indisposed to look truth in the face, and is timid in intellectual inquiry. We need not suppose the indisposition *congenital*, and if it be, Galton has in his *Hereditary Genius* already given a better explanation in the extermination and expatriation of the freethinking and inquiring members. We need not, as De Candolle and apparently Galton in this passage do, appeal to the inheritance of acquired characters. (p. 349.)

From the timidity in intellectual inquiry which results from rearing in a dogmatic creed, Galton turns to the relation between religion and science. "Can then religion and science march in harmony?" he asks; and his answer is of the following kind:

"The religious man and the scientific man have one great point in common, the devotion to an idea as distinct from a pursuit of wealth or social advancement. It is true that their methods are very different; the religious man is attached by his heart to his religion, and cannot endure to hear its truth discussed, and he fears scientific discoveries, which might in some slight way discredit what he holds more important than all the rest. The scientific man seeks truth regardless of consequences; he balances probabilities, and inclines temporarily to that opinion which has most probabilities in its favour, ready to abandon it the moment the balance shifts, and the evidence in favour of a new hypothesis may prevail. These, indeed, are radical differences, but the two characters have one powerful element in common. Neither the religious nor the scientific man will consent to sacrifice his opinions to material gain, to political ends nor to pleasure. Both agree in the love of intellectual pursuits, and in the practice of a simple, regular and laborious life, and both work in a disinterested way for the public good." (p. 349.)

Assuredly for Galton the ideal was the real! On p. 351 Galton again turns to this subject of acquired habits being transmitted hereditarily. He states that some acquired habits in dogs are certainly transmitted, but the number he tells us is small and "we have no idea of the cause of their limitation." Unfortunately he does not tell us what these habits are. It is not to be wondered at that Galton did not realise at first the full bearing of his theory of the continuity of the germ-plasm. Darwin like Lamarck had a belief in the transmission of some acquired characters, and Galton's close friend Herbert Spencer was a firm believer in the doctrine. Galton was at this time only feeling his way towards the conclusions that flowed from his hypothesis; he was beginning to be doubtful of the transmission of acquired characters, but he had yet to take the step of a full and complete denial. Thus after admitting that a few characters are acquired and transmitted by dogs[1], he writes:

"With men they are fewer still; indeed it is difficult to point out any one to the acceptance of which some objection may not be offered. Both M. de Candolle and Dr Carpenter have spoken of the idiocy and other forms of nervous disorders which beyond all doubt afflict the children of drunkards. Here then appears an instance based on thousands of observations at lunatic asylums and elsewhere in which an acquired habit of drunkenness, which ruins the will and nerves of the parent, appears to be transmitted hereditarily to the child. For my own part I hesitate in drawing this conclusion, because there is a simpler reason." (p. 351.)

The 'simpler reason' given by Galton is the toxic action of fluids saturated with alcohol on the reproductive elements of both sexes, and in the case of the mother on the unborn child. Perhaps the 'simplest reason' goes beyond Galton's! Now-a-days only asylum officers return 'Alcoholism' as a causal category of insanity. Those who have studied the pedigrees of the alcoholic know that nervous disorders are relatively as frequent in the ascendants as in the descendants; and the true explanation was given by De Quincey, when he said; "My friends thought drink drove me to insanity, but I knew that insanity drove me to drink." If we start with the drunkard, we find only too frequently mentally defective offspring. But if we start with complete pedigrees, we find stocks, whose members tend to nervous disorders, alcoholism, insanity and suicide. However, Galton's conclusion was a right one:

"We must not rely on the above-mentioned facts [alcoholism in parent and nervous disorder in offspring] as evidence of a once acquired habit being hereditarily transmitted."

The passage is, however, of great historical interest as it shows how Galton's theory of the distinction between the reproductive and the body cells was forcing him steadily to the conclusion that to modify heredity you must modify the germ cells, or at least an early stage of the embryo. Men like De Candolle, Spencer and Carpenter, nay, even Darwin himself, believed in the transmission of acquired characters; Galton, as far as I am aware, was the first to doubt it, and to assert that in the embryo of the offspring there is nothing represented that was not in the embryos of its parents. However crude now-a-days we may consider that statement to be, it really covers the continuity of the germ-plasm and the denial of the transmission of acquired characters. On the main point of difference between De Candolle and Galton, the present writer also finds Galton's criticism, as abstracted above, wholly justified. De Candolle's work, as he himself tells us, had been planned when he was a young man. His main idea seems to have then been to trace scientific achievement as arising from environmental causes—educational,

[1] I am not sure that Galton was not influenced by the aversion of Huggins' dog 'Kepler' to butchers' shops, an aversion shared by his relatives, and which some may assume to have originally been 'acquired.' There is a letter from Galton to Huggins on the subject of 'Kepler' dated Feb. 15, 1873, i.e. just before the *Fortnightly* paper. See our p. 66, second footnote.

religious, political institutions, family and national traditions, etc. Before his work was issued Darwin first and then Galton flared up on his horizon and the whole aspect of the country he was investigating appeared different. His book when it was written became a compromise between the old aspect and the new. It is, as Galton says, "deficient in method"—markedly so, I may add, from the statistical standpoint,—but nevertheless full of suggestion even to-day for more effective inquiry.

This somewhat lengthy digression on De Candolle is needful, not only because it casts light on both men's letters, but also on the influence Galton exercised over De Candolle, as well as the incentive the latter's book gave Galton's own closer inquiry, not only in *English Men of Science*, but also in the much later attempts he made to consider ability more generally in the material collected in this century, and partly published in *Notable Families*[1]. In this, however, Galton was merely extending, by aid of distributing *questionnaires*, the ideas and methods of his *Hereditary Genius*. Galton in the case of *English Men of Science* circularised a select list of 180 men of science—not all the members of the Royal Society. His schedule was a very lengthy one, and dealt not only with ancestry and education, but with many physical and mental characters classed in broad categories. As a rule he had about 90—100 cases with reliable information under each character heading. Perhaps the modern reader is most struck by the comparative paucity of data, or of statistical theory, which Galton thought needful for drawing conclusions. But actual investigation by more elaborate methods appears to confirm most of his results. He reached very often a conclusion, since proven, by what we may term a sound statistical instinct. I will illustrate this in two instances. He is considering whether in marriage "the love of contrast" does or does not "prevail over that of harmony[2]." He has only 62 cases of the statures of both parents, and he concludes by a very rough process that there is no sexual preference for *contrast* in height (pp. 31–32). I have worked out the correlation in stature between husband and wife for his data, and find it $+\cdot 19 \pm \cdot 08$, which shows that there probably existed a slight preference for harmony and this is well in accord with more recent determinations of the intensity of assortative mating. Again in 99 cases he had measures of the horizontal circumference of the head of men of science and 40 of these were men of great energy. From the data that of 13 men of 22″ or less circumference there were 10 of these energetic men, and that of 8 men of 24″ or more circumference there was only one of these very energetic men, Galton concludes that "although the average circumference of head among the scientific men is great[3]," there is a "cor-

[1] Issued in conjunction with Edgar Schuster, and re-issued recently in the *Galton Laboratory Publications*, Cambridge University Press.

[2] Galton's book contains the first tables for assortative mating in temperament, hair colour, figure and stature, although based on small numbers, and only reduced by considering percentages of 'harmonies' and 'contrasts.'

[3] This conclusion seems to be founded on the statement "that the average circumference of an English gentleman's head is $22\frac{1}{4}$ to $22\frac{1}{2}$ inches" (p. 99). If this be taken as 22·38″, Galton's value

relation[1]" between smallness of head and energy. Now if we distribute 40 cases uniformly among Galton's 99 individuals, we find:

| | Heads 22″ or less | Heads between 22″ and 24″ | Heads 24″ or more |
|---|---|---|---|
| Number ...................... | 13 | 78 | 8 (*A*) |
| Energetic men: observed ... | 10 | 29 | 1 (*B*) |
| „ „ uniformly distributed | 5·25 | 31·5 | 3·25 (*C*) |

A modern statistician would not be quite happy in asserting without further investigation that the owners of large heads were the less energetic. But actually the odds against (*B*) being a sample of (*C*) are about 50 to 1 and Galton's conclusion is reasonably justified. I have cited these results because, I think, Galton's statistical method does raise doubts in the minds of some readers as to its adequacy to support the superstructure[2]. I have been surprised myself on testing his data to find what a reasonable probability there is for the validity of his conclusions. After all the higher statistics are only common-sense reduced to numerical appreciation, and their deductions are what "les esprits justes sentent par une sorte d'instinct, sans qu'ils puissent souvent s'en rendre compte." Galton's judgments in his

for the scientists was 22·75″ ± 0·50, which is *not* a significant difference. I deduce these from Galton's pocket-book. The following are perhaps worth extracting:

| Subject | Head circumference in inches | Stature in feet and inches |
|---|---|---|
| Cayley ..................... | 23 | 5 7½ |
| Darwin..................... | 22¼ | 6 0 |
| Sir John Evans ........ | 22½ | 5 10 |
| Francis Galton ........ | 21⅛ | 5 9 |
| Grove ..................... | 24·0 | 6 0 |
| Sir Rowland Hill ...... | $23\frac{7}{16}$ | 5 7 |
| Hooker..................... | 21½ | 5 10½ |
| Huxley..................... | 23 | 5 10 |
| Jevons ..................... | 23 | 5 5½ |
| Clerk Maxwell.......... | 23½ | 5 4¼ |
| Owen ..................... | 23 | 5 11 |
| Marquis of Salisbury ... | 23 | 6 2 |
| Herbert Spencer .. ...... | — | 5 10 |
| Spottiswoode .......... | 22¾ | 5 11½ |
| Sylvester ................. | 22⅜ | — |

[1] This is the first occasion on which I have noticed Galton using this word, of course not yet with reference to the 'coefficient of correlation'.

[2] I feel considerable doubts myself as to the validity of the map of the distribution of birthplaces of 100 English men of science on p. 20. It may mean nothing more than a rough density of population map. The local distribution of each type of ability in the British Isles with reference to density of population would be well worth undertaking.

pre-statistical and early statistical days were essentially those of Laplace's 'esprit juste.'

Although Galton deals only with 180 selected names, he considers that the British Isles contain between the ages of 50 and 65 almost 300 men of this first class scientific status, and taking the male population between the same ages, he concludes that their frequency is about 1 in 10,000. He then proceeds to consider whether scientific distinction is more frequent than this among the relatives of scientific men. Of the above ratio, however, little use is made, because it includes men who lack the same education and opportunity. What Galton actually does is to take two groups of relatives of the scientific men, namely (1) grandparents and uncles, about 660, and (2) brothers and male cousins, about 1450, and inquire the number of distinguished individuals among these; and he finds their number much greater than in the case of men with like education and opportunity in the community at large, the latter ratio being derived from school and university data. The method, though rough, is probably adequate for Galton's purpose. It should be observed, however, that his environment being taken as approximately constant, Galton is really attempting to measure a 'partial' association, which is probably less intense than the total association[1]. However both the pedigrees provided and the statistics suffice to show that men of marked scientific ability come from able families.

Galton next proceeds to analyse the qualities on which his correspondents considered their success to have depended. In each case he points out how often these qualities were present in the parents, or relatives of the scientists. The chief qualities are (1) *Energy* much above the average—both physical and mental; (2) *Health*—"it is positively startling," Galton remarks, "to observe in these returns the strongly hereditary character of good and indifferent constitutions"; (3) *Perseverance*—also frequently reported in the parents; (4) *Practical business habits*—fully a half of those who possess these accredit one or both parents with the same faculty; (5) *Good Memory*—"heredity abundantly illustrated," about 30 cases of especial good memory, and 13 of poor memory; (6) *Independence of Character*—50 correspondents possess it in excess—strong evidence of its presence in the families of these scientists; (7) *Mechanical Aptitude*—found to exist largely in chemists, geologists, biologists and statisticians as well as in physicists and engineers. Galton then turns to religion and religious bias, having previously pointed out that the clergy are badly represented among men of science (pp. 25–26) or among their parents (p. 24). On this point it seems to me that such a question: "Has the religion taught in your youth had any deterrent effect on the freedom of your researches?" can scarcely be accurately answered by the subject himself. If he has cast off the religion of his youth, he may believe that he thinks freely; on the other hand if he is still a devout believer he would be unlikely to admit that his religious views hampered his scientific research. The scientific authors of the *Bridgewater Treatises* no doubt would have said their science was not the poorer for their religious bias, but we, who

[1] For example, if we endeavoured to measure the resemblance between brothers by taking sons of 6 foot fathers only, we should reach a value 40 °/₀ *too small.*

examine those treatises now, see that it most certainly was[1]. It does not seem to me therefore that much can be learnt from this section, except that a good many scientists of that day had discarded the religious faith of their youth, and more had lost all faith in any dogmatic theology.

Galton's third chapter is entitled "The Origin of Taste for Science," and here the bulk of the answers attributed the origin to inborn tendency. Out of 91 available answers, 56 said the taste was innate, only 11 said it was not innate and 24 were doubtful. On the other hand Galton does not dogmatically assert that innate taste of science is necessarily inherited. It is conceivable that the innate taste for science may arise from a combination of mental characters provided by otherwise unlike stocks[2]. Whatever the origin "a special taste for science seems frequently to be so ingrained in the constitution of scientific men, that it asserts itself throughout their whole existence." (p. 193.)

Other factors—fortunate accident, indirect opportunities, professional duties, encouragement of home or teacher, travel—are discussed by Galton, but are found to be rather auxiliary than primary sources of a love of science.

Of the advantages of travel there is a paragraph, which may be cited because, to a certain extent, it is autobiographical:

"Men are too apt to accept as an axiomatic law, not capable of further explanation, whatever they see recurring day after day without fail. So the dog in the yard looks on the daily arrival of the postman, butcher and baker as so many elementary phenomena, not to be barked at or wondered about. Travel in distant countries, by unsettling these quasi-axiomatic ideas, restores to the educated man the freshness of childhood in observing new things and seeking reasons for all he sees. I believe that a handsome endowment of travelling fellowships, thoroughly well paid with extra allowance for any special work allotted to their holders, given only to young men of high qualifications, and lasting for at least five years, would be money well bestowed in the furtherance of science[3]." (p. 219.)

Pp. 222–34 are some of the most interesting in the book, nay among the wisest that Galton ever wrote. They point out clearly some of our national difficulties, and the way that some have now been solved, and that others still seek solution. One is tempted to cite more than one has space for, especially when one remembers that the book can no longer be purchased!

"If we take a general survey of our national stock of capabilities and their produce, we see that the larger part is directed to gain daily bread and necessary luxuries, and to keep the

[1] Stokes stated that he had a strong religious bias and Huxley that he had a "profound religious tendency capable of fanaticism." Neither would admit that these characteristics interfered with their scientific work. Yet Stokes wrote his Burnett Lectures to demonstrate the evidence of design in the beneficial effects of light and Huxley wasted precious days that should have been given to science in pounding Gladstone and Semitic myths as to the creation!

[2] The author may, perhaps, claim an innate taste for science as he felt impelled to give up a profession to return to it; but he knows of no immediate ancestor or collateral with any scientific bent. On the other hand the names of some of his constituent stocks, Pearson, Beilby, Wharton, are not unfamiliar in science and distant links might be made if better data for testing the origin of innate scientific taste were not so easily accessible.

[3] The Cambridge and Oxford prize-fellowships are used, but not enough used, for travel. London is sadly wanting in anything of the kind, the only available travelling fellowships are the Albert Kahn. Five years work abroad is probably too long for all but young men of fortune or great push. The more modest are apt to be forgotten after a five years' absence—a common experience of young men who accept colonial or Indian scientific appointments!

great social machine in steady work. The surplus is considerable, and may be disposed of in various ways. Let us now put ourselves in the position of advocates of science solely, and consider from that point of view how the surplus capabilities of the nation might be diverted to its furtherance. How can the tastes of men be most powerfully acted upon, to affect them towards science?

The large category of innate tastes is practically beyond our immediate influence, but though we cannot increase the national store, we need not waste it, as we do now. Every instance in which a man having an aptitude to succeed in science is tempted by circumstances which might be controlled to occupy himself with subjects of less national value is a public calamity. Aptitudes and tastes for occupations which enrich the thoughts and productive powers of man are as much articles of national wealth as coal and iron, and their waste is as reprehensible. Educational monopolies, which offer numerous and great prizes for work of other descriptions, have caused enormous waste of scientific ability, by inducing those who might have succeeded in science to spend their energies with small effect on uncongenial occupations. When a pursuit is instinctive and the will is untaxed, an immense amount of work may be accomplished with ease." (pp. 222–3.)

"It is clear to all who have knowledge of the scope of modern science, that there exists an immense deal of national work which has to be performed, and which none but men of scientific culture are qualified to undertake. Scientific superintendence is required for all kinds of technical education, for statistical investigations of innumerable kinds, and deductions from them; for sanitary administration in the broadest sense; for agriculture, mining, industrial occupations, war, engineering. There is everywhere a demand for scientific assessors, who shall discover how to economise effort and find out new processes and fruitful principles. Professional duties generally ought to be more closely bound up with strictly scientific work than they are at present; and this requirement would tend to foster scientific tastes in minds which had little inborn tendency that way." (pp. 224–5.)

The reader, who remembers that this was written about 1873, before the foundation of most of our technical schools and engineering laboratories, and academic agricultural schools—to say nothing of the public health service—will grasp Galton's foresight. 'War' and 'statistical investigations of innumerable kinds'—both illustrations of the important future tasks of science—were chosen with surpassing aptness in 1873!

And again:

"It seems to me that the interpretation to be put on the replies we have now been considering, is that a love of science might be largely extended by fostering, not thwarting, innate tendencies, by the extension of scientific professional appointments and professorships, by assimilating in some cases the English system of teaching to that of the Scotch, and by creating travelling and other fellowships which shall enable their holders to view nature in various aspects, and to work with foreigners, whose habits of thought are fruitful in themselves but of a different kind to our own." (pp. 225–6.)

Among such fellowships Galton then demands the establishment of medical research fellowships.

"I appeal to capitalists, who know not what use, free from abuse, to make of their surplus wealth, to consider this want. They might greatly improve the practical skill of the English medical profession by affording opportunities of prolonged study. They might perhaps themselves reap some part of the benefit of it. A young medical man has now to waste the most vigorous years of his life in miserable routine work simply to obtain bread, until he has been able to establish his reputation. He has no breathing-time allowed him; the cares of mature life press too closely upon his student days to give him the opportunities of prolonged study that are necessary to accomplish him for his future profession." (pp. 226–7.)

How long after 1874 was the seed of that idea to lie buried, before the capitalist was forthcoming to fructify it by a valuable, if far from fully adequate, endowment?

There is one last citation which I must make from this chapter, because it has as much bearing on our nation to-day as it had fifty years ago; also it conveys Galton's views on those "social duties" which not a little fell to his lot:

"The influences which we have been considering are those which urge men to pursue science rather than literature, politics, or other careers; but we must not forget that there are deep and obscure movements of national life, which may quicken or depress the effective ability of the nation as a whole. I have not considered the reasons why one period is more productive of great men than another, my inquiry being limited, for the reasons stated in the first pages of this book, to one period and nation. But it may be remarked, that the national condition most favourable to general efficiency is one of self-confidence and eager belief of great works capable of accomplishment. The opposite attitude is indifferentism, founded on sheer uncertainty of what is best to do, or on despair of being strong enough to achieve useful results; a feeling such as that which has generally existed in recent years among wealthy men in respect to pauperism and charitable gifts. A common effect of indifferentism is to dissipate the energy of the nation upon trifles; and this tendency seems to be a crying evil of the present day in our own country. In illustration of this view, I will quote the following extract from a letter of one of my correspondents, who I should add is singularly well qualified to form a just opinion on the matter to which he so forcibly calls attention:—'The principal hindrance to inquiry and all other intellectual progress in the people of whom I see much, is the elaborate machinery for wasting time which has been invented and recommended under the name of "social duties." Considering the mental and material capital of which the richer classes have the disposal, I believe that much more than half the progressive force of the nation runs to waste from this cause'." (pp. 227–8.)

The evils pointed out by Galton have intensified rather than diminished since 1874, and especially since the Great War. Scientists have become more and more professional, that is there are fewer and fewer men of means, who pursue science for the pleasure of it. Science is now almost entirely one of the professional roads to a living. An examination of the testamentary dispositions published from day to day in the newspapers shows how little the meaning of science for national welfare, how little the social value of knowledge, have penetrated the minds of the wealthy. 'Conscience money' is amply provided for a variety of charities, many of which are with high probability anti-social in their effects. There is small doubt that salvation by 'good works' is still dominant in the minds of many, and 'good works' are still identified with charity by the majority of testators. A wealthy man will breed pheasants for sport, or dogs and pigs for prize points[1], but he would not spend a fraction of the money in striving to discover the laws of heredity, which might, if we had knowledge of them, aid not only national agriculture, but what Galton termed viriculture. The 'social duties' demand flowers, vegetables and fruit, and these a large establishment of gardeners and their underlings. But it is left to the florists to discover and make new varieties, and horticulture is part of the menial service of the establishment, not an intellectual pursuit of the proprietor; while his dog- and cattle-breeding follow narrow conventional lines, and their success is measured by the number of silver cups on a sideboard.

[1] Usually settled by the ignorant with a total disregard of the usefulness of the animal and in complete ignorance of its natural history.

The last chapter in Galton's book deals with education and endeavours from the replies of his correspondents to construct a scheme which would have favoured scientific development.

"As regards the precise subjects for rigorous instruction, the following seem to me in strict accordance with what would have best pleased those of the scientific men, who have sent me returns:—1. Mathematics pushed as far as the capacity of the learner admits, and its processes utilised as far as possible for interesting ends and practical application. 2. Logic (on the grounds already stated, but on those only). 3. Observation, theory and experiment, in at least one branch of science; some boys taking one branch and some another, to ensure variety of interests in the school. 4. Accurate drawing of objects connected with the branch of science pursued[1]. 5. Mechanical manipulation, for the reasons already given, and also because mechanical skill is occasionally of great use to nearly all scientific men in their investigations. These five subjects must be *rigorously* taught. They are anything but an excessive programme, and there would remain plenty of time for that variety of work which is so highly prized, as—ready access to books; much reading of interesting literature, history and poetry; languages learnt probably best during vacations, in the easiest and swiftest manner, with the sole object of enabling the learners to read ordinary books in them. This seems sufficient, because my returns show that men of science are not made by much teaching, but rather by awakening their interests, encouraging their pursuits when at home, and leaving them to teach themselves continuously throughout life. Much teaching fills a youth with knowledge, but tends prematurely to satiate his appetite for more. I am surprised at the mediocre degrees which the leading scientific men who were at the universities have usually taken, always excepting the mathematicians. Being original they are naturally less receptive; they prefer to fix of their own accord on certain subjects, and seem averse to learn what is put before them as a task. Their independence of spirit and coldness of disposition are not conducive to success in competition; they doggedly go their own way and refuse to run races." (p 257.)

It is in reading such a statement that one realises how much Galton gained by standing outside science as a profession, and one feels that science as a pursuit must always stand higher than science as a profession.

"If we class energy, intellect and the like, under the general name of ability, it follows that, other circumstances being the same, those able men who have vigour to spare for extra professional pursuits, will be mainly governed in the choice of them by the instinctive tastes of their manhood. The majority will address themselves to topics nearly connected with human interests; a few only will turn to science. This tendency to abandon the colder attractions of science for those of political and social life, must always be powerfully reinforced by the very general inclination of women to exert their influence in the latter direction. Again those who select some branch of science as a profession, must do so in spite of the fact that it is more unremunerative than any other pursuit. A great and salutary change has undoubtedly come over the feeling of the nation since the time when the leading men of science were boys, for education was at that time conducted in the interests of the clergy, and was strongly opposed to science. It crushed the inquiring spirit, the love of observation, the pursuit of inductive studies, the habit of independent thought, and it protected classics and mathematics by giving them the monopoly of all prizes for intellectual work.......This gigantic monopoly is yielding, but obstinately and slowly, and it is unlikely that the friends of science will be able, for many years to come, to relax their efforts in educational reform[2]. As regards the future provision

[1] There is a passage bearing on drawing on p. 142 worth citing: "There is an exact parallel between truthfulness of expression in speech and that of delineation in drawing. In the earliest sketch it is far better to be hard in outline than inaccurate. Subsequent touching up can smooth away the hardness; but there exists no proper material to work upon when there was carelessness in the first design."

[2] It is painful to read how little of their success many Victorian scientists attributed to their education. Darwin's education omitted mathematics, modern languages and all training in habits of observation and reasoning. He considered that all he had learnt of any value had

for successful followers of science, it is to be hoped that in addition to the many new openings in industrial pursuits, the gradual but sure development of sanitary administration and statistical inquiry may in time afford the needed profession. These and adequately paid professorships may, as I sincerely hope they will, even in our day, give rise to the establishment of a sort of scientific priesthood throughout the kingdom, whose high duties would have reference to the health and well-being of the nation in its broadest sense, and whose emoluments and social position would be made commensurate with the importance and variety of their functions." (pp. 258–60.)

Much of what Galton wished in 1874 to see achieved has since been done, although plenty remains to occupy fully the attention of educational reformers. It is singular, however, to note how little Galton's services to educational reform have been recognised, and yet in this book he is voicing the opinions of a very large section of the scientific men of that day; and these views filtered down through the press until they ultimately reached the politician. The last sentence but one appealing for development of sanitary administration and *statistical* inquiry finds Galton on common ground with Florence Nightingale—a link to which we shall return later. But alas! their dreams are still far from realisation; it is still held laughable to suggest that the statistician is a fundamental need, if we are to understand what makes for or mars the health and well-being of our nation in its broadest sense.

### B. DARWIN AND THE PANGENESIS EXPERIMENTS

As Galton's views on heredity brought him to a certain extent into conflict with De Candolle, so also they brought him at an even earlier date into a disagreement with Charles Darwin. At the end of 1869 as a result of his discussion of pangenesis in the Chapter entitled 'General Considerations' of *Hereditary Genius*, Galton determined to test experimentally Darwin's 'provisional' hypothesis. In that discussion Galton directly speaks of gemmules circulating in the blood (see our p. 113). Although Darwin read this book, I can find no trace of a letter at that date repudiating the idea of circulation in the blood being the essential method of transfer of gemmules. From December 1869 to June 1870 I find twelve letters of Galton to Darwin about the experiments on transfusion of blood. That Darwin answered some, perhaps all, of these letters is clear, but I have not succeeded in finding any replies. It is possible that after the letters to *Nature* of 1871, Galton destroyed them. At any rate in the list of Darwin letters prepared in 1896 by Galton himself none of these letters are referred to. All the Darwin rabbit letters that have survived are those which *followed* the publication of Galton's paper "Experiments in Pangenesis by Breeding from Rabbits of a pure variety, into whose circulation blood taken from other varieties had previously been largely transfused." This was read at the Royal Society on March 30, 1871[1]. These letters refer to a continuation of the experiments,

been self-taught and was due to his following up of an innate taste for science, and Galton expressed himself in much the same language: see our Vol. I, p. 12.

[1] *Royal Soc. Proc.* Vol. XIX, pp. 394–410, 1871.

also with negative conclusions, which results confirmatory of the thesis of his memoir Galton never to my knowledge published in detail. Those who read the letters below cannot doubt that Darwin knew the nature of the experiments, and knew that Galton was assuming that the 'gemmules' circulated in the blood. The whole point was to determine whether the hereditary units of a breed *A* could be transferred by transfusion of blood to members of a breed *B* and would 'mongrelise' the offspring conceived later by *B*. Was the 'blood' indeed as supposed in folk-language all over the world a true bearer of hereditary characters? That question is itself of importance, even apart from the question of Darwin's theory of heredity. But the publication of these letters has in this particular instance a deeper significance. It is a biographer's duty to illustrate the real strength of his subject's character, not merely to call it great. I know of no case in which a disciple's reverence for his master has exceeded that shown by Galton for Darwin in this matter. I doubt if any natures the least smaller than those of Darwin and Galton would have sustained their friendship unbroken, even for a day, after April 24th, 1871. I feel that the self-effacement of Galton in this instance is one of the most characteristic actions of his life; but it is not one that a biographer can disregard, however great his reverence for Darwin. Here are the letters extending from the start of the pangenesis experiments to nearly the time when Galton began to write his paper.

(1) 42, RUTLAND GATE, S.W. *Dec.* 11, 69.

MY DEAR DARWIN, I wonder if you could help me. I want to make some peculiar experiments that have occurred to me in breeding animals and want to procure a few couples of rabbits of marked and assured breeds, viz: *Lop-ear* with as little tendency to Albinism as possible. *Common Rabbits*, ditto. *Angora* albinos. And I find myself wholly unable to get them, though I have asked many people. Do you know anybody who has such things? I write without your book in reach, but you there especially mention a breeder of Angoras. Also you quote with approbation from Delaney's little book. Are either or both of those men accessible and likely to help? Pray excuse my troubling you; the interest of the proposed experiment—for it is really a curious one—must be my justification. Very sincerely yours, FRANCIS GALTON.

(2) 42, RUTLAND GATE, S.W. *March* 15, 70.

MY DEAR DARWIN, Very many thanks for the information and books. When I have got up the subject, I will write again, and will in the meantime take all care of the books.

I shall hope in a week from now to give you some news and by Saturday week definite facts about the rabbits. One litter [?doe] has littered to-day and all looks well with her. Two others towards the end of the week, viz: Wednesday and Saturday. I grieve to say that my most hopeful one was confined prematurely by 3 days having made no nest and all we knew of the matter was finding blood about the cage and the *head* of one of the litter. She was transfused from yellow and the buck also from yellow. Well the head was certainly much lighter than the head of another abortion I had seen, and was certainly *irregularly* coloured, being especially darker about the muzzle, but I did not and do not care to build anything about such vague facts and have not even kept the head. As soon as I know *anything* I will write instantly and first to you. For my part, I am quite sick with expected hope and doubt.

Ever very sincerely, F. GALTON.

It will be seen from Letters (1) and (2) that between Dec. 11, 1869 and March 15, 1870, Galton must by letter or verbally have communicated the purpose of his experiments to Darwin. He now speaks quite openly of the transfusion and its possible effect on the nature of the offspring.

(3) 42 RUTLAND GATE, S.W. *March* 17, 1870.

MY DEAR DARWIN, No good news. Bartlett assured me this morning that it was a popular prejudice that young rabbits might not be looked at, reasonable care being taken, so we opened 2 boxes and examined the litters. The first contained four dead young ones all true silver greys. One, however, has a largish light-coloured patch on its nose, but Bartlett tells me that this is not unusual with silver greys as the very tips of their noses are often white. However this patch is somewhat larger and there are faint hopes, I think, that it may prove more considerable than Bartlett believes. I have one more litter yet to come and hope to send you the result by Monday evening post. I have coupled a new pair and re-coupled the 2 does whose litters have failed, one of them with a more suitable mate, and expect the following results:

| Date of expected litter | Buck transfused from rabbit coloured as below | Doe transfused from |
|---|---|---|
| April 14 ...... | Hare-coloured | Hare-coloured |
| April 16 ...... | Yellow | Yellow |
| April 16 ...... | Black and white | Black and white |

The quantity of blood transfused was only 1·25 per cent. of the weight of the rabbits which is only the same thing as 30 oz. of blood to an ordinary man. I know this is a very small proportion of the whole amount of blood, but hope by a second operation on the old bucks and by improved operations on all the young ones to get a great deal more of alien qualities into their veins. Very sincerely yours, FRANCIS GALTON.

In a letter of Mrs Darwin's to her daughter Henrietta dated Down, Sat., Mar. 19 [1870] we read:

"F. [Father] is wonderfully set up by London, but so absorbed about work, etc. and all sorts of things that I shall force him off somewhere before very long. F. Galton's experiments about rabbits (viz. injecting black rabbit's blood into grey and *vice versa*) are failing, which is a dreadful disappointment to them both. F. Galton said he was quite sick with anxiety till the rabbits *accouchements* were over, and now one naughty creature ate up her infants and the other has perfectly commonplace ones. He wishes this exper[t] to be kept quite secret as he means to go on, and he thinks he shall be so laughed at, so don't mention......" *A Century of Letters*, Vol. II, p. 230.

(4) 42, RUTLAND GATE, S.W. *March* 22, 1870.

MY DEAR DARWIN. Another litter—this time of 4—and all of them are true silver greys.—Also, one of the does (mentioned in my last letter as transfused from a black and white) is dead.

My stud now stands as overleaf[1]. I call each silver grey by the name of the colour of the rabbit from which it has been infused. I also give the particulars of my first batch. You will see that there was much less variety in my pairs then, than there is now. I hope to try a new mode of transfusion upon a wholly new stock, taking younger rabbits and putting much more alien blood into them. Ever very sincerely yours, F. GALTON.

[There follows a list of transfusions into bucks and does of first and second batches.]

(5) 42, RUTLAND GATE, S.W. *March* 31, 1870.

MY DEAR DARWIN, Better news—decidedly better. I opened the hutches where the young rabbits are, this morning, and found now that the white patch on the nose of which I spoke had become markedly conspicuous and larger, but also that a white vertical bar had begun to

[1] I have not thought it needful to reproduce this table, as the details of the experiments are given in the paper as finally published.

appear in the forehead[1]. On going to the other litter, which I had never before got a proper view of, I found another young one with precisely similar marks. (The male parent was the same in both cases.) I have spent a most unsuccessful morning with new apparatus trying to inject more completely; but I have yet hopes of success by making some alterations.

I will return to you Naudin and the 2 pamphlets by *to-morrow's* book post. Very many thanks for them and for all the references. With great reluctance, I feel it would be too much for me to undertake the experiments. I am too ignorant of gardening, and, living in London with a summer tour in prospect, I don't see my way to a successful issue; but I hope to practise my eye and get some experience this year which may be of service next year or hereafter. I congratulate you about the Quagga taint. Once more about the rabbits, very many thanks for your hints, I will try more grey blood. Bartlett takes great interest and gives much care. Murie's assistant looks after the rabbits. Murie himself looks in now and then,

Very sincerely, F. GALTON.

Owing to the failure of Darwin's parallel letters we have no knowledge of what his hints were. The nature of the proposed plant-rearing experiments is equally unknown to us, but the suggestion may have remained in Galton's mind and have borne fruit in the sweet-pea experiments of a few years later. The Quagga taint[2] has close bearing on the present subject, for if a mother of breed *A* bore a child to a father of breed *B*, it seems likely that the 'gemmules' in the 'circulation' of the unborn child might pass into the mother's circulation and possibly affect a child born later to a father of her own breed *A*. The Quagga case, as indeed all instances up-to-date, of so-called telegony can now be dismissed from consideration. They depend essentially on (i) observation of variation within the pure breed not being sufficiently wide, or (ii) the assertions of kennel-men and others endeavouring to screen their responsibility for unplanned matings.

It is clear from this fifth letter that Galton was still hoping against the weight of accumulating facts for evidence that foreign 'gemmules' had been transfused with the blood.

(6) 5, BERTIE TERRACE, LEAMINGTON. *April* 8, 1870.

MY DEAR DARWIN, The white nose and vertical bar is, I find, of no importance. Bartlett was not accessible the day I found them out, but he has since told me they are common varieties, and I hear the same from Mr Royds, the rabbit-fancier and judge of poultry shows, from whom I bought them. Before leaving London last week I succeeded in infusing 2 per cent. of the rabbit's weight in alien blood, before I had only achieved 1·25 or 1/80th part which (on the supposition of Huxley that blood constitutes 1/10th of the whole weight of the body) is only 1/8th of the blood. In other words my transfusion, hitherto, has given only 1 great-grandparent of mongrel blood to the otherwise pure silver greys, and this is a very small matter. I do not like to risk another operation on the other jugular of my rabbits till after the forthcoming 3 litters, not till after I have had more success in the system of more abundant transfusion. I can do nothing with the blood in its natural state, it coagulates so quickly, so I defibrinise it. If I cannot ever succeed in transfusing as much into the rabbits as is necessary to make a fair experiment, I must go to larger animals, and try cross-circulation with big dogs.

[1] Pencil note against this word: 'white star'; Galton does not use the now common word 'flare.'

[2] See *Animals and Plants under Domestication*, Vol. I, pp. 403–4, 1st Edn. Vol. I, p. 345, Ed. 1875. Darwin believed absolutely in telegony and attributes it to the "diffusion, retention and action of the gemmules included within the spermatozoa of the previous male." *Animals and Plants*, 1st Edn. Vol. II, p. 388. Darwin's words seem to indicate that mere coition as apart from bearing offspring might produce telegony. The theory of telegony suggests that later born offspring should be more like the father than earlier born, but I have found no trace of this; see *R. S. Proc.* Vol. LX, pp. 273–83, 1896.

You are very kind in giving me so much valuable advice and so much encouragement.

Miss Cobbe's review is very characteristic. She has not, however, quite caught what I am driving at in religious matters and which—if the book shall be enough read to make it reasonable for me to do so—I shall express more clearly. Very sincerely yours, FRANCIS GALTON.

The religious views are probably those of the *Hereditary Genius*: see our p. 114. The review, entitled: 'Hereditary Piety[1],' by Frances Power Cobbe, will be found in the *Theological Review*, April, 1870. I do not know whether she was at this time a correspondent of Galton's, but she was so in 1877.

(7) 5, BERTIE TERRACE, LEAMINGTON. *April* 26/70.

MY DEAR DARWIN, Two more litters and no happy results, the young being all true silver greys. There ought to have been a third litter but the doe had not kindled. I shall next give a fresh infusion to every one of my old stock and hope to raise the proportion of alien blood in their bodies to at least 3 per cent. of their entire weight, or, say 30 per cent. of their entire blood.

I am obliged to defer all this for a week or two longer for my mother has been lying at the verge of death for a fortnight and I am wanted by her. She is now a trifle better and her illness—the result of bronchitis—may be less acute for a while and I may be able to get back to London. We have no reasonable hope that she will ever recover even a more moderate degree of health. Very sincerely yours, FRANCIS GALTON.

(8) 42, RUTLAND GATE, S.W. *May* 12, 1870 (written at the *Athenaeum*).

MY DEAR DARWIN, Good rabbit news! One of the latest litters has a white forefoot. It was born April 23rd, but as we did not disturb the young, the forefoot was not observed till to-day. The little things had huddled together showing only their backs and heads, and the foot was never suspected. The mother was injected from a grey and white and the father from a black and white. This, recollect, is from a transfusion of only 1/8th part of alien blood in each parent; now, after many unsuccessful experiments, I have greatly improved the method of operation and am beginning on the other jugulars of my stock. Yesterday I operated on 2 who are doing well to-day, and who now have 1/3rd alien blood in their veins. On Saturday I hope for still greater success, and shall go on...until I get at least one-half alien blood. The experiment is not fair to Pangenesis until I do.

We are for the time relieved from anxiety about my poor dear Mother, who suffered the agonies of death over and over again, but has strangely pulled through, and is now comfortable though very weak and seriously shaken. Very sincerely yours, FRANCIS GALTON.

The appearance of an 'orphan foot,' or even two, in normally whole-coloured animals purely-bred is a common event; but it is interesting to note how Galton seized any feature he could that supported mongrelisation, and thus the demonstration of the truth of 'pangenesis.' He discusses this white foot, pp. 402–3 of his paper, but, I think, might have dismissed it as he did the white noses and flare of some of his first batch of litters.

(9) 42, RUTLAND GATE, S.W. *June* 1st, 1870 (written at the *Athenaeum*).

MY DEAR DARWIN, Though I have no new litter to report, and shall have only one before the end of the month, I do not like to let more time go by, without heartily thanking you for your helpful and encouraging letter. I will not trouble you with details now, but simply say that I feel sure, unless some unexpected disaster to my stock should arise, that I shall have a very complete set of experiments finished before August. My bucks have been heavily re-transfused and I have a doe in the same state. Also I shall have all the combinations, extreme and intermediate, of pure and transfused bucks with pure and transfused does.

I find I cannot manage pigeons for want of a dove-cot, and dare not try dogs lest the Zoological Gardens should be alarmed by the noise and I should be extruded. But notwith-

[1] Reprinted in *Darwinism in Morals, and other Essays*, 1872.

standing this, I can assure you that I have the matter firmly in hand, and will be guided by the results, as to the extent of future work. Defibrinised blood is my salvation. I literally put into my silver greys during one operation as much blood as I can get from two rabbits each of the same size as the patient, and I have three bucks who have undergone two operations (but unluckily the earlier ones were far less successful). Very sincerely yours, FRANCIS GALTON.

(10) 42, RUTLAND GATE, S.W. *June* 25th, 1870.

MY DEAR DARWIN, A curious and, it may be, very interesting result delays my transfusion experiments. It is that 2, and I think all 3, of the does that had been coupled with the largely tranfused bucks prove *sterile*! Of course the sterility may be due to constitutional shock, or other minor matters, but, it *suggests* the idea that the reproductive elements are in the portion of the blood which I did *not* transfuse;—to wit the *fibrine*. In my earlier experiments, the blood was only partially defibrinised,—hence I was able to get a white leg; but in these later ones it was wholly defibrinised. It seems reasonable that the part of the blood which does most in the reparation of injuries should also be most rich in the reproductive elements. Of course I go on with the experiments with modifications of procedure....I wish I had more to tell you. I have transfused into 32 rabbits, in six cases twice over....

Very sincerely yours, FRANCIS GALTON.

The letters now break off, and the Galtons went to Paris on July 15th, intending to go to Switzerland; they did go to Grindelwald, but the declaration of war between France and Prussia led them to return. Here, after a stay at Folkestone, they paid visits to the Gurneys, at Julian Hill, at Leamington and at the Groves, reaching London only on October 17th (*L. G.'s Record*). On September 27th, George Darwin, however, wrote that his father sent his thanks for Galton's rabbit message and said that he was deeply interested in the success of the experiment. The nature of that experiment is clear, although Galton's letter detailing it appears to have perished; it is provided by Galton's paper itself; it was to cease defibrinisation, and it was done by establishing cross-circulation between the carotids, the great arteries of the neck.

"If the results were affirmative to the truth of Pangenesis, then my first experiments would not be thrown away; for (supposing them to be confirmed by larger experience) they would prove that the reproductive elements lay in the fibrine. But if cross-circulation gave a negative reply, it would be clear that the white foot was an accident of no importance to the theory of Pangenesis, and that the sterility need not be ascribed to the loss of hereditary gemmules, but to abnormal health, due to defibrinisation and, perhaps, to other causes also.

My operations of cross-circulation (which I call $x$) put me in possession of three excellent silver-grey bucks, and four excellent silver-grey does.......There were also three common rabbits, bucks, which were blood mates of silver-greys, and four common rabbits, does, also blood mates of silver-greys. From this large stock I have bred eighty-eight rabbits in thirteen litters, and in no single case has there been any evidence of alteration of breed. There has been one instance of a sandy Himalaya; but the owner of this breed assures me they are liable to throw them, and as a matter of fact, as I have already stated, one of the does he sent me did litter and throw one a few days after she reached me. The conclusion from this large series of experiments is not to be avoided, that the doctrine of Pangenesis, pure and simple, as I have interpreted it, is incorrect." (p. 404, *loc. cit.*)

Galton concludes that the gemmules are not independent residents in the blood; they either reside in the sexual gland itself, the blood merely forming nutriment to the growth, or they are merely temporary inhabitants of the blood and rapidly perish, so that the transfused gemmules perished before the period elapsed when the animals had recovered from their operations. Galton suggests that an experiment might be made—as the animals released from the operating table seemed little dashed in spirits, play, sniff and are ready to fight—to mate them at once.

"It would be exceedingly instructive, supposing the experiment to give affirmative results, to notice the gradually waning powers of producing mongrel offspring."

Galton clearly intended to continue the experiments; for a week after his paper was read he writes to George Darwin thanking him for a letter in which he had stated that his father was willing to take charge of eight of the rabbits[1]. Galton gives particulars about these eight young rabbits, how they should be mated and when the young should be returned to London for further operations.

"My paper will come out in the next number of the *R. Society Proceedings* and I will send your Father a copy with their pedigree marked." The *locus* for experimenting has, however, changed. "Though I shall not have my old excellent assistant Fraser, who sails this day week for Calcutta, I shall have the run of the University College Physiological Laboratory and shall be able, I believe, to conduct all the operations there with convenience greater than hitherto."

Again Darwin's letter is missing, but on April 25 Galton writes:

(11) 42, Rutland Gate, *April* 25, '71.

My dear Darwin, I am grieved beyond measure to learn that I have misrepresented your doctrine, and the only consolation I can feel is that your letter to 'Nature' may place that doctrine in a clearer light and attract more attention to it. I write hurriedly, as time is important to save the morning's post, in order to point out two passages which, I hope, in your letter to 'Nature' you will explain at length, so as to remove the false impression of Pangenesis under which I and probably others labour. In "Domestication of Animals etc." p. 374"......throw off minute granules or atoms, which *circulate* freely throughout the system......" And p. 379"...... the granules must be thoroughly diffused; nor does this seem improbable considering......the steady circulation of fluids throughout the body." (Is there not also a passage in which the words "circulating fluid" are used? I cannot hurriedly lay my hand on it, but believe it to exist.) Believe me—necessarily in great haste—Very sincerely yours, Francis Galton.

(12) 42, Rutland Gate, *May* 2/71.

My dear Darwin, I send a copy of the rabbit paper, in which I have marked the genealogy of the 6 little ones (p. 401).

You will see my reply in next week's 'Nature'. I justify my misunderstanding as well as I can and, I think, reasonably. The half plaintive end to the letter will amuse you. Very sincerely yours, Francis Galton.

I begin an entirely new and different series of experiments to-morrow.

One letter more before we come to the *Nature* correspondence. Darwin's and Galton's letters in *Nature* opened a general correspondence, in part of which Darwin was roughly handled and Galton wrote to him as follows:

(13) 42, Rutland Gate, *May* 12/71.

My dear Darwin, I have just seen ——'s not nicely conceived letter in 'Nature' on Pangenesis, and write at once to you, lest you should imagine that I in any way share the *animus* of the letter. I do not know him; at least, I have, perhaps twice only, had occasion to converse with him,—and what he says, certainly does not express my own opinion as expressed elsewhere and to others. I should not feel easy, if I did not disavow all share in it to you. Ever very sincerely, Francis Galton.

My new experiments are not hopeful—alas! I hope Pangenesis will get well discussed now.

[1] A postcard dated April 14th Down:—"The rabbits arrived safe last night and are lively and pretty this morning C. D."—seems to belong to this date.

Before we turn to the *Nature* letters, we must note one or two points, namely:

(*a*) Galton kept Darwin fully informed of the transfusion of blood experiments, and further stated their bearing on Pangenesis.

(*b*) Darwin clearly made throughout the experiments hints for their modification and extension even to other species.

(*c*) Galton's letters and paper are not compatible with Darwin having at any time warned him that the circulation of the blood was not a necessary factor in his own theory.

(*d*) Galton's words on p. 395 of his memoir cited by Darwin were too sweeping, but at the same time they were actually qualified by what he wrote on p. 404 that "the doctrine of Pangenesis, pure and simple, *as I have interpreted it*[1], is incorrect."

Letter of Charles Darwin in *Nature*, April 27, 1871.

"Pangenesis." In a paper, read March 30, 1871, before the Royal Society, and just published in the Proceedings, Mr Galton gives the results of his interesting experiments on the inter-transfusion of the blood of distinct varieties of rabbits. These experiments were undertaken to test whether there was any truth in my provisional hypothesis of Pangenesis. Mr Galton, in recapitulating "the cardinal points," says that the gemmules are supposed "to swarm in the blood." He enlarges on this head, and remarks, "Under Mr Darwin's theory, the gemmules in each individual must, therefore, be looked upon as entozoa of his blood," etc. Now, in the chapter on Pangenesis in my "Variation of Animals and Plants under Domestication," I have not said one word about the blood, or about any fluid proper to any circulating system. It is, indeed, obvious that the presence of gemmules in the blood can form no necessary part of my hypothesis; for I refer in illustration of it to the lowest animals, such as the Protozoa, which do not possess blood or any vessels; and I refer to plants in which the fluid, when present in the vessels, cannot be considered as true blood[2]. The fundamental laws of growth, reproduction, inheritance, etc., are so closely similar throughout the whole organic kingdom, that the means by which the gemmules (assuming for the moment their existence) are diffused through the body, would probably be the same in all beings; therefore the means can hardly be diffusion through the blood. Nevertheless, when I first heard of Mr Galton's experiments, I did not sufficiently reflect on the subject, and saw not the difficulty of believing in the presence of gemmules in the blood. I have said (Variation, etc., vol. ii, p. 379) that "the gemmules in each organism must be thoroughly diffused; nor does this seem improbable, considering their minuteness, and the steady circulation of fluids throughout the body." But when I used these latter words and other similar ones, I presume that I was thinking of the diffusion of the gemmules through the tissues, or from cell to cell, independently of the presence of vessels,—as in the remarkable experiments by Dr Bence Jones, in which chemical elements absorbed by the stomach were detected in the course of some minutes in the crystalline lens of the eye; or again as in the repeated loss of colour and its recovery after a few days by the hair, in the singular case of a neuralgic lady recorded by Mr Paget. Nor can it be objected that the gemmules could not pass through tissues or cell-walls, for the contents of each pollen grain have to pass through the coats, both of the pollen tube and embryonic sack. I may add, with respect to the passage of fluids through membrane, that they pass from cell to cell in the absorbing hairs of the roots of living plants at a rate, as I have myself observed under the microscope, which is truly surprising.

When, therefore, Mr Galton concludes from the fact that rabbits of one variety, with a large proportion of the blood of another variety in their veins, do not produce mongrelised offspring, that the hypothesis of Pangenesis is false, it seems to me that his conclusion is a little

[1] I have italicised these words to emphasise Galton's attitude.

[2] Note by the biographer. It would seem feasible to test the theory of pangenesis in the case of plants by considering the results obtained from the seeds of grafted and non-grafted plants of the same species.

hasty. His words are, "I have now made experiments of transfusion and cross-circulation on a large scale in rabbits, and have arrived at definite results, negativing, in my opinion, beyond all doubt the truth of the doctrine of Pangenesis." If Mr Galton could have proved that the reproductive elements were contained in the blood of the higher animals, and were merely separated or collected by the reproductive glands, he would have made a most important physiological discovery. As it is, I think every one will admit that his experiments are extremely curious, and that he deserves the highest credit for his ingenuity and perseverance. But it does not appear to me that Pangenesis has, as yet, received its death blow; though, from presenting so many vulnerable points, its life is always in jeopardy; and this is my excuse for having said a few words in its defence. CHARLES DARWIN.

Letter of Francis Galton in *Nature*, May 4th, 1871.

"Pangenesis." It appears from Mr Darwin's letter to you in last week's *Nature*, that the views contradicted by my experiments, published in the recent number of the "Proceedings of the Royal Society," differ from those he entertained. Nevertheless, I think they are what his published account of Pangenesis (Animals, etc., under Domestication, ii, 374, 379) are most likely to convey to the mind of a reader. The ambiguity is due to an inappropriate use of three separate words in the only two sentences which imply (for there are none which tell us anything definite about) the habitat of the Pangenetic gemmules; the words are "circulate," "freely," and "diffused." The proper meaning of circulation is evident enough—it is a re-entering movement. Nothing can justly be said to circulate which does not return, after a while, to a former position. In a circulating library, books return and are re-issued. Coin is said to circulate, because it comes back into the same hands in the interchange of business. A story circulates, when a person hears it repeated over and over again in society. Blood has an undoubted claim to be called a circulating fluid, and when that phrase is used, blood is always meant. I understood Mr Darwin to speak of blood when he used the phrases "circulating freely," and "the steady circulation of fluids," especially as the other words "freely" and "diffusion" encouraged the idea. But it now seems that by circulation he meant "dispersion," which is a totally different conception. Probably he used the word with some allusion to the fact of the dispersion having been carried on by eddying, not necessarily circulating, currents. Next, as to the word "freely." Mr Darwin says in his letter that he supposes the gemmules to pass through the solid walls of the tissues and cells; this is incompatible with the phrase "circulate freely." Freely means "without retardation"; as we might say that small fish can swim freely through the larger meshes of a net; now, it is impossible to suppose gemmules to pass through solid tissue without *any* retardation. "Freely" would be strictly applicable to gemmules drifting along with the stream of the blood, and it was in that sense I interpreted it. Lastly, I find fault with the use of the word "diffused" which applies to movement in or with fluids, and is inappropriate to the action I have just described of solid boring its way through solid. If Mr Darwin had given in his work an additional paragraph or two to a description of the whereabouts of the gemmules which, I must remark, is a cardinal point of his theory, my misapprehension of his meaning could hardly have occurred without more hesitancy than I experienced, but I certainly felt and endeavoured to express in my memoir some shade of doubt; as in the phrase, p. 404, "that the doctrine of Pangenesis, pure and simple, as I have interpreted it, is incorrect."

As I now understand Mr Darwin's meaning, the first passage (ii, 374), which misled me, and which stands: "......minute granules......which circulate freely throughout the system" should be understood as "minute granules......which are dispersed thoroughly and are in continual movement throughout the system"; and the second passage (ii, 379), which now stands: "The gemmules in each organism must be thoroughly diffused; nor does this seem improbable, considering......the steady circulation of fluids throughout the body," should be understood as follows: "The gemmules in each organism must be dispersed all over it, in thorough intermixture[1];

[1] In later editions of his book, Darwin replaced "circulate freely" by "are dispersed throughout the whole system" and he cancelled the words that this diffusion was not "improbable considering the steady circulation of fluids throughout the body." But elements "dispersed throughout the whole system" surely should have appeared in the blood. In a footnote to his later editions (1875, II, p. 350) Darwin admits that he should have expected to find gemmules in the blood "but this is no necessary part of the hypothesis."

nor does this seem improbable, considering......the steady circulation of the blood, the continuous movement, and the ready diffusion of other fluids, and the fact that the contents of each pollen grain have to pass through the coats, both of the pollen tube and of the embryonic sack." (I extract these latter addenda from Mr Darwin's letter.)

I do not much complain of having been sent on a false quest by ambiguous language, for I know how conscientious Mr Darwin is in all he writes, how difficult it is to put thoughts into accurate speech, and, again, how words have conveyed false impressions on the simplest matters from the earliest times. Nay, even in that idyllic scene which Mr Darwin has sketched of the first invention of language, awkward blunders must of necessity have often occurred. I refer to the passage in which he supposes some unusually wise ape-like animal to have first thought of imitating the growl of a beast of prey so as to indicate to his fellow-monkeys the nature of expected danger. For my part, I feel as if I had just been assisting at such a scene. As if, having heard my trusted leader utter a cry, not particularly well articulated, but to my ears more like that of a hyena than any other animal, and seeing none of my companions stir a step, I had, like a loyal member of the flock, dashed down a path of which I had happily caught sight, into the plain below, followed by the approving nods and kindly grunts of my wise and most respected chief. And I now feel, after returning from my hard expedition, full of information that the suspected danger was a mistake, for there was no sign of a hyena anywhere in the neighbourhood. I am given to understand for the first time that my leader's cry had no reference to a hyena down in the plain, but to a leopard somewhere up in the trees; his throat had been a little out of order—that was all. Well, my labour has not been in vain; it is something to have established the fact that there are no hyenas in the plain, and I think I see my way to a good position for a look out for leopards among the branches of the trees. In the meantime, Vive Pangenesis! FRANCIS GALTON.

In view of the previous correspondence lasting for nearly two years—referred to only in words which Darwin alone could appreciate: "followed by the approving nods and kindly grunts of my wise and most respected chief"—I think this letter of Galton's to *Nature* is one of the finest things he ever wrote in his life; it is few men who have such a great opportunity and use it so bravely. Vive Pangenesis!

Darwin may have saved his theory—for a time, but Galton saved by his restraint his own peace of mind. It suggests the spirit of the old Quaker David Barclay, his ancestor[1]:

Yet with calm and stately mien,
Up the streets of Aberdeen
Came he slowly riding...

It is certain that those who reverence Galton will appreciate what he did, and those who reverence both Galton and Darwin will rejoice that their friendship remained unbroken. Nay, not only seemed intensified, but *mirabile dictu* Darwin now took even an emphasised part in the blood transfusion experiments, which went on for another three years at least! The rabbits now passed to and fro between London and Down and several of Darwin's and Galton's letters exist[2]. I cannot help thinking that Darwin still thought some argument for Pangenesis might arise from this further

[1] For some account of this ancestor of Francis Galton, see Vol. I, p. 29.

[2] It is a grave misfortune that Darwin never put the *year* on any of these letters. Galton attempted but not very successfully to date them in 1896. When I wrote my *Francis Galton, A Centenary Appreciation* (University Press, Cambridge), I thought some of Darwin's rabbit letters referred to the first rabbit experiments, but I now feel sure this is not correct. I think I have got them into proper sequence with Galton's, and they all belong to the *second* and unpublished rabbit series.

work, otherwise it is hard to understand why the further work was carried out, especially why in association with Darwin, who had denied its bearing on Pangenesis.

There was evidently a good deal of correspondence, now sadly missing—which would have explained Darwin's views on these renewed experiments—during the summer of 1871. We shall now put before the reader the remainder of this somewhat fragmentary correspondence, using it as a frame for Galton's earlier work on heredity, which we shall discuss as it is referred to. Some few of the letters of Galton have been printed in the preceding chapter; others are omitted as merely referring to the arrangement of meetings in London or at Down.

(14) We are now in Yorkshire. (*Address*) 42, RUTLAND GATE, LONDON, *Sept.* 13/71.

MY DEAR DARWIN, I had proposed writing to you, in a few days' time, about the rabbits when I received your letter. First, let me thank you very much for the kind care you have taken of them. Secondly—I grieve to hear from you, that your holiday has not been so much of a success as you had hoped so far as health is concerned and, thirdly on my own part, I am glad to say, I am and have been particularly well (except only a boil inside the ear, which hurt badly for a few days).

To return to the rabbits:—Will you kindly prevent the bucks having any further access to the does, and make away with all the young except, say, 4 or 5 as a reserve in case of continued accident in the forthcoming series of operations. As soon as I return to town, towards the end of October, I will ask you to send me the old rabbits, and will begin at once to cross-circulate every one of them. My present assistant (a most accomplished young M.B. in medical science) has not the manipulative skill of my old friend and I fear I shall have an undue proportion of corpses, but there *must* be *some* successes out of the 3 does and 3 bucks that you have and the other 3 that I have.

Latterly, my whole heart has been in *rats*; white, old English black, and wild grey, which I have had Siamesed together in pairs, chiefly white and wild grey (for my stock of black is low), in a large number of cases—perhaps 30 or 40 pair. These have been fairly successful operations so far as the well-being and comfort of the animals is concerned, but unexpected, out-of-the-way accidents, are continually occurring. One pair died after 63 (about) days of [union] and injection into the body of the one passed into the other. I hope in this way to test Pangenesis better than by the cross-circulation for if even 1 *drop of blood per hour* passes from rat to rat, a volume equal to the entire contents of the circulation of either will be interchanged in 10 days, and this is equal in its effects to a pretty complete intermingling of the bloods. All crystalloids diffuse readily from rat to rat (as poisons) through the tissues, and as we know that eggs of entozoa are carried through the veins by the blood, it seems that a long continued Siamese union would be a valuable means of experiment.

We look forward with much pleasure to our return to town, to see your daughter in her new house. I do not think that I wrote myself, for my wife was writing to offer you, which I do now, my heartiest congratulations on the event. But, you must miss her.

Ever sincerely yours, FRANCIS GALTON.

(15) 42, RUTLAND GATE, S.W. *Nov.* 9 71.

MY DEAR DARWIN, I had not the least doubt but that I could have sent you before now definite results about my rabbits, but I cannot:—you must have patience with me and wait yet longer. The cold has killed one litter to which I had looked forward, and I have had a series of other mishaps not worth specifying, the result of which is that I have only one silver grey litter to go by—viz:—that of which I told you, which included a yellow one, slate grey on the belly, *with some white on his tail.* I should have thought this a great success but it may be pronounced a 'yellow smut'! Another result is that I have built a good serviceable little house for the rabbits in my own backyard and have all the best of them under my own eye, now. The litter that died from cold, *looked* very hopefully marked—but I think one cannot trust to, apparently, pied markings in very young silver greys. I will write again as soon as I have definite results; and when the little yellow fellow is somewhat older, he is now 6 weeks, I will get opinions about him. Very sincerely yours, FRANCIS GALTON.

*If* you can easily lay your hands upon Gould's Anthropology of N. America, I should be grateful for it.

(16) 42, RUTLAND GATE, S.W. *November* 21/71.

MY DEAR DARWIN, I am truly ashamed to have trespassed so long on your kindness, in keeping the rabbits, but until now, owing to a variety of causes (including an epidemic where the animals are kept), I could not ask for them back. Now, all is ready to receive them in University College and I should be much obliged if you would instruct your man to send them there. I enclose labels with the address:—Charles H. Carter, Museum, University College, Gower Street, London—to put on them. Mr Carter will receive them when they arrive. Please tell your man to keep the bucks and does separate and to write *bucks* on the hamper which contains them. Will you also let me know what I am indebted to you for their feed and keep, including a judicious 'tip' to your man. I am really most obliged to you, I should have been stranded in this experiment, without the help, because I have only 2 of my lot of rabbits alive and they are both out of condition and I doubt if one will live.

The College *shuts up at 5 in the afternoon* and nothing can be received after that hour. If that is too early for the carrier, what shall I do?—When may I expect them to arrive? My rats have died sadly, but owing to causes foreign to the effects of the operation. My last living pair, after being united nearly 3 months, were killed last week for the purpose of injection. Dr Klein kindly did it for me. One animal was injected with blue and the other with red, and *vascular union is proved*; but the connection was small, however Dr Klein thinks that with a more protracted connection the union would have been more complete. So I shall go on with vigour. Very sincerely yours, FRANCIS GALTON.

(17) 42, RUTLAND GATE, S.W. *November* 24/71.

MY DEAR DARWIN, The results are indeed most curious—You must kindly permit me to run down to you to-morrow (Saturday) for an hour or so, to see them and to fix what to do. I see my train would land me at Orpington at 11.12, so I suppose I should arrive at Down at about half past twelve. If however it should be a really wet day, I would postpone coming till Tuesday. You are indeed most kind to have taken all these pains for me and I sincerely trust the experiment may yet bear some fruit. I happened to be very unlucky with my Angora *transfusions* but there is no reason why they or the cross-circulation should not succeed and I will do my best to try it. Very sincerely yours, FRANCIS GALTON.

(18) 42, RUTLAND GATE, S.W. *Dec.* 2/71. (From *Athenaeum*)

MY DEAR DARWIN, The rabbits arrived quite safely and are in excellent condition. My man's letter to tell me of their arrival did not reach me till after post time last night or I should have written earlier. Once again, most sincere thanks for your kindness in taking care of them. Ever sincerely, FRANCIS GALTON.

*Jan.* 23rd [?1872] DOWN, BECKENHAM, KENT.

MY DEAR GALTON, The Rabbits have lost their patches and are grey of different tints, so you were right. They are quite mature now and ready to breed. We have put 2 does to a buck, for one more generation. Had you not better have the others soon, as we shall soon want space for the Breeders?

Have you seen Mr Crookes? I hope to Heaven you have, as I for one should feel entire confidence in your conclusion[1]. Ever yours sincerely, CH. DARWIN.

[1] I think this refers to Galton's investigations into spiritualism with Crookes (see our p. 63 *et seq.*). In *More Letters of Charles Darwin*, Vol. II, p. 443, there is a letter of Darwin to Lady Derby which reads: "If you had called here after I had read the article [probably Crookes' 'Researches in the Phenomena of Spiritualism,' *Quarterly Journal of Science*, 1874] you would have found me a much perplexed man. I cannot disbelieve Mr Crookes' statement, nor can I believe in his result. It has removed some of my difficulty that the supposed power is not an

(19) 42, RUTLAND GATE, S.W. *February* 1/72. (At *Athenaeum*)

MY DEAR DARWIN, If you can make it convenient to send, in separate hampers, 1 buck and 1 doe, I should be glad, as then my stock will be large enough to be above risk of accident. As for the others, pray do what you like with them. Would you send the pair, as before, addressed to—Dr Charles Carter, University College, Gower Street, and if you could kindly let a postage card be sent to him, to say when they might be expected, they would be the more sure to be immediately attended to. I grieve to say, that I find I must abandon the rats, as a task above my power to bring to a successful issue. I am most truly obliged for the care you have taken of the rabbits—I heartily wish, for my part, that I could have done more in the way of experiment than I have effected. Very sincerely yours, FRANCIS GALTON.

(20) 42, RUTLAND GATE, S.W. *May* 26/72.

MY DEAR DARWIN, I feel perfectly ashamed to apply again to you in my recurring rabbit difficulty, which is this: I have (after some losses) got three does and a buck of the stock you so kindly took charge of cross-circulated, and so have means of protracting the experiments to another generation, and of breeding from them and seeing if their young show any signs of mongrelism. They do not thrive over well in London, also we could not keep them during summer at our house, because the servants in charge when we leave could not be troubled with them. Is it possible that any of your men could take charge of them and let them breed, seeing if the young show any colour, then killing the litter and breeding afresh, 2 or 3 times over? I would most gladly pay even a large sum—many times the cost of their maintenance—to any man who would really attend to them. Can you help me? Ever sincerely yours, FRANCIS GALTON.

DOWN, BECKENHAM, KENT. *May* 27th. [1872?]

MY DEAR GALTON, We shall be very happy to keep the 4 rabbits and breed from them. I have just spoken to my former groom (now commuted into a footman) and he says he will do his utmost to keep them in good health. I have said that you would give him a present, and make it worth his while; and that of course adds to the expense that you will be put to, and I have thought that you would prefer doing this to letting me do so, as I am most perfectly willing to do.

If you will send an answer by return of post, I will direct our carrier, who leaves here every Wednesday night, to call on next Thursday *morning* at whatever place you may direct. Next week we shall probably be at Southampton for 10 days.

We have now got 2 litters from some of the young ones which you saw here; and my man says that in one litter there are some odd white marks about their heads; but I am not going again to be deluded about their appearance, until they have got their permanent coats.

Yours most sincerely, In haste for post, C. DARWIN.

(21) 42, RUTLAND GATE. *May* 28th, 1872.

MY DEAR DARWIN, You are indeed most kind and helpful and I joyfully will send the rabbits. But really and truly I must bear every expense to the full and will rely on your groom telling me, at the end; in addition to his present. The rabbits are none of them absolutely recovered, at all events the buck and 1 doe are not, but they will want no further attention in respect to what remains unhealed of their wounds. Two of the does are believed to be in kindle, having been left with the buck a fortnight and 10 days ago. I will tell

anomaly, but is common in a lesser degree to various persons. It is also a consolation to reflect that gravity acts at any distance, in some wholly unknown manner, and so may nerve-force. Nothing is so difficult to decide as where to draw a just line between scepticism and credulity. It was a very long time before scientific men would believe in the fall of aerolites; and this was chiefly owing to so much bad evidence, as in the present case, being mixed up with the good. All sorts of objects were said to have been seen falling from the sky. I very much hope that a number of men, such as Professor Stokes, will be induced to witness Mr Crookes' experiments."

It will be clear that at this time—after the Galton investigations but before Huxley's report (see our p. 67)—Darwin was endeavouring to retain an open mind.

Dr Carter to label and send all particulars with them and to mark their backs with big numerals in ink. The carrier should call at University College for them, asking the porter at the gate. I enclose a paper for him. Once again, with sincere thanks,

Ever yours, FRANCIS GALTON.

I have just corrected proofs of a little paper to be shortly read at the Royal Society on "Blood-relationship" in which I try to define what the kinship really is, between parents and their offspring. I will send a copy when I have one; it may interest you.

42, RUTLAND GATE, S.W. *May* 29/72.

MY DEAR DARWIN, May I lunch with you on Thursday and arrange about rabbits? We shall then be staying for 2 days in your neighbourhood at Mrs Brandram, Hayes Common.

Your letter reached me just before we were leaving town for a Saturday and Sunday visit, and I did not reply at once, waiting to be sure about our engagements. If I don't receive a post card at above address to say 'no,' I will come. Ever sincerely yours, FRANCIS GALTON.

The spiritualists have given me up, I fear. I can't get another invite to a séance.

42, RUTLAND GATE, S.W. *June* 4/72.

MY DEAR DARWIN, Thank you very much about the rabbits. I however sincerely trust you did not send your man all the way on purpose for them alone! Anyhow I feel I have put you to much trouble and can only repeat how greatly I am obliged.

Your criticisms on my paper are very gratifying to me, the more so that the question you put is one to which I can at once reply. You ask, why hybrids of the first generation are nearly uniform in character while great diversity appears in the grandchildren and succeeding generations[1]? I answer, that the diagram shows (see next page[2]) that only 4 stages separate the children from the parents, but 20 from their grandparents and therefore, judging from these limited data alone, (ignoring for the moment all considerations of unequal variability in the different stages and of pre-potence of particular qualities etc.,) the increase of the mean deviation of the several grandchildren (from the average hybrid) over that of the several children is as $\sqrt{20}$: or more than twice as great. The omitted considerations would make the deviation (as I am prepared to argue) still greater.

I will add the explanatory foot-note you most justly suggest, and should be very glad if you would let me have your copy back (I will return it) with marks to the obscure passages that I may try to amend them.

I found the [?writing] an uncommonly tough job; having to avoid hypothesis on the one hand and truism on the other and, again, the difficulty of being sufficiently general and yet not too vague. It is very difficult to draw a correct verbal picture in *mezzo-tint*, I mean by burnishing out the broad effects and not by drawing hard outlines.

Ever very sincerely yours, FRANCIS GALTON.

I have knocked every symbol out of my paper and wholly rearranged the diagrams etc., to make it less unintelligible. F. G.

A pleasant journey and rest to you all!

Galton's paper was read at the Royal Society on June 13th of this year, and we now turn to its examination.

We have seen (p. 114) that Galton as early as 1869 propounded a doctrine equivalent to the continuity of the germ-plasm and the non-in-

[1] In *Animals and Plants under Domestication*, 1st Edn. 1868, p. 400, Darwin writes "Crossed Forms are generally at first intermediate in character between their two parents; but in the next generation the offspring generally revert to one or both their grandparents, and occasionally to more remote ancestors." He then proceeds to explain this by latent gemmules, and had he been a statistician could have deduced at once a Mendelian quarter! He points out the triple character of the second generation of hybrids distinctly.

[2] I omit the diagram, as I have failed to interpret it and therefore cannot transcribe it properly.

heritance of acquired characters. This position, which is clear cut and fairly easily defensible, was I hold later obscured in his mind by two influences (*a*) the strong belief of Darwin in the inheritance of acquired characters, and (*b*) Darwin's doctrine of pangenesis. Both may be summed up in the single influence: an intense admiration for Darwin, which enforced an exaggerated respect for the authority of his judgment in individual instances. The doctrines of pangenesis and of the inheritance of acquired characters seem to me to have actually retarded Galton's progress and to have rendered his statement of his own views less clear than they otherwise would have been. I trace this influence particularly in his paper 'On Blood-relationship' of 1872[1]. This memoir would, I think, have given a sharp-cut theory had it not been darkened by the shadow of Darwin's views on heredity.

We will cite in regard to this the opening words of the 'Blood-relationship':

"I propose in this memoir to deduce by fair reasoning from acknowledged facts, a more definite notion than now exists of the meaning of the word 'kinship.' It is my aim to analyse and describe the complicated connection that binds an individual, hereditarily, to his parents and to his brothers and sisters, and therefore, by an extension of similar links, to his more distant kinsfolk. I hope by these means to set forth the doctrines of heredity in a more orderly and explicit manner than is otherwise practicable.

From the well-known circumstance that an individual may transmit to his descendants ancestral qualities that he does not himself possess, we are assured that they could not have been altogether destroyed in him, but must have maintained their existence in latent form. Therefore each individual may properly be conceived as consisting of two parts, one of which is latent and only known to us by its effects on his posterity, while the other is patent and constitutes the person manifest to our senses." (p. 394.)

Galton then proceeds to say that *both* these patent and latent elements in the parent give rise to the 'structureless elements' in the offspring. Now in the above sentences Galton clearly divides the 'structureless elements' of the parent into those which give rise to the somatic characters of the parent, and those which remain latent. At first sight we might suppose from the above definitions that Galton did not include latent elements similar to those which produced the somatic characters, but it appears from his remarks on p. 398 that he really did so, for he attributes on that page special features in the offspring corresponding to special features in the parents, not to the somatic characters in the parents, but to 'latent equivalents.' In other words, he considers that, in the bulk of cases, the correspondence in somatic characters between parent and child is not due to any influence of the somatic characters of the parent, but results from the latent elements of the parent. Thus Galton's 'latent elements' constitute absolutely the gametic elements of more modern notation. Had Galton gone at this time a stage further, and asserted that the somatic characters of the parent were only an index to the latent elements in him, and not directly associated with the bodily characters of the offspring, he would have reached an important principle. I hesitate to call that principle merely the continuity of the germ-plasm, for Galton saw a good deal further than anything contained in the word 'continuity' itself. He believed that both in the case

[1] *Proc. R. Society*, Vol. xx, pp. 394–402.

of the patent and the latent elements selection took place, so that not only are the somatic elements a selection of all possible somatic elements of an individual of the same ancestry, but the latent elements or germ-plasm were themselves a selection. This selection he termed 'class representation.' That the somatic or bodily characters are a selection is, of course, obvious; that the germ-plasm is selected also is extremely probable, but less easily demonstrated. Galton represented to himself the 'structureless elements' as a vast congeries of individual elements—like balls of a great variety of colours in a bag. A selection is made of these ('class representation') for the embryonic elements which by development become the adult elements, the somatic characters; that is the simple explanation of variation in the somatic characters of individuals of the same ancestry and reared under the same environment. Another selection from the same bag gives the germ-plasm of the individual on which his gametic characters depend, i.e. the possibilities of his descendants. Thus the continuity of the 'latent elements' or as we might say of the germ-plasm was in Galton's mind broken by continual selection. The 'class representation' of the somatic characters giving the phenomenon of visible variation, and the 'class representation' of the germ-plasm the variation of stocks or stirps.

Galton did not in this paper, I do not think he ever did, carry out his hypotheses to their legitimate conclusions. In the first place the two selections from our 'bag' cannot be treated as wholly independent; the somatic characters are not perfectly correlated with the gametic characters, but they are correlated with them, and as we descend to highly specialised races highly correlated with them. It would not be unreasonable to suppose that the somatic characters arise from a sub-selection of the gametic group, or from leaving a portion of this drawing 'on the table.' But the selection of the germ-plasm must lead to its simpler and simpler structure, especially in the case of unisexual reproduction. The course of evolution must on this hypothesis start with a highly complex germ-plasm and tend to break this up into simpler and simpler groups as generation by generation more elements are differentiated, i.e. organism differs from organism by having fewer and fewer common latent elements. We should see genera breaking up into species, species into local races, and ultimately races into stirps and possibly stirps into the merely ideal 'pure lines,' or organisms in the case of which it would be impossible to carry germ-plasm selection further for it would have become of one type only; the innumerable balls of immense variety in our bag would have been reduced to a single colour!

Darwin's natural selection acts only on evolution through the definite correlation of somatic and gametic characters. Galton's germinal selection, a random selection at the output of each new individual, must—if there be isolation—tend to produce species, races and sub-races. A pure race could only be one in which all latent elements were so substantially represented that there was little chance of a 'class representation' excluding any of them[1]. This is not the place to discuss at length the bearing of Galton's

[1] Purity of race might also be preserved by much intra-racial crossing.

germinal selection on the origin of species, or his 'class representation' on the origin of somatic variations. He did not press it himself to its legitimate conclusions, and probably did not see its full bearings on evolution. His general scheme from 'structureless elements' of parent to those of offspring is as follows:

| | | | | | | |
|---|---|---|---|---|---|---|
| Structureless Elements of Parent | through 'Class Representation' afford | Embryonic Elements | which by a development become | Adult Elements | which by a Second Selection contribute to | Structureless Elements of Offspring |
| | through 'Class Representation' afford | Latent Elements in Embryo | which by a development become | Latent Elements in Adult | which by a Second Selection contribute to | |

What I have termed a 'Second Selection' Galton terms 'Family Representation,' I think, on the ground that these selections produce the various somatic and gametic differences to be found in the members of the same family[1]. But it seems to me that it would be best simply to speak of first and second selections instead of 'class' and 'family' representations. Having put forward this scheme Galton now proceeds to express his grave doubts as to the adult elements' contributing anything or at least anything substantial to the 'structureless elements' of the offspring. He asserts that where the parents have a patent character that also exists in the latent form, i.e. in their gametic characters,

"I should demur, on precisely the same grounds, to objections based on the transmission of qualities to grandchildren being more frequent through children who possess those qualities than through children who do not; for I maintain that the personal manifestation is, on the average, though it need not be so in every case, a certain proof of the existence of some latent elements." (p. 399.)

In other words Galton is insisting on the somatic characters being only correlated with, or an index to, the gametic characters, and on the absence of complete association. He states that:

"the general and safe conclusion is, that the contribution from the patent elements [somatic characters of parent] is very much less than from the latent elements [gametic characters of parent]." (p. 399.)

And again:

"We see that parents are very indirectly and only partially related to their own children, and that there are two lines of connection between them, the one [adult latent elements] of large and the other [adult somatic elements] of small relative importance. The former is a collateral kinship and very distant, the parent being descended through two stages from a structureless source, and the child (as far as the parent is concerned) through five distinct stages from the same source; the other but unimportant line of connection is direct and connects the child with the parent through two stages." (p. 400.)

Galton even speaks of the 'structureless elements' that go to form the embryonic elements of the parents as going so far as heredity is concerned to "a nearly sterile destination."

Why did not Galton have the confidence at this time to say wholly sterile destination? I think there is not the least doubt that the *l'enfant*

[1] Of course Galton recognised the biparental contributions and in a second diagram shows the increased complexity.

*terrible* of Darwin still somewhat obscured his view[1]. Instead of 'Vive Pangenesis!' his cry ought to have been 'Pangenesis à la lanterne pour l'amour de la Science!' Galton could not swim absolutely against the current of gemmules flowing from the somatic organs to reinforce the germinal cells! He still thought that Darwin's insistence on the heredity of some acquired characters could not be the fabric of a dream[2]. He saw the light but authority was too great:

"We cannot now fail to be impressed with the fallacy of reckoning inheritance in the usual way, from parents to offspring, using those words in their popular sense of visible personalities. The span of the true hereditary link connects, as I have already insisted upon, not the parent with the offspring, but the primary elements of the two, such as they existed in the newly impregnated ova, whence they were respectively developed. No valid excuse can be offered for not attending to this fact, on the ground of our ignorance of the variety and proportionate values of the primary elements; we do not mend matters in the least, but we gratuitously add confusion to our ignorance, by dealing with hereditary facts on the plan of ordinary pedigrees—namely, from the *persons* of the parents to those of their offspring." (pp. 400–1.)

No Mendelian ever put more strongly than Galton thus did that somatic characters are no measure of gametic possibilities! Nay, Galton knew all about the fact that the second generation of hybrids shows more diversity than the first, but he did not call it, and perhaps rightly did not call it, 'segregation.'

"It is often remarked that the immediate offspring of different races resemble their parents equally, but that great diversities appear in the next and the succeeding generations.... A white parent necessarily contributes white elements to the structureless stage of his offspring and a black, black; but it does not in the least follow that the contributions from a true mulatto must be truly mulatto." (p. 402.)

Yet Galton—and after him the whole Biometric School—have been accused at random of asserting that all characters blend!

[1] The grave danger of Pangenesis was that it could, if by a very artificial mechanism, account for so much—rightly recognised or wrongly interpreted—phenomena; it therefore blocked the way to a simpler theory which, possibly truer to nature, could not account for the latter. Hence arose the controversy as to the inheritance of 'acquired characters' of later days, a slow process of getting rid of wrong interpretations. Lastly many phenomena which Darwin accounts for by the diffused gemmules of pangenesis can be equally well described by aggregated germinal units in the reproductive cells.

[2] Darwin even thought of the inheritance of insanity as that of an acquired character. Habits, mental instincts and even insanity modified the nerve-cells and were transmitted to the offspring by differentiated gemmules (*Animals and Plants...*, 1st Edn. Vol. II, p. 395). "No one who has attended to animals either in a state of nature or domestication will doubt that many special fears, tastes, etc., which must have been acquired at a remote period, are now strictly inherited": wrote Darwin in 1873 (*Nature*, Vol. VII, p. 281). While some instincts may have been developed by long ages of selection, "other instincts may have arisen suddenly in an individual and then been transmitted to its offspring independently both of selection and serviceable experience though subsequently strengthened by habit." Darwin then cites the case of Huggins' dog 'Kepler,' but it seems to me that there was far too little known of the ancestry of 'Kepler' in *all* lines to base any evidence for the inheritance of acquired characters in a certain family of dogs having an antipathy to butchers and their shops.

There is not the slightest doubt that 20 to 30 years hence we shall hear of nervous breakdowns attributed to 'shell-shocked' fathers of the Great War, and probably spoken of as instances of the inheritance of acquired characters. Investigation of the family history of cases of 'shell-shock' shows, however, that the bulk of these cases are associated with mentally anomalous stocks.

Galton concludes as follows, therein re-asserting the difference between somatic and gametic qualities, and at the same time the value of the statistical method:

"One result of this investigation is to show very clearly that large variation in individuals from their parents is not incompatible with the strict doctrine of heredity, but is a consequence of it wherever the breed is impure. I am desirous of applying these considerations to the intellectual and moral gifts of the human race, which is more mongrelised than that of any other domesticated animal. It has been thought by some that the fact of children frequently showing marked individual variation in ability from that of their parents is a proof that intellectual and moral gifts are not strictly transmitted by inheritance. My arguments lead to exactly the opposite result. I show that great individual variation is a necessity under present conditions; and I maintain that results derived from large averages are all that can be required, and all we could expect to obtain, to prove that intellectual and moral gifts are as strictly matters of inheritance as any purely physical qualities." (p. 402.)

It is curious that in the face of such a passage as this, there should still exist writers who have not grasped that the inheritance of the mental and moral qualities was a foundation stone of Galton's creed of life. His whole theory of inheritance was developed to account for supposed difficulties in this principle raised by his critics. And the principle itself—the equal inheritance of the psychical and physical characters—was the basis of his proposal to better the race of man by giving primary weight to his nature, and only secondary importance to his nurture. This paper of Galton's is now half-a-century old; I know of no earlier paper which pointed out so definitely the distinction between the somatic and gametic characters, which emphasised the continuity of the germ-plasm[1], which raised at the very least doubts as to the inheritance of acquired characters, which asserted that the personal or bodily characters of the offspring were not the product of those of the parents, and taught that the resemblance of father and son was really like that of brothers, for all were products of selected elements of a continuous germ-plasm. I feel that adequate credit has rarely been given by biologists to Francis Galton for these results, and there is no excuse for this neglect, for the paper in question was not published in an obscure journal, but in the proceedings of the foremost English learned society.

I can only hope that, however late in the day, this *Life* of Galton may aid in demonstrating the real parentage of certain now widely-current ideas.

We may now return to the rabbit correspondence.

9, ROYAL CRESCENT, MARINE PARADE, BRIGHTON. *August* 11/72.

MY DEAR DARWIN, The buck is quite well—the enclosed note just received explains everything. Now that Dr Carter has returned, he will see that all is rightly done. Will you kindly tell your servant to explain to the carrier? Very sincerely yours, FRANCIS GALTON.

[1] To show how opposed this was to Darwin's views I may cite the *Animals and Plants*..., 1st Edn. Vol. II, p. 383: "The reproductive organs do not actually create the sexual elements; they merely determine or permit the aggregation of the gemmules." "Use or disuse etc. which induced any modification in a structure should at the same time or previously act on the cells... and consequently would act on the gemmules" (p. 382). "Hence, speaking strictly, it is *not* the reproductive elements nor the buds, which generate new organisms, but the cells themselves throughout the body" (p. 374), i.e. by the production of gemmules which aggregate in buds or sexual elements.

42, RUTLAND GATE, *Nov.* 7/72.

MY DEAR DARWIN, Accept very best thanks for *Expression* which I have been devouring; you will, I am sure, receive numberless letters of hints corroborative of the points you make; even I could and will send some. But I write specially to say that if you care to send any more printed circulars of queries, I can dispose of three this very month most excellently for you. One by an expedition up the Congo, another by a man from the Zanzibar side into Africa and a third by a very intelligent German (English speaking) head of a missionary college on his way to my old country in Africa.

Would you have a short note sent me,—pray do not write yourself—about the rabbits.

Ever sincerely yours, FRANCIS GALTON.

P.S. You do not I think mention in *Expression* what I thought was universal among blubbering children (when not trying to see if harm or help was coming out of the corner of one eye) of pressing the knuckles against the eyeballs; thereby, reinforcing the orbicularis.

What a curious custom hand-shaking is and how rapidly savages take to it in their intercourse with Europeans.

I have a pamphlet of yours to send back.

DOWN, BECKENHAM, KENT. *Nov.* 8th. [1872?]

MY DEAR GALTON, I was going in a day or two to have written to you about the rabbits.

Those which you saw when here (the last lot) and which were then in the transition mottled condition have now all got their perfect coats, and *are perfectly true in character*. They are now ready to breed, or soon will be; do you want one more generation? If the next one is as true as all the others, it seems to me quite superfluous to go on trying.

Many thanks for your note and offer to send out the queries; but my career is so nearly closed, that I do not think it worth while. What little more I can do, shall be chiefly new work.

I ought to have thought of crying children rubbing their eyes with their knuckles; but I did not think of it, and cannot explain it. As far as my memory serves, they do not do so whilst roaring, in which case compression would be of use. I think it is at the close of a crying fit, as if they wished to stop their eyes crying, or probably to relieve the irritation from the salt tears. I wish I knew more about the knuckles and crying.

I am rejoiced that your sister is recovering so well: when you next see her pray give her my very kindest remembrances. My dear Galton, Yours very sincerely, CH. DARWIN.

What a tremendous stir-up your excellent article on 'Prayer' has made in England and America.

42, RUTLAND GATE, *Nov.* 15/72.

MY DEAR DARWIN, I have left your kind letter of ten days since unanswered, having some possible rabbit combinations in view which have ended in nothing. The experiments have, I quite agree, been carried on long enough. It would be a crowning point to them if your groom could get a prize at some show for those he has reared up so carefully, as it would attest their purity of breed. There is such a show, I believe, impending at the Crystal Palace. Enclosed is a £2 cheque. Will you kindly tip him with it for me, assuring him how indebted I feel for his attention. I don't know how I can repay *you*!

Would it not be worth while before abandoning the whole affair to get a litter from each of the available does, not with a view of keeping the young, but simply of seeing whether any are born mottled, and if not of then killing them? The reason being, that the mixed breed are so very apt to take wholly after one or the other ancestor, and one might get no other evidence of impure blood than a rare instance of a decidedly mongrel birth.

However I leave this quite in your hands, knowing that it means 5 or 6 weeks more trouble with the rabbits.

I read and re-read your *Expression* with infinite instruction and pleasure, and feel sure that its influence will soon be seen at the Royal Academy. Enclosed is a small addition to the note about the family on p. 34.

My sister Emma, I am rejoiced to say, is now at the seaside steadily mending in perfect quiet and in full hopes of complete restoration to health. I wish most heartily that yours was better. Ever sincerely yours, FRANCIS GALTON.

DOWN, BECKENHAM, KENT. *Dec.* 30 [1872].

MY DEAR GALTON, A young Mr Balfour, a friend of my son's, is staying here. He is very clever and full of zeal for [Biology]. He has been transplanting bits of skins between brown and white rats, in relation to Pangenesis! He wants to try for several successive generations the same experiment with rabbits. Hence he wants to know which colours breed truest. I have, of course, recommended silver greys. What other colour breeds true? Can you tell me? I think white or albinos had better be avoided. Do any grey breeds, of nearly the colour of the wild kind, breed true? Will you be so very kind as to let me hear? I much enjoyed my short glimpse of you in London. Ever yours, C. DARWIN.

DOWN, BECKENHAM, KENT. *Jan.* 4th [1873].

MY DEAR GALTON, Very many thanks for Fraser[1]: I have been greatly interested by your article. The idea of castes being spontaneously formed and leading to intermarriage is quite new to me, and I should suppose to others. I am not, however, so hopeful as you. Your proposed Soc[y] would have awfully laborious work, and I doubt whether you could ever get efficient workers. As it is, there is much of insanity and wickedness in families; and there would be more if there was a register. But the greatest difficulty, I think, would be in deciding who deserved to be on the register. How few are above mediocrity in health, strength, morals and intellect; and how difficult to judge on these latter heads. As far as I see within the same large superior family, only a few of the children would deserve to be on the register; and those would naturally stick to their own families, so that the superior children of distinct families would have a good chance of associating most and forming a caste. Though I see so much difficulty, the object seems a grand one; and you have pointed out the sole feasible, yet I fear utopian, plan of procedure in improving the human race. I should be inclined to trust more (and this is part of your plan) to experimenting and insisting on the importance of the all-important principle of Inheritance. I will make one or two minor criticisms. Is it not probable that the inhabitants of malarious countries owe their degraded and miserable appearance to the bad atmosphere, though this does not kill them; rather than to "economy of structure"? I do not see that an orthognathous face would cost more than a prognathous face; or a good morale than a bad one. That is a fine simile (p. 119) about the chip of a statue: but surely nature does not more carefully regard races than individuals, as [I believe I have misunderstood what you mean] evidenced by the multitude of races and species which have become extinct. Would it not be truer to say that nature cares only for the superior individuals and then makes her new and better races. But we ought both to shudder using so freely the word 'Nature' after what De Candolle has said.

Again let me thank you for the interest received in reading your essay.

Yours very sincerely, CH. DARWIN.

Many thanks about the rabbits: your letter has been sent to Balfour: he is a very clever young man, and I believe owes his cleverness to Salisbury blood.

This letter will not be worth your deciphering. I have almost finished Greg's *Enigmas*. It is grand poetry, but too utopian and too full of faith for me; so that I have been rather disappointed. What do you think about it? He must be a delightful man.

I doubt whether you have made clear how the families on the Register are to be kept pure and superior, and how they are in course of time to be still further improved.

I do not know whether Francis Balfour's experiments were ever pushed to their final conclusion, but if so, I have small doubt what that conclusion would be: A change of somatic character would not affect in a highly developed mammal the gametic characters, whatever arguments may be advanced from graft-hybrids[2]: Galton's own blood-transfusion experiments came to an end at this time. There are only two references that I have been able to find to the results of the second series, so much of which was

[1] This is the "Hereditary Improvement"; see our p. 117.

[2] *Animals and Plants under Domestication*, Vol. I, pp. 413—24, Vol. II, p. 360, 2nd Edn. 1875.

actually carried on at Down. The first is by Darwin himself in the footnote p. 350 (1875) of the 2nd edition of his *Animals and Plants...*:

"He [Mr Galton] informs me that subsequently to the publication of his paper he continued the experiments on a still larger scale for two more generations, without any sign of mongrelism showing itself in the very numerous offspring."

The second occurs in Galton's paper "A Theory of Heredity"[1] in a footnote, p. 342:

"I subsequently carried on the experiments with improved apparatus, and on an equally large scale, for two more generations."

Two slight footnote notices of what occupied much of Galton's time and energy for two or more years! But the result was really of value; it demonstrated that the blood was not a primary factor in heredity[2], and it weakened to an extent, perhaps hardly realised by Darwin, the probability of pangenesis. The misfortune was that Galton could not yet dismiss the whole mechanism of gemmules.

Differences, however, between the two men on this subject did not interfere for a moment with their warm friendship, and we next find Darwin giving Galton aid in two additional matters; the first is in answering his *questionnaire* concerning the nature and nurture of English men of science, and the second in growing sweet-peas—the inquiry which led to the conception of measuring correlation.

The answers which Galton received from his correspondents in the men of science inquiry are of extraordinary interest; they form brief auto-characterisations[3] by the leading scientific Victorians—Darwin, Hooker, Huxley, Spencer, Clerk Maxwell, Stokes and many others. The *questionnaire* was accompanied by a letter setting forth the scope of the inquiry. It runs

ANTECEDENTS OF SCIENTIFIC MEN.

42, Rutland Gate, London.

To Charles Darwin, Esq. In the pursuit of an inquiry parallel to that by M. de Candolle, I have been engaged for some time past in collecting information on the Antecedents of Eminent Men. My present object is to set forth the influences through which the dispositions of Original Workers in Science have most commonly been formed, and have afterwards been trained and confirmed. As a ready means of directing attention to the importance and interest of this inquiry, I append, overleaf, a reprint of a short review of the work of M. de Candolle, which I contributed to the 'Fortnightly Review' of March, 1873.

The result of my past efforts has clearly impressed upon me the fact that a sufficiency of data cannot be obtained from biographies without extreme labour, if at all; therefore, instead of imperfectly analysing the past, it seems far preferable to deal with contemporary instances, and none are more likely to appreciate the inquiry or to give correct information than Men of Science.

The number of persons in the United Kingdom who have filled positions of acknowledged rank in the scientific world is quite large enough for statistical treatment. Thus, the Medallists of the chief scientific societies; the Presidents of the same, now and in former years; those who have been elected to serve at various times on the Council of the Royal Society, and similarly,

[1] *Journal of the Anthropological Institute*, Vol. v, pp. 329–48, 1875.

[2] Even Darwin in his use of language was influenced by popular belief as the reader will find if he turns to the postscript of Darwin's letter of Jan. 4, 1873 on p. 176.

[3] Darwin's is reproduced at length in Francis Darwin's *Life and Letters of Charles Darwin*, Vol. iii, pp. 177–8.

the Presidents of the several sections of the British Association, form a body of little less than two hundred men, now living, a considerable portion of whom stand in more than one of the above categories. Other methods of selection give fifty or a hundred additional names.

Falling as you do within the range of this inquiry, may I ask of you the favour of furnishing me with information? If you should desire any portions of what you may send to be considered as private, they will be used in no other way than to afford material for general conclusions.

I send herewith a schedule which contains the questions to which I am seeking replies.

FRANCIS GALTON.

It would not I think be indiscreet to give in two notable instances the replies as to "special talents, as for mechanism, practical business habits, music, mathematics, etc.," also those on hereditary characteristics.

DARWIN.

Special talents, none, except for business, as evinced by keeping accounts, being regular in correspondence, and investing money very well; very methodical in my habits. Steadiness; great curiosity about facts, and their meaning; some love of the new and marvellous.

Somewhat nervous temperament, energy of body shown by much activity, and whilst I had health, power of resisting fatigue. An early riser in the morning. Energy of mind shown by vigorous and long-continued work on the same subject, as 20 years on the *Origin of Species* and 9 years on *Cirripedia*. Memory bad for dates or learning by rote; but good in retaining a general or vague recollection of many facts. Very studious, but not large acquirements. I think fairly independently, but I can give no instances. I gave up common religious belief almost independently from my own reflections. I suppose that I have shown originality in science, as I have made discoveries with regard to common objects. Liberal or radical in politics. Health good when young—bad for last 33 years.

*Father*. Practical business habits; made a large fortune and incurred no losses. Strong social affection and great sympathy with the pleasures of others; sceptical as to new things; curious as to facts; great foresight; not much public spirit; great generosity in giving away money and assistance. Freethinker in religious matters, great power of endurance.

*Mother*. Said to have been very agreeable in conversation.

HUXLEY.

Strong natural talent for mechanism, music and art in general, but all wasted and uncultivated. Believe I am reckoned a good chairman of a meeting. I always find that I acquire influence, generally more than I want, in bodies of men and that administrative and other work gravitates to my hands. Impulsive and apt to rush into all sorts of undertakings without counting cost or responsibility. Love my friends and hate my enemies cordially. Entire confidence in those whom I trust at all and much indifference towards the rest of the world. A profound religious tendency capable of fanaticism, but tempered by no less profound theological scepticism. No love of the marvellous as such, intense desire to know facts; no very intense love of my pursuits at present, but very strong affection for philosophical and social problems; strong constructive imagination; small foresight; no particular public spirit; disinterestedness arising from an entire want of care for the rewards and honours most men seek, vanity too big to be satisfied by them.

*Father*. A good musician and possessed a curious talent for drawing heads with pen and ink[1]. Impulsive but kindly; nothing otherwise remarkable.

*Mother*. Very impulsive and strong partizan; strong affections, marked religiosity and a constructive imagination worthy of a novelist. Physically and mentally I am far more like my mother than my father. Family generally, hot temper and tenacity of purpose; considerable power of expression in writing and speaking.

DOWN, BECKENHAM, KENT. *May* 28th, 1873.

MY DEAR GALTON, I have filled up the answers as well as I could; but it is simply impossible for me to estimate the degrees.

My mother died during my infancy and I can say hardly anything about her. It is so impossible for anyone to judge about his own character that George first wrote several of the answers about myself, but I have adopted only those which seem to me true.

[1] Inherited by his son: see *Life*, Vol. I, p. 4. The writer possesses a number of sketches by T. H. Huxley drawn on blotting paper and scraps of paper, probably at a committee meeting.

Now you may perhaps like to hear a few additional particulars about myself. I cannot remember the time when I had not a passion for collecting,—first seals, franks, then minerals, shells etc. As far as I am conscious, the one compulsory exercise during my school life which improved my intellect was doing Euclid, and this was *partly voluntary*.

At Edinburgh I do not think the lectures were of any service to me; but I profited as a naturalist by observing for myself marine animals.

At Cambridge getting up Paley's Evidences and Moral Phil. thoroughly well as I did, I felt was an admirable training, and everything else bosh.

My education really began on board the "Beagle."

I must add that my son Frank said he could safely give as my character, "sober, honest and industrious."

And now I want to ask you a question: if I had 50 men of 2 different nations, and for some reason could not measure all, if I picked out the 10 tallest of each nation, would their mean heights probably give an approximate mean between all 50 of each nation?

I hope you will get full answers to your queries, as I dare say the results will be interesting.

My dear Galton, Yours sincerely, CH. DARWIN.

42, RUTLAND GATE, S.W. *May* 30/73.

MY DEAR DARWIN, I am truly obliged by the Schedule. A few others are sent, many are promised and I have much hopes of useful statistical result in many ways. All I have thus far got confirms the belief that the families will be on the average very small. As for what the usual education will have been, I cannot yet guess.

In reply to your query about the 50, there seems—or it may be that I am stupid—that a word is omitted, displaced or somehow wrong, because the sense is not clear and I don't know how to interpret the meaning of the phrase "......would their mean heights probably give an approximate mean between all 50 of each nation," but the following will probably include what you want.

If nothing else could be assumed about the two nations than that the 10 tallest out of 50 taken at haphazard from $A$ had a mean height of $\alpha'$, and those from $B$ of $\beta'$, it would be impossible therefrom to deduce either:—

(1) $\alpha$ and $\beta$, the respective mean heights of the 50 $A$ and the 50 $B$ or

(2) the ratio of $\alpha$ to $\beta$.

But if you grouped the 10 tallest in either case according to their heights, that is, so many between 5′ 10″ and 5′ 11″, so many between 5′ 11″ and 6′ 0″ etc., it would be possible by comparing the run of these numbers with those of an ordinary Table of the Law of Error, to estimate approximately both (1) and (2).

10 is too small a number to be serviceable I should fear in this way;—100 ought to give excellent results; in any case the degree of regularity with which the numbers happened to run would be the measure of the probability of the accuracy of the results.

If you have any case you want worked out and would send me the figures I will gladly do it. Ever sincerely yours, FRANCIS GALTON.

For the year 1874 there are no letters. Darwin was ill in September 1873, Mrs Tertius Galton (Violetta Darwin) died in February 1874, Mrs Francis Galton was very ill in September and Galton himself at Christmas with "irregular gout and influenza." Darwin's eldest son George (later Professor Sir George Darwin) takes up the correspondence.

We return to-morrow to 42, RUTLAND GATE. *Nov.* 16/74.

MY DEAR GEORGE, Thank you kindly for your letter. My wife was alarmingly ill with a sudden vomiting of arterial blood, repeated during the night but fortunately never afterwards recurring. She was extremely weakened and unable to move out of bed for days, or out of the house where we were staying for weeks, but she has steadily mended and now 9 weeks have passed and she *is* almost and *looks* quite herself again[1]. We were staying with Judge Grove at the time, in a house he had taken in Dorsetshire for the shooting,—and his extreme kindness and that of all the family we can never forget.

I am rejoiced at the very good account you give of your health, and the good news of your Father. Somebody ought to make a fortune by "Drosera pills"—vegetable pepsin! The name would be capital. Poor Hooker,—what a frightful blow,—and a young family of girls wanting a mother.

We have been at Leamington for a fortnight and return home to-morrow. Previously we were at Bournemouth, when I renewed an acquaintance with H. Venn of Caius, who is great on "Chaucer." I wonder what your work now is. I saw your rejoinder in the *Quarterly* but not the original attack. I have alluded to your article on "Restrictions etc." in my book, which ought to be out soon. Ever yours, FRANCIS GALTON

GEORGE DARWIN, Esq.

42, RUTLAND GATE, S.W. *Jan.* 8/75.

MY DEAR GEORGE, Thanks for Lady R——'s letter, though her correspondent says little, and many thanks for your letter 3 or 4 days since.

That "curve of double curvature" was a sad slip for "curve of contrary flexure." The other point, I unluckily cannot answer, for I cannot get from the printer my copies of the paper and do not recall the passage or context. When we next meet I will tell you. Thank you much for the equation to the ogive.

Dr A. Clarke and nature have done me a world of good; my heart is set a going again and he quite withdraws a somewhat dispiriting diagnosis which he made when he first saw me. He told me of your diagram, on the facts of which I *most heartily* congratulate you.

On Thursday, Jan. 11th, there will be a Statistical Council when the papers will probably be arranged. If I get there, I will send a postcard to tell when your paper is to come in.

My twin papers come in and some are very interesting. J. Wilson of Rugby is a twin and sends me lots of addresses. I got a most curious letter from Lady E——, whose family abounds with twins, besides one treble and one quadruple birth. I feel saturated with midwifery and am haunted with imaginary odours of pap and caudle! You have real odours of pitch and tar.

Ever sincerely, FRANCIS GALTON.

GEORGE DARWIN, Esq.

42, RUTLAND GATE, LONDON. *April* 14/75.

MY DEAR DARWIN, George told me that you would very kindly have some sweet-peas planted for me, and save me the produce. I send them in a separate envelope with marked bags to put the produce in, and full instructions which I think your gardener will easily understand. I am most anxious to repair the disaster of last year by which I lost the produce of all my sweet-peas at Kew. With very many thanks, Yours very faithfully, FRANCIS GALTON.

*June* 2nd, 1875. (FONTAINEBLEAU, at present only.)

MY DEAR DARWIN, Thank you very much for your kind letter and information. It delights me that (notwithstanding the Frenchman's assertion) the large peas do really produce large plants, and that the extreme sizes sown (except *Q*) are coming up. I could not and did not hope for complete success in rearing all the seedlings, but have little doubt that the sizes that have failed may be supplemented by partial success elsewhere.

We have found Fontainebleau very pleasant and are now moving on via Neuchâtel, with some hope that George may, as he was inclined to do, hereafter fall in with us. He knows how to learn our address from time to time. My wife is already markedly better. With our united kindest remembrances to you all, Ever yours, FRANCIS GALTON.

It seems absurd to *congratulate* you on your election to the Vienna Academy, because you are a long way above such honours, but I am glad *they* have so strengthened their list by adding your name to it.

[1] The grave anxiety of a recurrence hung like a sword above the heads of the Galtons for many years. Mrs Galton's *Record* shows that from this time onward, till her death, she was more or less an invalid, in frequent pain, which limited largely her social activities.

42, RUTLAND GATE. *Sept.* 22/75.

MY DEAR DARWIN, In "Domestication," II, 253, you quote as a striking instance of *variation* a case communicated by Dr Ogle of 2 girl twins who had a crooked finger, no relative having the same. It happened, in my twin inquiries, that a case was sent me which is possibly or probably the same as your's—but which is a case of *reversion*. I send the particulars of this over leaf. You might think it worth while in the view of your 2nd Edition to ask Dr Ogle if his case *was* that of the Misses M——. I am not acquainted myself either with the Misses M—— or with Dr M——. Dr Gilchrist of the Crichton Institution, Dumfries, sent me Dr M——'s communication. We are only lately back in England and are not even yet settled in town. Will Frank kindly send me a line about the *sweet-peas*? With united kind remembrances to you all, Ever sincerely, FRANCIS GALTON.

I have been delighting in your "Insectivorous Plants."

Extract from a private letter to me, written by Dr F—— M——. (No address on this letter, but it is from Scotland and was enclosed by Dr Gilchrist of Dumfries.)

*The Misses M—— (twins Oct.* 16, *in* 1875)

"There is a congenital flexion at the second phalangeal joint of the little finger in each case, but the flexion is not so marked as to cause unsightliness or discomfort. I have ascertained that they inherited this peculiarity from their grandmother on the mother's side. The parents had no trace of it, nor any one of four brothers and three sisters!"

DOWN, BECKENHAM, KENT. *Sept.* 22nd, '75.

MY DEAR GALTON, I am particularly obliged for your letter, and will write to Dr Ogle. I think his case is different, and if you do *not* hear from me again, you will understand this to be the case.

I enclose a letter which when read kindly return to me.

With respect to the sweet-peas if you have time I think you had better come down and sleep here and see them. They are grown to a tremendous height and will be very difficult to separate. They ought to have been planted much further apart. They are covered with innumerable pods. The middle rows are now the tallest. Three of the plants are very sickly and one is dead. The row from the smallest peas are still the smallest plants. See what I say in "Var. under Dom." Vol. II, p. 347, about the peculiar properties of plants raised from the *small* terminal peas of the pods.

I am surprised and very much pleased at your liking my "Insectivorous Plants." I hope that your tour has done you much good. My dear Galton, Yours very sincerely, CH. DARWIN.

42, RUTLAND GATE. *Sept.* 24th, 1875.

MY DEAR DARWIN, We have stayed on in town another day so I have got from the Royal Society and send herewith Parts XIV and XV of the *Revue Scientifique* which contain the part of Claude Bernard's lectures which you wished to see. I have put pencil X at pages 324, 325, 327, 352 (in each case on the 2nd column of the page)[1]. These are the principal passages. Please send the pamphlets back when done with, to the Royal Society, *as returned by me*. Also I return the slips from *Nature* (Romanes) with many thanks.

Overleaf I send a note about the continuation of my Pangenesis experiments. I see I made a great mistake about the number of generations when we spoke yesterday. There were only 3 generations operated on, on *both* sides. I don't care to claim cases in which a great grand-*son* was matched with a grand-*daughter* as an additional generation. Besides, the cases were few.

Very sincerely yours, FRANCIS GALTON.

*Nov.* 2nd [1875]. DOWN, BECKENHAM, KENT.
RAILWAY STATION, ORPINGTON ON S.E.R.

MY DEAR GALTON, I hear from George that you are going to write on inheritance and therefore I think it worth telling you that Huxley does not at all believe in Balbiani's views and statements. He says he published some years ago some strange facts and then went right round and gave them all up. I send you Wedderburn's note and a pamphlet by him which will amuse you and which need not be returned. Yours very sincerely, CH. DARWIN.

[1] The pencil crosses may still be traced in the Royal Society copy of the *Revue Scientifique* 1874, witness to the fact that great men are not always great enough to obey library regulations!

42, RUTLAND GATE, S.W. *Nov.* 3/75.

MY DEAR DARWIN, It was truly kind of you, to write me with your own hand, a note of warning about Balbiani; but I do not use his statements in any way, in my forthcoming memoir which is to be read next Tuesday at the Anthropological Society.

The general line of it is this:

First I start with the 4 postulates, in favour of which you have so strongly argued, and which may reasonably be now taken for granted:—

1. Organic units in great number.
2. Germs of such units in still greater number and variety (existing *somewhere*).
3. That undeveloped germs do not perish; but multiply and are transmissible.
4. Organisation wholly depends on mutual affinities.

From these 4 postulates, I logically deduce several results, one of which is the importance and almost the necessity of double parentage in all complex organisations, and consequently of sex.

Then I argue that we must not look upon those germs that achieve development as the main sources of fertility; on the contrary, considering the far greater number of germs in the latent state, the influence of the former, i.e. of the personal structure, is relatively insignificant. Nay further, it is comparatively sterile, as the germ once fairly developed is passive; while that which remains latent continues to multiply. From this follows:—

(1) The extremely small transmissibility of acquired modifications (to which I recur).

(2) The fact that exceptional gifts are sometimes barely transmissible (here the sample was over rich and drained the more fecund residue).

(3) The fact of some diseases skipping one or more generations; (here the supposition is made of the germs of those diseases being peculiarly gregarious, hence the general outbreak of them leaves but a small residuum which has not strength to break out in the next generation, but being husbanded in a latent form, there multiplies and recovers strength to break out in the next or in a succeeding generation).

Next, I go into the question of affinities and repulsions, which I put as necessarily numerous and many-sided (while professing entire ignorance of their character) and I argue thence, a long period of restless unsettlement in the newly fertilised ovum, accompanied as we know it to be, with numerous segregations and segmentations in each of which the dominant germs achieve development, while the residue is segregated to form the sexual elements. But I argue, that as our experience of political and other segregations shows that they are never perfect, we are justified in expecting that numerous alien germs will be lodged in every structure and that specimens of all of them will be found in almost all parts of the body. In this way, I account for the reproduction of lost parts, etc., as well as for the inheritance of all peculiarities that had been congenital in an ancestor.

I then consider the cases of inheritance of what had been non-congenital in an ancestor, but acquired by him. I show that the deduction usually made, that the structure reacts on the sexual elements, is not justified by the evidence of adaptivity of race, *when* this depends on conditions which *act equally* on all parts of the body. My reason is, that since the same agents (viz. the germs) are concerned both in growth and in reproduction, the conditions that would modify the one, would simultaneously modify the other; hence they would be collaterally affected and the apparent inheritance is not a case of inheritance at all, in the strict sense of the word. Nay the progress may begin to vary under changed conditions *sooner* than the parent (as in the hair or fleece of the young of dogs and sheep, transported to the tropics).

As regards Brown-Séquard's guinea-pigs;—if I rightly understand and am informed of his experiment, it is open to fatal objection. The guinea-pigs that were operated on appear to have been kept separate from the rest. If so, we should expect the young sometimes to have convulsive attacks from mere imitation, just as we should expect of children brought up in a ward of epileptic patients, or among hysterical people (revivals, dancing mania etc.). Besides, there is not the least evidence that the mutilation of the spinal marrow, on which the parental epilepsy primarily depended, was inherited. I also disparage much other evidence of the inheritance of acquired modifications, leaving but a very small residue to accept. For this residue, I account by supposing the germs thrown off by the structure during its regular reparation, to frequently find their way into the circulation and some of these occasionally to reach the sexual elements and to become lodged and naturalised there, either by finding an unoccupied place or by dislodging others, like immigrants into an organised society, coming from a foreign country. Thus I account both for the fact, and for the great rarity and slowness of the inheritance of acquired modifications.

In conclusion, I restate a former definition, that I gave of the character of the relationship between parent and child, which I make out to be, not like that which connects a parent nation and its colonists, but like that which connects the *representative government* of the parent nation with the representative government of the colonists; with the further supposition, that the government of the parent country is empowered to nominate a small proportion of the colonists.

I have now, so far as the limits of a letter admit, made a clean breast of my audacity in theoretically differing from Pangenesis:—

(1) In supposing the sexual elements to be of as early an origin as any part of the body (it was the emphatic declarations of Balbiani on this point that chiefly attracted my interest) and that they are not formed by aggregation of germs, floating loose and freely circulating in the system, and

(2) In supposing the personal structure to be of very secondary importance in Heredity, being, as I take it, a *sample* of that which is of primary importance, but not the thing itself.

If I could help, even in accustoming people to the idea that the notion of Organic Germs is certainly that on which the true theory of Heredity must rest, and that the question now is upon details and not on first principles, I should be very happy. Ever yours, FRANCIS GALTON.

Thanks for the letter on the Hindoo family, which I will keep, and for the pamphlet on the wholesale execution of weakly people, which I return by book post.

*Nov.* 4th [1875]. DOWN, BECKENHAM, KENT.
RAILWAY STATION, ORPINGTON, S.E.R.

MY DEAR GALTON, I have just returned from London where I was forced to go yesterday for Vivisection Commission.

I have read your interesting note and am delighted that you stick up for germs. I can hardly form any opinion until I read your paper *in extenso*. I have modified parts of the Chapter on Pangenesis which is now printing, and have allowed that the gemmules may, or perhaps do, multiply in the reproductive organs. I write now as I fancy that you have not read B.-Séquard's last paper, in which he gives 17 or 13 (I forget which) instances of deficient toes on the *same* foot in the offspring of parents, which had gnawed off their own gangrenous toes owing to the sciatic nerve having been divided.

You speak of "almost the necessity of double parentage in all complex organisations." I suppose you have thought well on the many cases of parthenogenesis in Lepidoptera and Hymenoptera and surely these are complex enough.

I am very glad indeed of your work, though I cannot yet follow all your reasoning.
In haste, Most sincerely yours, C. DARWIN.

DOWN, BECKENHAM, KENT. [? *Nov* 4, 1875[1].]
RAILWAY STATION, ORPINGTON, S.E.R.

MY DEAR MR GALTON, My father thought you might care to have the reference to Brown-Séquard's paper. There is a good résumé of all his observations in the 'Lancet,' Jan. 1875, p. 7.
Yours very sincerely, FRANCIS DARWIN.

[1] The reader will note with amusement the *complete* omission of date—the inheritance in an intensified form of a habit peculiar not only to Charles Darwin but also to Mrs Darwin. I only know one letter to which Darwin did put a date, it is the following written to his aunt Violetta Galton, Francis Galton's mother.

*July* 12, 1871. DOWN, BECKENHAM, KENT.

MY DEAR AUNT, I am very much obliged to you for your great kindness in writing to me with your own hand. My sons were no doubt deceived, and the picture-seller affixed the name of a celebrated man to the picture for the sake of getting his price. Your note is a wonderful proof how well some few people in this world can write and express themselves at an advanced age. It is enough to make one not fear so much the advance of age, as I often do, though you must think me quite a youth! With my best thanks, pray believe me with much respect, Your affectionate nephew, CHARLES DARWIN.

This letter so gracefully suggestive of both Violetta Darwin and Charles Darwin deserves to be put on record.

42, RUTLAND GATE. *Nov.* 5/75.

MY DEAR DARWIN, Three proofs reached me from the *Contemporary Review* of my 'Theory of Heredity,' so I can spare one, and as I know you like to mark what you read, do not care to return it. I hope it will make my meaning more clear. The remarks printed as a note on p. 5, but which I ought to have put in the text, will meet what you wrote about the Hymenoptera.

I am most obliged for what you tell me about Brown-Séquard; I did not know of it, and will hunt up the passage to-day. (Thanks for the reference, received this morning.)

I should be truly grateful for criticisms which might enable me to modify or make clear before it is too late. Ever yours, FRANCIS GALTON.

What a nuisance this modern plan is, of sending proofs in sheet, and not in strip. One can't amend freely.

The paper which Galton sent Darwin is entitled 'A Theory of Heredity.' This memoir was in type for the *Contemporary Review* in November 1875[1], and was read before the Anthropological Institute in the same month. It was revised and printed in the *Journal of the Anthropological Institute* (Vol. v, pp. 329–48), and it is to this issue that we shall refer. The paper follows generally the lines of the 'Blood-relationship' of 1872, except that it still more definitely discards 'Pangenesis' and casts still further doubt on the heredity of acquired characters, and modification of offspring characters by the use or disuse of the same characters in the parent. The paper therefore marks a further stage in Galton's dissent from Darwin's theory and Darwin's views. Galton writes as follows:

"The facts for which a complete theory of heredity must account may conveniently be divided into two groups; the one refers to those inborn or congenital peculiarities that were also congenital in one or more ancestors, the other to those that were not congenital in the ancestors, but were acquired for the first time by one or more of them during their lifetime, owing to some change in their conditions of life.

The first of these two groups is of predominant importance, in respect to the number of well-ascertained facts that it contains, many of which it is possible to explain, in a broad and general way, by more than one theory based on the hypothesis of organic units. The second group includes much of which the evidence is questionable or difficult of verification, and which, as I shall endeavour to show, does not, for the most part, justify the conclusion commonly derived from it. In this memoir I divide the general theory of heredity into two parts, corresponding respectively to these two groups. The first stands by itself, the second is supplementary and subordinate to it." (pp. 329–30.)

After noting that Darwin, in the chapter on Pangenesis in the *Animals and Plants...*, had given the most elaborate epitome then extant of the many varieties of facts which a complete theory of heredity must account for, Galton states that his conclusions will differ essentially from Darwin's, and continues:

"Pangenesis appears more especially framed to account for the cases which fall in the second of the above-mentioned groups[2], which are of a less striking and assured character than those in the first group, and it will be seen that I accept the theory of Pangenesis with considerable modification, as a supplementary and subordinate part of a complete theory of heredity, but by no means for the primary and more important part." (p. 330.)

[1] It appeared in that *Review* in the following month. It was published also in the *Revue Scientifique*, T. x, pp. 198–205, 1876.

[2] Later on p. 347 Galton says that Pangenesis over-accounts for the facts of acquired modifications and reparations.

Galton next defines the word 'stirp' to

"express the sum total of the germs, gemmules or whatever they may be called, which are to be found, according to every theory of organic units, in the newly fertilised ovum—that is in the early pre-embryonic stage—from which time it receives nothing further from its parents, not even from its mother, than mere nutriment[1]......This word 'stirp,' which I shall venture to use, is equally applicable to the contents of buds, and will, I think, be found very convenient, and cannot apparently lead to misapprehension."

We now pass to the essential features of Galton's theory, which corresponds far more closely than Darwin's to modern ideas, indeed it is often difficult to say how much modern ideas have taken from Galton—without acknowledgment of the source.

The stirp is the organised aggregate of organic units, or germs. The personal structure develops by selection out of a small portion of these units, and the sexual elements of the new individual are generated by the residuum of the stirp. There is no free circulation of gemmules from the cells to be aggregated in the sexual organs. When the somatic elements are being formed from the stirp any segmentation may contain 'stray and alien gemmules,' and many of these may become entangled and find lodgment in the tissue. When these gemmules are lodged in great variety, the somatic cells are really reproductive cells and thus Galton would account for the replacement of a lost limb in the lower animals, or the reparation of simple tissues in the higher ones. The *selection* of organic units to form the somatic characters of the individual from the whole host in his stirp Galton looks upon as of the highest importance. He considers that a sort of struggle for place goes on among the innumerable germs of the stirp, and those germs which are most frequent or have certain intrinsic qualities[2] will be most successful. He considers that this continual selection leads ultimately in unisexual reproduction to the elimination of necessary units and so to degeneration; sex, he argues, is not primary, but a result of the advantage of a more primary double parentage, which lessens the chance of one or more of the needful species of germs in the stirp disappearing by selection[3]. Galton even goes so far as to suggest that where an excess of germs has been withdrawn from the stirp to form a marked character, for example, great ability or even a pathological state, there will be an absence of these germs in the residue, which goes to form the new sexual element, and he accordingly accounts in this way for the offspring of a man of genius having small ability, or again

[1] Galton (p. 341) very aptly remarks that if pangenetic gemmules circulated freely through the system, there can be little doubt that they would reach the body of an unborn child. Thus the paternal gemmules in that body would be dominated by an invasion of maternal gemmules with the final result that an individual would transmit maternal peculiarities far more than paternal ones; "in other words people would resemble their maternal grandmothers very much more than other grandparents, which is not at all the case."

[2] The "dominant germs" are "those that achieve development." (p. 341.)

[3] "There is yet another advantage in double parentage, namely that as the stirp whence the child sprang is only half the size of the combined stirps of his two parents, it follows that one-half of his possible heritage must have been suppressed. This implies a sharp struggle for place among the competing germs, and the success, as we may infer, of the fitter half of their numerous varieties." (p. 334.)

for diseases skipping a generation. This selecting out of germs will not occur in animals of pure breeds for their stirp contains only one or a very few varieties of each species of germ, so that the selection will contain all, and thus the offspring resemble their parents and one another.

"The more mongrel the breed, the greater is the variety of the offspring." (p. 336.)

To this principle, however, Galton adds a limitation, the stirp cannot be indefinitely increased in complexity, because there is a limit to the space it occupies. There is a finite, if great, number of varieties of germs, and of the individual germs in each variety.

"Thus in the gradual breeding out of negro blood, we may find the colour of a mulatto to be the half, and that of a quadroon to be the quarter of that of his black ancestors; but as we proceed further, the sub-division becomes very irregular; it does not continue indefinitely in the geometrical series of one-eighth, one-sixteenth, and so on, but is usually present very obviously or not at all, until it entirely disappears." (p. 335.)

Turning now to the germ which has developed into a somatic cell, Galton questions whether it does produce gemmules at all—at any rate its fertility is far less than that of the latent germ. Influences acting on the somatic cells of the parent are only slightly or not at all represented in the like somatic cells of the offspring. He considers at some length instances of inherited mutilations and of acquired characters, and thinks they may be reasonably looked upon as a 'collection of coincidences.' Even if there are real cases of changes in the somatic cells of the parents influencing the somatic characters of the offspring, Galton would but admit that occasionally gemmules are thrown off by somatic cells, which find their way into the circulation and ultimately obtain a lodgment in the already constituted sexual elements. Such a process is, however, independent of and subordinate to the causes which mainly govern heredity (pp. 347–88). Even to the last Galton did not wholly give up Pangenesis, for Darwin had accepted Brown-Séquard's epileptic guinea-pigs, yet as Galton remarked:

"It is indeed hard to find evidence of the power of the personal structure to react upon the sexual elements that is not open to serious objection." (p. 345.)

Finally I may cite:

"The hypothesis of organic units enables us to specify with much clearness the curiously circuitous relation which connects the offspring with its parents. The idea of its being one of direct descent, in the common acceptation of that vague phrase, is wholly untenable, and is the chief cause why most persons seem perplexed at the appearance of capriciousness in hereditary transmission. The stirp of the child may be considered to have descended directly from a part of the stirps of each of its parents, but then the personal structure of the child is no more than an imperfect representation of his own stirp, and the personal structure of each of the parents is no more than an imperfect representation of each of their own stirps." (p. 346[1].)

Such a modern idea as that parents are only conduit-pipes for the germplasm of their stocks is fully expressed by Galton with better limitation, and with fuller suggestiveness, both in this paper and in that on Blood-

[1] From the modern biometric standpoint the association is 'correlational' not causal.

relationship, than in much current literature. It is only the terminology and the fact that Galton was not a professional biologist which have deprived him of the credit due to him as the discoverer or inventor of what we now term the 'continuity of the germ-plasm.' Might not that theory, Galton modestly suggests, be substituted with advantage for that of pangenesis?

DOWN, *Nov.* 7th [1875].

MY DEAR GALTON, I have read your essay with much curiosity and interest, but you probably have no idea how excessively difficult it is to understand. I cannot fully grasp, only here and there conjecture, what are the points on which we differ—I daresay this is chiefly due to muddle-headiness on my part, but I do not think wholly so. Your many terms, not defined "developed germs"—"fertile" and "sterile" germs (the word 'germ' itself from association misleading to me), "stirp,"—"sept," "residue" etc. etc., quite confounded me. If I ask myself how you derive and where you place the innumerable gemmules contained within the spermatozoa formed by a male animal during its whole life I cannot answer myself. Unless you can make several parts clearer, I believe (though I hope I am altogether wrong) that very few will endeavour or succeed in fathoming your meaning. I have marked a few passages with numbers, and here make a few remarks and express my opinion, as you desire it, not that I suppose it will be of any use to you.

(1) If this implies that many parts are not modified by use and disuse during the life of the individual, I differ from you, as every year I come to attribute more and more to such agency.
(2) This seems rather bold, as sexuality has not been detected in some of the lowest forms, though I daresay it may hereafter be.
(3) If gemmules (to use your own term) were often deficient in buds I could but think the bud-variations would be commoner than they are in a state of nature; nor does it seem that bud-variations often exhibit deficiencies which might be accounted for by absence of the proper gemmules. I take a very different view of the meaning or cause of sexuality.
(4) I have ordered Fraser's Mag. and am curious to learn how twins from a single ovum are distinguished from twins from 2 ova. Nothing seems to me more curious than the similarity and dis-similarity of twins.
(5) Awfully difficult to understand.
(6) I have given almost the same notion.
(7) I hope that all this will be altered. I have received new and additional cases, so that I have now not a shadow of doubt.
(8) Such cases can hardly be spoken of as very rare, as you would say if you had received half the number of cases which I have.

I am very sorry to differ so much from you but I have thought that you would desire my open opinion. Frank is away; otherwise he should have copied my scrawl.

I have got a good stock of pods of Sweet Peas, but the autumn has been frightfully bad; perhaps we may still get a few more to ripen.

My dear Galton, Yours very sincerely, CH. DARWIN.

A. R. Wallace took a different view as to what Galton had achieved in a letter of the following spring.

THE DELL, GRAYS, ESSEX. *March* 3rd, 1876.

DEAR MR GALTON, I return your paper signed. It is an excellent proposal. I must take the opportunity of mentioning how immensely I was pleased and interested with your last papers in the *Anthrop. Journal*. Your 'Theory of Heredity' seems to me most ingenious and a decided improvement on Darwin's, as it gets over some of the great difficulties of the cumbrousness of his Pangenesis. Your paper on Twins is also wondrously suggestive.

Believe me, Yours very faithfully, ALFRED R. WALLACE.

F. GALTON, Esq.

42, RUTLAND GATE. *Nov.* 8/75.

MY DEAR DARWIN, Alas! Alas!—and I had taken such pains to express myself clearly, and I see what I mean, so clearly!

I was most obliged for the Brown-Séquard reference in the *Lancet*, and will certainly alter the paragraph. His non-publication of the papers, even in abstract, read by him at the British Association in 1870, had given me additional fear that there was something wrong.

All the other points you refer to in your letter, I will do what I can about: i.e., make clearer, answer, or amend; but it is too late to make more than small alterations in the proof.

Thank you for reference and offer to send Panum, but I have a description of his results, so far as I want them, in C. Dareste (*Ann. Sc. Naturelles* [*Zoologie*, T. XVII], 1862, 'Sur les œufs à double germe,' p. 34).

In my 'Fraser' article there is a most unlucky and absurd collocation of words, which I heartily hope no critic will seize upon, for which I simply can't account except in the supposition of badly scratching out in the MS., and variously altering some passage. It is about 'double yolked eggs' and 'simple germs'. I ought never to have passed it in proof; but there it is.

The twins born in one chorion,—never mind whether 2 amnions or not,—is Kleinwächter's dictum which he fortifies by numerous modern German authorities; Kiwisch being the only one who, it appears, still talks of fusion of membranes. I also noted the remark in the Catalogue of the Museum Coll. Surgeons "Teratology" that twins in one chorion are *probably* (I think that was the word) derived from 2 germinal spots on one ovum.

If you care to see Kleinwächter, I could send it you.

Very sincerely yours, FRANCIS GALTON.

42, RUTLAND GATE, S.W. *Nov.* 10/75.

MY DEAR GEORGE, I got my back Statistical Society publications last night and have read your cousin-paper with very great interest[1]. You certainly have exploded most effectually a popular scare. Would it be profitable to make any "probable error" sort of estimate of your results, which should eventuate in some such form as this: "The injurious effects of first-cousin marriages, measured in such and such ways, cannot exceed so and so, and probably do not exceed so and so"?

You ought to found a fortune upon your discovery,—Thus: there are, say, 200,000 annual marriages in the kingdom, of which 2,000 and more are between first cousins. You have only to print in proportion, and in various appropriate scales of cheapness or luxury:

"WORDS of Scientific COMFORT
and ENCOURAGEMENT
To COUSINS who are LOVERS"

then each lover and each of the two sets of parents would be sure to buy a copy; i.e. an annual sale of 8,000 copies!! (Cousins who fall in love and don't marry would also buy copies, as well as those who think that they *might* fall in love.)

I read my "Theory of Heredity" at the Anthropological last night, when up got a mad spiritualist who orated, and then offered to address the meeting on the subject as a medium; the spirit speaking through his lips. (This was not accepted.)

Ever sincerely yours, FRANCIS GALTON.

GEORGE DARWIN, Esq.

*Nov.* 10th. Night. [1875]

DOWN, BECKENHAM, KENT.
RAILWAY STATION. ORPINGTON. S.E.R.

MY DEAR GALTON, I have this minute finished your article in Fraser and I do not think I have read anything more curious in my life. It is enough to make one a Fatalist, I am in a passion with the *Spectator* who always muddles if it is possible to muddle. But after all he does not write so odiously as I did in my letter, which you received so beautifully. I should be glad

[1] *Journal of the Royal Statistical Society*, Vol. XXXVIII, pp. 153–82. I may perhaps be permitted to add the word of warning that the danger of cousin marriage is *not* a popular scare. Any patent or latent defect is certain to be emphasised by cousin marriage as of course any good characteristic.

to be convinced that the obscurity was *all* in my head, but I cannot think so, for a clear-headed (clearer than I am) member of my family read the article and was as much puzzled as I was. To this minute I cannot define what are "developed," "sterile" and "fertile" germs. You are a real Christian if you do not hate me for ever and ever.

I shall try you when we come to London in a month or six weeks time, as I want to ask a question about averages, which can be asked in a minute or two, but would fill a long letter.

Yours very sincerely, CH. DARWIN.

P.S. As soon as I am sure that no more pods of Sweet Peas will ripen, I will send all the bags in a box per Railway to you.

42, RUTLAND GATE, S.W. *Nov.* 26/75.

MY DEAR DARWIN, How can I thank you sufficiently for the trouble you have taken with the peas, which arrived last night in beautiful order. You must let me know, when we next meet, if there is anything I owe you for payments of any kind connected with them; Will you, in the meantime, give the enclosed 10/- (I send an order made out in *your* name) to the gardener from me? and tell him that I am much obliged for his care.

Ever yours, FRANCIS GALTON.

Romanes has told me much of his wonderfully interesting results with the Medusae.

*Dec.* 18th [1875] (Home on Monday).

MY DEAR GALTON, George has been explaining our differences. I have admitted in new Edit. (before seeing your essay) that perhaps the gemmules are largely multiplied in the reproductive organs; but this does not make me doubt that each unit of the whole system also sends forth its gemmules. You will no doubt have thought of the following objection to your view, and I should like to hear what your answer is. If 2 plants are crossed, it often or rather generally happens that every part of stem, leaf—even to the hairs—and flowers of the hybrid are intermediate in character; and this hybrid will produce by buds millions on millions of other buds all exactly reproducing the intermediate character. I cannot doubt that every unit of the hybrid is hybridised and sends forth hybridised gemmules. Here we have nothing to do with the reproductive organs. There can hardly be a doubt, from what we know, that the same thing would occur with all those animals which are capable of budding and some of those (as the compound Ascidians) are sufficiently complex and highly organised.

Yours very sincerely, CH. DARWIN.

42, RUTLAND GATE. *Dec.* 19/75.

MY DEAR DARWIN, The explanation of what you propose does not seem to me in any way different on my theory, to what it would be in any theory of organic units. It would be this:

Let us deal with a single quality, for clearness of explanation, and suppose that in some particular plant or animal and in some particular structure, the hybrid between white and black forms was exactly intermediate, viz: grey—thenceforward for ever. Then a bit of the tinted structure under the microscope would have a form which might be drawn as in a diagram, as follows:—

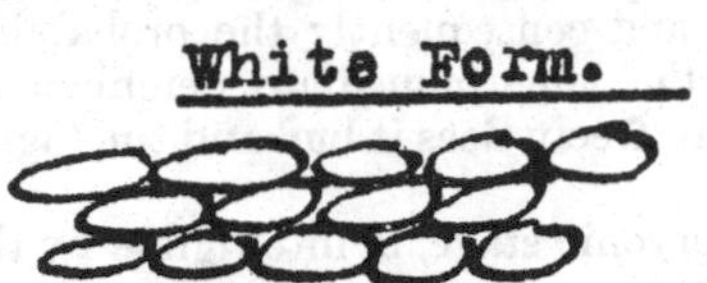


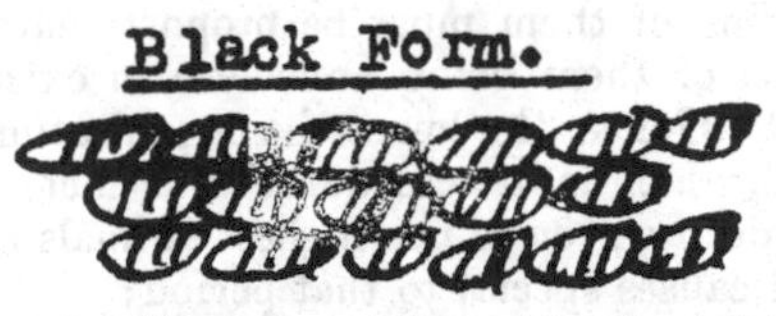


whereas in the hybrid it would be either that some cells were white and others black, and nearly the same proportion of each, as in (1) giving *on the whole* when less highly magnified a

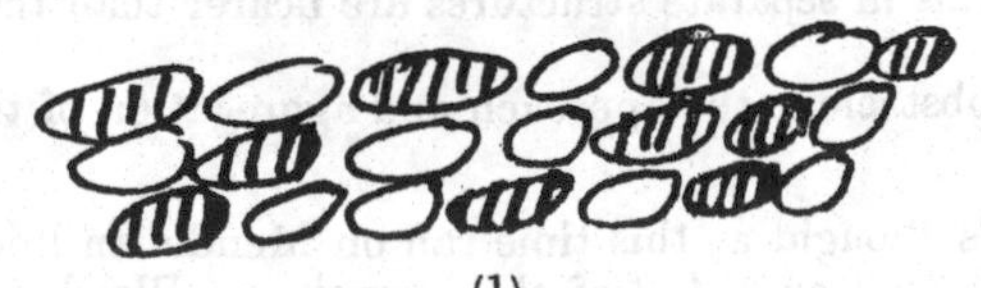
(1)

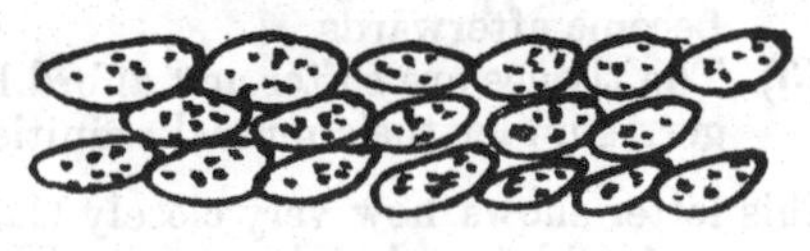
(2)

uniform grey tint,—or else as in (2) in which *each cell* had a uniform grey tint.

In (1) we see that each cell had been an organic unit (quoad colour). In other words, the structural unit is identical with the organic unit.

In (2) the structural unit would not be an organic unit but it would be an organic *molecule*. It would have been due to the development, not of one gemmule but of a group of gemmules, in which the black and white species would, on statistical grounds, be equally numerous (as by the hypothesis, they were equipotent).

The larger the number of gemmules in each organic molecule, the more *uniform* will the tint of greyish be in the different units of structure. It has been an old idea of mine, not yet discarded and not yet worked out, that the number of units in each molecule may admit of being discovered by noting the relative number of cases of each grade of deviation from the mean greyness. If there were 2 gemmules only, each of which might be either white or black, then in a large number of cases one-quarter would always be quite white, one-quarter quite black, and one-half would be grey. If there were 3 molecules, we should have 4 grades of colour (1 quite white, 3 light grey, 3 dark grey, 1 quite black and so on according to the successive lines of "Pascal's triangle"). This way of looking at the matter would perhaps show (*a*) whether the number in each given species of molecule was constant, and (*b*), if so, what those numbers were[1]. Ever very faithfully yours, FRANCIS GALTON.

42, RUTLAND GATE, *Dec.* 22/75.

MY DEAR GEORGE, I have never supposed otherwise than that the gemmules *breed abundantly* all over the body, though I look upon them merely as *local parasites*, so to speak, that live, multiply and die in great multitudes in the places where they are lodged, though occasionally some of them may be detached and drifted along with the circulation, and so find their way to the sexual elements—as was explained in the second part of the paper.

It is by the *abundance* of all sorts of them, in every part of the body, that I accounted in my paper for the reproduction of mutilated parts, and other specified phenomena, adding: "It would much transcend my limits if I were to enter into these and kindred questions, but it is not necessary to do so, for it is sufficient to refer to Mr Darwin's work, where they are most fully and carefully discussed, and to consider while reading it whether the theory I have proposed could not, as I think it might be, substituted with advantage for Pangenesis."

[I have not the *Contemporary Review* by me and cannot give the page of the extract. My copy is merely a revise, paged from 1 onwards. It is in the 12th page of the revise.]

In this passage, I meant to include propagation by buds. You will see in the preceding page an allusion to the way in which the scattered alien germs "thrive and multiply."

Now for the application of all this: wherever in a plant developed out of a bud or seedling, (no matter which, for the 'stirp' is similar in both cases) the alien, localised germs happen to be congregated in sufficient number and varieties to form material for a fresh stirp, there will be a tendency to produce a bud. Structural conditions, such as those found at the parts where buds usually shoot, must of course be helpful in forwarding this tendency.

The advantage of my theory appears to be this:—

By Pangenesis, we should expect *all* animals, however highly organised, to throw out buds.

By my theory, I argue that where the animals are complex, the variety of germs concerned in the making of them must be proportionately great, and consequently the probability of a complete set of them being anywhere in existence, in the same immediate neighbourhood, is diminished. Hence, the lower the organisation, the more freely does it bud and the higher ones do not bud, which is in accordance with fact.

The budding, even of the highest animals in the embryonic stage, is intelligible by the joint action of 3 causes special to that period:

(1) The differentiation is less complete, and germs destined to be separated are then together.

(2) The embryo being small, the alien germs in separate structures are nearer than they become afterwards.

(3) The tissues are softer and afford less obstacle to the approach and aggregation of the germs under their mutual affinities.

[1] This letter shows how very closely Galton's thought at this time ran on Mendelian lines. The passage should be taken in conjunction with that on p. 402 of the memoir on Blood-relationship. See our pp. 170–4 and compare p. 84.

I hope I have answered fully enough, and much regret that I misunderstood the question, as put in your Father's letter, and have given you both unnecessary trouble. I am eager to receive criticisms—even adverse ones. Ever yours, FRANCIS GALTON.

About your Father's plants and the statistics of growth:—In cases where not only the *one* biggest of each sort, but the two or three biggest were measured, the uncertainty of the relative values of the moduli of variability of the two sorts would be materially diminished.

42 RUTLAND GATE, LONDON. *Jan.* 30th, 1876.

MY DEAR GEORGE, I was very glad to hear good news of you from Litchfield, who dined with us a few days back; (but not with your sister, I am sorry to say, as she was not then well). Strachey was nearly going yesterday to look after your map frame, possibly he did after all (he asked me to join him but I was engaged). He thought of taking it bodily away. Never did a thing hang so long in hand as this, but I am powerless to help. I can't understand it, as Strachey is so energetic in much that he undertakes and does it so well.

I got a letter from Glaisher a short time back about my "exponential ogive" whereof he much approves, name and all, and he gives me a compact expression for it, in terms of his "error function." I enclose a copy of part of what he says. In working out your Father's plant statistics, it occurred to me that it would be uncommonly convenient to calculate an exponential ogive table, which I did, and since receiving Glaisher's letter I sent it to him to see if he could get it properly recalculated for me directly from his formula. You see,—by knowing *any* two ordinates, you know the whole curve and can at once get the value of any other ordinates in it. I need not bother you with particulars about the table, further than that it gives ordinates from 1 to 50 in an ogive of 100 places, from 1 to 50 in an ogive of 1,000 places, ditto 10,000, 100,000 and a million. So that all goes into a page.

But I could not make out anything by its means about those data concerning your Father's self- and crop-fertilised plants in which only the biggest were measured. Their "run" was too irregular. I could get no two trustworthy ordinates. The ignorance of the number of plants in the row did not so much matter, because one knew it within limits and could find what the result would be for those limits; between which the real result must lie, and these were not extravagantly far apart.

We have had astonishing fogs in this part of London, that is going up from here to Hyde Park corner. I never saw one thicker than yesterday. Your friend Cookson, whom I met walking this morning, told me that in one place he could only see three flagstones off. I suppose you have glorious sunshine in Malta.

Tyndall's lecture about Bacteria was a great success and seems to have utterly smashed the adherents of Bastian. I conclude from the theory that the physiological reason of immortality in the next life is that there are no Bacteria in the pure air of heaven!—nothing to cause corruption.

I send reprints of my twins and theory of heredity (revised); one of the twin papers is new and so is the last paragraph about the *cuckoo* in the one that was in Fraser, which if you care to look at may interest you. Romanes' paper has been selected as the Croonian Lecture of the Royal Society for the year; a well-deserved honour. There seems to be an epidemic in the learned societies. Not only the Linnean, but now the Anthropological has got into such a state, and the respectable Athenaeum is all in a boggle about its future trustee to replace Lord Stanhope. Pray remember me very kindly to your brother and with my wife's best regards. Ever yours, FRANCIS GALTON.

GEORGE DARWIN, Esq.

2, BRYANSTON ST. [1877]

MY DEAR G[ALTON]. I have just bethought me, that I received a French essay a few months ago on the effects of the conscription on the height of the men of France and on their liability to various diseases which rendered them unfit for the army, due to the weaker men left at home propagating the race. He shows, I think rightly, that no one hitherto had considered the problem in the proper light. I forget author's name,—and where published.

Do you know this essay? and would you care to see it. I suppose that I could find it, but I think I have not yet catalogued it. It seemed to me a striking essay.

Ever yours sincerely, CH. DARWIN.

42 RUTLAND GATE, *Jan.* 12/77.

MY DEAR DARWIN, Thanks very many: When you come across the essay I should be very glad to see it. I know of a curious *Swiss* memoir, something apparently to the same effect, in which the author says that the Swiss yeomen are very apt to leave their homestead to a sickly son, knowing that he will not be called out on service, nor tempted to take service abroad in any form, but will stay at home and look after the property. Consequently the Swiss landed population tend to deteriorate.

I will try hard to put in practice your valuable hints about making my lecture as little unintelligible and dull as may be and have hopes of succeeding somewhat. George has *most kindly* taken infinite pains to the same end. Ever sincerely yours, FRANCIS GALTON.

CHARLES DARWIN, Esq.

DOWN, *Feb.* 11th. [1877?]

MY DEAR G. The enclosed is worth your looking at. It was sent me from N. Zealand as the writer thought we should not in England see Tickner's Life! I should think T. was to be trusted, and if so case very curious. It makes me believe statement about inherited handwriting. I shall never work on inheritance again. The extract need not be returned.

Ever yours, C. DARWIN.

I do hope Mrs Galton is pretty well again.

42, RUTLAND GATE, *Feb.* 22/77.

MY DEAR DARWIN, By this book post I return Tickner's book with many thanks (after keeping it an unconscionable time, but I knew you did not want it and it was useful to refer to to me).

About the deaf and dumb men speaking with Castilian etc. accent, according to their teachers, I cannot help thinking it sufficiently explained by their imitation of the actions of the lips etc. of the teachers. I have tried in a looking-glass, and it seems that I mouth quite differently when I speak broad Scotch; again, last year, I was trying some experiments with Barlow's 'logograph' and the traces were greatly modified under different conditions of cadence.

Let me, before ending, heartily congratulate you on the German and Dutch testimonial of which I see a notice in to-day's *Times*, and take the opportunity of wishing you many, very many happy returns of the birthday. Ever sincerely yours, FRANCIS GALTON.

My wife is convalescent and already walks out a little.

42, RUTLAND GATE, *May* 24/78.

MY DEAR DARWIN, The enclosed "Composite Portraits" will perhaps interest you. The description of them is in this week's *Nature* (p. 97). You will see that I have there published the letter you kindly forwarded to me from Mr Austin of New Zealand (to whom I am now about to write a second time). Together with the villain's (absit omen!) I send 3 of our own family ancestors which I have had made, and for which you may care to find some place somewhere. The original portraits are in the possession of Reginald Darwin and are those of our uncle Sir Francis Darwin and of our great-grandfather and of our great-great-grandfather respectively[1] (as you will find written on their backs). These take the Darwin family back for 2½ centuries. There seems to be a great deal of the Darwin type in William Darwin b. 1655.

I hear vague rumours of your wonderful investigations in the growth etc. of plants, and am eager for the time when they shall be published. Ever sincerely yours, FRANCIS GALTON.

DOWN, BECKENHAM, KENT. *March* 22/79.

MY DEAR GALTON, Dr Krause has published in Germany a little life of Dr Eras. Darwin, chiefly in relation to his scientific views; and to do our grandfather honour, my brother Eras. and myself intend to have it published in English. I intend to write a short preface to it, chiefly for the sake of contradicting the chief of Miss Seward's calumnies; and this I can do from having a letter from your aunts written at the time, and from my father's correspondence with Miss Seward. But I further intend to add a few remarks about our grandfather. Can you aid me with any information or documents?

[1] See Plates VI and LXII in Vol. I and remarks on p. 243.

PLATE XV

(i)

Dr Erasmus Darwin,
"Physician, Philosopher and Poet."

(ii)

Sir Francis Darwin,
"Physician, Traveller and Naturalist."

To illustrate the influence of the Collyer blood in modifying the Darwin strain. From miniatures in the possession of Mr Darwin Wilmot.

Sir Francis Sacheverell Darwin, "Physician, Traveller and Naturalist," son of Elizabeth Collier (Mrs Chandos-Pole) and godfather of Francis Galton. From a drawing in the possession of Mr Darwin Wilmot.

I have one nice and curious letter to Miss Howard which I will publish. Also many letters to Josiah Wedgwood and to the famous Reimarus, but I doubt whether any of these will be worth publishing. Do you know whether there are any letters in the possession of any members of the family which might be worth publishing; and could you take the trouble to assist me by getting the loan or copies of them?

Several years ago I read the memoirs of your Aunt Mrs Schimmelpenninck and so far as I can remember many of the stories about Dr Darwin seemed very improbable. Did you ever hear your mother speak of this book, and can you authorise me to contradict any which are injurious to his good name? I am sure you will forgive me for troubling you on this head as we have a common interest in our grandfather's fame. Yours very sincerely, CHARLES DARWIN.

Saturday, 6, QUEEN ANNE STREET.

MY DEAR GALTON. If it would not bore you, can you come to luncheon here on Monday at 1 o'clock; as it will be my best chance of seeing you. I have been extremely sorry to hear that you have not been well of late and that you are soon going abroad.

Yours very sincerely, CH. DARWIN.

*April* 30 [1879]. DOWN, BECKENHAM.

Many thanks. The extract will come in capitally. You are vy. good to take so much trouble. Mrs Sch.[1] received all safe, and shall soon be returned. I much enjoyed my talks with you. C. D.

The following letter probably has reference to Elizabeth Collier's birth[2], and may possibly aid in the final solution of the difficulty as to her origin.

DOWN, BECKENHAM, KENT. *June* 8 [1879].

MY DEAR GALTON, Many thanks for your note. I have lately been staying with my sister, Caroline, and she says my memory is in error about the mysterious visitor. She believes his name was Brand, and that it was in the time of Colonel Pole; I cannot but doubt about the latter point. My sister feels pretty positive that the gentleman stayed at the house of a neighbour (name forgotten) and never visited Mrs Pole or Mrs Darwin, but sent her respectful and very friendly messages. Nevertheless she was never at ease till he had left the country. Thanks for all your help. I have fixed our photograph of Dr D. Ever yours, C. DARWIN.

P.S. If you should come across Dr Lauder Brunton, see if he has anything more to communicate about Dr D. for I shall soon go to press.

42, RUTLAND GATE, *Nov.* 12/79.

MY DEAR DARWIN, It was with the greatest pleasure that I received and read your biography of Dr Darwin.

What a marvel of condensation it is, and how firmly you lay hold of facts that had long been distorted and ram them home into their right places.

The biography seems to me quite a new order of writing, so scientifically accurate in its treatment. The many passages you quote are curiously modern in their conception and—(Excuse this horrid paper which folds the wrong way) simple in expression (considering his average style). I still can't quite appreciate the flaw in his mind which made it possible for him to write so very hypothetically for the most part, while at the same time his strictly scientific gifts were of so high an order. There seems to be an unexplained residuum, even after what you quote from him, about the value of hypotheses. I see you have mentioned me twice, very kindly—but too flatteringly for my deserts. How you are *down* upon Mrs Schimmelpenninck and Miss Seward[3]!

[1] Mrs Schimmelpenninck, Galton's aunt: see Vol. I, p. 54.

[2] See Vol. I, p. 21.

[3] I think Galton had a truer appreciation of Erasmus Darwin than possibly his cousin had,—a better historical perspective,—and with all their faults of exaggeration the ladies in question did give something of the 'atmosphere,' which Charles Darwin's portrait lacks. That portrait is wanting, in the full characterisation of a many-sided figure; we can only give reality to it by a study of Erasmus Darwin's own works, local gossip about him and the public opinion of his day—

I now, with fear and trembling lest you should finally vote me a continued bore, venture to enclose copies of some queries I have just had printed and am circulating, after having obtained by personal inquiries a good deal of very curious information on the points in question. I venture to ask you more particularly, because the 'visualising' faculty of Dr Darwin appears to have been remarkable and of a peculiar order and it is possible that yours, through inheritance, may also be similarly peculiar. It is perfectly marvellous how the faculty varies, and moreover some very able men intellectually do not possess it. They do their work *by words*, I am in correspondence with Max Müller about this, who is an *outré* "nominalist."

Very sincerely yours, FRANCIS GALTON.

Thanks for Bowditch (children's growth) which you kindly sent me.

*Nov.* 14th [1879]. DOWN, BECKENHAM, KENT.
RAILWAY STATION, ORPINGTON. S.E.R.

MY DEAR GALTON, I have answered the questions, as well as I could, but they are miserably answered, for I have never tried looking into my own mind. Unless others answer very much better than I can do, you will get no good from your queries. Do you not think that you ought to have age of the answers? I think so, because I can call up faces of many schoolboys, not seen for 60 years, with *much distinctness*, but now-a-days I may talk with a man for an hour, and see him several times consecutively, and after a month I am utterly unable to recollect what he is at all like. The picture is quite washed out.

I am *extremely* glad that you approve of the little life of our grandfather; for I have been repenting that I ever undertook it as work quite beyond my tether. The first set of proofsheets was a good deal fuller, but I followed my family's advice and struck out much.

Ever yours very sincerely, CHARLES DARWIN.

QUESTIONS ON THE FACULTY OF VISUALISING[1].

For explanations see the other side of this paper.

The replies will be used for *statistical purposes only* and should be addressed to:—
FRANCIS GALTON, 42, RUTLAND GATE, LONDON.

| *Questions.* | *Replies.* |
|---|---|
| 1. Illumination. | Moderate, but my solitary breakfast was early and morning dark. |
| 2. Definition. | Some objects quite defined, a slice of cold beef, some grapes and a pear, the state of my plate when I had finished and a few other objects are as distinct as if I had photos before me. |
| 3. Completeness. | Very moderately so. |
| 4. Colouring. | The objects above-named perfectly coloured. |
| 5. Extent of field of view. | Rather small. |
| *Different kinds of Imagery.* | |
| 6. Printed pages. | I cannot remember a single sentence, but I remember the place of the sentences and the kind of type. |
| 7. Furniture. | I have never attended to it. |
| 8. Persons. | I remember the faces of persons formerly well-known vividly, and can make them do anything I like. |
| 9. Scenery. | Remembrance vivid and distinct and gives me pleasure. |
| 10. Geography. | No. |
| 11. Military Movements. | No. |
| 12. Mechanism. | Never tried. |

and I would add, an examination of the innumerable paintings of him from various aspects. He was in no sense a bloodless man, but clearly a man of many crotchets and peculiarities of temperament. I have had the privilege of examining a considerable number of Erasmus Darwin's letters and papers, and feel that his true characterisation remains to be drawn. The final portrait will not be that of Schimmelpenninck, but again not that of Charles Darwin. Meanwhile I find my imagination persists in coupling the supposed extremes: Samuel Johnson and Erasmus Darwin!

[1] For the nature and occasion of these questions the reader must consult Chapter XII.

| *Questions.* | *Replies.* |
|---|---|
| 13. Geometry. | I do not think I have any power of the kind. |
| 14. Numerals. | When I think of any number, printed figures rise before my mind; I can't remember for an hour 4 consecutive figures. |
| 15. Card-playing. | Have not played for many years, but I am sure should not remember. |
| 16. Chess. | Never played. |
| *Other senses.* | |
| 17. Tones of voices. | Recollection indistinct, not comparable with vision. |
| 18. Music. | Extremely hazy. |
| 19. Smells. | No power of vivid recollection, yet sometimes call up associated ideas. |
| 20. Tastes. | No vivid power of recalling. |

*Signature of Sender and Address.* CHARLES DARWIN, Down, Beckenham. (Born *Feb.* 12th, 1809.)

*April* 7, 1880. DOWN, BECKENHAM, KENT.

MY DEAR GALTON, The enclosed letter and circular may perhaps interest you, as it relates to a queer subject. You will perhaps say: hang his impudence. But seriously the letter might possibly be worth taking some day to the Anthropolog. Inst. for the chance of some one caring about it. I have written to Mr Faulds telling him I could give no help, but had forwarded the letter to you on the chance of its interesting you.

My dear Galton, Yours very sincerely, CH. DARWIN.

P.S. The more I think of your visualising inquiries, the more interesting they seem to me.

42, RUTLAND GATE, *April* 8/80.

MY DEAR DARWIN, I will take Faulds' letter to the Anthro. and see what can be done; indeed, I myself got several thumb impressions a couple of years ago, having heard of the Chinese plan with criminals, but failed, perhaps from want of sufficiently minute observation, to make out any *large* number of differences. It would I think be feasible in one or two public schools where the system is established of annually taking heights, weights etc., also to take thumb marks, by which one would in time learn if the markings were as persistent as is said. Anyhow I will do what I can to help Mr Faulds in getting these sort of facts and in having an extract from his letter printed. I am so glad that my 'visualising' inquiries seem interesting to you. I get letters from all directions and the metaphysicians and mad-doctors have been very helpful.

Very sincerely yours, FRANCIS GALTON.

Our united kindest remembrances to you all.

Galton communicated Dr Faulds' letter to the Anthropological Institute; the original is now before me, and it is inscribed, "Addressed to Charles Darwin, Esq. and communicated by F. Galton." Apparently that body did not publish it as they certainly ought to have done. Many years afterwards it was discovered in their archives. Its non-publication, however, was not of such importance as it might have been, for on Oct. 28, 1880, a very full letter from Dr Faulds appeared in *Nature* covering the same ground. To this matter we shall return later.

42, RUTLAND GATE, *July* 5/80.

MY DEAR DARWIN, Best thanks for sending me *Revue Scientifique* with Vogt's curious paper, which I return with many thanks. The passage you marked for me makes me sure that he would give help of the kind I now want and I will write to him. (De Candolle and another Genevese, Achard by name, have already kindly done much.)

I send an advance copy of those "Visualised Numerals" of mine, not to trouble you to re-read what you know the pith of already, but because of the illustrations at the end and also for the chance of your caring to see there the confirmation from other sources (I find that the editor has cut out all Bidder's remarks on this point—which I much regret) of what Vogt says about the left hand executing with facility *in reverse* what is done by the right hand. I made

Bidder scribble flourishes with pencils held in both hands simultaneously and the reflexion of the one scrawl in a mirror was just like the other picture seen directly.

I have just published in *Mind* something more about mental imagery, and when I get my reprints I will send one in case you care to glance at it.

Enclosed is a reference that might be put among your Dr Erasmus Darwin papers, in the event of having again to revise the 'Life'. I had not a notion, until I began to hunt up for the reference, how much he had considered the subject of mental imagery, or the very striking experiment in Part I, Section XVIII. 6 (which in my edition of 1801 is in Vol. I, p. 291), which shows that he himself possessed the faculty in a very marked manner.

We came back after a very successful Vichy visit[1]; my Wife improved at once on getting there, but for my part I have since been unlucky, and am only just out of bed after a week's illness of the same kind as Litchfield's long affair—this partly accounts for bad handwriting. With kindest remembrances to you all from us both and from my sister Emma who is now with us for a few days, Ever sincerely yours, FRANCIS GALTON.

42, RUTLAND GATE, Monday morning, *March* 7/81.

DEAR DARWIN, About Worms[2]:—I have waited for an opportunity of verifying what I told you about the effect of heavy soaking rain, *when it suddenly succeeds moderate weather*, in driving the worms from their holes to the gravel walks, where they crawl for long distances in tortuous courses, and where they die. It has been very frequently observed by me in Hyde Park, and this morning I have again witnessed it in a sufficiently well-marked degree to be worth recording.

It rained heavily on Saturday night last, after a spell of moderate weather. Unluckily I was not in the Park on Sunday till near 1 h. by which hour the birds had had abundant time to pick up the worms. Still, dead worms were about and their tracks were most numerous. On Sunday (last night) it again rained heavily and I was in the Park at 10 h. The tracks were not nearly so numerous as they had been on Sunday morning, but more dead worms were about. I began counting, and found they averaged 1 to every 2½ paces (in length) of the walk, the walk being 4 paces and a trifle more in width.

Walking on, I came to a place where the grass was swamped with rain-water on either side of the raised gravel path, for a distance of 16 paces. In those 16 pace-lengths I counted 45 dead worms.

On not a few previous occasions when I have been out before breakfast, I have *under the conditions already mentioned* seen the whole of the walks strewn with worms almost as thickly as were the 16 pace-lengths just described. The worms are usually very large. I rarely notice dead worms on the paths at other times. Ever sincerely yours, FRANCIS GALTON.

I shall be very curious to learn about the effects of the red light as against those of a strongly actinic colour.

DOWN, BECKENHAM, KENT. *March* 8th [1881].

MY DEAR GALTON, Very many thanks for your note. I have been observing the *innumerable* tracks on my walks for several months, and they occur (or can be seen) only after heavy rain. As I know that worms which are going to die (generally from the parasitic larva of a fly) always come out of their burrows, I have looked out during these months, and have usually found in the morning only from 1 to 3 or 4 along the whole length of my walks. On the other

[1] Both the Galtons enjoyed Vichy and visited it yearly from 1878 to 1881.

[2] Miss Margaret Shaen tells me that she first met Francis Galton at Down, when Darwin was studying earthworms. "They had much talk together on the subject, Mr Galton getting most eager in trying to picture to himself exactly how the worms drew things into their holes to close them up. Mr Darwin was then experimenting with little bits of paper like this ◁, laying them near the worm holes, and finding them drawn down by the point. I remember Mr Galton trying to do the like with his pocket pencil, i.e. to draw the paper down inside his pencil case. I am pretty sure he was keen to test the worms perception of angles by altering the sides of the triangle, getting them more equal to see if the worms would still detect the smallest of the angles and draw that one in. I don't know if Mr Darwin did try any such experiments." See *The Formation of Vegetable Mould through the Action of Earthworms*, pp. 14, 85–95.

hand I remember having in former years seen scores or hundreds of dead worms after heavy rain. I cannot possibly believe that worms are drowned in the course of even 3 or 4 days immersion; and I am inclined to conclude that the death of sickly (perhaps with parasites) worms is thus hastened. I will add a few words to what I have said about their tracks, after stating that I found only a very few dead ones. Occasionally worms suffer from epidemics (of what nature I know not) and die by the million on the surface of the ground.

Your ruby paper answers capitally, but I suspect that it is only by dimming the light, and I know not how to illuminate worms by the same intensity of light, and yet of a colour which permits the actinic rays to pass. I have tried drawing the angle of damp paper through a small cylindrical hole, as you suggested, and I can discover no source of error. Nevertheless I am becoming more doubtful about the intelligence of worms. The worst job is that they will do their work in a slovenly manner when kept in pots, and I am beyond means perplexed to judge how far such observations are trustworthy.

Ever my dear Galton, Yours most sincerely, CH. DARWIN.

42, RUTLAND GATE, *Oct.* 9/81.

MY DEAR DARWIN, Pray accept my best thanks for the worm book, which I have read, as I read all your works, with the greatest interest and instruction. I wish the worms were not such disagreeable creatures to handle and keep by one, otherwise they would become popular pets, owing to your book, and many persons would try and make out more concerning their strange intelligence. Once again very best thanks and believe me,

Ever sincerely yours, FRANCIS GALTON.

DOWN, BECKENHAM. *March* 22nd [1882]

MY DEAR GALTON,—I have thought that you might possibly like to read enclosed which has interested me somewhat, and which you can burn.—I have been on the sick-list, but am improving. Ever my dear Galton, yours very sincerely, CH. DARWIN.

Such, a month before his death, was the last letter of Darwin to Galton.

42, RUTLAND GATE, *March* 23/82.

MY DEAR DARWIN, Best thanks for the American article, which is certainly suggestive, where paradoxical. It is delightful to find that virtue mainly resides in large and business-like families, fond of science and of arithmetic! It eminently hits off the character of your own family and in some fainter degree of my brothers and sisters, and of all Quakerism.

I hope you are quite well again. With our kindest remembrances,

Ever yours, FRANCIS GALTON.

DOWN, Thursday, 20th *April* 1882.

DEAR MR GALTON, My mother asks me to write to you and tell you of my dear father's death. He died yesterday afternoon about 4. He was taken ill in the middle of Tuesday night and remained in a great state of faintness, suffering terribly from deep nausea and a most distressing sense of weakness. He was conscious till within a ¼ hr. of his death. He gradually became more and more pallorless and at last became suddenly worse. I cannot help saying how often I have heard him speak with affection of you[1], Yours affectionately, FRANCIS DARWIN.

I forgot to say what I especially meant to, that my mother bears it wonderfully, she is very quiet and calm.

[1] Mrs Litchfield, Darwin's daughter, tells me that her Father had a great admiration for Galton's acuteness and she has also a memory of her Father saying what fun Galton was. Miss Elizabeth Darwin recalls a visit of Galton when they were all children, and his talking of mesmerising them, but it was not attempted in case it should frighten them. After Miss Henrietta Darwin's marriage, Galton told her he was sure he could mesmerise her, but that it would not be good for her. In his *Memories*, p. 80, Galton tells us that he learnt the art in Austria during his undergraduate days, and mesmerised some 80 persons, but "it is an unwholesome procedure, and I have never attempted it since." By experiment, however, he demonstrated that the exercise of will power by the operator is unnecessary, it is a purely subjective operation.

The following letter to his sister, Miss Emma Galton, is not only of historical interest, but portrays the intense reverence Galton felt for his cousin:

42, RUTLAND GATE, *April* 22/82.

DEAREST EMMA, I feel at times quite sickened at the loss of Charles Darwin. I owed more to him than to any man living or dead; and I never entered his presence without feeling as a man in the presence of a beloved sovereign. He was so wholly free of petty faults, so royally minded, so helpful and sympathetic. It is a rare privilege to have known such a man, who stands head and shoulders above his contemporaries in the science of observation. When the news came on Thursday I went to the Royal Society which met that day and arranged that a request should be telegraphed to the family by the President in the name of the Royal Society asking if they would consent to an interment in Westminster Abbey, to which I have some reason to believe the Dean (who is abroad) would in no way object. If so the funeral would be attended by deputations from all the learned societies. I wrote to Lord Aberdeen, who fully consents on behalf of the Geographical, and who has written accordingly. I was absolutely engaged all yesterday (till after dinner hour even), and could not learn progress. I hope the first wishes of the family may yield and that Charles Darwin may be laid by the side of Newton as the two greatest Englishmen of Science. I had a brief letter from Frank Darwin on Thursday with nothing however in it that was not in the next day newspapers. It was evidently angina. The world seems so blank to me now Charles Darwin is gone. I reverenced and loved him thoroughly. Ever affectionately, FRANCIS GALTON.

On April 26th Darwin was buried in Westminster Abbey[1]; the funeral card runs, "Wednesday April 26th 1882, at 12 o'clock precisely. Admit the Bearer at eleven o'clock to the Jerusalem Chamber." Galton walked in the procession[2], and on the same evening wrote to his sister, Miss Emma Galton, as follows:

42, RUTLAND GATE, *April* 26/82.

DEAREST EMMA, The great ceremony in the Abbey is over. The whole "family" of scientific men were there, a great and imposing gathering. No ostentation but great from its intrinsic worth. The Duke of Argyll and Wallace were the two end pall-bearers, Huxley and Canon Farrar were together, thus all shades of opinion and station were merged. It was touching to see the blind Postmaster-General [Fawcett] led past the coffin. Several past Cabinet Ministers were also present. They had asked me to find out Canon Farrar's views, wishing to have some prominent ecclesiastic, especially one connected with Westminster Abbey, as a

[1] It is noteworthy, perhaps, that Galton on Dec. 27, 1881 had sent a note to the *Pall Mall Gazette* urging the stringent enforcement of rigid sanitary conditions of burial in the case of interments in the Abbey.

[2] Galton was also at Lord Tennyson's funeral and these ceremonies in the Abbey impressed him with the existence of a great failure on such occasions. The solemn procession up the nave to the chancel was not visible to the bulk of the congregation in the transepts. Galton in a letter to *The Times* May 25, 1898 writes: "My own seat was in a good position, but I saw nothing of the distinguished persons who formed the procession except the foreheads of two of the pall-bearers who were of exceptional stature, whose well-known names I need not specify. All the others were sunk wholly out of sight in a trough of crowded humanity. It is a sad waste of effort and opportunity to so mal-organise a great spectacle that its most imposing feature proves to be invisible to the great majority of those who come to see it." Galton's solution was a slightly raised causeway from choir to chancel. It may be objected that we go to honour the dead and not to see a spectacle. But this is not wholly true; it is the spectacle which impresses itself on the multitude and makes them realise, perhaps for the first time, the national value of the great dead. They go to hear and they go to see that their memories and their imaginations may be indelibly impressed. A solemn national funeral repercusses in wider circles than are ever reached by the acts or words of a national hero during life. It sets even the inert inquiring.

Miss Emma Galton, from a photograph taken in the 'fifties.

pall-bearer and he (Farrar) entered most cordially into the wishes of the family. He offered to act as a pall-bearer either in or without his robes, as desired. He is to preach next Sunday on Darwin at the Abbey and tells me that he wishes to make such amends as he can for the reception formerly given by the Church party to Darwin's works, and we have talked over some points for the sermon.

Reginald Darwin was there and Emma Wilmot and Cameron Galton and H. Bristowe. The family party was so large that most of the ladies (including Louisa) and about half of the men were placed in the seats by the altar rails else the procession would have been too long. H. Bristowe and I walked together. Louisa will write more details. The newspapers will give a much fuller account. The service was not particularly touching; it never is in the Abbey; it is more like the ceremonial of *giving a University Degree*.

I got a card for Erasmus to attend with the family and telegraphed to him to Loxton thinking it *possible* that owing to his admiration of Darwin's works he might like to come, but he declined.

Mrs Darwin is very composed now.

I feel this is a worthless and heartless sounding letter, but as I said the feeling promoted by the ceremony is *not* a solemn one but rather the sense of a national honour and glory.

Ever affectly, FRANCIS GALTON.

The words of the anthem, taken from Proverbs iii. 13, 15, 16 and 17 "Happy is the man that findeth wisdom, and the man that getteth understanding.... Her ways are ways of pleasantness, and all her paths are peace," were aptly chosen, as also the anthem of Handel at the grave-side: "His body is buried in peace, but his name liveth evermore." The ceremony did not strongly appeal, however, to Darwin's Quaker-minded cousin; for him the restful burial in the little churchyard of Claverdon Leys thirty years later seemed indeed appropriate. The next day, April 27th, Darwin's daughter, Miss Elizabeth Darwin, wrote to Galton's sister Emma:

"We have had a great deal of sympathy and it is soothing to feel how many appreciated our dear Father's goodness. He always had a very real affection for your brother and took great pleasure in his company."

On the same day appeared a letter by Galton in the *Pall Mall Gazette*:

The Late Mr Darwin: A suggestion

SIR,—Next Sunday numerous congregations will expect some honourable recognition of the character and works of Charles Darwin. Let me suggest to clergymen generally that they should substitute on that day the 'Benedicite' for the more usual 'Te Deum,' as many of its noble verses are pointedly appropriate to what they would probably wish to say afterwards from the pulpit:—

O all ye Works of the Lord, bless ye the Lord : praise him, and magnify him for ever.
O all ye Green Things upon the Earth, bless ye the Lord : praise him, and magnify him for ever.
O ye Whales, and all that move in the Waters, bless ye the Lord : praise him, and magnify him for ever.
O all ye Fowls of the Air, bless ye the Lord : praise him, and magnify him for ever.
O all ye Beasts, and Cattle, bless ye the Lord : praise him, and magnify him for ever.
O ye holy and humble Men of heart, bless ye the Lord : praise him, and magnify him for ever.

In pursuance of the same idea, let me add that a stained glass window in Westminster Abbey, symbolising these and other verses of the same canticle in its several panels, would be a beautiful monument to the memory of Charles Darwin, and quite in harmony with the surroundings. It would afford a desired opportunity for other countries to share in the erection of a memorial without merging their several contributions indistinguishably into one, as each country might contribute a separate panel. I suggest this window in addition to, and not in substitution of, any bust or tablet that may hereafter be decided upon, and towards all of which I, for one, am prepared to subscribe liberally. I am, Sir, Your obedient servant, F. G.

It was, perhaps, too generous an idea to expect in 1882 that an 'evolution' window could, even in Westminster Abbey, replace the old 'creation' window based upon its neolithic myth. But the time may yet come when the national mausoleum shall contain not only the ashes of the nation's great dead, but some appropriate witness to those living embers of the mind which entitled them to their final resting-place. Galton strongly believed in and generously supported all projects of perpetuating the memory of the worthy dead. It was exhibited not only in the case of Darwin, but in several other instances. Thus in the monument he put up to Erasmus Darwin in Lichfield Cathedral[1], in his support of the Speke memorial and his desire to see it extended to embrace other African pioneers (see our p. 25 *ftn.*), and again in the substantial aid he gave to the Oxford Weldon memorial. I have no doubt fuller investigation would lead to the discovery of other instances[2].

But for Darwin, Galton's affection and reverence were unlimited. Within three weeks of the former's death he wrote to Darwin's son George as follows:

42, RUTLAND GATE, *May* 16th, 1882.

MY DEAR GEORGE, You may be glad to hear that the memorial to your father was fairly started this afternoon and very shortly the letters to foreigners will be sent and notices in the papers will appear. A Sub-Cmte. of the executive Cmte. has only now to fix a few details. I was very sorry to have missed you when you called, as there is much I should like to have heard about you all. I am very glad that your Mother bears up so well.

I wanted too, to speak to you (as I have to Spottiswoode) about getting together available illustrations and memorial scraps of all kinds for a book of mementos for the Royal Society (like those of Priestley—do you know them?). There ought to be a picture of the 'Beagle' if one is procurable and copies (small) of *all* the pictures and photographs. You are no doubt collecting all available information of his early life before his contemporaries and seniors shall have passed away. Every month is precious. I do wish somebody had done this many years ago for Dr Erasmus Darwin. If omitted, this want is soon irrevocable. When you are next in Town pray come to us. Ever yours, FRANCIS GALTON.

Talking once to the husband of one of the greatest of Victorian women, about the loss of a great friend—to whose learning and scholarship I owe whatever love I may possess for accurate investigation—he remarked:

"It is difficult to measure what the mental development of an individual loses and what it gains by the death of a friend of dominant personality."

The words seemed to me then harsh and unsympathetic, but I have learnt with the years the element of truth in the experience expressed by them. That truth is not wholly appropriate to the friendship of Galton with Darwin; the latter was only thirteen years Galton's senior, but those years, and Galton's unlimited reverence for intellectual power did, as in the

[1] See Note at the end of this Chapter.

[2] One other instance I can indeed refer to from letters in my possession. He was the prime mover in the scheme for obtaining a portrait of Sir Joseph Hooker. There are numerous letters to Galton approving and enclosing subscriptions, and the letter of Hooker to Galton is worthy of being preserved elsewhere than in an autograph book where I found it:

ROYAL GARDENS, KEW, *May* 15/80.

MY DEAR GALTON, Your kind letter announces a most unexpected honour, and a crowning one. I only wish I could feel that I was worthy of it. I am quite at Mr Collier's disposal and very pleased to find that he is the selected artist. Very sincerely Yours, JOS. D. HOOKER.

42. Rutland Gate
London S.W.
Dec 9. 1859.

My dear Darwin

Pray let me add a word of congratulation on the completion of your wonderful volume, — to those which I am sure you will have received from every side. I have laid it down in the full enjoyment of a feeling that one rarely experiences after boyish days, of having been initiated into an entirely new province of knowledge which, nevertheless, connects itself with other things in a thousand ways.

I hear you are engaged on a second Edition. There is a trivial error in p. 68. about Rhinoceroses, which I thought I might as well point out, & have taken advantage of the same opportunity to scrawl down ½ a dozen other notes which may or may not, be worthless to you.

With our united kind regards to yourself & Mrs Darwin
Believe me very sincerely yours
Francis Galton

Francis Galton's Letter to Darwin on the publication in 1859 of *The Origin of Species*.

PLATE XIX

The first study at Down, the room in which *The Origin of Species* was written. Photographed in his Father's lifetime by Major Leonard Darwin.

case of Pangenesis, unconsciously shackle the free development of Galton's own ideas. Galton would never have admitted such an aspect of the friendship. To him Darwin was the man who freed him from superstition and directed his life-work into new channels[1]; but nevertheless the onlooker may note, what individual actors cannot apprehend, for he like the dramatist sees the play as a whole. Be this as it may, undoubtedly the year 1882 marked an epoch in Galton's career[2]. As Mrs Galton records, Darwin's death "cast

[1] Galton did not only acknowledge this in the memorable letter to Darwin himself in 1869 (see Vol. I, Plate II) but most gracefully in the speech he made at the Royal Society dinner after receiving the gold medal in 1886. I will cite a portion of it:

"The ethnological aspects of geography now [1860] began to attract me more than the physical ones. It was about this time that the fact dawned on scientific men that the key to the origin of society among civilised nations and to many of their unexplained customs was to be found in the habits of contemporary barbarians. I can assure you, as a specialist in heredity, that I am not speaking without reason when I say that qualities which I seem to have inherited through two of my grandparents gradually yielded precedence to those that I certainly inherited from the other two. Recollect, please, that this medal is awarded to me for 'statistical inquiries into biological phenomena.' I can account fully both for the statistics and the biology. You must please allow me the pleasure of dissecting myself. On my father's side, I know of many most striking, some truly comic, instances of statistical proclivity. I have in my possession many pounds weight of ruled memorandum books severally allotted to almost every conceivable household purpose, which belonged to an aged female relative who died years ago. I also reckon at least five other remarkable instances of a love of tabulation within two degrees of kinship of myself. Again, as regards biology, I am sure there is a similarity between the form of the bent of my mind and that of my mother's father, Dr Erasmus Darwin. The resemblance chiefly lies in a strong disposition to generalise upon every-day matters that commonly pass unnoticed. I have myself attempted some of the very inquiries to which he had drawn attention, in complete unconsciousness that he had done so. It was owing to this hereditary bent of mind that I was well prepared to assimilate the theories of Charles Darwin when they first appeared in his 'Origin of Species.' Few can have been more profoundly influenced than I was by his publications. They enlarged the horizon of my ideas. I drew from them the breath of a fuller scientific life, and I owe more of my later scientific impulses to the influences of Charles Darwin than I can easily express. I rarely approached his genial presence without an almost overwhelming sense of devotion and reverence, and I valued his encouragement and approbation more, perhaps, than that of the whole world besides. This is the simple outline of my scientific history." (*The Times*, Dec. 1, 1886.)

[2] Galton's last tribute probably to Darwin was paid at the Darwin-Wallace celebration of the Linnean Society on July 1st, 1908. The present writer saw him to and from the meeting and knew that he was feeling unwell; his few words were a great effort. After thanking the President for his kind remarks, Galton turned to the main point on which he felt our generation's gratitude to Darwin should be keenest—the freedom Darwin gave us from theological bondage: "You have listened to-day to many speakers and I have little new to say, little indeed that would not be a repetition, but I may add that this occasion has called forth vividly my recollection of the feelings of gratitude that I had towards the originators of the then new doctrine which burst the enthraldom of the intellect which the advocates of the argument from design had woven round us. It gave a sense of freedom to all the people who were thinking of these matters, and that sense of freedom was very real and very vivid at the time. If a future Auguste Comte arises who makes a calendar in which the days are devoted to the memory of those who have been the beneficent intellects of mankind, I feel sure that this day, the 1st of July, will not be the least brilliant." *The Darwin-Wallace Celebration...by the Linnean Society of London*, 1908, pp. 25–6.

It is characteristic of Francis Galton that it was not the enormous influence of Darwin on the biological sciences that he thought of in the first place, but the emancipation of the human intellect from its centuries-old neolithic traditions—the common gain of the average man, only indirectly affected by the spread of scientific knowledge—that he wished to see emphasised.

a deep gloom" over her husband; but it was followed by his most productive decade. Interests in psychological and in statistical investigations had originated well before this date, but as our following chapters will show they now became predominant and displaced to a large extent the more biological aspect of the inquiries which we have associated in the second half of this chapter with Darwin. The philosopher of Down was no longer there either to check error or to restrain imagination. The miniature of Darwin remained on the writing-table, but rather as a symbol of method, than to suggest the warning voice of the revered master:

*Ignoramus, in hoc signo laboremus!*

NOTE I. ON THE MONUMENT TO ERASMUS DARWIN ERECTED BY FRANCIS GALTON IN LICHFIELD CATHEDRAL, 1886.

About the time when the question of a monument to Charles Darwin in Westminster Abbey was being raised, Galton determined to commemorate the grandfather of both in Lichfield Cathedral, and obtained the permission of the Dean and Chapter for the erection of a memorial medallion. This was executed by E. Onslow Ford. See our Plate XIX. The work of Krause and Charles Darwin on the life and ideas of Erasmus Darwin had drawn the attention of Galton again to his grandfather, and he was more than inclined to revise the opinion he had expressed to de Candolle in 1882 (see our Vol. I, p. 13). Perhaps what weighed much with Galton were the lines from the preface to the *Zoonomia.*

The great Creator of all things has infinitely diversified the works of his hands, but has at the same time stamped a certain similitude on the features of nature that demonstrate to us that the whole is one family of one parent.

There is not a doubt, I think, that Erasmus Darwin anticipated Lamarck in propounding a doctrine of evolution based upon the inheritance of acquired characters, and that he recognised a unity of origin for all forms of life. It was with this impression strong upon him that Galton made his first draft for the Lichfield inscription. It ran as follows:

In memory of Erasmus Darwin, M.D., F.R.S., Physician, Philosopher, and Poet; Author of *Zoonomia, Botanic Garden,* &c.; Earliest propounder of the Theory greatly elaborated by his more distinguished grandson, Charles Darwin, which ascribes to the operations of animals and plants, prompted in the first instance by their individual needs, the secondary and higher function of modifying through inheritance by various indirect and slow though certain methods, the forms and instincts of their respective races, in increasing adaptation to the habits of each and to their physical surroundings and thus of furthering the development of organic nature as a whole.

This inscription certainly accords with Erasmus Darwin's view, if it does not lay as much stress on the element of 'will' as Erasmus did. It was, perhaps, not incompatible with Charles Darwin's opinion that at least some acquired characters are inherited. Galton sent it to Huxley for criticism and Huxley replied with the following characteristic note:

Will you have patience to look at my article on "Origin" in "Lay Sermons" written in 1860?

4, Marlborough Place, Abbey Road, N.W. *Oct.* 12, 1886.

My dear Galton, I have read Krause afresh and I can only say that if no better case is to be made out for Erasmus Darwin my verdict in the article "Evolution" in the Encyclopaedia Britannica—(not the part which I advised Mr Gladstone to read)

"Erasmus Darwin, though a zealous evolutionist, can hardly be said to have made any real advance on his predecessors"

seems to me to express the exact truth.

That Erasmus Darwin anticipated Lamarck's central idea is perfectly true—indeed he was much more consistent and logical than Lamarck for he saw the necessity of extending his view to the vegetable kingdom, and his conception of heredity and desire in plants, was the necessary consequence of the general theorem that adaptation has been brought about by desire. But this central idea of Lamarck is exactly the most worthless part of the "Philosophie Zoologique." Krause expressly admits that Erasmus knew nothing about the struggle for existence as a selective agent. Again it is clear that Erasmus had no notion of variation in Charles Darwin's sense, see p. 183. "The original living filament" is "excited into action by the necessities of the creatures, which possess them and on which their existence depends."

The principles of Charles Darwin if I understand him rightly were these—Two are matter of fact, namely:

1. Animals and plants vary within limits which cannot be defined and from causes of which we have no knowledge.
2. Of these variations some give rise to forms better adapted to cope with existing conditions and some to forms worse adapted—The latter tend to extinction; the former to supersession of the original or at any rate to coexistence with it. The last is speculation.
3. The interaction of variation with conditions thus understood is a *vera causa* competent to account for the derivation of later from earlier forms of life.

This is the Darwinian faith "which except a man keep whole and undefiled without doubt" he shall be made a bishop—If you can find the smallest inkling of it in Erasmus—I cave in.

Ever yours very truly, T. H. Huxley.

Of Galton's reply to Huxley I find the following 'rough draft.'

*Oct.* 16/86. 42, Rutland Gate, S.W.

My dear Huxley, Thank you much for your full letter. I see that the proposed epitaph on Dr Erasmus Darwin must be abandoned, both because it is founded on Krause's views, which are debatable, and because it expresses those views in a way open to misapprehension. Would you kindly tell me if anything strikes you as objectionable in an epitaph written in the following sense? And if you could suggest any well-sounding phrase as well, I should be truly obliged.

In memory of Dr E. D. Author of etc., etc. Gifted with a vivid imagination, uncommon power of observation, an indefatigable love of research, original and prescient in his views, he was among the first who occupied himself with the topics that were subsequently explored by his grandson, Charles Darwin, to whom he transmitted many of his characteristics.

This I know is bald in the extreme, but if the intended sense is decided on one must hope by much touching up to make it read more suitably. Thinking it might be convenient I copy out some "pièces justificatives" on the opposite page (slightly shortening phrases and sometimes transposing). Ever sincerely yours, Francis Galton.

Galton's *pièces justificatives* contain the views of Charles Darwin as to Erasmus' mental powers, some citation of Krause's opinions and extracts from Erasmus Darwin's writings showing that he only attributed variation in *part* to desire or will, but in part also to accidental or spontaneous causes. Galton also indicates that Erasmus had clearly stated that the final result of the contest between males was that the strongest and most active animal propagated the species and thus it became improved.

*Oct.* 18, 1886. 4, MARLBOROUGH PLACE, ABBEY ROAD, N.W.

MY DEAR GALTON, I have nothing to say against your new form of epitaph. Indeed I think it is quite just. I am not quite sure, however, whether a 'topic' can be 'explored.' But it is easier to criticise than to suggest something better. Charles Darwin used to scold me for not appreciating the 'Vestiges' (that flimsiest of books) so you may judge how tenderly he looked at his grandfather's work. He had the noble weakness of thinking too much of other people's doings and too little of his own. I wish none of us had worse sins to be forgiven.

Ever yours very sincerely, T. H. HUXLEY.

Did Erasmus Darwin possess vivid imagination, incessant activity and energy of mind, great originality of thought, prophetic spirit both in science and the mechanical arts, had he the true spirit of the philosopher—all which characteristics Charles Darwin ascribed to him,—or did he contribute little or nothing to posterity? Personally I see no truer way of answering that question than pointing to the achievements of his grandchildren, his great grandchildren and great great grandchildren. Their scientific worth and mechanical ingenuity, their originality of thought and energy of mind suggest that Charles Darwin's judgment was not in this case the product of a 'noble weakness.' But those who have had a friend in the least approaching Darwin in largeness of mind and charm of character will fully appreciate Huxley's criticism on the first of Galton's epitaphs which ventured to name Erasmus in the same breath as his own hero Charles. On our Plate XX the reader will find the final 'epitaph,' and its phraseology may obtain for him a deeper meaning in view of this correspondence between Galton and Huxley.

## NOTE II. CONTINUATION OF THE GALTON AND DE CANDOLLE CORRESPONDENCE, 1879–1885.

In order not to break the thread of Chapters XI and XII, I have placed here the continuation for six years of the above correspondence. It is not an unfitting place for it, because the letters not only provide further evidence of Galton's admiration for Darwin, but indicate how occupied his mind was during these years with psychological problems. The "visualised numerals," and the "visions of the sane" are discussed in Chapter XI, and composite photographs ("photographies cumulées") in Chapter XII.

GENÈVE, 8 *nov.* 1879.

MON CHER MONSIEUR, J'ai lu avec beaucoup de plaisir les opuscules que vous avez bien voulu m'envoyer. Les photographies *cumulées* m'ont paru curieuses. Elles serviront probablement dans plusieurs cas et pour diverses recherches de médecine, physiologie etc. On peut en inférer aussi certaines conséquences utiles dans les arts. Je me propose, par ce motif, de montrer vos *generic images* dans une séance de notre société des arts où se trouvent toujours des dessinateurs, des photographes, etc. Si vous publiez d'autres essais, je vous serai très obligé de me les envoyer avant le mois de janvier prochain.

L'aspect de vos criminels confirme ce que je lisais hier dans un article scientifique signé de Parville dans le *Temps*, sur les crânes de 36 assassins français. Leurs crânes étaient larges d'une tempe à l'autre.

Les observations psychométriques dont vous décrivez les résultats me semblent bien difficiles à faire. Cependant je sais, par expérience, qu'on arrive en se donnant de la peine à constater des faits de cette nature. J'ai aussi observé des choses analogues pendant la nuit, et ce que j'ai conclu appuie votre opinion par des approchements assez intéressants.

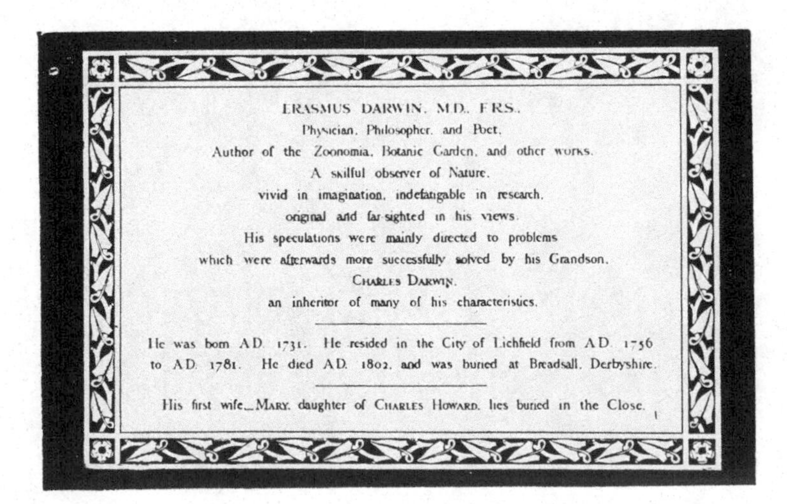

ERASMUS DARWIN, M.D., F.R.S.
Physician, Philosopher, and Poet,
Author of the Zoonomia, Botanic Garden, and other works.
A skilful observer of Nature,
vivid in imagination, indefatigable in research,
original and far-sighted in his views.
His speculations were mainly directed to problems
which were afterwards more successfully solved by his Grandson,
CHARLES DARWIN,
an inheritor of many of his characteristics.
He was born A.D. 1731. He resided in the City of Lichfield from A.D. 1756
to A.D. 1781. He died A.D. 1802, and was buried at Breadsall, Derbyshire.
His first wife, MARY, daughter of CHARLES HOWARD, lies buried in the Close.


Monumental Tablet to Erasmus Darwin (executed by Onslow Ford) and erected by his grandson Francis Galton in Lichfield Cathedral, 1886.

Pendant deux années, à l'âge de 71 et 72 ans, je me suis appliqué à saisir au moment du réveil la nature de mes rêves. Avec une ferme volonté on y parvient. Les détails s'oublient très vite, mais on peut noter dans sa tête si le rêve se rapporte à des choses anciennes ou récentes et à des choses dont on s'était occupé la veille ou auxquelles on n'avait pas pensé depuis longtemps. Voici ce que j'ai trouvé:

1°. très souvent dans la soirée j'avais parlé de quelque personne ou objet qui a servi de point de départ à un rêve. Quelquefois on en avait parlé devant moi, ou j'avais lu à haute voix le nom ou celui de l'objet. Rarement une lecture des yeux produisait cette conséquence. Le nom ou le mot était devenu source de quelque association involontaire d'idées, comme dans vos observations, mais les déductions étaient errantes et souvent absurdes. Des conversations ou lectures de quelques jours antérieurs, même je crois de 24 heures, ne conduisaient pas à des rêves.

2°. (et ceci encore plus certain): *jamais* je n'ai rêvé à des choses qui m'avaient causé ou récemment ou autrefois de vives inquiétudes ou une vive émotion. Quoique ma carrière ait été assez uniforme j'ai éprouvé des chagrins, j'ai eu des soucis qui m'empêchaient de dormir. J'ai assisté à des scènes révolutionnaires qui m'irritaient au plus haut degré. Ma vie a été exposée plusieurs fois dans des courses de montagne, etc. Or ces évènements ne se sont jamais présentés dans mes rêves, non plus que des plaisirs anciens très vifs.

3°. J'ai rêvé à des personnes mortes depuis longtemps, mais presque toujours je me suis souvenu que j'en avais parlé dans la soirée ou qu'une liaison d'idées analogue avait existé. Par exemple je rêvais souvent être avec mon père, mort en 1841: il me semblait le voir, causer avec lui, sur des affaires scientifiques avec beaucoup de suite et de raison, mais je travaille tous les jours dans la bibliothèque de mon père, je consulte ses ouvrages etc. C'est dans le courant habituel de mes idées. Tout cela confirme, par une autre voix, vos réflexions de la p. 7.

Les idées qu'on a la nuit, quand on est réveillé et qu'on ne peut pas s'endormir, confirment également ce que vous dites. Dans cet état on a (du moins moi) deux ou trois idées très précises qui reviennent couramment. Ce sont des idées qui vous préoccupent depuis quelques jours: probablement une inquiétude sur un de vos proches, un procès, etc., ou quelque lettre difficile à rédiger, quelque discours à faire etc. Comme on ne voit rien, d'autres causes d'idées n'existent pas et celles qui vous dominent ont une force extraordinaire. Un peu d'agitation nerveuse qui empêche de dormir augmente cette vivacité des idées nocturnes. J'ai été si souvent frappé de leur netteté que je m'étais fait fabriquer une ardoise, avec règle mobile, pour pouvoir écrire dans mon lit certaines phrases, certaines divisions d'un sujet qui m'apparaissaient tout-à-coup et que le lendemain je ne pouvais plus retrouver. L'appareil n'est pas assez commode pour l'employer, aussi quelquefois j'allume une lumière pour noter ce qui vient de m'apparaître dans ces nuits d'insomnie. Pour un écrivain l'absence de sujets de distractions me paraît une cause essentielle de succès. Je ne comprends pas du tout ceux qui étudient ou rédigent en se promenant dans la campagne.

Voilà, mon cher Monsieur, des observations dont vous ferez ce que vous voudrez. Je n'ai pas l'intention de les publier. Si vous voulez en parler dans quelque note je n'ai pas d'objection, sans cependant vous le demander. Agréez, je vous prie, l'assurance de mes salutations les plus dévouées. ALPH. DE CANDOLLE.

42, RUTLAND GATE, LONDON. *April* 11/80.

MY DEAR SIR, Thank you very much for the kind efforts you have made to procure me information about the visualised numerals. They have caused M. Achard to send me his numeral forms and some interesting accompanying remarks, which I have added to my collection. Perhaps the enclosed reprint may interest you, if you have not by chance already seen it, as it gives a recent account of the facts and some remarks upon them which I think you will find to answer in part the very reasonable doubts you suggest in your kind letter, which I received last night.

I do not think these forms of any value to those who see them, nor that they should be cultivated,—but they strike me as exceedingly curious and instructive survivals of the earliest mental processes of a child. They are specially interesting because of the reasons given in the enclosed reprint, they have been invented by the child himself, but I will not write what you will more rapidly read in print.

It has been a most amusing but somewhat discouraging experience to find how very many wise men are, as it were, vexed and put out by finding that other people have real undeniable gifts that they do not themselves possess a vestige of, and are inclined in consequence to discredit the inquiry. M. Antoine d'Abbadie who sees these number-forms clearly, kindly questioned for me several of his colleagues of the Académie des Sciences and came to just the same result that I did. It was therefore with some wicked feeling of triumph that I collected and marched off with[1], to the evening meeting of the Anthropological Institute, six good men including persons well known to science, who were prepared to describe their number-forms and who did so very effectively. I am now busy on a more generally interesting part of the subject of Mental Imagery. Believe me with thanks, and warm acknowledgement of the kind interest you have so often shown in my work, Very faithfully yours, FRANCIS GALTON.

I am afraid our mutual friend Mr Bentham has felt the gloom and severity of this past winter, for he does not look well, and complains about himself.

To M. ALPHONSE DE CANDOLLE.

42, RUTLAND GATE, LONDON. *June* 5/82.

MY DEAR SIR, Thank you *much* for your interesting brochure[2] on Ch. Darwin, analysing the causes that contributed to his success. It has been a great satisfaction in all our grief at his loss, to witness the wide recognition of the value of his work. He certainly, as you say, appeared at a moment when the public mind was ripe to receive his views. I can truly say for my part that I was groaning under the intellectual burden of the old teleology, that my intellect rebelled against it, but that I saw no way out of it till Darwin's 'Origin of Species' emancipated me. Let me, while fully agreeing with the views expressed in the pamphlet in all important particulars, supply a few minor criticisms which it might be well to mention.

(1) As to the pecuniary fortune of Darwin, I think the phrases "moyenne pour l'Angleterre etc."—"la maison modeste..." [pp. 12–13] hardly convey the right idea. I should think that his fortune was much more considerable—say upwards of £5000 a year, before his brother's death in 1881, and subsequently larger. The house was maintained in thoroughly substantial and costly comfort,—but when the particulars of the Will are published, which I suppose they soon will be, we shall know.

(2) "Les descendants du poète phys[iologiste] p. 12...ont lu certainement de bonne heure les ouvrages de leur aïeul." I am almost certain of the contrary in every case except Ch. Darwin, (and I doubt in his case whether he had). To myself the florid and now ridiculed poetry was and is intolerable and the speculative physiology repellent. I had often taken up the books and could never get on with them. Canning's parody "the loves of the triangles" quite killed poor Dr Darwin's reputation. It just hit the mood of the moment and though my mother never wearied of talking of him, his life was to me like a fable only half believed in. That much the same was the case with some of Charles Darwin's sons, I can, I think, affirm.

(3) George Darwin "déjà connu par de bons mémoires de statistique" [p. 13]. Probably you may not know his present *very high* position as a mathematical astronomer, who has revealed the past history of the planetary system, in a most unexpected way. His works are spoken of in the presidential address of the Royal Society etc. as *massive* works. They are only slowly becoming known, being exceedingly laborious mathematical work of a kind that is within the practice of very few men indeed, but by them cordially recognised as commensurate in originality and importance with that of Laplace. His calculations depend on the "viscosity" of all solid bodies on the yielding of their *substance* to a tidal action, and most unexpected results came out, which bind under one scheme a large variety of astronomical phenomena.

When I received your pamphlet, it so happened that your name had just been on my lips in respect to quite another matter, in which you were at one time much interested and which is now being taken up here. It is a question of *cumulative* temperature on vegetation. I have been since the beginning one of the members of the council to whom a large annual grant is entrusted by Government to carry on the systems of Forecasts in land and ocean meteorology and we are endeavouring to give weekly data that may be of direct use to agriculturists. In reply to questions that we circulated as to the best form for that purpose, frequent mention was made of the cumulative values of heat. We have accordingly been investigating the probability of calculating these values in units of 'day-degrees' viz. (1) cumulative effect of heat derived from 1 Fahr. of temperature acting during 24 hours, or of (2) acting during 12

[1] After giving them all a good dinner, *Memories*, p. 271.

[2] "Darwin aïeul, considéré au point de vue des causes de son succès et l'importance de ses travaux" tiré des *Archives des Sciences de la Bibliothèque Universelle*, Tome VII. *May* 1882.

hours, and so on. The result is that it is quite feasible to do so, with fair approximation, on the data of registered days' maxima and minima, and accepting any arbitrary base-line above which the accumulative temperature is to be reckoned. We can easily give 2 or more *slices* of the diurnal curve; that is to say the cumulative values between 3 or more arbitrary temperatures.

My colleagues ask me to inquire of you whether you happen lately to have again attended to the subject or whether you have any suggestions to make that might help us, in addition to what you have already published and which we find to be thoroughly appreciated by some of our correspondents?

It is rather out of our line to do so, but we might perhaps, if it were thought essential, get experiments made on the cumulative effects of temperature on some forms of vegetation—say the cereals—but probably sufficient information for our purpose already exists. We can measure cumulative effects of *sunshine, rain,* and *temperature* and could measure that of *evaporation* under any one definite condition, but it is a question whether the latter limitation would not render the results of little general service. I should be greatly obliged for a reply to the above question. Believe me, my dear Sir, very faithfully yours, FRANCIS GALTON.

To M. ALPHONSE DE CANDOLLE.

GENÈVE, 14 *Janvier* 1884.

MON CHER MONSIEUR, Je prépare une seconde édition de mon volume de *l'Histoire des sciences et des savants* qui est épuisé depuis longtemps et que les libraires me demandent. Pour cela je fais grand usage de vos *Englishmen of Science* et du volume récent des *Inquiries into Human Faculties* qui contient beaucoup d'articles curieux. Nous suivons la même méthode, celle d'observer, et quand on le peut, de compter pour comparer, par conséquent nous devons nous appuyer l'un l'autre et nous risquons bien peu d'être en opposition. Permettez-moi de vous demander quelques informations sur des savants anglais.

Pourriez-vous me dire quelles étaient les positions ou professions des pères du célèbre Zoologiste *Owen* nouvellement créé K.C.B., de Sir George Airy et de Sir George Wheatstone? Je n'ai pas pu le savoir d'après les dictionnaires biographiques à ma portée.

Je présume que Sir William Thomson, né à Belfast, fils d'un professor de mathématiques, était un protestant, d'une famille écossaise ou anglaise établie en Irlande. Est-ce exact?

Le caractère de votre illustre cousin Charles Darwin est si honorable, si éminent sous plusieurs rapports, que j'aimerais connaître sur lui certains détails d'une valeur même secondaire. Par exemple, avait-il une disposition naturelle aux *arts du dessin*? et à la *musique*? Rien ne l'indique dans ses ouvrages.

Je ne sais pas s'il faut lui attribuer une *imagination forte.* Beaucoup de personnes le lui reprochaient, parce qu'elles ne comprenaient pas la valeur de ses observations et déductions, et qu'il leur plaisait de dire qu'il se livrait à de pures hypothèses. Pour moi qui ai reconnu très vite la sagesse de son esprit et de sa prudence, je ne sais pas si ces qualités avaient occupé la place entière de l'imagination, ou s'il faut admettre que même avec beaucoup de vigueur de raisonnement il avait beaucoup d'imagination.

L'aîné de mes petits-fils, Raymond de Candolle, né Anglais, élevé à Rugby et qui vient d'entrer à Trinity College, Cambridge, voit les chiffres disposés en séries, et certains chiffres plus apparents que d'autres dans son esprit. C'est le seul cas de ce genre dans ma famille.

Toujours, mon cher Monsieur, votre très dévoué, ALPH. DE CANDOLLE.

42, RUTLAND GATE, LONDON. *Jan.* 27/84.

MY DEAR SIR, I delayed answering until I had an opportunity of talking over the questions you put about Darwin, with his very intelligent daughter.

He did *not* draw, he had not a good ear for music, but was much affected by it, sometimes to tears. He had naturally, (excuse the word which I know you detest! but I mean 'innately') a very emotional disposition, which was repressed by his habits of hard thinking, but always ready to burst out. Thus his delight in the scenery of a tour about the English Lakes a few years ago, had all the freshness and eagerness of that of a boy. However his nature could not be called aesthetic. As regards imagination I hardly know whether I understand the word in your sense, nor indeed if I have any definition of my own. I know that his faculty of mental imagery was once vivid and had become diminished, both from what he distinctly told me and from corroborative evidence. But that he ever was deceived by imagination I should think most unlikely, as he was so remarkably veracious.

He may be said to have studied veracity as the highest of arts. If imagination is cited in the sense of living in an ideal world of day dreams and poetry, I understand he was very fond of poetry as a boy but his interest in it faded by disuse. His scientific imagination in the sense of the power of envisaging abstract ideas, and living among them, and interesting himself with them was obviously great, on the evidence of his works. I am sorry not to be able to give you at present much information about the other men of whom you ask.

*Wheatstone* was an artisan, a mere workman originally. He took much interest in my inquiries and helped me in any way I asked, *except* as to his own history.

*Airy* promised to send me details, but eventually did not. His parents on both sides came of substantial farmers, solid men, of local notoriety. A certain disposition to dominate in argument is a strongly marked hereditary characteristic on the maternal (Biddell) side.

*Owen* I forget at the moment; they were low rather than high-middle-class. In a few days I shall be in the way of reviving recollections and will write again.

*Thomson* (Sir W.) I think you are quite right but here also I will write again.

Excuse this imperfect letter, but I am on the point of going to the country for a while and thought it best to write before going rather than after my return. I am very glad indeed that you are about to issue a new edition of your admirable volume. Let me say about the Darwin family that 4 of the 5 sons have achieved a very considerable reputation here. *George* the Plumian Professor at Cambridge is looked upon as one of the ablest of the rising men in mathematical physics. He has made a great mark already and is rapidly rising in repute. *Frank* who lectures at Cambridge on Botany was invited to be a candidate for the Professorship of Botany at Oxford, with a certainty of election, but for domestic reasons he refused. *Horace* has set up, in conjunction with a friend, a laboratory at Cambridge for the manufacture of high class scientific instruments. He is most ingenious as a mechanical originator. *Leonard* the Captain of Engineers, is one of the most scientific of his standing in that scientific corps. In the entrance examination he was first of all the candidates.

Thank you for telling me of your son's "number-form." I feel a little wicked delight at the fact occurring in your own family, because I recollected that you were at first somewhat sceptical of the reality of that curious tendency.

You will be glad to hear that we have begun at the Meteorological Office to publish *accumulated* temperatures in units of "day-degrees," counting (1) from Jan. 1, (2) from the first of the current week. General Strachey has worked out a beautiful method of obtaining them approximately, from the data of the daily maxima and minima, and a monthly (and probably a local) constant. I will tell them at the office to send you one of our new sheets as a specimen. It will give me the greatest pleasure if at any time I can be of service to you in obtaining information. Yours very sincerely, FRANCIS GALTON.

Our friend Mr Bentham continues *very* weak, but he has no organic malady.

To M. ALPHONSE DE CANDOLLE.

42, RUTLAND GATE, LONDON. *Oct.* 17/84.

MY DEAR SIR, I have read and re-read your new edition of the "Histoire des Sciences" with great interest and instruction, and trust you will appreciate my attention to even the briefest criticism by the improved handwriting of this letter in deference to what you justly say (and said before) at the bottom of p. 541. It is very singular how closely in many respects, our lines of inquiry run side by side. I shall be very curious indeed to see how far my own data will confirm yours in the 'nouvelles recherches,' but doubt much whether they will show the effect of heredity to be so strong, especially, for example, in myopism. Your appraisement of the several faculties, and selection of the faculties most convenient to be appraised, falls in very closely with an effort I lately indicated in the "Fortnightly" and am now making, to find out the best data by which the appraisement may be swiftly and fairly made. It has struck me that the masters and mistresses of schools might be able to indicate some often recurrent events in ordinary school life, which evoke different conduct in different children, and that by statistics of their conduct on these occasions, some fair guide to their habits and therefore to their character *at the time being* might be obtained. I should be greatly obliged for any hints that your experience may have suggested, how to appraise these qualities.

I venture to send you the "Fortnightly" of which I speak, not only on account of that article, but because of a very curious one in it about the Jews, by L. Wolf, which I feel sure will interest you in connection with your remarks on p. 174. I think however, that Mr Wolf overstates his case. We have arranged to talk the matter over and he will show me his data. It strikes me that the Jews are specialised for a *parasitical* existence upon other nations, and that there is need of evidence that they are capable of fulfilling the varied duties of a civilised nation by themselves.

I see that you still adhere to your view of the influence of the parental feelings at the time of conception, on the child. Could not that be experimentally tested upon such animals as rabbits? and possibly upon guinea pigs? The cause of fear might be the exhibition of a weasel or ferret. I cannot conclude without expressing my sincere pleasure at the way in which you have spoken of my inquiries. It is one of the pleasantest feelings to know that one is in intellectual sympathy with others. Believe me very sincerely yours, FRANCIS GALTON.

To M. ALPHONSE DE CANDOLLE.

GENÈVE. 27 *Oct.* 1884.

MON CHER MONSIEUR, Je suis très heureux de penser que vous approuvez mes dernières recherches sur l'hérédité! A peine cependant elles étaient publiées que je voyais d'autres considérations qui méritaient examen et avanceraient nos connaissances. J'espère qu'elles se trouveront dans votre prochain volume.

Il n'y a pas de doute que les maitres et maîtresses d'écoles pourraient faire des observations intéressantes sur les enfants, mais il faut vouloir observer, et le temps manque souvent à des personnes aussi fatiguées par leurs fonctions. Si on leur *ordonne* de faire telle ou telle observation elle se fait moins bien que si cela vient de leur propre volonté.

Les écoles dont on pourrait tirer le plus grand parti sont, par exemple, les écoles polytechniques, normale, navale, etc., en France. On y entre, à un âge déterminé, après des examens qui classent les individus. Les élèves sont soumis au même régime de nourriture, etc. Ensuite on les classe à la sortie. On pourrait donc constater si, dans des conditions égales, les plus habiles ont été les plus ou les moins robustes, les blonds ou les bruns, les fumeurs ou les nonfumeurs, les fils d'hommes instruits ou d'autres, etc. Il faudrait dans une de ces écoles qu'un médecin ou un des administrateurs eût l'idée de faire ces comparaisons, voilà le difficile, quoique deux années fussent suffisantes.

Votre article 'Measurement of Character' a l'avantage de provoquer les réflexions et suscitera peut-être de bonnes observations. Je vous remercie de me l'avoir envoyé!

L'article de Mr Wolff sur les Juifs me paraît un peu trop flatteur à leur égard. Si certaines préscriptions du Talmud sont bonnes pour la santé, il me paraît bien douteux qu'elles soient suffisamment observées par les fidèles, comme le prétend l'auteur. Les injonctions ne les ont pas rendus généralement propres. Ils sont moins propres que beaucoup de chrétiens auxquels la religion ordonne de mépriser leur corps, de penser surtout à l'âme et à la vie future. Je ne puis croire que les séparations des époux pendant des périodes aussi longues que celles dont parle Mr Wolff, soient réellement observées, surtout les mariages étant précoces, et si on les observe il doit y avoir un palliatif dans une polygamie plus ou moins admise. Les anciens Juifs étaient régulièrement polygames; ensuite les servantes ont probablement servi de complément à la femme légitime et aujourd'hui que se passe-t-il chez les Israélites? Je voudrais d'autres témoignages que ceux de Mr Wolff, par exemple le dire de médecins. Il y a beaucoup d'actrices juives et de modèles pour les peintres. La prévoyance habituelle chez la race et leurs mariages précoces ont probablement plus d'importance pour régler les mœurs que les préscriptions de la loi religieuse. L'article de Mr Wolff m'a appris du reste bien des choses que j'ignorais et ses réflexions sont souvent très justes.

Vous avez bien raison de dire que les Juifs sont adaptés à la vie parasite. C'est une bonne définition des faits. Il faut dire qu'on les a forcés à cette vie exceptionnelle. Si les difficultés étaient complètement levées pour eux, ils changeraient peut-être. D'Israeli a été un homme d'état égal à beaucoup de plus distingués.

C'est un professeur de physiologie Juif qui me proposait d'exécuter avec moi des expériences sur l'effet de l'alcoolisme, de la peur, etc., sur les produits dans les lapins, ou les cobayes, ou les chiens. J'avoue que ces sortes d'expériences me répugnent, mais je crois qu'elles prouveraient ce que je suppose: que l'état momentané des parents influe sur les produits. On l'a vu maintes fois pour l'alcoolisme et je viens de lire dans une revue très sérieuse (*Revue d'hygiène*, Octobre 1884 p. 875) ce qui suit:

"Notre ami, le Prof. Layet, de Bordeaux, avait à faire (au congrès d'hygiène de Lattaeie) un rapport sur la *restriction volontaire de la natalité du point de vue de ses conséquences humanitaires et sociales.*" Selon lui: "Au point de vue moral, elle favorise l'illégitimité, les 9 départements qui ont le moins de naissances légitimes ont aussi le coefficient d'illégitimité le plus élevé....L'accomplissement incomplet d'une fonction pervertit les excitations au lieu de les éteindre. L'habitude de la restriction amène une perturbation du système nerveux des conjoints; *les enfants nés par erreur dans ces conditions se ressentent de la perturbation nerveuse qu'a présidé à leur conception.* Les aliénés sont plus nombreux dans les départements où les époux ont le moins d'enfants."

"M. Lunier, de Paris, est convaincu de l'influence de la restriction volontaire sur le développement de l'état névrosique, particulièrement de l'hystérie et de la névralgie iléo-lombaire chez les femmes, et en cite de nombreuses observations nouvelles. La suppression du droit d'aînesse fait que le mal s'étend comme une tache d'huile dans les campagnes." Certainement dans les pays, comme France, où la loi assure un partage égal entre les enfants, beaucoup de femmes et de maris ont *peur* d'une 3ième, 4ième ou 5ième grossesse. L'effet de cette peur sur certains enfants est probable. Me voici sur un sujet bien scabreux. Il faut mieux s'arrêter et vous prier, mon cher Monsieur, de me croire toujours votre très dévoué ALPH. DE CANDOLLE.

GENÈVE. 29 *Sept.* 1885.

MON CHER MONSIEUR, Vous avez eu la bonté de m'envoyer votre savant article sur l'hérédité de la stature et je m'empresse de vous en remercier.

La taille (soit stature) est bien un des meilleurs caractères à étudier au point de vue de l'hérédité. Il est précis et d'une nature simple qui s'adapte à la statistique. Sur le continent nous avons une quantité de données relatives à la taille des jeunes gens appelés forcément au service militaire, mais rien sur l'hérédité de ce caractère. On a constaté souvent qu'un changement dans l'alimentation d'une population modifie la taille moyenne. Ainsi, dans les départements français, la stature s'est quelquefois élevée quand le maïs a été substitué au sarrasin ou au seigle. Inversement le canton de Berne en Suisse avait une belle population, forte et de taille plutôt élevée, mais les paysans se sont mis depuis 50 ans à distiller de l'eau de vie de pommes de terre. Leurs femmes et même leurs enfants en boivent. Il en est resulté un affaiblissement très visible de la race, et vu le nombre de scrofuleux, goîtreux, idiots, etc., il est probable que la taille moyenne a diminué. Nous n'avons pas de document ancien qui permette de le bien constater, mais je vous signalerai une publication récente du Bureau fédéral de Statistique fort intéressante pour les recherches dont vous vous occupez. Elle est intitulée: *Résultats de la visite sanitaire des recrues en automne* 1884, par le Bureau de Statistique fédéral, 31 pages en 4, Berne 1885, Orell, Fussli et C^ie^, éditeurs à Zurich.

On trouve dans ce cahier beaucoup de faits sur la taille, le thorax, l'acuité de la vue, la profession, etc., des recrues acceptées ou refusées pour le service militaire. Le texte est en français. Il est de la plume de M. Kummer, chef du Bureau, homme très exact et judicieux.

J'ai sur ma table un volume—malheureusement en allemand—qui contient le résumé de tout ce qu'on a réuni sur la proportion des sexes et le nombre des naissances en divers pays, chez l'homme et dans les animaux et les plantes. Le titre est: *Die Regulierung des Geschlechtverhaltnisses bei der Vermehrung der Menschen, Tiere und Pflanzen, von Carl Dusing, Dr Phil. Jena* 1884. On rendrait service en traduisant cet ouvrage en anglais ou en français. Cependant plusieurs parties, et naturellement les tableaux, sont intelligibles pour les personnes qui ne savent pas bien l'allemand.

La proportion énorme des enfants mâles quand le père est âgé m'a extrêmement frappé. Quand un de vos Lords voudra avoir des fils il faut lui conseiller de se marier tard! La nutrition qui modifie la proportion des sexes dans les produits paraît différente de ce qu'on suppose communément. Sur ce point les moyennes obtenues pour les animaux domestiques, selon la nourriture préalable des parents, sont curieuses et positives. Les vétérinaires en Allemagne ont réuni beaucoup de faits. J'espère que vous avancez dans vos relevés des nombreuses observations que vous avez faites. Ça sera un travail bien intéressant. Les publications de la Société d'anthropologie anglaise n'arrivent pas ici. Auriez-vous la bonté de me donner le titre et le prix du journal de cette société? Je trouverais peut-être moyen de le faire acheter par notre Société de Lecture, sorte d'institution littéraire qui reçoit déjà des publications françaises et allemandes sur l'anthropologie. Recevez, mon cher Monsieur, l'assurance de mes sentiments très dévoués.

ALPH. DE CANDOLLE.

PLATE XXI

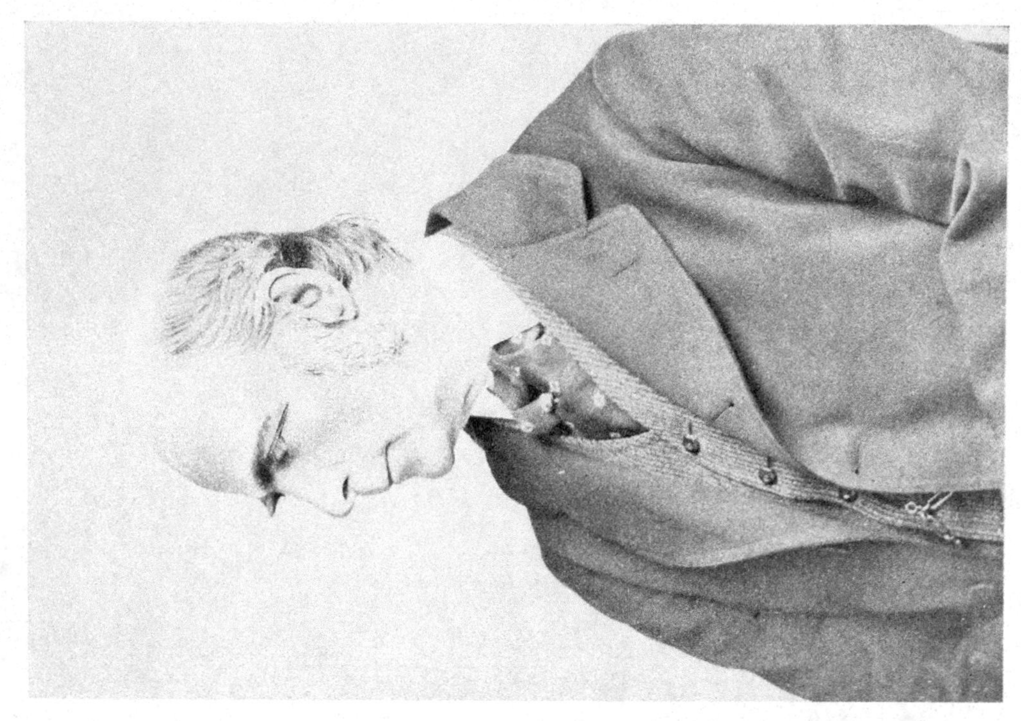

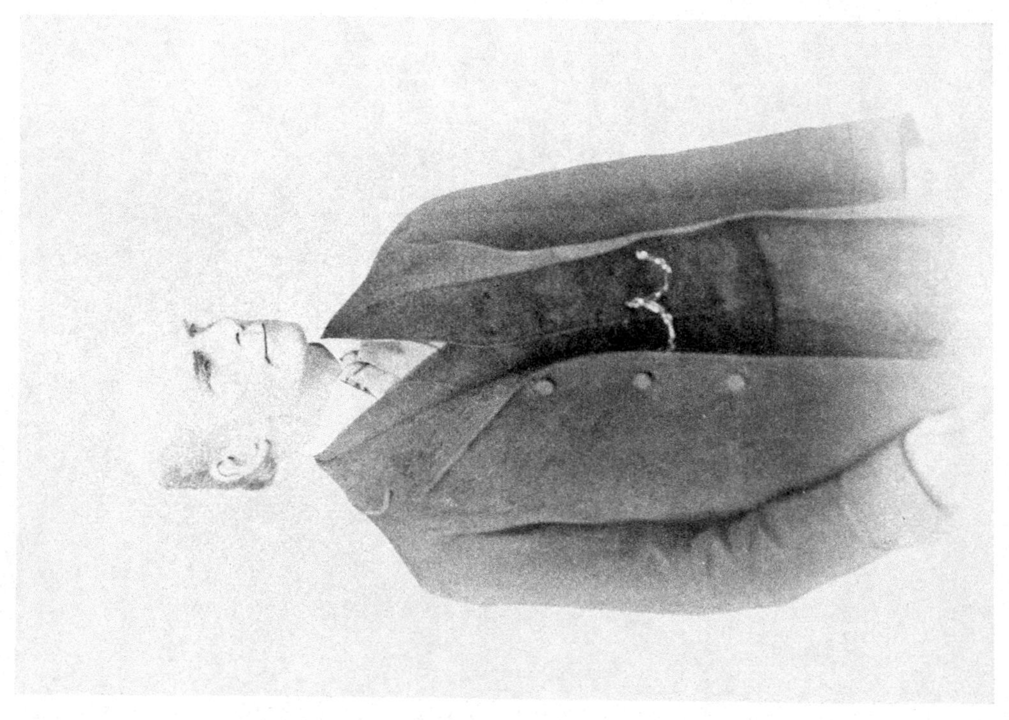

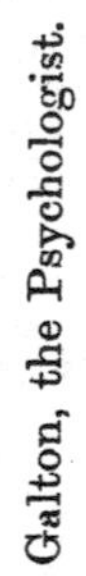

Galton, the Psychologist.

# CHAPTER XI

## PSYCHOLOGICAL INVESTIGATIONS

**"While recognising the awful mystery of conscious existence and the inscrutable background of evolution, we find that as the foremost outcome of many and long birth-throes, intelligent and kindly man finds himself in being. He knows how petty he is, but he also perceives that he stands here on this particular earth, at this particular time, as the heir of untold ages and in the van of circumstance. He ought therefore, I think, to be less diffident than he is usually instructed to be, and to rise to the conception that he has a considerable function to perform in the order of events, and that his exertions are needed. It seems to me that he should look upon himself more as a freeman, with power of shaping the course of future humanity, and that he should look upon himself less as the subject of a despotic government, in which case it would be his chief merit to depend wholly upon what had been regulated for him, and to render abject obedience."**

**FRANCIS GALTON, *Inquiries into Human Faculty*, 1883.**

*Introductory.* We have marked the transition of Galton's mind from interest in geographical to interest in anthropological studies. But once deeply interested in physical anthropology, he very soon grasped that the superficial anthropometric characters were no adequate index to the real man himself. Probably to the day of his death he would have been unwilling to admit that the size of a man's head had no real prognostic value as a measure of his intelligence. But he gradually came to the conclusion that the static anthropometric superficial characters afforded little index to a man's mentality, and from the middle of the seventies onwards Galton's thoughts turned more and more to the psychometric side of anthropology. He thus grew to have less and less faith in any superficial or bodily measurements being of psychological importance. He did not, I think, consider whether the dynamic anthropometric characters were more closely related than the statical to mental efficiency; indeed the measurement of the correlation between the physiological functioning of the various organs of the body and its psychical activities is a problem of quite recent days; and we stand only at its threshold as far as scientific—by which I understand quantitative—solution goes. Galton was, however, among the first, if not absolutely the first, in this country to insist that anthropometry cannot make real progress without psychometric observation and experiment. He was the first to insist upon the importance of experimental psychology—and he approached the subject from the standpoint of the anthropologist. It is perfectly true that Germans were working at experimental psychology at least as early as Galton. Wundt reversed Galton's process and passed with doubtful success from psychology to anthropology[1]. But it seems to me that the work of the two men was wholly independent and that Galton was the pioneer of experimental psychology in this country. Indeed very little real progress was possible in this new science without the aid of Galton's correlational calculus, and psychologists not only owe Galton a great debt for his suggestive experiments and actual apparatus, but also for those mathematical methods which are now the commonplace tools of psychological investigation.

I do not speak without careful examination of the facts, when I claim for Galton a pioneer position in experimental psychology in Great Britain. His *Inquiries into Human Faculty and its Development* appeared in 1883, but it was a *résumé* of work which had occupied Galton for at least seven years previously, and if we include folk psychology, for twelve years[2]. Galton's notebooks and queries to himself and friends begin as early as 1876, and one docket is inscribed by himself "Psychometric Inquiries 1876."

In March, 1883, Galton printed and issued a four-page pamphlet in the preparation of which he had the aid of the late Professor G. Croom-Robertson[3]. Galton opens with the statement that:

"I am endeavouring to compile a list of instruments suitable for the outfit of an Anthropometric Laboratory, especially those for testing and measuring the efficiency of the various mental and bodily powers. The simplest instruments and methods for adequately determining the delicacy of the several senses are now under discussion. After these shall have been disposed of, the next step will be to consider the methods of measuring the quickness and the accuracy of the Higher Mental Processes. Any information you can give, or suggestions that you can make, will be thankfully accepted."

The remainder of the pamphlet deals with the measurement of sensitivity, giving an analysis of the facts of sensation, and a programme of what has to be measured in (I) Skin-sensation, (*a*) Temperature, and (*b*) Touch, (II) Sight, (III) Hearing, (IV) Smell, (V) Taste, and (VI) the so-called muscular sense. Much of this is of course very familiar now. But it led Galton himself to devise various instruments for testing skin-sensation, hearing, smell, etc. As the pamphlet states, having the facts clearly before us, we must next "proceed to consider the most suitable apparatus to afford the measurements (or other tests) suggested by the several paragraphs." This pamphlet was followed by a proposal to hold an exhibition of psycho-

[1] Compare the great difference in value between Wundt's *Psychologische Studien* and his *Völkerpsychologie*.

[2] As evidenced by correspondence in the Galton Laboratory. The first published paper was that on the *Whistles* of 1877, and the *Composite Portraits* and *Generic Images* followed in 1878 and 1879 respectively.

[3] Galton's friendship with Croom-Robertson began in 1876, when the latter was just starting *Mind*. Galton had sent him two of his papers on Heredity, and Croom-Robertson said they should not be overlooked in the second issue of that Journal. He also asked Galton for psychological contributions. "There was no one to whose intelligent cooperation I then owed more than Professor Croom-Robertson (1842–1892) of University College. His genius and temperament were of the most attractive Scottish type—exact, sane and very genial.... He was a thorough friend whose death left a void in my own life that has never been wholly filled." *Memories*, p. 267.

metric instruments which was circulated among the leading English psychologists[1]. The noteworthy fact that resulted was that very little apparatus of the kind existed in England, and practically none had been invented there. One distinguished psychologist wrote to Galton:

"I regard you as a public benefactor and only wish I could be of more use to you. For some time I have been intending to get together some psychophysical apparatus but the difficulty has been to get the money. Just as that difficulty was to some extent surmounted I found myself committed to a biggish piece of literary work which will take all my time for some months to come.

One of the first things I meant to do was to write to you and ask to be allowed to see some of your apparatus; that I shall now be able to do when this exhibition comes off. I expect you know a great deal more about the whole thing than I do. I may, however, mention two or three books and papers in which apparatus has been described."

And then follows a list of references, almost entirely to German papers.

"But I am afraid in saying all this to you I am making myself very offensive, sending slack to Newcastle. However you must forgive me, if you will, and believe that I am only anxious to be of use to you if I can."

These sentences seem to suggest that in 1884 a leading psychologist could recognise Galton as a pioneer. The same authority, writing in 1911, says:

"The position I think is this: Galton deserves to be called the first Englishman to publish work that was strictly what is now called Experimental Psychology, but the development of the movement academically has, I believe, in no way been influenced by him."

Possibly it would have been better for English psychology had it been more influenced by Galton. We should then have had an original English School of Psychology, not handicapped by German dominance. But no one can to-day examine American and English psychological papers without recognising that their chief superiority over German and French work lies in the adoption of Galton's correlational calculus. It has given them a methodology far superior to that of their continental competitors, and on

[1] 42, Rutland Gate, S.W. *Jan.* 28/84.

Dear Mr —— I have undertaken to arrange and exhibit at the large forthcoming Health Exhibition a suitable outfit for an Anthropometric Laboratory. Its object would be to afford means of defining and measuring personal peculiarities of Form and Faculty, more especially to test whether any given person, regarded as a human machine, was at the time of trial more or less effective than others of the same age and sex. Again, to show by means of testings repeated at intervals during life, whether the rate of his development and decay was normal. The apparatus should refer to:—

1. Ordinary weighing and measuring, spirometer, colour of hair and eyes, etc.
2. Muscular action,—strength, estimate of range of motion, right and left-handedness, steadiness of hand, etc.
3. Effectiveness of the various senses, duration of impressions, after-images, reaction-times, waxing and waning stimuli, etc.
4. Higher mental processes, judgment of lengths, angles, memory of eye and ear. Elementary judgments, maps and apparatus.

(This is but a brief offhand and not a well methodised description, but it will serve for the present.)

Have you any special apparatus that you would allow me to exhibit in your name? Either the apparatus itself, a picture of it, or any hints from which I could have apparatus made? I should be most *grateful* for any hints. Very faithfully yours, Francis Galton.

this account alone it is impossible to assert that experimental psychology in our universities has been in no way influenced by Galton. There are not wanting signs also, that academic psychology may awake to a truer sense of what Galton achieved in this field and will cease, while adopting his calculus, to disregard both his apparatus and observational work. Four years before Galton started his exhibition he was, however, collecting his information and distributing his schedules. The following letter to Professor James Ward written in 1880 will indicate how Galton was then working on visualised numbers:

42, RUTLAND GATE, *Feb.* 9/80.

DEAR MR WARD, What a charming, interesting and full letter you have sent me. I wish Pythagoras was in reach of the penny post that I might send him a schedule. But failing that, please tell me if I rightly catch your explanation. Is it that Pythagoras who (to use your numerical equivalents) always visualised "a clever boy" whenever there was any question of the number five,—five men, five shields, five dinners etc.,—came to think that the "clever boy" was more of a reality than the men, shields or dinners? I could better understand that "numbers are the *μίμησιν* of things" than the converse way in which it is put by him. Will you kindly write and tell me?......

The association between number and colour has, I find, to be criticised rather closely to be sure that it has not a trivial origin. A young lady of apparently more than average ability had astonished her Father by an accidental allusion to these things. He told me of it and I questioned her. One very decided association was red with "million"; she told me she thought it due to the play of the word "*vermillion.*" Another correspondent (indeed 2 or 3 I think) speaks of much the same thing as regards letters. One wrote to me this morning saying that *e* was always green; but he believed this due to the *ee* in the word. But there is no doubt that blue has a calming effect and red an irritating one, for the Italian mad-doctors find an advantage in putting their irritable patients in a room lighted with blue light, and their apathetic ones under a red light.

As regards the preference for particular numbers, it would indeed be a curious inquiry. I had some experience more than once in that myself;—thus in getting census returns of age, 30 is a favourite answer, there are a paucity of 29's and 31's and a superabundance of 30's. Also in meteorological readings a tendency of that kind shows itself. The Hebrew 40, etc., are similar cases. (In my own family 16 was an habitual noun of indefinite magnitude, I could not conceive why.) Your suggestion, however, throws much light on the usual causes for preference.

I quite understand your ⁙ etc. in the sense you mentioned, but I see that I have somewhat bungled in the use of it notwithstanding. It often seems to me that there is a perverse demon, who somehow makes one write or do differently to what one intended to do. I can recall one gross error that I once made in pure defiance of my better judgment; it seemed temporarily sent to sleep while the hand wrote. A poor excuse!

What you say about your rudimentary diagram of figures is doubly interesting. It helps to show continuity between total absence and full existence and it is the first clear account I have received of *motor* sensation associated with number. The absence of these has hitherto astonished me, because my own representations are eminently motor. I can't think of "gratitude" without mentally acting the part of a grateful man, etc.

I really think there will turn up as you suppose, some facts bearing on teaching arithmetic. A girl of French parentage (the father is a mathematician settled in England) had her system sent me by her Father, together with his own. It shows clearly the influence of the French *names* for numerals. With many thanks for all you have sent.

Yours faithfully, FRANCIS GALTON.

## A. PSYCHOMETRIC INSTRUMENTS

I think it will be best in dealing with Francis Galton's psychometric work to start with some account of the instruments devised by him. In 1876 there was exhibited at South Kensington a "Special Loan Collection of Scientific Apparatus," and in connection with this exhibition a series of conferences was held in the month of May. At these conferences discussions on various subjects took place, largely in relation to the instruments exhibited. Spottiswoode was President of the section of Physics, and among the Vice-Presidents were De la Rue, Helmholtz, Tyndall and Sir William Thomson (Lord Kelvin). On May 19, one of the subjects for discussion was "The Limits of Audible Sound," and among other papers Galton gave an account of his "Whistles for determining the upper limits of audible sound in different persons[1]."

Galton notes that the number of vibrations perceived of a "closed pipe" or whistle depends upon its length. Accordingly he alters its length by a screwed plug at the closed ends; the number of turns and part turns of the screw are registered on a scale fixed to the walls of the whistle and on the screw head. The pitch of the screw is 25 to the inch. Hence one turn of the head shortens the tube by $\frac{1}{25}$th of an inch and the head of the screw being divided into ten parts it is possible to shorten the whistle by $\frac{1}{250}$ of an inch with perfect ease. Now the velocity of sound in ordinary conditions of temperature and pressure being 13,440 inches per second, the note of the whistle may be found by dividing 13,440 by four times the length in inches, i.e. by $4n \times \frac{1}{250}$, where $n$ is the reading on the scale, or $840,000/n$ is the number of vibrations per second[2]. For example, if the screw be set at 10, there are 84,000 vibrations a second, if at 70, 12,000, while a setting of 120 denotes 7000 vibrations per second. This rule of course applies only to strictly longitudinal vibrations. Galton very properly observes that it ceases to apply when the length of the tube is less than one-and-a-half times its diameter. When the tube is reduced to a shallow pan, it is the transverse vibrations which are all important. The necessity of preserving a fair proportion between diameter and length, led Galton to reduce the bore of his tube in some cases to a very minute dimension. On this account he considered that his whistles could not be relied on for vibrations of more than 14,000 to the second.

Galton notes than when the limits of audibility for a given person are reached "the sound usually gives place to a peculiar sensation, which is not sound but more like dizziness, and which some persons experience in a high degree." He further remarks that young people hear shriller sounds than

[1] *South Kensington Museum Conferences held in connection with the Special Loan Collection of Scientific Apparatus*, 1876. *Physics and Mechanics Volume*, p. 61. Published by Chapman and Hall. Galton's account was reproduced with some introductory matter in a pamphlet entitled "Galton's Whistles," issued by Tisley and Co., who manufactured the whistles commercially. They are still manufactured, but not by this firm's successors.

[2] The original gives $84,000/n$ by a slip.

older people, and cites a Dorsetshire proverb "that no agricultural labourer who is more than forty years old, can hear a bat squeak." He distinguishes between the sharpness of hearing and the hearing of high notes, and indicates that the position of the whistle—opposite to the auricular orifice—may be of importance.

Dalby 'the aurist' had already used one of Galton's whistles for diagnosis, and Galton himself had tried experiments with them on all kinds of animals at the Zoological Gardens and on insects. He put one of his whistles at the end of a hollow walking-stick which had a bit of india-rubber piping under the handle, brought the stick as near as was safe to the animal's ear, and when it was accustomed to it, squeezed the tube, and observed whether it pricked its ears. If it did, it probably heard the whistle. Cattle and ponies, much more than horses, hear high notes. If you pass through the streets of a town, working the walking-stick whistle, all the *little* dogs turn round, but it does not seem to have any effect on the large ones.

"Of all creatures I have found none superior to cats in the power of hearing shrill sounds. It is perfectly remarkable what a faculty they have in this way....... You can make a cat, who is at a very considerable distance, turn its ear round by sounding a note that is too shrill to be audible by any human ear."

Galton attributes this faculty in cats to natural selection, differentiating them so that they can hear the shrill notes of mice and other animals they need to catch. Some of Galton's audience at the conference heard the high notes of his whistles, others failed to catch them at all. Among the former was Alexander J. Ellis, translator of Helmholtz's *Lehre von der Tonempfindungen*, who stated that he heard all the high notes perfectly.

It is clear that very useful work might be done to-day by testing the members of families and forming pedigrees for cases in which there is a faculty for hearing very high notes, and probably Galton's whistles would be an adequate means of investigation. I do not remember ever seeing a frequency curve for a large general population of the limit of audibility[1].

An addendum to the above paper on whistles was contributed to *Nature*[2] by Galton in March, 1883. He notes that while his little whistle, set at ·14 of an inch, would give about 24,000 vibrations per second if air were puffed through it, the vibrations will be some 86,500 a second if hydrogen be used, because the number of vibrations per second is inversely proportional to the square root of the specific gravity of the gas blown through, and hydrogen is thirteen times lighter than air. Galton tested first with coal-gas, the specific gravity of which is not much more than half that of common air. He found that a length of ·13 of the whistle gave him personally no audible note for air; but he heard the note at ·14; he could for coal-gas get no audible note at ·24. Galton suggests that the whistle-lengths at limit of audibility, being as ·14 to ·25, or as ·56 to 1, are nearly in the ratio of 60 to 1, or the specific gravities. But if the audibility depends on the period and not the square of the period, ·56 to 1 should be as the *square roots*

[1] Galton's published data do not really provide material for such a curve (see our p. 221).

[2] Vol. XXVII, p. 491, corrected Vol. XXVIII, p. 54.

of the specific gravities. The experiment may possibly indicate that the subject appreciated the notes not by their number of vibrations, but by their energy.

As some persons can hear a musical note with the air whistle set at much less than 14, it may be concluded that 173,000 vibrations per second are possible with a hydrogen whistle.

"Mr Hawksley is making for me an apparatus with small gas bag for hydrogen pure or diluted, and an india-rubber ball to squeeze to enable hydrogen to be used with the whistle when desired. The whistle is fixed to the end of a small india-rubber tube in order to be laid near the insect whose notice it may be desired to attract."

Galton thought it possible that some insects may hear notes quite inaudible to man and he proposed to put this to the test of experiment. I do not know of any report on the results of experiments with this hydrogen whistle on insects. The difficulty for fieldwork, as apart from laboratory experiment, would be the transport of the hydrogen.

From Hearing Galton turned his attention to the "muscular sense," or rather to that combination of senses which tests by lifting weights what difference, if any, there is between them[1]. Galton, adopting Weber's law, took his weights in geometrical progression, i.e. as

$$WR^0,\quad WR^1,\quad WR^2,\quad WR^3, \text{ etc.}$$

He chose $W = 1000$ grains and $R = 1020$ grains and had ten varieties taking $R$ to the powers:

$$0,\ 1,\ 2,\ 3,\ 3\tfrac{1}{2},\ 4\tfrac{1}{2},\ 5,\ 6,\ 7,\ 9,\ 12.$$

He made his weights by charging cartridge cases with shot and closing in the usual way with a wad. If the weights be numbered with the power of $R$, Galton obtained a series of triplets of the following kind:

| Just Perceptible Ratio | Grade of Sensibility | Sequence of Weights |
|---|---|---|
| 1·020 | I | 1, 2, 3 |
| 1·030 | I½ | 2, 3½, 5 |
| 1·040 | II | 3, 5, 7 |
| 1·050 | II½ | 2, 4½, 7 |
| 1·061 | III | 0, 3, 6 |
| 1·071 | III½ | 0, 3½, 7 |
| 1·082 | IV | 1, 5, 9 |
| 1·093[2] | IV½ | 0, 4½, 9 |
| 1·104 | V | 2, 7, 12[2] |
| [1·115 | V½ | 0, 5½, 11][3] |
| 1·126[2] | VI | 0, 6, 12 |

Galton chose his lowest weight ($WR^0$) so that it gave a decided sense of weight, and his highest so that it could be handled without sense of fatigue. The test consisted in placing the weights in each series in correct order of

[1] "An Apparatus for Testing the Delicacy of the Muscular and other Senses in different Persons." *Journal of the Anthropological Institute*, Vol. XII, pp. 469–75.

[2] Corrected values; these are errors in the original paper.

[3] Interpolated to complete series, but not available with Galton's original ten weights.

magnitude. The grade, beyond which the order was not correctly given, measured the muscular sensitivity of the individual. Galton emphasised the fact that beyond true appreciation, the correct order might be given by chance in, perhaps, one or another case.

The important points here are: (i) How far the sense measured is touch and how far muscular appreciation. In Galton's method of handling even inertia might be a factor of the appreciation[1] (pp. 473–4). (ii) Galton assumes the geometrical law, and this plays a large part in his later work. (iii) He does not suppose with Weber that $W$ and $R$ vary from individual to individual. He assumes a sort of population average value for $W$ and $R$. I am by no means sure that his purpose could not have been accomplished with equal effectiveness by taking the first weight the same in each triplet (or quartet) and making the others proceed, not by equal ratios, but by equal differences; in fact his geometrical series, except in the lowest grades of sensitivity, are very approximately arithmetic series.

Galton remarks:

"Blind persons are reputed to have acquired, in compensation for the loss of their eyesight, an increased acuteness of their other senses. I was therefore curious to make some trials with my test apparatus, and I was permitted to do so on a number of boys at a large educational blind asylum, but found that although they were anxious to do their best, their performances were by no means superior to those of other boys. It so happened that the blind lads who showed the most delicacy of touch, and won the little prizes I offered to excite emulation, barely reached the mediocrity of the sighted lads of the same ages, whom I had previously tested. I have made not a few observations and inquiries, and find that the guidance of the blind depends mainly on the multitude of collateral indications, to which they give much heed, and not in their superior sensitivity to any one of them. Those who see do not care for so many of these collateral indications, and habitually overlook and neglect several of them. I am convinced also, that not a little of the popular belief concerning the sensitivity of the blind is due to occasional exaggerated statements that have not been experimentally verified." (p. 475.)

So Galton destroyed another of the beliefs, which are only held because men in general have been too sluggish to test their truth experimentally.

In a footnote added in March of the following year, 1883[2], Galton endeavoured to distinguish between the sense of touch and the sense of muscular effort. He supposes the test object held in the palm of the hand, palm uppermost, while the back of the extended hand rests on a broad and padded stirrup, connected by a string with fixed pulleys and a counterbalance weight. There is then no muscular effort to support the weight, and the hand can distinguish easily between the localised pressure of the weight on the palm and the "soft and broad pressure" of the stirrup on the back of the hand. The counterbalance is then removed and the "operator" experiences at once the muscular efforts necessary to support the weight and distinguishes it from the mere pressure of the weight on the palm. I believe Galton was the first investigator anywhere to measure muscular sensitivity by the discrimination of weight boxes.

As Galton's anthropometric measurements of sensitivity and of physique

[1] In the Anthropometric Room of the Galton Laboratory four not three weights are used for each test. Each weight consists of a circular tin box loaded with shot, and is lifted by the thumb and two fingers *without rocking*.

[2] The paper was read Nov. 14, 1882.

increased in range he invented further instruments. Thus we find, in 1889, an "Instrument for testing the perception of differences of Tint[1]." This might be described as a double wedge photometer, one photometer being set by the examiner and the other by the examinee, who endeavours to match the known tint set by the examiner. Actually Galton got over the expense of wedge photometers by using sheets of coloured glass, each rotating on a horizontal wheel on the same axis, and which could thus be set at any angle to the examinee's line of sight; a rotation of either wheel caused the light from an illuminated screen to pass through a greater thickness of the coloured glass. For the measurement of white light Galton replaced the sheets of coloured glass by gratings. The whole apparatus was extremely simple; the examinee, with his head screened from the light, looked through a slit into a horizontal tube blackened inside, at the other end of which were two windows, with outlook on an illuminated screen. Inside the tube in front of the two windows were placed the two wheels carrying the examiner's and the examinee's photometric sheets of glass diametrically, one was controlled by the examiner and the other by the examinee, and the former recorded the angular settings of both. The difference was a measure of the goodness of the colour matching. The great advantage of the instrument over a wedge photometer system lies not only in economy, but in the power it gives the experimenter of changing his colours[2]. Some disadvantage arises from the varying amount of light reflected from sheets at varying angles.

Another instrument exhibited at the same time[3], but the details of which belong to an earlier period, was a pendulum for "Determining Reaction-Time." This consisted of a fairly massive seconds pendulum, which could be released at an angle of 18° from the vertical; during its descent it gave a light-signal by brushing against a very light and small mirror which reflected a light off or onto a screen, or on the other hand it gave a sound-signal by a light weight being thrown off the pendulum by impact with a hollow box[4]. The position of the pendulum at either of these occurrences is known. The position of the pendulum, when the response is made to the signal, is obtained by means of a thread stretched parallel to the axis of the pendulum by two elastic bands above and below and in a plane perpendicular to that of the motion of the pendulum. This thread moves freely between two parallel bars in a horizontal plane, and pressing a key causes the bars to clamp on the thread, just, for illustration, as the bars of a parallel ruler might close on the thread. This determines the response-position of the pendulum, the motion of which is not suddenly checked by the clamping of the thread, owing to the elastic bands. The horizontal bars are just below

[1] *Journal of the Anthropological Institute*, Vol. XIX, pp. 27–29, 1889.

[2] I have recently had such a piece of apparatus constructed in the Biometric Laboratory for testing personal equation. Some mechanical difficulty arose in bringing the two coloured windows adequately close together for reasonable comparison. I surmounted this by aid of a prism of Iceland spar, the image of the ordinary ray of one could be juxtaposed to the image of the extraordinary ray of the other, the other images being cut off by a diaphragm. The slight colouring of the border of the extraordinary image was found negligible. [3] *Ibid.* p. 28.

[4] A similar pendulum, adjustable to any time of oscillation, in the Galton Laboratory, just touches a delicately-balanced hammer which falls on a bell.

a horizontal scale which is 800 mm. below the point of suspension of the pendulum. Galton provides a table for reading off the distances along the scale from the vertical position of the pendulum in terms of the time the pendulum takes from the vertical position to the position in which the thread is clamped. The reaction time is thus ascertainable on the assumption that the time from pressing the key to the mechanical clamping of the thread is negligible compared with the reaction time. It would however be easy to correct for this, if we arranged occasionally for the pendulum to work the clamping key itself, and so ascertained the time of clamping independently of the living being's reaction and response.

Another instrument designed by Galton was intended to measure the rapidity of a blow, or indeed the rate of movement of any limb[1]. The principle of the mechanism is that the limb is attached by a string to a light mechanism which draws in the string at a faster rate than the limb moves. The motion of the string is checked when the limb reaches its full extension, but a light weight on a platform continues to rise freely and measures by the height it reaches the velocity of the platform (and of the string) when the string was checked. The whole scheme is indicated in the accompanying diagram. *AB* is a stretched india-rubber band; in the actual machine as worked this was much longer than indicated in the diagram. *BC* is a thin steel wire to which the conical platform *D* is firmly attached, an ivory cylinder *E* rests on the platform and runs loosely on the steel wire. When the platform *D* is checked, *E* goes forward with *D*'s final velocity and this velocity is measured on the scale behind *BC* according to the height to which *E* rises. A string passes from the vertex of *C* and is wound round and ultimately fastened to one wheel *F* of a differential pulley. Another string is wound round the second wheel *G* of the differential pulley and ultimately fastened to it; the remainder of this string is made horizontal by being carried over a small pulley. To the

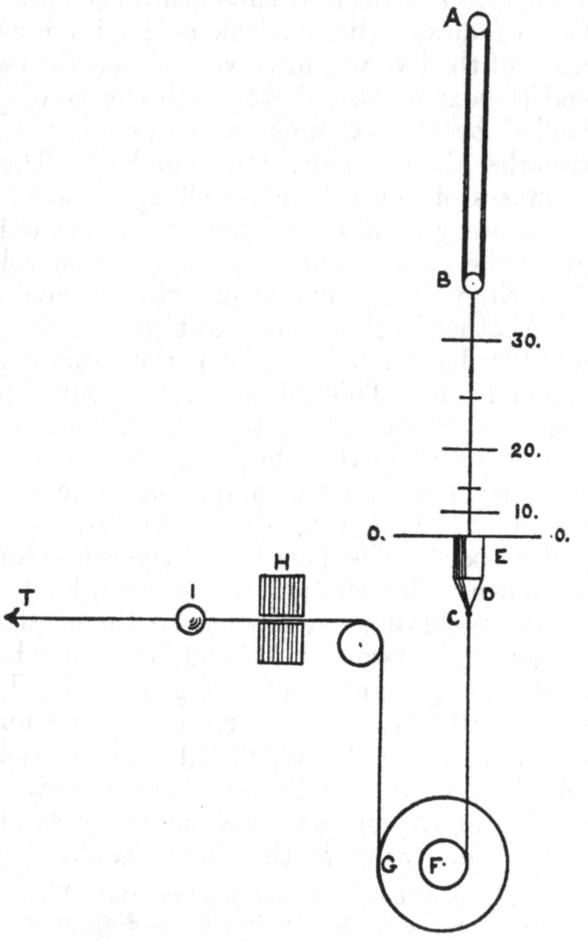


[1] *Journal of the Anthropological Institute*, Vol. xx, pp. 200–204, 1890. "A new Instrument for Measuring the Rate of Movement of the various Limbs."

horizontal portion a small india-rubber ball $I$ is firmly attached, which rests against the fixed buffer $H$, when the top of the cylinder $E$ is at the zero of the scale. The ball $I$ must be against $H$ when the limb is fully extended. $T$ represents the direction of the string carried to the moving limb. The differential pulley reduces the motion to $\frac{1}{3}$ that of the moving limb. The height $h$ in inches reached by $E$ for a given velocity $w$ of $I$ on impact is $0 \cdot 0207 \times w^2$, which allows of an easy graduation of the scale.

To measure the velocity of a blow Galton places the examinee with his back to a wall, and he strikes at a long feather, so adjusted that (i) when the fist reaches the feather the india-rubber ball strikes the buffer. Care must be taken (ii) that the wrist is not bent, and (iii) that the extended horizontal limb is in the horizontal line of the string: the free end of the string is attached to the fist. The machine requires vertical and horizontal adjustments to allow of the fulfilment of (i) and (iii). Galton states that the instrument had worked successfully in his laboratory. It neglects the resistance of the air on the small bead or cylinder $E$, together with the possible friction of the steel wire, and the additional acceleration due to the pull of the string on the limb; all these are however very secondary factors, and might, were it necessary, be allowed for. As usual with Galton's apparatus, the constituents are of the simplest character, and any man with a modest mechanical knowledge could rig up such an instrument.

A different arrangement for measuring the velocity of a blow was used by Galton in 1882. It is figured by Galton in his account of his first Anthropometric Laboratory[1], to which we shall refer later, but the instrument was discarded as it was liable to be injured if the blow was not a straight one, and occasionally in that case the experimentee injured himself! The instruments were chiefly devised by Galton himself, and included:

(*a*) For Hearing: both *Acuteness* and the *Highest audible Note.*

Under sound we first reach a point on which Galton was rather insistent, namely that the sensitivity of women, the fineness of touch, of hearing, of taste, etc., was not greater than that of men, although the contrary had been often asserted. Galton tested with four of his whistles, giving 20,000, 30,000, 40,000 and 50,000 vibrations a second[2], with the following results:

| Sex | Ages | Percentage of cases in which the undermentioned numbers of vibrations were perceived as a musical note | | | | Number of cases |
|---|---|---|---|---|---|---|
| | | Number of vibrations per second | | | | |
| | | 20,000 | 30,000 | 40,000 | 50,000 | |
| Males ... | 23—26 | 99 | 96 | 34 | 18 | 206 |
| | 40—50 | 100 | 70 | 13 | 1 | 317 |
| Females... | 23—26 | 100 | 94 | 28 | 11 | 176 |
| | 40—50 | 100 | 63 | 8 | 1 | 284 |

[1] "On the Anthropometric Laboratory at the late International Health Exhibition." *Journal of the Anthropological Institute*, Vol. XIV, pp. 205–212, 1885.

[2] I feel some doubt as to the accurate standardisation of these whistles.

On this Galton remarks: "It will be seen here, as in every other faculty that has been discussed, the male surpasses the female[1]." Elsewhere Galton writes[2]:

"The trials I have as yet made on the sensibility of different persons confirm the reasonable expectation that it would, on the whole, be highest among the intellectually ablest. At first owing to my confusing the quality of which I am speaking with that of nervous irritability, I fancied that women with delicate nerves who are distressed by noise, sunshine, etc., would have acute powers of discrimination. But this I found not to be the case. In morbidly sensitive persons, both pain and sensation are induced by lower *stimuli* than in the healthy, but the number of just perceptible grades of sensation between the two is not necessarily altered.

I found, as a rule, that men have more delicate powers of discrimination than women, and the business experience of life seems to confirm this view. The tuners of pianofortes are men, and so, I understand, are the tasters of tea and wine, the sorters of wool, and the like. These latter occupations are well salaried, because it is of the first moment to the merchant that he should be rightly advised on the real value of what he is about to purchase or to sell. If the sensitivity of women were superior to that of men, the self-interest of merchants would lead to their being always employed, but as the reverse is the case the opposite supposition is likely to be the true one."

The suggestion here made was worth consideration, but only limited weight can be given to it, when we consider how many callings at that date were closed to women, without their being really unfitted for them. Greater stress must, however, be placed upon Galton's actual observations such as those just recorded for the audibility of high notes. At a later date[3] Galton made experiments on the sensitivity of men and women with regard to their discrimination in touch, using as an aesthesiometer a pair of dividers applied to the nape of the neck. He found that women were *superior* to men in tactile sensibility in the ratio of about 7 to 6. Galton's result has been confirmed by many later investigators. He also shows in the same paper that women are more variable in sensitivity of touch than men. He dealt with 932 males and 377 females, and worked by the method of median and quartiles. There are irregularities in the tabled data, however, which suggest some anomalies in the recorder's (Sergeant Randall's) method of measurement; they are probably inadequate to influence the main results.

Thus Galton's original generalisation was too sweeping. If we look to the evolutionary standpoint and indulge for a moment in hypotheses, we might suppose natural selection endowed the hunter and warrior with great sensitivity in the matter of sight and sound, while sensitivity to touch after capture may well have played a part in the surrender of the female and successful mating in a much earlier stage of living forms than the human[4].

(*b*) For Sight: *Keenness of Vision*, measured by an ingenious arrangement, one size of type, diamond, only being used, and the specimen cards, all

[1] *Loc. cit.* pp. 278, 286.

[2] *Journal of the Anthropological Institute*, Vol. XII, pp. 472–3, 1883.

[3] "The Relative Sensitivity of Men and Women," *Nature*, Vol. L, p. 40, 1894 (May).

[4] One may reasonably recognise female sensitivity to touch in the play of tail, rubbing of fur, and other excitatory actions of the male dog in courtship.

fastened square to the line of sight at distances 7″, 9″, 11″ and so on up to 41″. The curve of the frame along which the test blocks are placed was actually an equiangular spiral. *Colour sense.* A series of bars packed closely with coloured wools wound round their centres, and the examinee had to place pegs against such of the bars as had any shade of green wound round them.

*Judgment of the Eye: As regards Length.* A first bar is shifted along until a pointer is considered to bisect it, and a second bar until a pointer is considered to trisect it. A hinged lid in both cases screens a scale on the top of the bar, which has a central fiducial mark and $\frac{1}{100}$th graduations of its whole length on either side. *As regards Perpendicularity*[1]. A bar rotates about a screened pivot on a horizontal table; this bar must be set perpendicular to a line drawn on the table. When set, a lid is raised, and a protractor rendered visible on which the difference of the setting and of true perpendicularity can be read off.

(*c*) Instruments for measuring *Sense of Touch* were also exhibited but not used. Some years afterwards Galton adopted as aesthesiometer dividers applied to the nape of the neck[2].

(*d*) Later Galton dealt with the *Sense of Smell*, and in the Galton Laboratory we still use his method and his very bottles! The tests consist: (*a*) in sorting out by smell from a number of bottles those having the same contents and (*b*) in placing in order a number of bottles having various intensities of smell of the same material.

(*e*) In the test of the *Eye and Hair Colours* Galton used artificial glass eyes respectively dark blue, blue, grey, dark grey, brown grey (green, light hazel), brown, dark brown, black. He also used standard samples of hair: flaxen, light brown, dark brown, black, and three shades of red: fair red (golden), red, dark red (chestnut auburn). He was certainly among the first to introduce standard scales of this kind, and, what is more, to realise the difficulty of reproducing them. Such eye and hair scales are common enough now, but were by no means so in 1882, yet the difficulty remains of reproducing them accurately even when manufactured by one firm. The glass eyes of two standard scales are found not to have the same amount of pigment in them, and the spun glass silk used for standard hair scales not always the same

[1] Galton terms it "judgment of squareness," but I think such a name is better reserved for another sort of test which I have personally used. A number of rectangles, not diverging widely from squares in both directions and containing one true square, are given in confused order to the examinee and he is asked to give the number of the rectangle he considers square. In the same way a number of ellipses differing slightly from a circle are given, and he is asked to choose the circle; of course in both cases without correcting glasses. By giving each member of an audience a slip of paper as he enters, and throwing ellipses and rectangles on the screen by a lantern, I have been able to measure the astigmatism of 400 or 500 persons in a few minutes, and thus find not only the average astigmatism but the frequency distribution of astigmatism at the same time. The method was suggested to me by the contour of the dome of St Paul's, which always seems to me to have its major axis vertical, and to look ungraceful, until I rotate my head to the horizontal position, when it becomes gracefully proportioned. I found several of my friends thought the minor axis of the dome vertical. This "judgment of squareness" of course involves the error of judgment as well as astigmatism, but the latter is, I think, the chief contributory factor.

[2] See our p. 222.

amount of dye. Galton felt keenly the need for a standard and permanent set of colours, and made a suggestion on this point of great value. In 1869 he had been struck by the great variety of permanent colours which are produced for mosaic work. He had been over the *Fabbrica* of mosaics attached to the Vatican and seen their 25,000 numbered trays or bins of coloured mosaic. He realised at once the opportunity thus afforded not only for the establishment of a general colour scale in this country, but, as the mosaics were manufactured for the representation of human figures among other things, for skin, hair and eye-colour scales for anthropometric purposes[1].

On Feb. 3, 1870, Galton sent the following letter to the *Science and Art Department, South Kensington*. I cite from a rough draft in the *Galtoniana*:

"Certain scientific inquiries in which I am engaged have brought forcibly before my notice the great desideratum of being able to obtain an accepted standard scale of colours, by reference to which a person's meaning might be expressed with precision whenever he desired to designate a particular hue or tint. The exhibition of such a standard would fall, I venture to say, most legitimately within the province of the South Kensington Museum, and I will now show how very easily and efficiently this desideratum might be supplied. In the *Fabbrica* of mosaics at the Vatican in Rome there are no less than 25 thousand trays or bins, numbered consecutively, and each filled with cakes of mosaic material, each separate bin being devoted to a different colour. The workers on the mosaics in the *Fabbrica* send, as they require, to the superintendent of this department for so many pounds weight from such and such specified bins, the colours they want being solely expressed by the numbers attached to the bins. I have read cursory accounts of this large and most remarkable factory and I have visited it myself as an ordinary though much interested sightseer, but I cannot find any full description of its management either in the Art Library of the South Kensington Museum, or elsewhere. However it may be taken for granted that the facts of the case are substantially as I have stated them.

Now I beg to propose that the authorities of the South Kensington Art Department should make application to the Pope for mosaic tablets containing in order specimens of each of their 25 thousand bins to be suspended in the Museum for the purpose of reference as a standard of colour."

Galton then proceeds to discuss the space that such a scale of colour would occupy; if each fragment of mosaic were $1'' \times \frac{1}{2}''$, the space required would be about ten square yards. Supposing we arranged our tablets in series of 10 in file and 10 in rank, we should have for 20 rows deep, a length of about 52 feet for the scale. For square specimens $\frac{1}{2}'' \times \frac{1}{2}''$, which would probably be adequate, with 40 rows deep, the length of the scale would be about seven yards. Galton continues:

"It might be disposed as a frieze running along the wall at a height convenient for reference, the bits of mosaic perhaps arranged in tablets of 100 containing 10 ranks and 10 files, with dark lines at the 5th division each way for convenience of immediately ascertaining the number appertaining to each several bit.......

The *Fabbrica* at the Vatican is maintained by the Papal Government solely for the purpose of mosaics for public buildings in the Roman States and for making gifts to foreign potentates. Presents of art works are given in this way that required, I am afraid to say how many separate pieces of material for their construction and that have demanded the lifetime of a skilled artist for their completion. But the series of tablets of which I speak would be far more easily made.

[1] Many years after Galton's suggestion Professor von Luschan's useful mosaic skin colour scale came into existence. I have also procured mosaics from the *Hof-Fabrik* in Berlin and formed permanent scales for coat colour in mammals. Galton's proposal was a most fruitful one, and it is to be regretted that it was never carried out in its entirety.

They would be built up as readily as a wall is built with bricks. Even if it occupied a man a whole day to make a single tablet (10 × 10), the entire affair would fill less than a year of his time.

It is not to be supposed that the Vatican scale of colour is scientifically regular in the interval between the several graduations, neither have I reason to believe that scrupulous pains have been taken to keep the tints and hues of each bin identical in their character for consecutive centuries, or even for shorter periods; but this at least is certain: that the series is as minute and as comprehensive as it is possible to be; that it exists in the most durable of all materials, that it would be exceedingly useful to England to possess such a scale, that it might be had almost for the asking and that it would be a highly interesting and ornamental adjunct to the South Kensington Museum.

It might well be a subject for the subsequent consideration of the authorities of South Kensington whether they should not select by means of the large amount of skill and science at their disposal say one tenth of the Vatican series to create what might be called a South Kensington scale of colours, and distribute identical copies of it in mosaic, which would occupy a space according to the above calculation of less than 10 feet × 1 foot, among the art schools of the United Kingdom."

In a postscript sent two days later Galton suggested that to avoid difficulty and delay in Rome, it might be adequate to ask for rough specimens with their numbers from every bin and let the grinding to the required size be done in England, where the machinery to do it was better and more accessible than in Italy. Galton's letter was written on Feb. 3, the correspondence from the Museum up to May 16th is a series of letters saying that the subject "will receive consideration." After which date Galton, I presume, gave up asking for an answer to his letter! *Sixteen* years later (1886) Galton returned to his suggestion impressed by the fading of the original paintings of Broca for skin tints[1], and by a further brief stay in Rome where he had again visited the Vatican factory and made further inquiries. He now found that there were 40,000 bins of mosaics, and of these 10,752 were classified; they occupied 24 cases in each of which were 16 rows of 28 samples. The flesh tints appropriate to European nations were about 500 in number, so that the Vatican factory provided ample material for the selection of a series of tints such as anthropologists desired. Topinard, Galton stated, was preparing a new scale of only five or six tints for hair colour to be correlated with Broca's numbers, the latter's original tints having changed colour. Galton had asked for a copy of this new scale in order to match it by mosaics; he had promised to provide the cost, and he suggested that such scales in mosaics should be circulated among anthropological institutes and museums. He now adds that it may not be possible to get such mosaics from the Vatican factory to judge by a former experience, but they could possibly be obtained elsewhere. He then refers to his proposal of 1870 to the South Kensington authorities and states that Mr Odo Russell—later Lord Ampthill—our semi-official representative at the Vatican (till 1870) was ultimately asked to inquire as to the feasibility of carrying out the scheme,

"but the price asked by the Papal Government was altogether excessive, and so the matter dropped. Now, however, resulting not improbably from my then abortive suggestions, I find that such samples are being produced. I saw one set in process of being made." (p. 146.)

[1] *Journal of the Anthropological Institute*, Vol. XVI, p. 145, 1886. "Notes on Permanent Colour Types in Mosaic."

Galton exhibited cakes of Roman enamel suitable for anthropometric standards of colour[1]. And then in this century we have the idea carried out by a German with German made mosaics, and no one gives Galton credit for originating the idea!

I cannot trace that Galton got either the simple Topinard hair scale, or Broca's scale reproduced in mosaics. Before the war I found that painted scales sent to Berlin were very speedily and accurately matched in mosaics.

It seems in place here to summarise a further paper of a later date "On recent Designs for Anthropometric Instruments" in so far as it deals with the subject of our present chapter[2]. The pioneer work of Galton is here recognised to the full. His instruments had passed from South Kensington to Cambridge, and an anthropometric laboratory had been opened there. Messrs Horace Darwin and Dew Smith were improving old and devising new anthropometric instruments in Cambridge, and a good deal of this paper concerns their work.

A Japanese professor had sent Galton money to provide an outfit for an anthropometric laboratory in Tokio. Professor Giuseppe Sergi wished to add to his anthropological cabinet a set of instruments suitable for school work, and desired Galton to select a list for him. Topinard, one of the leading French anthropologists, wrote:

"I have written nothing as yet concerning physiological instructions to travellers, awaiting a convenient moment for doing so. I am disposed to take directly your system, and will ask to have all your apparatus sent to me. We possess no samples of colours for hair and eyes beyond the polychromatic table of Broca, which the Anthropological Institute employs, but I am about to undertake new work of this kind, and intend shortly to have some samples made; but not many of them, probably five for eyes and five for hair. My present difficulty is to select the exact shades and tints; if you have yourself made any such sets, I should be much obliged if you would let me have one." (p. 4.)

The impetus given by Galton's anthropometric laboratories was indeed universal, and he was admitted then to have led the way in this matter, an admission which has been almost overlooked since.

Among matters which concern us in this chapter are standards for hair and eye colours. Here Galton directly suggests "glass spun by a glass blower for comparison with hair." Thus before 1886 he had proposed sets of standard glass eyes, mosaics for skin colour and spun glass for hair; all three of these suggestions have been carried out in this century—by Germans—in the well-known eye-scale of Professor R. Martin, in Professor von Luschan's skin-scale and Professor G. Fischer's glass-silk hair scale. Thus the best of what we can do *now*, was suggested by Galton twenty to thirty years earlier.

Horace Darwin showed (i) a very simple chronograph designed by Francis Galton, (ii) an instrument for measuring the relative sensitiveness of the eye to various colours, designed at the suggestion of Galton, and (iii) an instrument for testing an individual's keenness in distinguishing small differences

[1] The originals are not in the *Galtoniana*, and I have not succeeded in tracing them.
[2] *Journal of the Anthropological Institute*, Vol. XVI, pp. 2–9, 1886.

PLATE XXII

Francis Galton in holiday garb; taken at Vichy, August, 1878, when aged 56 years. The Galtons were at Vichy again in 1880: see our p. 196.

in the pitch of a musical note, presumably designed entirely by Darwin himself. All these instruments are very simple in character and ought not to be overlooked by the anthropometrician. In particular the chronograph is very ingenious. Darwin thus describes it:

"A wooden rod is supported at its upper end by a detent, and can be released at will. The rod then falls freely in space passing through, a hole in a fixed diaphragm. A weight in the form of a ring larger than the hole in the diaphragm, rests on a collar near the top of the rod. Thus, after rod and weight together have fallen a definite distance, the weight is caught by the diaphragm and makes the signal sound, while the rod still continues to fall. On hearing the signal sound the person to be tested presses down a lever, thereby releasing a spring clamp which grips the falling rod firmly. The interval of time between the signal sound and this operation is measured by the space the rod has fallen through, and is read at once in hundredths of a second from graduations on the rod itself." (p. 9.)

Lastly, we may note that early in 1890 the Royal Society appointed a committee of which Lord Rayleigh was Chairman, Captain Abney, Secretary, and Sir George G. Stokes, then President of the Society, Brudenell Carter, Church, Evans, Michael Foster, Dr Farquharson, Galton and W. Pole were members. The average attendance was six to eight and Galton appears to have attended with the greatest regularity. At the fifth meeting he presented a memorandum as to testing colour blindness, (i) "under not dissimilar conditions to those in which signals are seen by sailors and engine drivers," (ii) in which the attendant does not know the colour being exhibited, and (iii) the subject indicates the colour not by its name, but by turning a thumb and finger piece attached to each colour box, with rough side up for red, smooth side up for green, and into an intermediate position for neutral tinted colour. The whole takes place in the dark, and the subject's answers are ascertained by examining the colour-boxes later. There were to be nine colour-boxes, three for red, three for green, three for neutral tint; each series with one, two and three thicknesses of glass. A good many other methods of measuring colour sense were described, and much evidence taken. I do not know whether Galton's apparatus ever came into practical use: like all his instruments, it was very simple, the light being provided by a policeman's "bull's eye."

Galton's investigation of mental characters led him directly to the Weber-Fechner Law of the geometrical mean. Such a law appears directly opposed to the Gaussian hypothesis that the arithmetic mean gives the best "medium," i.e. the most probable or modal value of a series of observations. Galton accordingly proposed the following problem: Assuming the geometrical mean and not the arithmetical mean to give the best "medium," what is the mathematical form of the frequency distributions[1]? Galton seems to have held that not only in tint and length judgments[2], but in many

[1] *R. Soc. Proc.* Vol. XXIX, pp. 365–6, 1879. "The Geometric Mean in Vital and Socia Statistics."

[2] "Three rods" Galton writes "of lengths $a$, $b$, $c$ if taken successively in the hand appear to differ by equal intervals when $a:b::b:c$ and not when $a-b=b-c$" (p. 366). I have made a number of individual tests on myself, but my judgment supports an arithmetic not a geometric mean in the case of the three rods. I once asked between 200 and 300 individuals to select on

sociological categories, the geometric mean would dominate the frequency distributions:

"My purpose is to show that an assumption which lies at the basis of the well-known law of "Frequency of Error" (commonly expressed by the formula $y = e^{-h^2x^2}$) is incorrect in many groups of vital and social phenomena, although that law has been applied to them by statisticians with partial success and corresponding convenience." (p. 365.)

By "vital phenomena" Galton here refers to those assumed to be governed by the Weber-Fechner law; as illustrations of "social phenomena" he cites growth of population following a geometrical increase, or increase of capital in a business which is proportional to its size.

"In short, sociological phenomena, like vital phenomena, are as a general rule subject to the condition of the geometric mean." (p. 367.)

That many sociological phenomena do lead to markedly skew distributions is I think a point of very great importance, and Galton's attention had soon been drawn to it. It is, however, very questionable whether the theory of the geometrical mean is the only, or a wide enough avenue of approach.

Galton put the matter in the hands of Mr (now Sir) Donald MacAlister, who deduced the frequency distribution at once[1], on the assumption that the logarithms of these vital and sociological variates would obey the frequency of error-curve. I am unaware of any comprehensive investigation being ever undertaken to test the "goodness of fit" of this geometric mean curve to actual observations. MacAlister gives no numerical illustration, and I do not think Galton ever returned to the topic. It would still form the subject of an interesting research, but I fear the Galton-MacAlister curve would be found wanting. See *Biometrika*, Vol. IV, pp. 193 *et seq*.

### B. PSYCHOMETRIC OBSERVATIONS AND EXPERIMENTS

Perhaps the most significant evidence of how Galton's mind was turning from physical to psychical anthropometry is to be found as early as 1877, in his "Address to the Anthropological Department of the British Association," at the Plymouth meeting of that year[2]. He there made, what for that

a tint scale a tint exactly intermediate between two tints $A$ and $B$, which actually contained 1/10 and 9/10 of black. The geometric mean would have given the mode at 3/10; it was actually about 7/10. This was confirmed by a second series of guesses. It is possible that the eye measured the amount of white not of black in the tint shades.

[1] "The Law of the Geometric Mean." *R. Soc. Proc.* Vol. XXIX, pp. 367 *et seq.*, 1879. The curve is $y = y_0 \frac{h}{\sqrt{\pi}} \frac{e^{-h^2(\log x/a)^2}}{x}$

[2] There is an historically very instructive series of letters which were interchanged between Galton and Huxley preserved in the *Galtoniana*, regarding the foundation of the "Department" of Anthropology in 1866. Huxley was president of Section D Biology, from which had sprung the "Departments" of Physiology and Anthropology, and he practically nominated all the officers of all three branches and Botany as well. "I think I mentioned to you that I proposed to ask Humphry to be President of the Physiol. Department and Wallace to take charge of the gentle Anthrop's. Both have consented."..."X. is the one man who won't do for any office in division Anthropology! *Dix mille fois, non!* Rolleston would go into convulsions at the mere rumour, and I confess that the less often that young gentleman comes in my way—the

time was a bold proposal, that all anthropologists should turn for a time from physical anthropology and study prevalent types of human character and temperament. He points out how it has now become possible to inquire by exact measurement into certain fundamental qualities of the mind; the new science of what has been termed Psychophysics shows that the difference in the mental qualities of man and man admits of being gauged by a suitable scale. Galton further suggested that mental qualities such as 'personal equation' and its basis in reaction time should be measured with a view of correlating them with temperament and external physical characters. Among other things he suggests the classification of individuals by the time they occupy in forming a judgment. He notes that the interval of time between the perception of a signal and the recording of it by tapping a key, is modified when there are alternative signals *A* and *B*, and the recording of *A* is to be done by the right and of *B* by the left hand. An interval is required to discriminate between the signals and between the hands. In such a way the individual time in forming a judgment can be to some extent measured. Galton compares the advance of that day in the measurement of mental characters with the numerical measurement by the thermometer of heat and cold in the days of old. As Dr John Beale wrote to Boyle in 1663:

"If we can discourse of heat and cold in their several degrees so as we may signify the same intelligibly...it is more than our forefathers have taught us to do hitherto."

The pity is that so much psychometric apparatus is far more expensive than thermometers! If we can, however, obtain a group with differentiated mental characters, how shall we ascertain the external physical features most commonly associated with its members? And here Galton turns, I think for the first time, to photography for assistance.

He suggests, in the first place, a standard form of photography in which by the aid of three mirrors, a direct three-quarter face, and reflected profile, full-face and top of head aspects would be obtained on the same plate at the same time. Unfortunately he does not describe adequately the positions of these mirrors, and I have been unable to determine them. I can get by reflection *norma facialis, norma lateralis* and *norma verticalis* (as they are termed in craniometry), but then the direct aspect appears to be a three-quarter occipital view! Galton next makes what I believe is his first announcement as to composite photography; that is the method he proposes of ascertaining whether those with differentiated mental characters have differentiated physical features. He writes:

"Having obtained drawings or photographs of several persons alike in most respects but differing in minor details, what sure method is there of extracting the typical characteristics from them? I may mention a plan which had occurred both to Mr Herbert Spencer and myself, the principle of which is to superimpose optically the various drawings and to accept the aggregate result. Mr Spencer suggested to me in conversation that the drawings reduced to the same scale might be traced on separate pieces of transparent paper and secured one upon

sweeter my temper is likely to be.—He is such a choice specimen of the Snob scientific." X. is dead now, without leaving his impress on science, but the term Huxley found in his wrath to characterise the young gentleman is perhaps worthy of preservation.

another, and then held between the eye and the light. I have attempted this with some success. My own idea was to throw faint images of the several portraits, in succession, upon the same sensitised photographic plate. I may add that it is perfectly easy to superimpose optically two portraits by means of a stereoscope and that a person who is used to handling instruments will find a common double eye-glass fitted with stereoscopic lenses to be almost as effectual and far handier than the boxes sold in shops[1]." (pp. 9–10.)

Thus was launched the first idea of composite photographs. But Galton very rarely made a suggestion without already having applied it himself[2], and in this case he had chosen as his subject "the criminals of England who have been condemned to long terms of penal servitude for various heinous crimes."

He had formed his own views on "the ideal criminal." He has three peculiarities of character: (i) his conscience is almost deficient, (ii) his instincts are vicious, and (iii) his power of self-control is very weak. His instincts determine the description of his crime, and the absence of self-control may be due to ungovernable temper, to sensual passion, or to mere imbecility.

Galton as a biologist is very cautious in his discussion of "vicious instincts." He says:

"The subject of vicious instincts is a very large one: we must guard ourselves against looking upon them as perversions, inasmuch as they may be strictly in accordance with the healthy nature of the man, and being transmissible by inheritance, may become the normal characteristics of a healthy race, just as the sheep-dog, the retriever, the pointer, the bull-dog have their several instincts. There can be no greater popular error than the supposition that natural instinct is a perfectly trustworthy guide, for there are striking contradictions to such an opinion in individuals of every description of animal. All that we are entitled to say is, that the prevalent instincts of each race are trustworthy, not those of every individual. A man who is counted as an atrocious criminal by society, and is punished as such by law may nevertheless have acted in strict accordance with his instincts. The ideal criminal is deficient in qualities that oppose his vicious instincts; he has neither the natural regard for others which lies at the base of conscience, nor has he sufficient self-control to enable him to consider his own selfish interests in the long run. He cannot be preserved from criminal mis-adventure, either by altruistic or by intelligently egoistic sentiments." (pp. 11–12.)

Having defined the mental characters which he considers peculiar to the criminal Galton next proceeded to investigate how far these peculiarities are correlated with the physical characters, in particular with the physiognomy. He divided his criminals into three main groups taking in all cases the photographs[3] only of men sentenced to long terms of penal servitude; the groups were (*a*) Murder, Manslaughter and Burglary, (*b*) Felony and Forgery, (*c*) Sexual offences. Galton believed that by continually sorting the photographs in tentative ways certain natural classes began to appear, some very well marked, and that the proportion of these in the three crime-

[1] Galton's double eyeglass with stereoscopic lenses is in the *Galtoniana*.

[2] I think an exception must be made to this rule in the case of the four aspects on our photographic plates referred to above. But I may have overlooked some possible mirror arrangement, and if not we have to remember that Galton's address was prepared in great haste, for he had been suddenly called upon to occupy the chair owing to the ethnologist who would otherwise have presided being debarred by illness.

[3] Identification photographs of the Home Office provided by the Surveyor-General of Prisons, Sir Edmund Du Cane.

groups was significantly different. If this were substantiated, the composite photographs of the three crime-groups should be markedly differentiated. The reader who has studied carefully the above account will appreciate what Galton was seeking in the composite photograph. He looked upon the mental traits as "transmissible by inheritance," he held that the physical traits were also inherited, and he was searching to divide man up into varieties in which the physiognomic characters should be indices of the mental traits. The common inheritance of both was fundamental to his idea.

"The Anthropologist has next to consider the life-history of those varieties, and especially their tendency to perpetuate themselves, whether to displace other varieties and to spread, or else to die out. In illustration of this, I will proceed with what appears to be the history of the criminal class. Its perpetuation by heredity is a question that deserves more careful investigation than it has received; but it is on many accounts more difficult to grapple with than it may at first sight appear to be. The vagrant habits of the criminal classes, their illegitimate unions, and extreme untruthfulness, are among the difficulties. It is, however, easy to show that the criminal nature tends to be inherited; while, on the other hand, it is impossible that women who spend a large portion of the best years of their life in prison can contribute many children to the population. The true state of the case appears to be that the criminal population receives accessions from classes who, without having strongly marked criminal natures, do nevertheless belong to a type of humanity that is exceedingly ill suited to play a respectable part in our modern civilisation, though well suited to flourish under half-savage conditions, being naturally both healthy and prolific. These persons are apt to go to the bad; their daughters consort with criminals and become the parents of criminals." (pp. 13–14.)

Galton then cites the now famous Jukes family[1], of which an account had been published in the preceding year.

"I have alluded to the Jukes family in order to show what extremely important topics lie open to inquiry in a single branch of anthropological research and to stimulate others to follow it out. There can be no more interesting subject to us than the quality of the stock of our countrymen and of the human race generally, and there can be no more worthy inquiry than that which leads to an explanation of the conditions under which it deteriorates or improves[2]." (p. 15.)

The genealogy of other "criminal" families published since, confirms Galton's views, but his call to scientific criminology met with little response for nearly thirty years. Even to the present day English anthropologists do not seem to grasp that a study of the *mental* varieties of their own race may be of more importance than recording the discovery of another Romano-Briton or the funereal trappings of an Egyptian monarch.

From the time of this paper onwards for several years Galton worked hard at composite photographs. There has been on the whole a great deal of unjustified disappointment in regard to them. This has largely arisen from a misunderstanding of what was expected from them, and a neglect of Galton's purpose in suggesting their use. That purpose is quite evident from this first paper: It was to ascertain whether men's mental characteristics were *intraracially* correlated with their facial characteristics. The fact that

[1] *Thirty-first Annual Report of the Prison Association of New York*, 1876.

[2] In these words Galton definitely lays down the principle that anthropology is not a mere antiquarian investigation, but is essentially occupied with some of the most urgent of our present social problems.

intraracial groups markedly differentiated in mental characters do not give markedly differentiated composite photographs, should not be considered merely negative and disappointing. It should have been interpreted as a most valuable anthropometric result, namely that mental characters are not highly correlated with external physical characters. That conclusion is confirmed by modern research on quite different lines; there is little or no correlation between human mentality and external anthropometric characters. I am fully aware that this result cuts directly at the whole of popular belief in physiognomy and phrenology and of the old anatomical ideas of craniometry. But this principle statistically demonstrated will stand, and composite photographs pointed at an earlier date in the same direction. The characters of the mind, the workings of the brain depend in the main upon commissures and linkages, matters of a far more subtle nature than the shape of the brain case. Whether the efficiency of the mind is more closely correlated with the physiological processes of the body, i.e. with its dynamic qualities, than with its static properties is another question, still *sub judice*. But one fundamental result of Galton's introduction of psychometry into anthropometric measurements has been to demonstrate the very small relation of mentality to external bodily characters. It is from this standpoint that Galton's composite photographs did and may still do useful work.

It may be argued that the American Indian, the Negro and the Western European have as markedly divergent and individual mental characters as they have divergent and individual physical characters (see our p. 81), and that both are inherited within these races of men. That there is *interracial* correlation between mental and physical attributes goes without saying as long as races are inbred. Each race simply transmits its own mentality and its own physique, but that is no proof of a high *intraracial* correlation between the two. Any geneticist knows how relatively easy it is to separate the mental and superficial characters of one breed by crossing it with another, much easier than it is to combine the forelimbs of one breed with the hindlimbs of a second; the simple reason being the relatively high correlation of the two members[1]. Goring has shown[2] that the average criminal is not differentiated markedly from the normal man by his physical characters; in England at any rate he is not the physically anomalous being of the Lombrosian school of criminologists.

The non-differentiation in a markedly significant manner of the composites of groups selected by mental characters contained a fundamental scientific fact, which has had to wait many years for us to grasp its full significance, and will possibly have to wait more years still for its general popular recognition.

[1] In breeding several hundred dogs from crosses of Pekingese and Pommeranians, there has only been one instance in which it might be supposed that a Pekingese forelimb was combined with a Pommeranian hindlimb; but it has been quite possible to obtain a pointed muzzle and chocolate coat combined with the strong mental individuality of the Pekingese. I feel certain that a differentiation by mental qualities of our hybrids would not on composite photography reproduce Pommeranian and Pekingese external characters.

[2] *The English Convict, A Statistical Study.* By Charles Goring, M.D., H.M. Stationery Office.

There have been many unconsidered opinions expressed about composite photography; they may be chiefly summed up in the view expressed by a well-known zoologist at the British Association Meeting at Plymouth in 1898[1]. He said that he had never been able to see the scientific value of the composite photograph. It represented the haphazard obliteration of one element by another. To which Galton fitly replied that the value of the composite photograph was that it brought out what was common to all the components, while eliminating that which was exceptional.

We shall postpone all further discussion of Galton's work on Composite Photography until the following chapter, but that work is only interpretable when we remember its psychological origin: Galton was inquiring into the extent to which mentality is associated with physiognomy.

In 1879 Galton published his first investigation into the working of his own mind. It was issued in two forms differing a good deal in detail. The first paper, entitled "Psychometric Facts," appeared in the *Nineteenth Century*[2], and the second paper, with the title "Psychometric Experiments," in *Brain*[3]. The two articles were independently written, the latter being the more statistical. The latter opens with the words:

"Psychometry, it is hardly necessary to say, means the art of imposing measurement and number upon operations of the mind, as in the practice of determining the reaction-time of different persons. I propose in this memoir to give a new instance of psychometry, and a few of its results. They may not be of any very great novelty or importance, but they are at least definite, and admit of verification; therefore I trust it requires no apology for offering them to the readers of this Journal, who will be prepared to agree in the view, that until the phenomena of any branch of knowledge have been submitted to measurement and number, it cannot assume the status and dignity of a science." (p. 148[4].)

Galton divides thought into two main categories. In the first category ideas present themselves by association with some object newly perceived by the senses, or with previous ideas. In the second such of these associated ideas, as happen to be germane to the topic on which the mind is set, are fixed by attention. Galton's investigation applied entirely to the first category, the automatic arising of ideas by association; they come of their own accord and cannot, except in indirect and imperfect ways, be compelled to come. The inquiry dealt with the rate at which these associated ideas come; their sameness and their difference, and the periods of life in which they were originally formed. He remarks that the experiments were "exceedingly trying and irksome, and that it required much resolution to go through with them, using the scrupulous care they demanded." This it is easy for the reader to verify; I have personally tried it on Galton's actual test list of words; my chief difficulty being the reluctance of associated ideas to appear, and their utter triviality compared with Galton's experience. As Galton himself says:

"When we attempt to trace the first step in each operation of our mind, we are usually baulked by the difficulty of keeping watch, without embarrassing the freedom of its action.

[1] *Times* Report of Section D, September 10th, 1898.

[2] March, 1879, pp. 425–33. [3] Vol. II, pp. 149–57.

[4] The last sentence was adopted many years ago as the motto of the *Biometric Laboratory.*

The difficulty is much more than the common and well-known one of attending to two things at once. It is especially due to the fact that the elementary operations of the mind are exceedingly faint and evanescent, and that it requires the utmost painstaking to watch them properly....... My method consists in allowing the mind to play freely for a very brief period, until a couple or so of ideas have passed through it, and then while the traces or echoes of those ideas are still lingering in the brain to turn the attention upon them with a sudden and complete reawakening; to arrest, scrutinise them, and to record their exact appearance; afterwards I collate the records at leisure, and discuss them and draw conclusions." (p. 150.)

Galton's first experiment was a leisurely walk of 450 yards down Pall Mall, on an occasion when he felt himself unusually capable of the kind of effort required. He reckoned that 300 objects caught his eye, although he never allowed his mind to ramble.

"It was impossible for me to recall in other than the vaguest way the numerous ideas that had passed through my mind; but of this, at least I was sure, that samples of my whole life had passed before me, that many bygone incidents, which I never suspected to form part of my stock of thoughts, had been glanced at as objects too familiar to awaken the attention. I saw at once that the brain was vastly more active than I had previously believed it to be, and I was perfectly amazed at the unexpected width of the fields of its everyday operations."

After an interval of some days in which he kept his mind from dwelling on his first experiences, Galton took a second experimental walk. He was struck as before by the variety of ideas that presented themselves, but his admiration for the activity of the mind was reduced by the observation that there was a great deal of repetition in his thought. He next devised an experiment for testing these associations and repetitions. He selected a list of 75 suitable words and sitting at a table with a stop-watch, started it on exposing a word of which he was previously ignorant. He waited till the word called up two directly associated ideas and then stopped the watch and recorded these ideas. The second associated idea was always derived from the word itself and not from the first associated idea, for he kept his attention firmly concentrated on the word itself. Sometimes he only got one associated idea; sometimes three or four occurred together and he was able to record them, but as a rule he only managed to record two with precision. Galton went through the 75 words on four occasions at intervals of a month, "but it was a most repugnant and laborious work, and it was only by strong self-control that I went through my schedule according to programme."

The total number of associated ideas was 505, and took 660 seconds to form; or at the rate of about 46 per minute or 2755 in an hour[1]. Of the 505 ideas, however, 29 occurred in all four trials, 36 in three, 57 in two and 107 in one trial only. Galton concluded therefore that reiterated association, even under the very different conditions of place and time of his experiments, was a much more marked feature than he had anticipated. He held from the proved number of faint and barely conscious thoughts and from the proved iteration of them, that the mind is perpetually travelling over familiar ways without the memory retaining any impression of its excursions.

"My associated ideas were for the most part due to my own unshared experiences, and the list of them would necessarily differ widely from that which another person would draw

[1] There were 13 cases of "puzzle" in which nothing sufficiently definite occurred in the maximum of time, 4 seconds, allowed for each test.

up who might repeat my experiments. Therefore one sees clearly, and I may say one can see *measurably* how impossible it is in a general way for two grown-up persons to lay their minds side by side together in perfect accord. The same sentence cannot produce precisely the same effect on both, and the first quick impressions that any given word in it may convey, will differ widely in the two minds." (p. 157.)

Galton was able in 124 cases of associated ideas to determine the period of life at which they became associated with the word. His results may be thus abstracted:

*Associations formed at following periods of Life.*

| | No. | Percentages | | | | |
|---|---|---|---|---|---|---|
| | | Total | Four times | Three times | Twice | Once |
| Boyhood and Youth ... | 48 | 39 | 10 | 9 | 7 | 13 |
| Subsequent Manhood ... | 57 | 46 | 8 | 7 | 5 | 26 |
| Quite Recent Events ... | 19 | 15 | 0 | 3 | 1 | 11 |
| Total ... | 124 | 100 | 18 | 19 | 13 | 50 |

The greater fixity of the earlier associations is clear as well as the fact that half the associations date from the period of life before leaving college. Associations are largely fixed in childhood and adolescence, but I do not think it necessarily follows as Galton seems to suggest that early *education* has a large effect in fixing our associations. The result may flow from mental plasticity, or the unstocked condition of the mental storehouse of youth.

Lastly Galton divides the original words into three classes, and the associated ideas into four.

The original words:

(i) were capable of mental images, as 'abbey,' 'aborigines,' 'abyss.'

(ii) represented actions or states of mind as 'abasement,' 'abhorrence,' 'ablution.'

(iii) formed more abstract notions as 'aptness,' 'ability,' 'abnormal.'

The associated ideas were:

(*a*) Sense imagery, chiefly visual.

(*b*) Histrionic, the mind visualised itself acting a part.

(*c*) Merely verbal associations as names of persons.

(*d*) Verbal associations as in phrases and quotations.

Galton gives the following analysis:

*Per cent. nature of Associated Ideas.*

| No. | Nature of Words | Sense Imagery | Histrionic | Verbal Associations | | Per cent. |
|---|---|---|---|---|---|---|
| | | | | Persons | Phrases and Quotations | |
| 26 | Capable of Mental Images | 43 | 11 | 30 | 16 | 100 |
| 20 | Actions or States of Mind | 32 | 33 | 13 | 22 | 100 |
| 29 | Abstract Notions ... ... | 22 | 25 | 16 | 37 | 100 |

Of these results Galton writes that they

"have forcibly shown to me the great imperfection in my generalising powers; and I am sure that most persons would find the same if they made similar trials. Nothing is a surer sign of high intellectual capacity than the power of quickly seizing and easily manipulating ideas of a very abstract nature. Commonly we grasp them very imperfectly, and hold on to their skirts with great difficulty. In comparing the order in which the ideas presented themselves, I find that a decided precedence is assumed by the Histrionic ideas, whenever they occur; that verbal associations occur first and with great quickness on many occasions, but on the whole they are only a little more likely to occur first than second; and that Imagery is decidedly more likely to be the second, than the first of the associations called up by a word. In short, gesture-language appeals the most quickly to our feelings." (pp. 161–2.)

"Perhaps the strongest of the impressions left by these experiments regards the multifariousness of the work done by the mind in a state of half-unconsciousness, and the valid reason they afford for believing in the existence of still deeper strata of mental operations, sunk wholly below the level of consciousness, which may account for such mental phenomena as cannot otherwise be explained[1]. We gain an insight by these experiments into the marvellous number and nimbleness of our mental associations, and we also learn that they are very far indeed from being infinite in their variety. We find that our working stock of ideas is narrowly limited, but that the mind continually recurs to them in conducting its operations, therefore its tracks necessarily become more defined and its flexibility diminished as age advances." (p. 162.)

There can be little doubt that Galton broke new ground in these papers both as to substance and method. But they produced little repercussion among English psychologists; not improbably because it is an easier task to experiment on another's mind than on one's own mind.

Galton's work of 1879 undoubtedly turned his thoughts to Mental Imagery, and he issued in November of that year a schedule containing Questions on the Faculty of Visualising[2]. On the data obtained from this questionnaire Galton published in *Mind* for July, 1880, a paper entitled: "Statistics of Mental Imagery[3]." The scope of this paper was twofold: namely to indicate how very varied is the intensity of visualising in the male members of the English Race and to indicate how Galton's method of ranking or of percentiles (see our Chapter XII) could be applied to such psychometric statistics.

[1] In the *Nineteenth Century* (p. 433) Galton writes: "The unconscious operations of the mind frequently far transcend the conscious ones in intellectual importance. Sudden inspirations and those flashings out of results which cost a great deal of conscious effort to ordinary people, but are the natural outcome of what is known as genius, are undoubtedly products of unconscious cerebration. Conscious actions are motived, and motives can make themselves attended to, whether consciousness be present or not. Consciousness seems to do little more than attest the fact that the various organs of the brain do not work with perfect ease or cooperation. Its position appears to be that of a helpless spectator of but a minute fraction of a huge amount of automatic brain work."

[2] Galton suggested the morning's breakfast table as an object for visualisation and requested answers to the following questions: (1) Illumination? (2) Definition? (3) Completeness? (4) Colouring? (5) Extent of Field of View? He then turned to various concrete examples of visualisation and asked his examinees to state whether they could visualise (6) Printed pages? (7) Furniture? (8) Persons? (9) Scenery? (10) Geography? (11) Military Movements? (12) Mechanism? (13) Geometry? (14) Numerals? (15) Card Playing? (16) Chess? There is no doubt that the answers he received under (14) were the original source of his later work on "Visualised Numerals."

[3] Vol. v, pp. 301–18.

Galton confesses that the first results of his inquiry amazed him. He had begun by questioning friends in the scientific world, because he thought they were the most likely persons to give accurate answers, and to his astonishment most of the men of science replied that mental imagery was unknown to them.

"They had no more notion of its true nature than a colour-blind man, who has not discerned his defect, has of the nature of colour. They had a mental deficiency of which they were unaware, and naturally enough supposed that those who were normally endowed were romancing." (p. 302.)

The members of the French Institute exhibited a like incredulity as to the reality of the visualising faculty. On the other hand in general society Galton found many men and women with the power of visualising. He was thus compelled to the conclusion that, whatever its cause might be, scientific men as a class have feeble powers of visual representation.

"My own conclusion is, that an over-readiness to perceive clear mental pictures is antagonistic to the acquirement of highly generalised and abstract thought, and that if the faculty of producing them was ever possessed by men who think hard, it is very apt to be lost by disuse. The highest minds are probably those in which it is not lost, but subordinated, and is ready for use on suitable occasions. I am, however, bound to say, that the missing faculty seems to be replaced so serviceably by other modes of conception, chiefly I believe connected with the motor sense, that men who declare themselves entirely deficient in the power of seeing mental pictures can nevertheless give lifelike descriptions of what they have seen, and can otherwise express themselves as if they were gifted with a vivid visual imagination. They can also become painters of the rank of Royal Academicians." (p. 304.)

Galton data were collected from 100 adult men, of whom 19 were Fellows of the Royal Society, three times as many more of distinction in other kinds of intellectual work, and the remainder of less note. He had also returns from 172 Charterhouse boys who had been interested in the matter by their Science Master Mr W. H. Poole. The whole of the original material—with much that Galton collected later for a new edition of the *Inquiries into Human Faculty*—is in the *Galtoniana*, and would be well worth working up by more modern statistical methods than were available in 1879.

What Galton does in this paper is to arrange the answers to each of his questions—vividness of imagery, colour representation, extent of field of mental view—in ranks by order of intensity, for his 100 adult males and for two groups of the Charterhouse boys: *A* for the upper classes, *B* for the five lower classes of the school. When the material was ranked Galton cited the Highest, the first Suboctile, the first Octile, the first Quartile, the Median, the last Quartile, the last Octile, the last Suboctile and the Lowest Answers. The intensities exhibited by the two Charterhouse groups at the various selected ranks were very similar, and the adult males were not very dissimilar from these, but they did not form as regular a series as the boys. They were avowedly not members of a true statistical group:

"being an aggregate of one class of persons who replied because they had remarkable powers of imagery and had much to say, of another class of persons, the scientific, who on the whole are very deficient in that gift, and of a third class who may justly be considered as fair samples of adult males." (p. 312.)

The reader of the paper will certainly realise—probably for the first time—how very varied is the power of mental imagery among individuals. But unless the reader is very familiar with the process of 'ranking' he is unlikely to extract at once from that system such results as that: 12 per cent. of persons see the mental image as vividly as the real thing, 12 per cent. only recall colours by a special effort for each, and more than 6 per cent. have a larger field of mental than of normal view, i.e. can see more than a hemisphere, all the faces of a die at once or the three walls of a room, and even the fourth simultaneously by an effort. It may be doubted whether the ranking scheme was best adapted to attract attention to a most interesting investigation. The paper concludes with a few observations on "visualised numerals."

I had frequently been puzzled by a number of lantern slides in the *Galtoniana*, which besides giving various phenomena associated with mental imagery provided illustrations of Bushman, Eskimo and palaeolithic drawings and carvings. They undoubtedly belonged to some public lecture, but there was nothing in the three lists of papers prepared by Galton himself to indicate that this lecture was ever published, nor was there any statement on p. 339 of the *Inquiries into Human Faculty* to say that the "1880 Mental Imagery, *Fortnightly Review*; *Mind*" referred to practically distinct papers. They are, however, distinct, and although the *Fortnightly*[1] paper, entitled "Mental Imagery," does not cover the whole ground of the lantern slides, there is little doubt that it contains a great deal of the substance of the lecture to which they belonged. The lecture was certainly one on "Mental Imagery," and, although it was not published *in extenso*, the *Fortnightly* probably contained the substance of it. There is little doubt that both slides and *Fortnightly* paper deal with the matter of Galton's popular lecture at the Swansea Meeting of the British Association in 1880. According to *L. G.'s Record* that meeting was attended by Galton and his wife. Mrs Galton makes no reference to the lecture, nor have I discovered any manuscript of it in the *Galtoniana*. It is possible that it was needful to cut out a good deal of the material of the lecture from the *Fortnightly* article as it would not be intelligible without illustrations.

The paper commences with what Galton himself calls vague physiological considerations concerning the difference between a sensation received by the optic nerve and transmitted to the brain, and a mental image where the sequence of events would occur in the *reverse* order, there being the propagation of a central impulse from the brain towards the optic nerve. This reverse process can be so vigorous that the mental image is vivid, and may in certain cases amount to a hallucination. These considerations

"justify us in ascribing the marked differences in the quality, as well as the vividness, of the mental imagery of different persons, to the various degrees in which the several links of a long nervous chain are apt to be affected." (p. 313.)

Galton states that

"his purpose is to point out the conditions under which mental imagery as above defined is

[1] *Fortnightly Review*, Vol. XXVIII, N.S. pp. 312–24, September, 1880. The *Mind* paper is entitled "Statistics of Mental Imagery." (*Brit. Assoc. Report*, 1880, p. lxxviii.)

most useful, and the particular forms of it which we ought to aim at developing, and I shall adduce evidence to show that the visualising faculty admits of being educated, although no attempt has ever yet been made, as far as I know, to bring it systematically and altogether under control." (p. 313.)

Galton applies his "ogive curve" here, as in *Mind*, and concludes that "the medium quality of mental imagery among Englishmen may be briefly described as fairly vivid but incomplete." Owing to the flatness of the curve between the quartiles, our author holds that it should be feasible to educate the faculty among the great majority of men to the degree in which it manifests itself without any education at all in at least one person out of sixteen, i.e. to the suboctile value, where the image is firm and clear. I must confess that I do not feel convinced of the great "educability" of the general population in visual imagery, by the rather slender evidence Galton gives from the *École Nationale de Dessin* in Paris, and from an eminent engineer, who had great visual faculty in form, and acquired it by practice in colour also (pp. 322–3). The visual faculty may be largely innate in such selected populations as engineers and artists, and may merely need exercising. It is difficult also to reconcile Galton's view that education could span the gap from lower quartile to upper suboctile with his statement in the following sentences:

"The visualising faculty is a natural gift, and like all natural gifts, has a tendency to be inherited. In this faculty the tendency to inheritance is exceptionally strong, as I have abundant evidence to prove, especially in respect to certain rare peculiarities, of which I shall speak [number forms and colour associations], and which when they exist at all, are usually found among two, three, or more brothers and sisters, parents, children, uncles and aunts and cousins[1]." (p. 314.)

From families Galton turns to races, and while admitting the difficulty in civilised races of the modification by education considers that the French possess the visualising factor in a high degree, noting their power of prearranging ceremonials and fêtes, and their genius for tactics and strategy, which show that they are able to foresee effects with unusual clearness. Their phrase "figurez-vous" or "picture to yourself," he says, seems to express their dominant mode of perception. Galton next turns to uncivilised races and stresses the cave drawings of the Bushmen of South Africa. He considers that the drawings of uncivilised races are largely the products of "mental imagery." This he justifies from a letter to himself from Dr Mann, of the Cape, who in 1860 observed a Bushman lad at work:

"He invariably began by jotting down upon paper or on a slate, a number of isolated dots which presented no connection or trace of outline of any kind to the uninitiated eye, but looked like the stars scattered promiscuously in the sky. Having with much deliberation satisfied himself of the sufficiency of these dots, he forthwith began to run a free bold line from one to the other, and as he did so the form of an animal—horse, buffalo, elephant or some kind of antelope—gradually developed itself. This was invariably done with a free hand, and with such unerring accuracy of touch that no correction of a line was at any time attempted. I understood from this lad that this was the plan which was invariably pursued by his kindred in making their clever pictures." (p. 316.)

[1] Galton states that the fact that scattered members of the same family had number forms was often discovered for the first time by his own inquiries.

Galton concludes from Dr Mann's account that a drawing by this method would be impossible if the artist had not a clear image[1] of the animal in his mind's eye. He refers also to the engravings of mammoth, elk and reindeer on bone by the men of the ice-age as illustrating the same visualising faculty. His argument would have been much strengthened had the cave-drawings of palaeolithic man been known in 1880, for these must have been made in semi-obscurity without the presence of the model being possible.

Among other illustrations of the visual imagery of the uncivilised races Galton cites the Eskimo performances, in particular, a chart drawn from memory of the coast from Pond's Bay to Fort Churchhill[2], a straight line distance of more than 1100 miles, which the draughtsman must at one time or another have visited in his canoe, and which was in remarkable accordance with the Admiralty Chart of 1870 (p. 316).

Galton next turns to number forms and colour associations, that is colours associated with numbers, letters or more particularly vowels. He had formed a collection of hundreds of such cases, not only from English, but from American, French, German, Italian, Austrian and Russian correspondents. He points out how in many cases the visualising faculty is not under control, the first acquired image of any scene holds its place, and cannot be subsequently corrected. Many persons find no difficulty in recalling faces uninteresting to them but are powerless to summon up the looks of dear relatives lost to them (p. 319). Galton gives an amusing experiment he made with a young lady and a philosophising friend. Both he accosted with the words: "I want to tell you about a boat." The young lady immediately visualised a rather large boat pushing off from the shore, filled with gentlemen and with ladies dressed in blue and white. The philosopher said that the word 'boat' called up no definite visual image, for he at once exerted himself to hold his mind in suspense, refusing to think of any particular boat, with any particular freight from any particular point of view. Galton suggests that:

"A habit of suppressing mental imagery must therefore characterise men who deal much with abstract ideas; and as the power of dealing easily and firmly with these ideas is the surest criterion of a high order of intellect, we should expect that the visualising faculty would be starved by disuse among philosophers, and this is precisely what I have found on inquiry to be the case." (p. 319.)

Galton points out that while our readings with mental visualisation may be dangerous it is equally inadvisable to starve this power. He suggests that if the boat-experience had been carried a stage further, the speaker saying: "the boat was a four-oared racing boat, it was passing quickly just in front of me, and the men were bending forward to take a fresh stroke," the listener ought to have had a definite picture well before his or her eyes. It ought to have the distinctness of a real four-oar going either to the right

[1] Later in the paper (p. 322) Galton refers to the rare power of throwing a mental image on to a sheet of white paper and holding it fast there while it is outlined with a pencil. He considers the Bush-boy had something of this faculty.

[2] Lat. 73° to lat. 58° 44′. The chart was published on p. 224 of *Captain Hall's Journals* issued by the U.S. Government in 1879.

or left, in short to be a generic image of a four-oar formed by a combination into a single picture of many sight-memories of such boats. "I argue," he writes, "that the mind of a man whose visualising faculty is free in its action forms these generalised images of its own accord out of its past experiences" (p. 320).

Galton states the forms of the visualising faculty which he thinks ought to be aimed at in education:

"The capacity of calling up at will a clear, steady and complete mental image of any object that we have recently examined and studied. We should be able to visualise that object freely from any aspect; we should be able to project any of its images on paper and draw its outline there; we should further be able to embrace all sides of the object simultaneously in a single perception, or at least to sweep all sides of it successively with so rapid a mental glance as to arrive at practically the same result. We ought to be able to construct images from description or otherwise, and to alter them in whatever way we please. We ought to acquire the power of combining separate, but more or less similar images into a single generic one. Lastly we should learn to carry away pictures at a glance of a more complicated scene than we can succeed at the moment in analysing[1]." (p. 322.)

A final point which Galton makes in this extraordinarily interesting paper is that the will cannot render vivid a faint image; its action is negative being limited to the suppression of what is not wanted and would confuse:

"It cannot create thought, but it can prevent thoughts from establishing themselves which lead in a false direction; so it keeps the course clear for a logical sequence of them. But if appropriate ideas do not come of their own accord, the will is powerless to evoke them. Thus we forget a familiar name, it is impossible to recall it by force of will. The only plan in such a case is to think of other things, till some chance association suggests the name. The mind may be seriously dulled by over-concentration, and it will only recover its freshness by such change of scene and occupation as will encourage freedom and discursiveness in the flow of ideas." (p. 324.)

The paper concludes with the extract which we have cited in the fuller form from the *Inquiries into Human Faculty* (see our p. 211).

Galton's investigation of visualised numerals or number forms sprang directly from his inquiries as to mental imagery. Several of his correspondents referred to their "number forms," i.e. the schemes in which they visualised the numerals from 1 to 200, or in some cases to a thousand or even a million. Closely allied to these number forms were arrangements of months of the year and of the days of the week. Others visualised in much the same way the years of their life and even the centuries of history. Not a few of these "forms" were associated with colours or shading. Galton collected both before and after the publication of his *Inquiries into Human Faculty* large quantities of these forms, and there is very ample correspondence with regard

[1] About this Galton writes: "A useful faculty, easily developed by practice, is that of retaining a mere retinal picture. A scene is flashed upon the eye; the memory of it persists, and details which escaped observation during the brief time when it was actually seen may be analysed and studied at leisure in the subsequent vision." This point needs very full investigation. Personally I have tried in vain to get any detail of scene or action, which I had not individually taken in on the occurrence. I feel grave doubts whether the details "which escaped observation" would not be supplied later because they were probable accompaniments, and to give evidence in a court of law of what happened by aid of such a visualising faculty would be for me a very real danger.

to them. Much of the data has never been published; Galton continued to collect for a revised edition of the *Inquiries* which he never issued. In particular there is a docket dealing with heredity in number forms, and he accumulated much evidence to show—not that the particular number forms—but that the tendency to visualise numbers runs in families. Before he discussed the matter in his *Inquiries into Human Faculty*, he published two memoirs on the topic. The first, entitled "Visualised Numerals," appeared in *Nature* Jan. 15, 1880[1]. And the second, with the same title, in the *Journal of the Anthropological Institute*, being a paper read on March 9, 1880[2]. At this time Galton had collected eighty such number forms and he found that about one person in thirty adult males and one in fifteen adult females possessed a number form. Among children they appeared to be more frequent, but were less fixed and distinct and tended to fade away with age. The 'form,' Galton considered, was of an older date than that at which a child began to learn to read, and represented his mental processes at a time of which no other record remains (*J. A. I.* p. 93 and especially *Nature*, p. 495). The 'forms,' he held, were the most remarkable existing instances of what has been termed "topical memory," the establishment of an association between position and the thing to be remembered; a link emphasised by teachers of mnemonics when they advise speakers to associate mentally the corners of a room with the different topics of a speech they are about to deliver. Discussing the relative frequency of number forms in the two sexes Galton writes:

> "I have been astonished to find how superior women usually are to men in the vividness of their mental imagery and in their powers of introspection...... I find the attention of women, especially women of ability, to be instantly aroused by these inquiries. They eagerly and carefully address themselves to consider their modes of thought, they put pertinent questions, they suggest tests, they express themselves in well-weighed language and with happy turns of expression, and they are evidently masters of the art of introspection. I do not find any peculiar tendency to exaggeration in this matter either among women or men; the only difference I have observed between them is that the former usually show an unexpected amount of intelligence, while many of the latter are unexpectedly obtuse. The mental difference between the two sexes seems wider in the vividness of their mental imagery and the power of introspecting it than in respect to any other combination of mental faculties of which I can think." (*Nature*, p. 252.)

The paper read before the Anthropological Institute was not only fuller than that in *Nature*, but was of special interest because Mr George Bidder, Colonel Yule, the Rev. G. Henslow, Mr (now Sir) Arthur Schuster, and others each described their own number forms. It would seem that these gentlemen were unaware, until Francis Galton began his inquiry, that there was anything unusual in the possession of a number form. This experience I also have had not infrequently, when I have found a person with a number form; he seemed to suppose everybody had a number form, and to be rather incredulous when I asserted that this was not so. Galton himself, it is of interest to note, did not visualise numerals. He writes:

[1] Vol. XXI, pp. 252–3, 323, 494–5. [2] Vol. X, pp. 85–102. The copies of both these memoirs in the *Galtoniana* contain the names of the various contributors of number forms.

"Another general experience is that the power of seeing vivid images in the mind's eye has little connection with high or low ability, or any other obvious characteristic, so that at present I am often puzzled to guess from my general knowledge of a friend, whether he will prove on inquiry to have the faculty or not. I have instances in which the highest ability is accompanied by a large measure of this gift, and others in which the faculty appears to be almost wholly absent. It is not possessed by all artists, nor by all mathematicians, nor by all mechanics, nor by all men of science. It is certainly not possessed by all metaphysicians, who are too apt to put forward generalisations, based solely on the experiences of their own special way of thinking, in total disregard of the fact that the mental operations of other men may be conducted in very different ways to their own." (*Nature*, p. 252.)

And again:

"Although philosophers may have written to show the impossibility of our discovering what goes on in the minds of others, I maintain an opposite opinion. I do not see why the report of a person on his own mind should not be as intelligible and trustworthy as that of a traveller upon a new country, whose landscapes and inhabitants are of a different type to any which we ourselves have seen. It appears to me that inquiry into the mental constitution of other people is a most fertile field for exploration, especially as there is much in the facts adduced here, as well as elsewhere, to show that original differences in mental constitution are permanent, being little modified by the accidents of education[1], and that they are strongly hereditary." (*Nature*, p. 256.)

Our Plates (XXI and XXII) give specimens of number forms. The *Galtoniana* contains many more and further slides of a certain number of coloured ones which do not appear in the published papers.

The next paper we reach was given as a Friday evening discourse at the Royal Institution, May 13, 1881[2]. It is entitled: "The Visions of Sane Persons." The object of this lecture was to show the unexpected prevalency of a visionary tendency among persons who form a part of ordinary society. Visions, illusions, hallucinations are stages of the same mental phenomenon, and may grade in intensity up to the star of Napoleon I or the *daimon* of Socrates and ultimately link up with a touch of madness.

Galton commences his lecture with referring in succession to:

(*a*) *Number Forms.* "Strange visions for such they must be called, extremely vivid in some cases, but almost incredible to the great majority of mankind," who are inclined to set them down as fantastic nonsense.

(*b*) *The Association of Colour with Sound.* The persistence of colour association with sound is fully as remarkable as that of Number Forms with numbers; generally it is concerned with the vowel sounds, and it is not a mere general colour, but a very distinct tint of that colour, which is associated with the given sound. The association is permanent, but very arbitrary, no two persons agreeing in their distribution of tints to sounds.

(*c*) *Association of Words with visualised Pictures.* Sometimes this curious fantasy occurs in a vague fleeting way, but occasionally the pictures are strangely vivid and permanent. Thus in Mrs Haweis' mind the interrogation

[1] This sentence since visualisation is part of the mental constitution does not seem wholly in accord with Galton's view that it should be possible by education to raise the intensity of that faculty in the general population so that the present grade at the upper suboctile should represent that of the lower quartile of the new population.

[2] Published also in the *Fortnightly Review*, June, 1881; *Proceedings, Royal Institution*, Vol. IX, pp. 644–55, 1882.

'What?' always excited the idea of a fat man cracking a long whip. And such pictures are the regular concomitants of the words and go back as long as memory is able to recall.

(*d*) *Pictures in the Field of View, when the eyes are closed, or in perfect darkness.* Many persons appear to have this kaleidoscopic change of forms, if they simply close their eyes and wait; thus Galton himself had these forms to a slight extent, but too fugitive to describe or draw. The Rev. George Henslow had them in a marked degree, and Goethe apparently also[1].

(*e*) *Phantasmagoria.* A common form of vision is the appearance of a crowd of phantoms hurrying past like men in the street. They are occasionally seen in broad daylight, but generally come to a person in bed, after putting the candle out and preparing to sleep, but by no means yet asleep. Galton reports that he knew three scientific men of eminence who had such phantasmagoria in one form or another[2]. Galton concludes with actual hallucinations occurring to sane people in good working health corresponding to the familiar hallucinations of the insane.

"I have," he writes, "a sufficient variety of cases to prove the continuity between all the forms of visualisation, beginning with an almost total absence of it and ending with a complete hallucination. The continuity is, however, not simply that of varying degrees of intensity, but of variations in the character of the process itself, so that it is by no means uncommon to find two very different forms of it concurrent in the same person. There are some who visualise well and who also are seers of visions, who declare that the vision is not a vivid visualisation, but altogether a different phenomenon. In short if we call all sensations due to external impressions '*direct*,' and all others '*induced*,' then there are many channels through which '*induction*' may take place, and the channel of ordinary visualisation in the persons just mentioned is different from that through which the visions arise." (p. 649.)

"It is remarkable how largely the visionary temperament has manifested itself in certain periods of history and epochs of national life. My interpretation of the matter, to a certain extent, is this—that the visionary tendency is much more common among sane people than is generally suspected. In early life it seems to be a hard lesson to an imaginative child to distinguish between the real and visionary world. If the fantasies are habitually laughed at and otherwise discouraged, the child soon acquires the power of distinguishing them; any incongruity or non-conformity is quickly noted, the vision is found out and discredited, and is no further attended to. In this way the natural tendency to see them is blunted by repression. Therefore, when popular opinion is of the matter-of-fact kind, the seers of visions keep quiet; they do not like to be thought fanciful or mad, and they hide their experiences, which only come to light through inquiries such as those that I have been making. But let the tide of

[1] The present writer has them somewhat vividly, first colour patterns, then floral devices, succeeded by the abrupt appearance of highly characteristic faces, corresponding to no individuals known to him, and with traits emphasised to caricature.

[2] I do not know whether Galton would have classed under vision or phantasmagoria another form of visualisation which comes to the present writer without any willing or power of control. Waking in the morning he lies on his back and looks eyes wide open at the empty white washed ceiling. In a varying number of seconds it will become closely covered with written matter in long narrow columns. It is never print, but has finely made, heavy black vertical letters, as those of a medieval MS. *Hortulus animae.* The words although apparently on the ceiling and of normal size are perfectly clear and legible, but on attempting to read them only a word, here or there, will be grasped before the whole script either vanishes, or changes. The author recently caught two words widely apart 'mathematics' and 'faithful' in the vision. He can well imagine that more easily moved natures, unaware of the frequency of such phantasmagoria, might by pondering on them intensify them and read from them supernatural messages directing their conduct, thus crossing the border line between sanity and insanity.

PLATE XXIII

Miss Millicent Adèle Galton (Mrs Bunbury). Died in 1883.
From a photograph taken in the 'sixties.

opinion change and grow favourable to supernaturalism and the seers of visions come to the front. It is not that a faculty previously non-existent has been suddenly evoked, but that a faculty long smothered in secret has been suddenly allowed freedom to express itself, and it may be to run into extravagance owing to the removal of reasonable safeguards." (p. 655.)

We may consider here Galton's last published experimental investigation on introspection. In 1884, the year after the appearance of the *Inquiries into Human Faculty*, he issued in *Mind*[1] a paper entitled: "Free-will, Observations and Inferences." The experiment was actually made in 1883, "during the somewhat uneventful but pleasant months of a summer spent in the country[2]." Galton explains his aims in the following words:

"The cases appear rare in which any of the numerous writers on Free-will have steadily, and for a long time together, watched the operations of their own mind whenever it was engaged in such an act, and discussions on Free-will have certainly been much more frequent than systematic observations of it. Consequently for my own information, I undertook a course of introspective inquiry last year; it was carried on almost continuously during six weeks, and has been proceeded with, off and on, for many subsequent months. As the results were not what I expected and as they were very distinct, I publish them, of course on the understanding that I profess to speak only of the operations of my own mind. If others will do the same, we shall be hereafter in a position to generalise.

My course of observation was that, whenever I caught myself engaged in a feat of what might fairly be called Free-will, I checked myself and recalled the antecedents and noted any circumstances that might have influenced my decision and forthwith wrote down an account of the whole transaction. After I had collated several notes I found that the variety of processes to be observed was small: I therefore discontinued my notes, but maintained the observations, until I felt satisfied that I could describe as much of what goes on in my own mind as falls within the ken of its consciousness.

I may say that, after some preliminary maladroitness had been overcome, I did not find the task difficult, nor even irksome; not nearly so much as in other introspective inquiries I have made. It is true that facility in any kind of introspection is difficult to acquire; it depends on the establishment of a habit something like that of writing in the midst of [other] avocations. When the latter has once been attained, the writer recovers the thread of thought that has been dropped at each interruption, and rarely finds it broken. So it is with introspection." (p. 406.)

Galton at once discards acts of 'Will' as distinguished from free-will as they are usually automatic; tenacity of purpose does not denote free-will, and is not usually considered to be a high order of psychical activity[3].

[1] Vol. IX, pp. 406–13.

[2] The Galtons were "done up by London whirl and grief for Mr Spottiswoode's death and the funeral in the Abbey, July 5th, and I became so unwell at the Jenkinsons that we began our summer outing at Boscombe and Bournemouth and spent a pleasant month meeting pleasant people. All the time I was on starvation diet. Then we went to Newton Abbot near Torquay and visited Totnes and Dartmouth and Torquay; also a pleasant time, and with nice dry weather such as one seldom enjoys in England. Still I prefer a foreign climate and think it suits my tiresome ailment better." *L. G.'s Record.*

Galton's sister Adèle Bunbury died on Dec. 31st and Montagu Butler's first wife Georgina during this year, so that the Galtons lost three close connections in the year following Darwin's death. Francis Galton wrote an obituary notice of Spottiswoode for the Royal Geographical Society (*Proceedings N. M. S.*, 1883, Vol. v, pp. 489–91) and concluded it with the words: "his name will assuredly take its place in the national memory as one of those upon whose ability, moral character, and resolute work, the credit of the English nation is mainly founded." Spottiswoode and Galton had been joint Honorary Secretaries of the Royal Geographical Society and intimate friends for many years.

[3] "As obstinate as a mule" or more vulgarly "as obstinate as a pig" are cited by Galton to express his meaning.

Again, appetites as motives automatically direct the will, and these cases may be disregarded, nor did Galton trouble about cases in which two motives of the same kind were in conflict and the greater prevailed. "There is no more anomaly in these than there is in the heavier scale-pan of a balance descending." Galton ultimately associated the possible cases of free-will with cases of irresolution, and to his great surprise found that not more than about one such case arose in the day. "All the rest of my actions seemed clearly to lie within the province of normal cause and consequence."

Galton classified these classes of "irresolution" into three categories:

(i) Each of two alternative plans grew less attractive the longer it was looked at, and so the mind "swung to and fro incapable of wholly fixing itself on either."

(ii) A fitfulness in the growth of a desire to change one's condition or occupation. "The resolution was delayed until a considerable rise of the new desire corresponded with a sudden fall of the old one." Galton illustrates this by the daily act of waking up and rising in the morning.

(iii) A change of Ego. An Ego which wants to continue staying comfortably in bed, and an Ego with a faint voice preaching the merits of early rising.

"To this I may give intellectual assent, but before it is possible for me to will to rise the Ego that is subsisting in content must somehow be abolished and a transmigration must take place into a different Ego, that of wide-awake life."

The mind may be shifted into a new position of stable equilibrium, by such a small matter as a twig tapping against the window.

"I suspect that much of what we stigmatise as irresolution is due to our Self being by no means one and indivisible, and that we do not care to sacrifice the Self of the moment for a different one. There are, I believe, cases in which we are wrong in reproaching ourselves sternly, saying, 'The last week was not spent in the way you now wish it had been' because the Self was not the same throughout. There is room for applying the greatest happiness of the greatest number, the particular Self at the moment of making retrospect being not the only one to be considered." (p. 409.)

Galton next turns[1] to what he terms 'incommensurable motives,' cases in which

"the one that was not the most keenly felt, nor gave the greatest pleasure in any sense of the word, emerged triumphant." (p. 409.)

He argues that the 'apparently' stronger psychical motive may not be the physiologically stronger motive, had we an exact cognisance of the battles

[1] He gives the following illustration: "An imperious old lady, infirm and garrulous, called at my house just as I had finished much weary work and was preparing with glee for a long walk. Hearing that I was at home, she dismissed her carriage for three quarters of an hour, so I was her prisoner for all that time. As she talked with little cessation, I had full opportunity for questioning myself on the feeling that supported me through the infliction. The response always shaped itself in the same way, 'social duties may not be disregarded; besides this is a capital occasion for introspection.'" Galton comments: "Leaving aside the last clause of the reply we see here...how a keen desire may wither under the influence of something about which our consciousness is scarcely exercised; some one of the many habits, whose quiet and firm domination gives a steadiness and calm to mature life that children cannot comprehend." (p. 410.)

in our brain; the appearance of relative strength is deceptive. Galton draws attention to the startling spontaneity with which some of the ideas that determine the will seem to arise. He suggests a subconscious chain of ideas a part of which suddenly comes into consciousness and may dominate the will. "Most of our ideas are partially shaped when they are first consciously perceived, and frequently they are fully shaped."

Those who with closed eyes witness a whole series of transformations not called up by any act of will of their own, and of which they cannot change the sequence by any conscious effort, will be prepared to consider favourably Galton's view that every form of sudden presentation, every new idea, has an analogous source to these visual ones.

"Moreover, as the imagination works in obscure depths out of the usual ken of consciousness, there seems reason for supposing that the 'something' upon which it works may in most cases be equally beyond its view. It is also certain that those who introspect, and those who study the genesis of dreams, succeed in discovering plain causes for numerous images and thoughts that had seemed to have arisen spontaneously. If these explanations are correct, as I feel assured they are, we must understand the word 'spontaneity' in the same sense that a scientific man understands the word 'chance.' He thereby affirms his ignorance of the precise causes of an event, but he does not in any way deny the possibility of determining them. The general results of my introspective inquiry support the views of those who hold that man is little more than a conscious machine, the larger part of whose actions are predictable. As regards such residuum as there may be, which is not automatic and which a man however wise and well informed could not possibly foresee, I have nothing to say, but I have found that the more carefully I inquired, whether it was into the facts of hereditary similarities of conduct, into the life-histories of very like or very unlike twins, or now introspectively into the processes of what I should have called my own Free-will, the smaller seems the room left for the possible residuum." (pp. 412–3.)

Galton would have been the last to claim finality for his conclusions, but his investigation raises many points of interest, and like so much of his psychological work emphasises the wide field of subconscious mental activity springing at odd intervals into consciousness. This Galton compares with the sudden and silent appearance of the head of a seal above the surface of still water and its just as sudden and silent disappearance, the observer being yet aware that the seal has been continuously active in a manner unperceived below the surface.

Three other psychological experiments on himself were made by Galton, but the results were not published. He refers to them in his *Memories*[1]. In the first, made in his youthful days, he was guided by a passionate desire to subjugate the body to the spirit, and determined that the will should replace automatic acts. He applied this to breathing, and every breath was submitted to the will. The normal power of breathing was dangerously interfered with and he felt as if he should suffocate, if he ceased to will. He had a terrible half-hour in which by slow and irregular steps the lost automatic power was recovered. Secondly Galton determined to gain some of the commoner feelings of Insanity. He adopted the plan of investing everything he met with the imaginary attributes of a spy. Galton found the experiment only too successful; in the course of a morning stroll by the time he had

[1] Pp. 276–7.

walked from Rutland Gate to the Green Park cabstand in Piccadilly every horse even on the stand seemed watching him, either with pricked ears, or else disguising its espionage. Hours elapsed before the uncanny sensation wore off and Galton said that he could only too easily re-establish it.

In his third experiment Galton strove to gain an insight into the abject feeling which a savage has for his fetish or idol, and he fixed on the grotesque figure on *Punch's* wrapper, and made believe in Punch's possession of divine attributes, and his mighty power to reward or punish men according to their treatment of him. The experiment gradually succeeded, and for a long time he retained for *Punch's* image a large share of the feelings that a barbarian has for his idol and learnt "to appreciate the enormous potency they might have over him."

Personally I have been much puzzled by the resurrection in modern days of the mascot, and by the apparent depth of feeling in some minds with regard to mascots; re-reading Galton's experiment with *Punch*, explained to my unimaginative mind how easily such reversions to fetishism may arise in the case of more emotional natures among modern men.

These three experiments aptly illustrate what serious endeavours Galton made to understand and appreciate the workings of his own and other men's minds.

C. INQUIRIES INTO HUMAN FACULTY AND ITS DEVELOPMENT, 1883.

This is the third of the larger works of Francis Galton, but it differs to some extent from the earlier two in being more completely a summary of the memoirs of the preceding ten to twelve years[1]. It is true there is a good deal added, but there is a considerable amount omitted, and those omissions to some extent may lead the reader of the book to suppose the conclusions based on less substantial evidence than a reader of the memoirs would have before him. On this account I have considered it best to discuss the memoirs at length, and in this section merely to supplement the earlier sections of this and those of the following chapter by drawing attention to novel points.

Writing of the memoirs he had published since the appearance of *Hereditary Genius* in 1869 Galton says:

"They may have appeared desultory when read in the order in which they appeared, but as they had an underlying connection it seems worth while to bring their substance together in logical sequence into a single volume. I have revised, condensed, largely rewritten, transposed old matter, and interpolated much that is new; but traces of the fragmentary origin of the work still remain, and I do not regret them. They serve to show that the book is intended to be suggestive, and renounces all claim to be encyclopaedic. I have indeed, with that object, avoided going into details in not a few cases where I should otherwise have written with fullness, especially in the anthropometric part. My general object has been to take note of the

[1] Of the twenty-two memoirs on which the work is based seventeen have been already considered in this or earlier chapters, four will be dealt with in Chapter XII and one in Chapter XIII. For the titles of these memoirs: see Appendix, pp. 338–9 of the work. Three memoirs on composite photography (including that on 'Generic Images'), the memoir on the fertility of Town and Country populations, that on Test Weights and that on Galton's Whistles, together with the questionnaire on visualising, are reproduced on pp. 340–80 of the Appendix.

varied hereditary faculties of different men, and of the great differences in different families and races, to learn how far history may have shown the practicability of supplanting inefficient human stock by better strains, and to consider whether it might not be our duty to do so by such efforts as may be reasonable, thus exerting ourselves to further the ends of evolution more rapidly and with less distress than if events were left to their own course.......I thought it safer to proceed like the surveyor of a new country, and endeavour to fix in the first instance as truly as I could the position of several cardinal points." (pp. 1–2.)

It is clear from this passage that Galton recognised he was a pioneer. He was, indeed, the first to grasp that if evolution be the true doctrine of the development of living forms, then it is desirable for rational man to take stock of his varieties, mental and physical, to measure their evolutionary value, and to throw himself into sympathy with the changes Nature foreshadows for his kind. The intention of Galton's work is to touch on various topics more or less connected with the cultivation of race or, as he puts it (p. 24), with "eugenic" questions[1]. Galton proposes to tell us the range of qualities found in man and therein must lie man's possibility of improvement. Is it not a *religious* duty of the men of to-day to leave their race better than they found it? Or, as Romanes phrased Galton's idea: Is it not man's high prerogative to cooperate with the unknown Worker in promoting the great work? The world was not ripe for such a doctrine in 1883, and, needless to say, it raised theological ire. *The Guardian* published a thoroughly hostile review from which I cite a few sentences:

"The author cannot even refrain from trespassing upon the territory of those with whom he is at issue, a territory which for him is not matter, which cannot be seen, or touched or measured or weighed—and so cannot be proved (by his method of proof) to exist. We are henceforth to apply ourselves to elicit the '*religious* significance' of the doctrine of evolution; whether if we substitute for religious *anti-religious*, Mr Galton would be able to demonstrate any difference in the meaning conveyed by the words he uses we take leave to doubt."

Speaking of Galton's remarks on the herd (see our p. 74) the critic writes:

"A small tribe is sure to be slaughtered or enslaved; a large one falls to pieces through its own 'unwieldiness.' It must be 'either deficient in centralisation or straightened for food or both.' 'Self-reliant individuals' are required; but neither too few nor too many. The importance of gregarious instincts in savage life is fully set forth; but they are not equally important to 'all forms of savage life.' Natural selection tends to give one leader 'and to repress superabundant leaders.' As we have been taught before, this wonderful law of natural selection creates and destroys, reduces and enlarges, raises and represses, originates and annihilates."

Galton, as we know, discussed only the *objective* efficacy of prayer (see our pp. 115–17), and the critic cites his words with the comment we give following them:

[1] "That is, with questions bearing on what is termed in Greek *eugenes*, namely, good in stock, hereditarily endowed with noble qualities. This and the allied words, *eugeneia*, etc., are equally applicable to men, brutes and plants. We greatly want a brief word to express the science of improving stock, which is by no means confined to questions of judicious mating, but which, especially in the case of man, takes cognizance of all influences that tend in however remote a degree to give to the more suitable races or strains of blood a better chance of prevailing speedily over the less suitable than they otherwise would have had. The word *eugenics* would sufficiently express the idea; it is at least a neater word and a more generalised one than viriculture, which I once ventured to use." (pp. 24–5; see our p. 110.) Thus the name for the science of eugenics was invented just forty years ago.

"'We simply look to the main issue—Do sick persons who pray or are prayed for, recover on the average more rapidly than others?'

'I have discovered hardly any instance in which a medical man of repute has attributed recovery to the influence of prayer.' 'The universal habit of the scientific world to ignore the power of prayer is a very important fact.'

Is this a fact at all? What evidence has Mr Galton to bring forward in support of this outrageous assertion concerning the scientific world[1]?"

And again:

"A nation, he informs us, ought not to hold together by purely gregarious instincts, 'a mob of slaves clinging to one another through fear,' it should consist of 'vigorous self-reliant men, knit to one another by innumerable ties,' and as he ought to have added, well versed in the new doctrines of evolution and determined to destroy their weaker brethren in obedience to the great law of the survival of the fittest in the struggle for existence. Instead of wasting his time upon the records of the past and preparing for a future state, the new animal man is to 'awake to a fuller knowledge of his relatively great position, and begin to assume a deliberate part in furthering the great work of evolution.' It is his 'religious duty,' says Mr Galton, to do this 'deliberately and systematically.' This is the practical outcome of the new philosophy for the new animal—the only religious duty he has to perform in the new Cosmos[2]."

I have cited these passages—very characteristic of the ecclesiastical feeling of that day—to show how the anti-Darwinian *odium theologicum* was within a year of Darwin's death transferred to his cousin, because, going farther than Darwin, he had seen that if the doctrine of evolution through heredity and natural selection be true, then man ought to use this principle as any other natural law to raise his kind. The thoughts and purposes of the Deity, Florence Nightingale held, are only to be discovered by the statistical study of natural phenomena, and both Francis Galton and Florence Nightingale believed that application of the results of such study was the *religious* duty of man. Are we any nearer to-day than the theological world was in 1883 to a true appreciation of that position? Are Dean Inge and Canon Barnes average representatives of the modern Church, or is their grade, as Galton would have put it, somewhere about the "suboctile"? We sadly fear that Father Wasmann, Mr G. K. Chesterton, and Herr Bumüller would more nearly reproduce the median theological mind of to-day. In 1883 it was probably Romanes alone who recognised the fact that Galton was virtually marking out the lines of what may be appropriately called a new religion.

"We have of late had so many manufactures of this kind that the market is somewhat glutted, and therefore it is very doubtful how far this new supply will meet with an appropriate demand; but we can safely recommend Mr Galton's wares to all who deal in such commodities as the best which have hitherto been turned out. They are the best because the materials of their composition are honesty and commonsense without admixture with folly or metaphor[3]."

After this slight indication of the reception the publication of Galton's work met with, I turn to its contents. The earlier pages discuss material

[1] Yet surely Galton was merely stating a universal experience! What chance of publication by a recognised scientific society would a memoir have if the author, describing the sequence of any physical or vital phenomena, added: "but according to my experience the sequence is modified in $x\,^{\circ}/_{\circ}$ of cases by the power of prayer"? We must go back to Cuvier practically to find breaks in the sequences of natural phenomena directly attributed in a "scientific" memoir to theocratic intervention.

[2] *The Guardian*, July 4, 1883, p. 1001.

[3] *Nature*, Vol. XXVIII, p. 98, Mar. 31, 1883.

with which the reader of the present volume is familiar; they deal chiefly with the anthropometric characters, with the variety in features, with the type face as reached by composite portraiture, with the healthy, the diseased and the criminal. Galton then reiterates his view on the influence of town life (see our pp. 123–25). He next turns to a very important matter, which had been thrust on his attention when dealing with English men of science, namely *Energy*.

"Energy is the capacity for labour. It is consistent with all the robust virtues, and makes a large practice of them possible. It is the measure of fullness of life; the more energy the more abundance of it; no energy at all is death; idiots are feeble and listless. In the inquiries I made on the antecedents of men of science no points came out more strongly than that the leaders of scientific thought were generally gifted with remarkable energy, and that they had inherited the gift of it from their parents and grandparents. I have since found the same to be the case in other careers. Energy is an attribute of the higher races, being favoured beyond all other qualities by natural selection. We are goaded into activity by the conditions and struggles of life. They afford stimuli that oppress and worry the weakly, who complain and bewail, and it may be succumb to them, but which the energetic man welcomes with a good humoured shrug, and is the better for in the end.

The stimuli may be of any description: the only important matter is that all the faculties should be kept working to prevent their perishing by disease. If the faculties are few, very simple stimuli will suffice. Even that of fleas will go a long way. A dog is continually scratching himself, and a bird pluming itself, whenever they are not occupied with food, hunting, fighting, or love. In those blank times there is very little for them to attend to besides their varied cutaneous irritations. It is a matter of observation that well washed and combed domestic pets grow dull; they miss the stimulus of fleas[1]." (pp. 25–6.)

Galton further remarks that it does not follow that because men are capable of doing hard work that they like doing it. Some may fret if they cannot let off their superfluous steam, but others need a strong stimulus such as wealth, ambition or passion to compel them to action.

"The solitary hard workers, under no encouragement or compulsion except their sense of duty to their generation, are unfortunately rare among us." (p. 26.)

"It may be objected that if the race were too healthy and energetic there would be insufficient call for the exercise of the pitying and self-denying virtues, and the character of men would grow harder in consequence. But it does not seem reasonable to preserve sickly breeds for the sole purpose of tending them, as the breed of foxes is preserved solely for sport and its attendant advantages. There is little fear that misery will ever cease from the land, or that the compassionate will fail to find objects for their compassion; but at present the supply vastly exceeds the demand; the land is overstocked and overburdened with the listless and the incapable.

In any scheme of eugenics, energy is the most important quality to favour; it is, as we have seen, the basis of living action, and it is eminently transmissible by descent." (p. 27.)

Galton next deals with sensitivity, describing his weight-lifting and whistle test for touch and sound. Speaking of discrimination by the senses, he remarks on the limitation of language to express various degrees of difference by what we now term broad categories. He writes:

"We inherit our language from barbarous ancestors, and it shows traces of its origin in the imperfect ways by which grades of difference admit of being expressed. Suppose a pedestrian is asked whether the knapsack on his back feels heavy. He cannot find a reply in two words

[1] The humour of this passage quite escaped one critic. Otherwise he might have realised that Galton's production in the critic's case of "cutaneous irritations" was a most useful stimulus against the critic himself growing dull.

that cover more varieties than (1) very heavy, (2) rather heavy, (3) moderate, (4) rather light, (5) very light[1]. I once took considerable pains in the attempt to draw up verbal scales of more than five orders of magnitude, using those expressions only that any cultivated person would understand in the same sense; but I did not succeed. A series that satisfied one person was not interpreted in the same sense by another." (p. 33.)

The general aim of this section of Galton's work is to show that the range of sense discrimination in man is wide, that delicate discrimination is an attribute of a high race, and that it is not, as some have supposed, necessarily associated with nervous irritability.

The author next emphasises the importance of family anthropometric registers, a matter to which we shall shortly return. We may note his concluding remarks:

"The investigation of human eugenics—that is, of the conditions under which men of a high type are produced—is at present extremely hampered by the want of full family histories, both medical and general, extending over three or four generations. There is no such difficulty in investigating animal eugenics, because the generations of horses, cattle, dogs etc. are brief, and the breeder of any such stock lives long enough to acquire a large amount of experience from his own personal observation. A man, however, can rarely be familiar with more than two or three generations of his contemporaries before age has begun to check his powers; his working experience must therefore be chiefly based upon records. Believing, as I do, that human eugenics will become recognised before long as a study of the highest practical importance, it seems to me that no time ought to be lost in encouraging and directing a habit of compiling personal and family histories. If the necessary materials be brought into existence, it will require no more than zeal and persuasiveness on the part of the future investigator to collect as large a store of them as he may require." (pp. 44–5.)

Then follows a discussion of statistical methods, in particular of the "ogive curve" (see our Chapter XII); it is followed by a study of character (see our pp. 268–271), by a discussion of the criminal (see our pp. 229–231) and the insane, and their heredity; and then we have the salient points of the paper on gregarious and slavish instincts reproduced (see our pp. 72–74). Galton next turns to the great variation in the visualising power of man and summarises, and to some extent expands, the memoirs we have already discussed (see our pp. 236–45). He refers in more detail to blindfolded chess-players, who play several games at once, and notes cases of orators mentally reading manuscript in making speeches.

"One statesman has assured me that a certain hesitation in utterance, which he has at times, is due to his being plagued by the image of his manuscript speech with its original erasures and corrections. He cannot lay the ghost, and he puzzles in trying to decipher it.

Some few persons see mentally in print every word that is uttered; they attend to the visual equivalent and not to the sound of the words, and they read them off usually as from a long imaginary slip of paper, such as is unwound from telegraphic instruments. The experiences differ in detail as to size and kind of type, colour of paper, and so forth, but are always the same in the same person." (p. 96.)

Galton next deals at some length with the visualising power of uncivilised races; he notes that Bushmen and Eskimo are an exception to the rule

[1] Five categories are usually adequate for the statistician; he has unfortunately often to content himself with three. But if seven are desirable then not unreasonable results may be obtained by such a system as (1) extremely heavy, (2) heavy, (3) rather heavy, (4) medium, (5) rather light, (6) light, (7) extremely light,—provided personal equation is considered, eliminated, or standardised.

PLATE XXIV

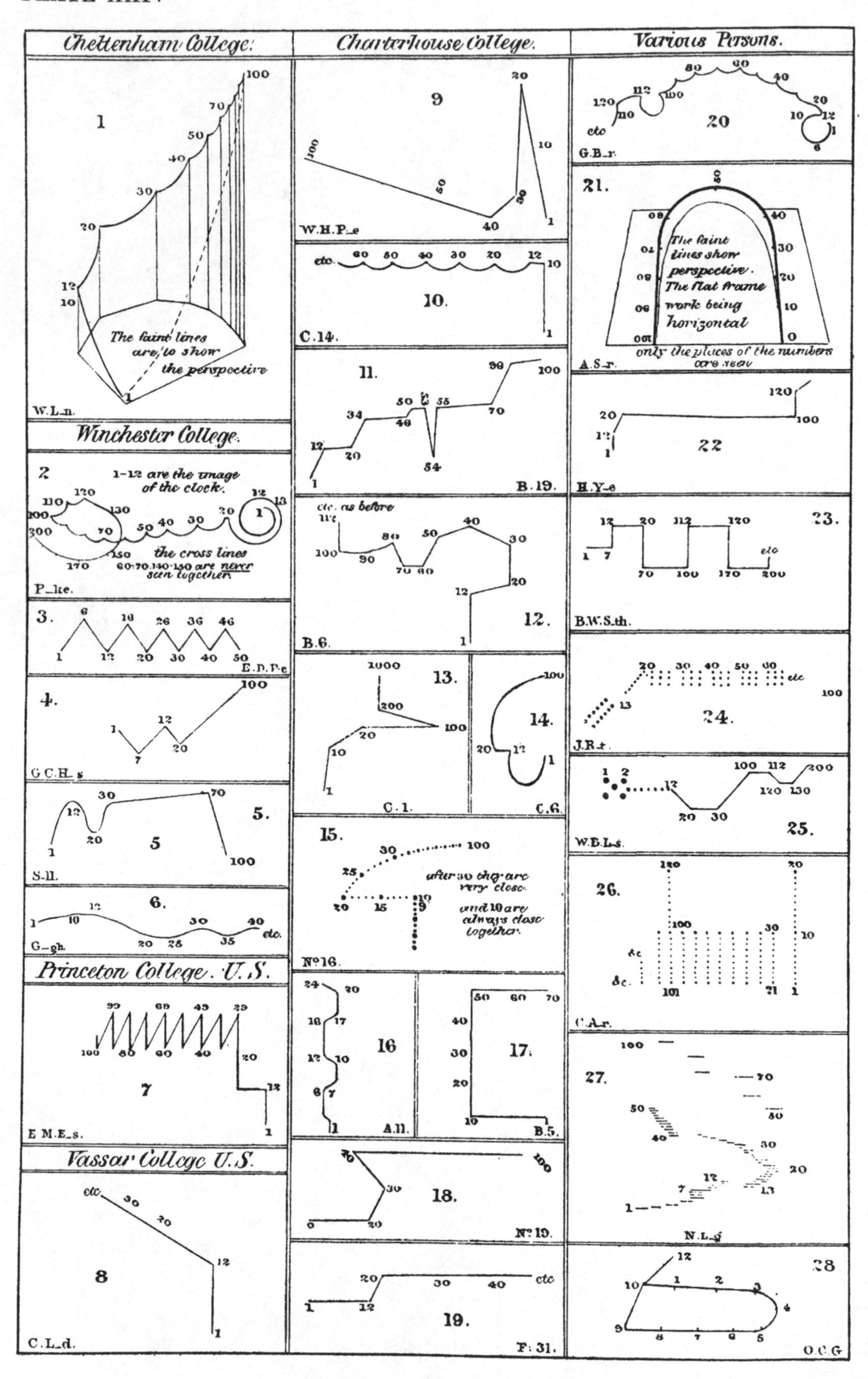

Cheltenham College.
Charterhouse College.
Various Persons.
The faint lines are to show the perspective
W.L_n.
W.H.P_e
G.B_r.
C.14.
A.S_r.
Winchester College.
1-12 are the image of the clock.
the cross lines 60-70.140-150 are never seen together
P_ke.
B.19.
H.Y_e
The faint lines show perspective. The flat frame work being horizontal
only the places of the numbers are seen
E.D.F_e
B.6.
B.W.S_th.
G.C.H_s
C.1.
C.6.
J.R_t.
S_ll.
after 30 they are very close
and 10 are always close together
W.B.L_s.
G_gh.
No. 16.
C.A_r.
Princeton College. U.S.
E.M.E_s.
A.11.
B.5.
Vassar College U.S.
No. 19.
N.L_g
C.L_d.
F. 31.
O.C.G.

EXAMPLES OF NUMBER FORMS.

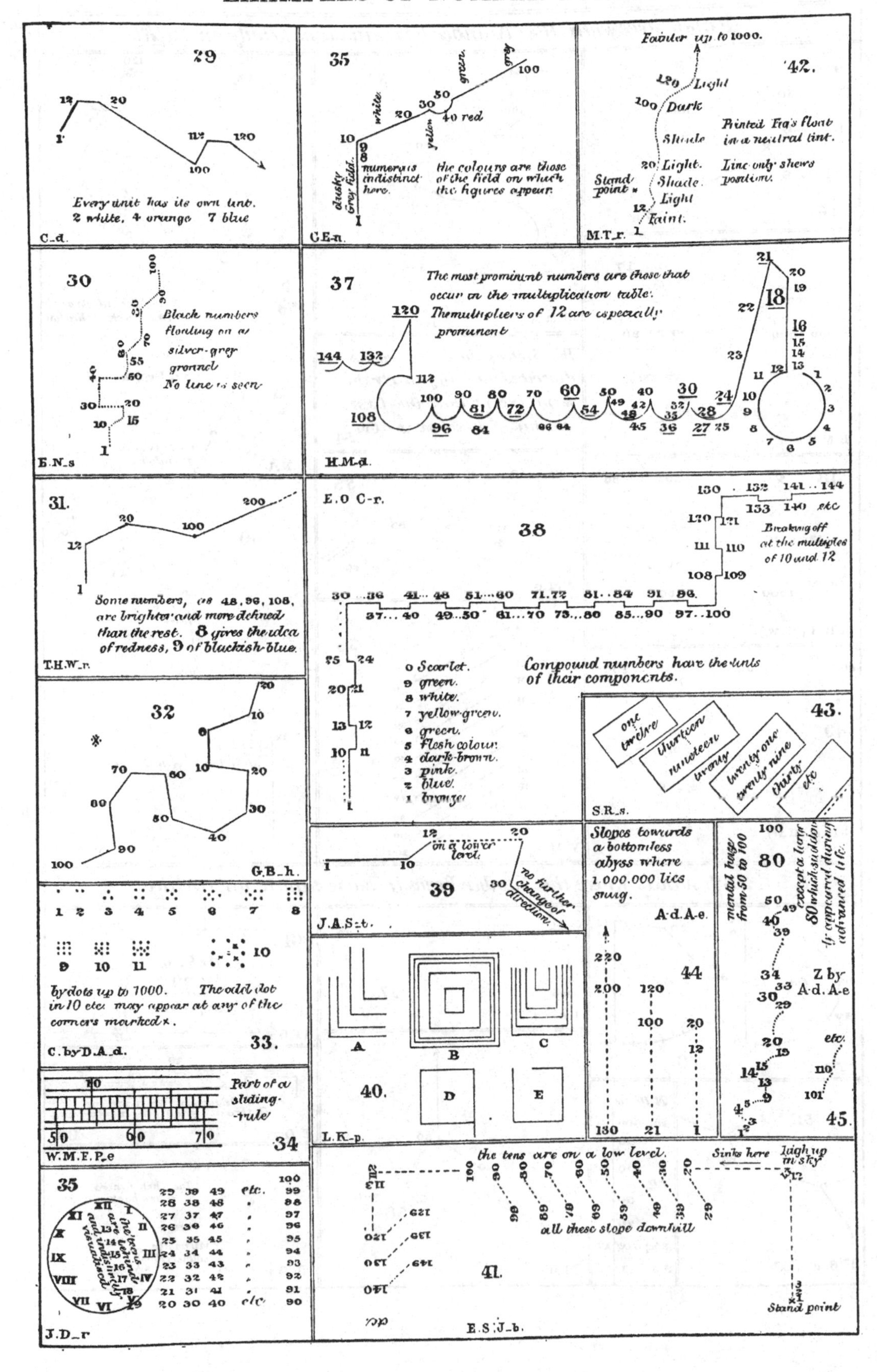

29
Every unit has its own tint.
2 white, 4 orange 7 blue
C_d.
35
numerals indistinct here.
the colours are those of the field on which the figures appear.
C.E_n.
Fainter up to 1000.
42.
Light
Dark
Shade
Light.
Shade.
Light
Faint.
Stand point
Printed Fig's float in a neutral tint.
Line only shews position.
M.T_r.
30
Black numbers floating on a silver-grey ground.
No line is seen.
E.N_s
37
The most prominent numbers are those that occur in the multiplication table.
The multipliers of 12 are especially prominent.
H.M_d.
31.
Some numbers, as 48, 96, 108, are brighter and more defined than the rest. 8 gives the idea of redness, 9 of blackish-blue.
T.H.W_r.
E.O C-r.
38
Breaking off at the multiples of 10 and 12
Compound numbers have the tints of their components.
0 Scarlet.
9 green.
8 white.
7 yellow-green.
6 green.
5 Flesh colour.
4 dark-brown.
3 pink.
2 blue.
1 bronze.
32
G.B_h.
43.
one twelve
thirteen nineteen twenty
twenty one twenty nine
thirty etc
S.R_s.
on a lower level.
no further change of direction.
39
J.A.S_t.
Slopes towards a bottomless abyss where 1.000.000 lies snug.
A.d.A_e.
44
mental haze from 50 to 100
except a line 80 which suddenly appeared during advanced life.
Z by A.d.A_e
45.
by dots up to 1000. The odd dot in 10 etc. may appear at any of the corners marked ×.
C. by D.A_d.
33.
40.
L.K_p.
Part of a sliding-rule
34
W.M.F.P_e
35
the teens are behind and indistinctly visualised.
etc.
J.D_r
the tens are on a low level.
Sinks here
high up in sky
all these slope downhill
41.
Stand point
E.S.J_b.

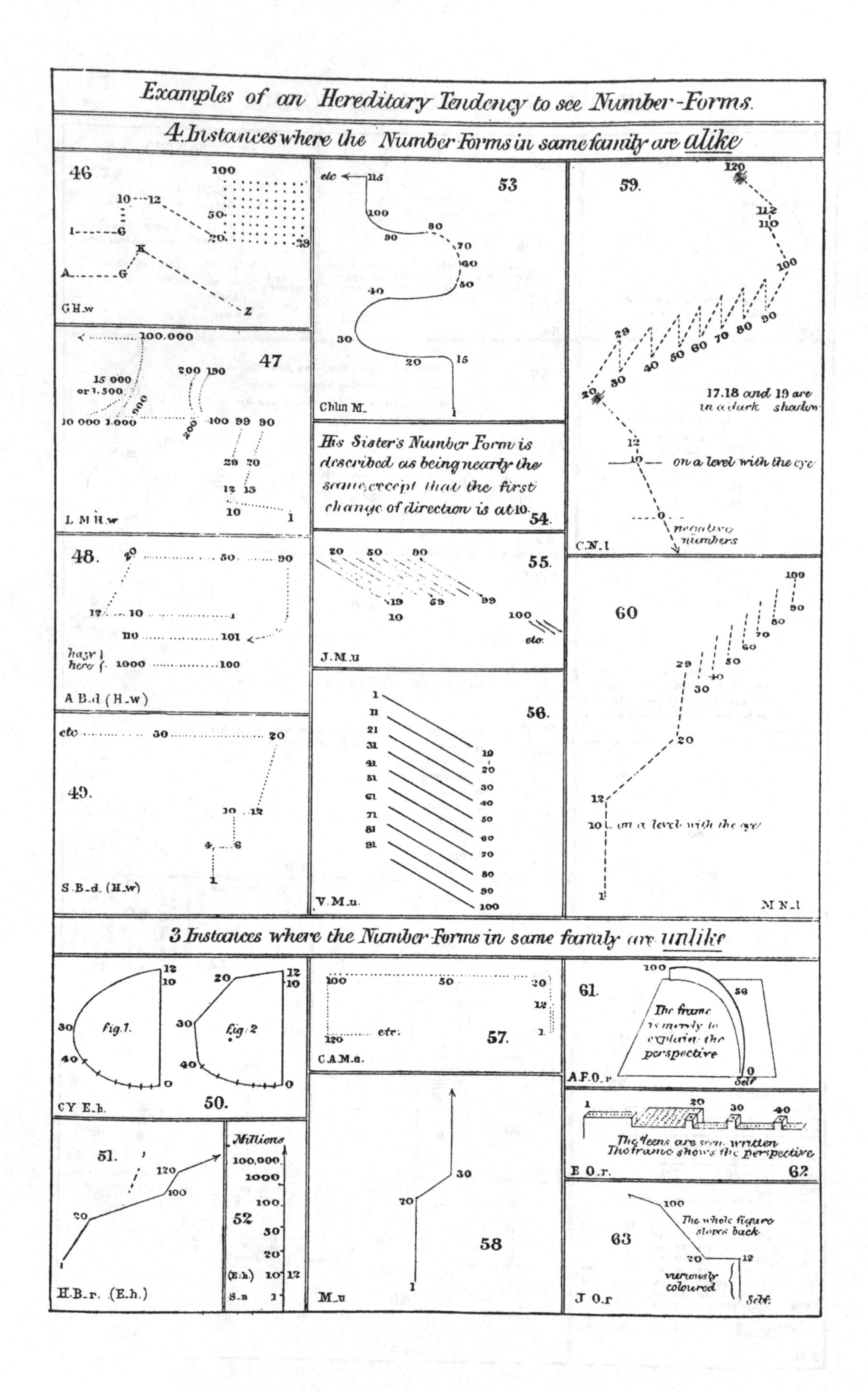

Examples of an Hereditary Tendency to see Number-Forms.
4 Instances where the Number Forms in same family are alike
46
GH.w
47
L M H.w
48.
A B.d (H.w)
49.
S.B.d. (H.w)
53
Chun M.
His Sister's Number Form is described as being nearly the same, except that the first change of direction is at 10.
54.
55.
J.M.u
56.
V.M.u.
59.
17.18 and 19 are in a dark shadow
on a level with the eye
negative numbers
C.N.1
60
on a level with the eye
M N.1
3 Instances where the Number Forms in same family are unlike
fig.1.
fig.2
CY E.h.
50.
57.
C.A.M.u.
61.
The frame is merely to explain the perspective
Self
A.F.O.r
The teens are seen written
The frame shows the perspective
E O.r.
62
51.
H.B.r. (E.h.)
Millions
52
(E.h)
S.n
58
M.u
63
The whole figure slopes back
variously coloured
Self.
J O.r

and quotes from his paper in the *Fortnightly* of September 1880 (see our pp. 238–41), and in concluding the subject expands the last paragraph of that paper into his final expression of opinion:

"There can, however, be no doubt as to the utility of the visualising faculty when it is duly subordinated to the higher intellectual operations. A visual image is the most perfect form of mental representation whenever the shape, position and relations of objects in space are concerned. It is of importance in every handicraft and profession where design is required. The best workmen are those who visualise the whole of what they propose to do, before they take a tool in their hands. The village smith and the carpenter who are employed on odd jobs require it no less for their work than the mechanician, the engineer and the architect. The lady's maid who arranges a new dress requires it for the same reason as the decorator employed on a palace, or the agent who lays out great estates. Strategists[1], artists of all denominations, physicists who contrive new experiments, and in short all who do not follow routine, have need of it. The pleasure its use can afford is immense. I have many correspondents who say that the delight of recalling beautiful scenery and great works of art is the highest that they know; they carry whole picture galleries in their minds. Our bookish and wordy education tends to repress this valuable gift of nature. A faculty that is of importance in all technical and artistic occupations, that gives accuracy to our perceptions, and justness to our generalisations, is starved by lazy disuse, instead of being cultivated judiciously in such a way as will on the whole bring the best return. I believe that a serious study of the best method of developing and utilising this faculty[2], without prejudice to the practice of abstract thought in symbols, is one of the many pressing desiderata in the yet unformed science of education." (pp. 113–4.)

Galton next passes to "Number Forms" and gives here the fullest account that he has provided of them, although in no way comparable with the range of his collected material. He publishes three plates of "Number Forms" and a fourth plate showing some typical associations of numbers with colours. He also indicates that some persons associate character with numerals, but rarely, except in the case of 12, to which most pay great respect, is there any agreement in the characterisation. Thus 3 may be a "treacherous sneak," a "feeble edition of 9," "a good old friend" and "delightful and amusing." There is no agreement as to the *sex* of numbers, although Galton himself imagined that the even numbers must *of course* be male (p. 144).

He then refers to the very strong evidence he had collected for the hereditary character, not of particular number forms, but of the tendency to visualise numbers. He next turns to colour associations and describes them at considerable length (pp. 145–54). He emphasises the fact that while to the ordinary man these associations of colour with letters or numbers appear equally "wild and lunatic[3]," no two colour visionaries agree in their schemes, and one seer is scandalised and almost angry at the heresies of another!

[1] Napoleon I seems to have held that men who formed mental pictures (*tableaux*), no matter what their intellect, courage and knowledge, were unfit to command. *Maximes de Guerre et Pensées*, No. 73.

[2] Galton, notwithstanding his evidence for the hereditary character of this faculty, yet held that it could be developed by training, and cited Légros' old teacher Lecoq de Boisbaudran, who had developed at the *École Nationale de Dessin* in Paris a complete training in visualisation. It can, no doubt, where it exists be developed by practice, but it may be questioned whether it can be originated in an individual without it, any more than musical sense or mechanical ingenuity can be developed in those in whom they are not innate.

[3] The complexity of some of the colour schemes as shown on Galton's Plate IV is marvellous; that plate required 14 colour stones to produce it lithographically, and therefore, fascinating as it is, I cannot reproduce it here!

As I have remarked, Galton describes and figures only a small part of his material, but enough to succeed

"in leaving a just impression of the vast variety of mental constitution that exists in the world, and how impossible it is for one man to lay his mind strictly alongside that of another, except in the rare instances of close hereditary resemblance." (p. 154.)

The next section of the work is entitled *Visionaries*, and consists substantially of the material we have discussed in our *résumé* of the "Visions of Sane Persons" (see our pp. 243–45). The essential point is the frequency with which the automatic construction of fantastic figures takes place, and their continued sequence without control of the volition. The transition of such visions to hallucinations was regarded by Galton as only a matter of the intensity of nerve excitement, which might be produced by ill-health, brain-storms or drugs. The following section of the book under discussion is termed "Nurture and Nature."

"Man," writes our author, "is so educable an animal that it is difficult to distinguish between that part of his character which has been acquired through education and circumstance and that which was in the original grain of his constitution." (p. 177.)

Galton considers that the character of a nation may not change, but a different phase or mood of it may become dominant owing to some accident causing the special representatives of that phase to be for a time national leaders.

"The love of art, gaiety, adventure, science, religion may be severally paramount at different times." (p. 178.)

Now follows a passage which I think must be cited as a whole, for it needs some consideration:

"One of the most notable changes that can come over a nation is from a state corresponding to that of our past dark ages into one like that of the Renaissance. In the first case the minds of men are wholly taken up with routine work, and in copying what their predecessors have done; they degrade into servile imitators and submissive slaves to the past. In the second case some circumstance or idea has finally discredited the authorities that impeded intellectual growth, and has unexpectedly revealed new possibilities. Then the mind of the nation is set free, a direction of research is given to it, and all the exploratory and hunting instincts are awakened. These sudden eras of great intellectual progress cannot be due to any alteration in the natural faculties of the race, because there has not been time for that, but to their being directed to productive channels. Most of the leisure of the men of every nation is spent in a round of reiterated actions; if it could be spent in continuous advance along new lines of research in unexplored regions, vast progress would be sure to be made. It has been the privilege of this generation to have had fresh fields of research pointed out to them by Darwin, and to have undergone a new intellectual birth under the inspiration of his fertile genius." (pp. 178–9.)

The comparison of the Darwinian movement with that of the Renaissance is a very apt one. But in neither case was it the "mind of the nation" which was set free. The movement in Germany, for instance, merely transferred the masses of the people physically and mentally from one bondage to a second, and where the new ideas did reach them they became symbols of an economic revolt, as in the Peasants' Rebellion, rather than marks of great intellectual progress. So it has been with the Darwinian doctrines, they did just reach and interest the more thinking working men in the seventies, but they have ceased to have any meaning for the bulk of the population to-day; its problems are essentially economic, and it will accept as an intellectual faith any doctrine which apparently offers better economic conditions[1]. The error is that we assume great progress in the intellectual views of the leaders of thought in a nation always corresponds to some mental development, or to some change in the culture of the mass of the people.

Galton concludes this section (p. 182) by stating that while we know "that the bulk of the respective provinces of nature and of nurture are totally different, and although the frontier between them may be uncertain," yet "we are perfectly justified in attempting to appraise their relative importance."

Our author now turns to *Associations*. He writes:

"The furniture of a man's mind chiefly consists of his recollections and the bonds that unite them. As all this is the fruit of experience, it must differ greatly in different minds according to their individual experiences. I have endeavoured to take stock of my own mental furniture in the way described in the next chapter, in which it will be seen how large a part consists of childish recollections, testifying to the permanent effect of many of the results of early education." (p. 182)......"The character of our abstract ideas, therefore, depends to a considerable degree on our nurture." (p. 183.)

I think in these remarks Galton does not allow adequately for the difference in receptivity in the material educated. Galton and his brothers, Darwin and Erasmus, had very similar early nurtures, but what made the elder brothers merely country gentlemen in ideas and habits, and the younger brother a foremost man of science of his day? Surely it was a differentiated receptivity, which caused Francis to store his mind—from practically the same environment—in a wholly different way; and there can be little doubt that this receptivity, which stored experiences wholly otherwise than his brothers did, was an innate faculty, a result of nature not of nurture. Again, many lads had the training of a classical school and of a university, precisely as Charles Darwin had, but their receptivity was very different in its selection from his, and the result left them largely mediocrities. That basal distinction was one of nature. Again, it is not only the selective action in storing experiences, but the manner in which the brain associates them, which is important. I cannot think, therefore, that because Galton in his *Psychometric Experiments*[2] found many of his associations were from early childhood that this denotes a large part played by nurture in mental efficiency. I think the effectiveness of the brain in summoning fitting associations is recognised by Galton in the following section of his book entitled *Antechamber of Consciousness*. Here he writes:

[1] A recent talk with a Russian Soviet professor from Moscow threw some light on the idealist views of the Soviet leaders. The results of modern science were to be broadcasted among the people, and the ecclesiastics who opposed this were to be removed; it was to be science for the people as against theological bondage, but the new scientific faith was to be associated with an economic revolution, which would benefit the masses. There certainly is a philosophic reading of history—what we might term an anthropological sense—in this combination. And I await with greater interest and more understanding the outcome of these idealistic scientists and politicians!

[2] He reproduces largely his papers on this subject (see our pp. 233–36) in pp. 185–203 of the *Inquiries*.

"When I am engaged in trying to think anything out, the process of doing so appears to me to be this: the ideas that lie at any moment within my full consciousness seem to attract of their own accord the most appropriate out of a number of other ideas which are lying close at hand, but imperfectly within the range of my consciousness. There seems to be a presence-chamber in my mind where full consciousness holds court, and where two or three ideas are at the same time in audience, and an antechamber full of more or less allied ideas, which is situated just beyond the full ken of consciousness. Out of this antechamber the ideas most nearly allied to those in the presence-chamber appear to be summoned in a mechanically logical way, and to have their turn of audience. The successful progress of thought appears to depend—first, on a large attendance in the antechamber; secondly, on the presence there of no ideas except such as are strictly germane to the topic under consideration; thirdly, on the justness of the logical mechanism that issues the summons. The thronging of the antechamber is, I am convinced, altogether beyond my control; if the ideas do not appear, I cannot create them, nor compel them to come. The exclusion of alien ideas is accompanied by a sense of mental effort and volition, whenever the topic under consideration is unattractive, otherwise it proceeds automatically, for if an intruding idea find nothing to cling to, it is unable to hold its place in the antechamber and slides back again." (pp. 203–4.)

Galton's analysis suggests the importance of (i) the selective action of the brain in storing ideas drawn from experiences, and of (ii) its efficiency in associating these ideas. In both these faculties it seems to me that we are dealing with an innate quality of the brain, which distinguishes two brothers reared under the same environment, or two youths educated in the same way in the same school and the same university. It is impossible to reproduce here the whole of Galton's suggestive thought in this section of his work on the *Antechamber of Consciousness*; we must refer the reader to the work itself. One further citation of a characteristic kind may be given:

"Extreme fluency and a vivid and rapid imagination are gifts naturally and healthfully possessed by those who rise to be great orators or literary men, for they could not have become successful in those careers without them. The curious fact already alluded to of five editors of newspapers being known to me as having phantasmagoria, points to a connection between two forms of fluency, the literary and the visual. Fluency may be also a morbid faculty, being markedly increased by alcohol (as poets are never tired of telling us), and by various drugs, and it exists in delirium, insanity, and states of high emotion. The fluency of a vulgar scold is extraordinary." (pp. 205–6.)

Galton's next section is entitled "Early Sentiments" (pp. 208–16), and in it he endeavours to show that

"the power of nurture is very great in implanting sentiments of a religious nature, of terror and of aversion, and of giving a fallacious sense of their being natural instincts." (p. 216.)

He states that:

"The models upon whom the child or boy forms himself are the boys or men whom he has been thrown amongst, and whom from some incidental cause he may have learnt to love and respect. The every-day utterances, the likes and dislikes of his parents, their social and caste feelings, their religious persuasions are absorbed by him; their views or those of his teachers become assimilated and made his own."... "He is born prepared to attach himself as a climbing plant is naturally disposed to climb, the kind of stick being of little importance." (p. 208.)

It seems to me that Galton overlooks here the fact that "slavish acceptance" is very frequently an inherited character. The child accepts the first thing placed before it, not necessarily because it is the first thing or comes from its parents, but because it lacks desire to inquire for itself. Galton asserts that mere chance of birthplace makes religion a matter of accident, but it is not chance, but mental inertia which leads many persons to retain the religion of their childhood without further inquiry. I cannot lay the stress Galton does on the danger of dogma or sentiments instilled in early youth becoming ingrained in the character. One sees too many young people now-a-days who have changed the religious and social faiths of their childhood, to lay exaggerated stress in this respect on nurture. It may have been true 40 years ago that:

"In subjects unconnected with sentiment, the freest inquiry and the fullest deliberation are required before it is thought decorous to form a final opinion; but whenever sentiment is involved, and especially in questions of religious dogma, about which is more sentiment and more difference of opinion among wise, virtuous and truth-seeking men than about any other subject whatever, free inquiry is peremptorily discouraged. The religious instructor in every creed is one who makes it his profession to saturate his pupils with prejudice." (p. 210.)

Whatever the religious instructor of to-day may say or do, I think it would have small effect on the youth of to-day! They have won their freedom, or, perhaps, it were truer to say, it has been won for them, and in my experience they think and choose for themselves both their social and their religious creeds. Those that do not, fail, not so much from prejudices inculcated by parents and pastors, as from intellectual inertia, which the careful observer will probably recognise as the really vital contribution of the parents to their offspring[1]. Still we may well agree with Galton that

"there are a vast number of foolish men and women in the world who marry and have children, and because they deal lovingly with their children it does not at all follow that they can instruct them wisely." (p. 210.)

Galton points out that the wisest men of all ages may have led upright and consistent lives and been honoured by a wide circle for their unselfish furtherance of the public good, but that they have belonged to many races, and have been claimed by many dogmatic faiths (pp. 211–13).

Conscience is next dealt with and it is stated that it arises from two sources (*a*) inheritance, and (*b*) early training. Ethnologists have shown that conscience varies from race to race and age to age; it is partly transmitted by inheritance in the way and under the conditions suggested by Darwin[2]:

"The value of inherited conscience lies in its being the organised result of the social experience of many generations, but it fails in so far as it expresses the experience of generations whose habits differed from our own. The doctrine of evolution shows that no race can be in perfect harmony with its surroundings; the latter are continually changing while the organism of the race hobbles after, vainly trying to overtake them. Therefore the inherited part of conscience cannot be an infallible guide, and the acquired part of it may, under the influence of dogma, be a very bad one. The history of fanaticism shows too clearly that this is not only a theory but a fact. Happy the child, especially in these inquiring days, who has been taught a religion that mainly rests on the moral obligations between man and man in domestic and national life, and which, so far as it is necessarily dogmatic, rests chiefly on the proper interpretation of facts about which there is no dispute,—namely, on those habitual occurrences which are always open to observation, and which form the basis of so-called natural religion." (p. 212.)

[1] Discussing recently with a friend whether Galton's views applied to the young people of to-day, I mentioned a number of them known to us both who had certainly thought for themselves. The reply came: "Yes, but *they* have minds," and not till the words were out of the mouth did the speaker realise that the case had been given away.

[2] *The Descent of Man*, 1871, Vol. I, p. 102, etc.

Terror, Galton asserts, is early learnt and he refers to the manner in which gregarious animals learn it from each other. In man, he mentions the inculcation in medieval times by preachers and artists of a belief in the horrible torments of the damned, and suggests that as torture was practised in the judicial proceedings of those days and was considered an appropriate attribute of the highest authority, so there appeared no inconsistency in a supremely powerful ruler, however beneficent, making the freest use of it. Aversion, like terror, is easily taught, and Galton points to the ideas concerning clean and unclean of Jews and Mussulmans. He even notes that his sojourn in the East during a very receptive stage of his life (see our Vol. I, Chap. VI) had impressed upon him the nobler aspects of Mussulman civilisation (see Vol. I, p. 207 and II, p. 28), and that he then adopted some of their aversions, even 40 years later looking upon his left hand as unclean (p. 216).

Whatever present-day readers may feel as to the power of nurture in implanting dogmatic belief, or in creating terrors and aversions in the mind of the child which it is not able thereafter to cast off, there is no doubt that the theological readers of 1883 were vastly incensed by Galton's book and gave expression to it in a series of hostile criticisms.

Galton's final answer[1] to the problem of the relative strengths of Nurture and Nature is based, as we have indicated elsewhere (pp. 126–29), on the "History of Twins." This subject occupies pp. 216–43 of his treatise, but it is unnecessary to repeat its conclusions here. He finishes his section by reference to the small effect that nurture has on the nature of the young cuckoo.

Then follows a reproduction of the memoir of 1865 on the *Domestication of Animals* (see our pp. 70–72). Galton claims that the facts cited show the small power of nurture against adverse natural tendencies. By this he means that every wild animal has practically had its chance of being domesticated, but that nurture has in the great bulk of cases failed to achieve domestication. Those who fail, sometimes only in one small particular, are destined to perpetual wildness so long as their race continues. "As civilisation extends they are doomed to be gradually destroyed off the face of the earth as useless consumers of cultivated produce." Galton infers that because very slight differences may make domestication impossible in related species, so very slight differences in the natural dispositions of human races may either lead them irresistibly to a certain career or make that career impossible (p. 271). Galton's next section is entitled: *Possibilities of Theocratic Intervention* and here again he commits the unpardonable offence of trespassing fearlessly upon the territory of those with whom he is at issue (see our p. 249); I fear this practice is rather the rule in the case of warfare, which is not unusually carried even into the enemy's camp. Be this as it may, Galton replies to the criticism that it is idle to compare the intensities of nature and nurture, because these may not be the only influences at work. There is the possibility of theocratic interference either on the Deity's own initiative or as a response to prayer. Galton endeavours

[1] "There is no escape from the conclusion that nature prevails enormously over nurture when the differences of nurture do not exceed those commonly found among persons of the same rank of society and in the same country." (p. 241.)

to show that there is only one mode of theocratic interference, which could upset the statistical comparison of the relative intensities of nature and nurture. He illustrates his point by supposing a caretaker tending a large number of silkworms of various breeds and fed in different ways, and that an observer watched his proceedings as well as he could, but only during the day-time and through a telescope. Now the caretaker might have a custom, of which the observer was ignorant, of feeding the silkworms in various ways during the night, and Galton asks how this would affect the statistical conclusions. He suggests four possibilities and considers in each case that the caretaker's unobserved interference would not affect the statistical conclusions based upon classifications by nature and by nurture. But, I think, the reason of this is that Galton supposes the caretaker to pay attention in his secret proceedings only to race (i.e. nature) and to the day feedings of these races (i.e. nurture). What if he thought nothing about race or day feedings, but classified his worms by some characteristic of the individuals? Suppose he fed them differentially so as to bring all worms up to practically the same size and colour, which might be the very characteristics by which the observer had classified respectively for nurture and breed? Clearly no comparison of the effects of nurture and nature would be possible, and by less complete changes the observer might be led to very false conclusions. Further, this would be done without the caretaker knowing, as Galton supposes in his fifth alternative, that he was watched and, because he objected to being watched, devising plans to deceive the observer. There is no necessity to suppose

"the homologue would be a god with the attributes of a devil, who misled humble and earnest inquirers after truth by malicious artifice." (p. 275.)

There is in fact no need to appeal to Milton's God, who could be moved to laughter by man's quaint attempts to understand his works[1].

Surely the problem is of a different kind. Either theocratic interference is perpetual and consistent, in which case it is as definite as any law of nature, and cannot be distinguished from it, and will not alter statistical results; or, it is occasional and capricious, in which case statistical samples taken under apparent sameness of physical environment will give divergent results. The general stability of statistical ratios, like the general fulfilment of prediction from so-called physical laws of nature, is the best argument against occasional and capricious theocratic interference. On pp. 277–94 Galton repeats his statistical arguments (see our pp. 115–17) against the "Objective Efficacy of Prayer." He expands to some extent his earlier arguments:

"The cogency of all these arguments is materially increased by the recollection that many items of ancient faith have been successively abandoned by the Christian world to the domain

[1] *Paradise Lost*, Bk VIII, ll. 70 *et seq.*:

"Or if they list to try
Conjecture, he his fabric of the Heavens
Hath left to their disputes, perhaps to move
His laughter at their quaint opinions wide
Hereafter, when they come to model Heaven,
And calculate the stars."

of recognised superstition. It is not two centuries ago, long subsequent to the days of Shakespeare and other great men whose opinions still educate our own, that the sovereign of this country was accustomed to lay hands on the sick for their recovery, under the sanction of a regular Church service, which was not omitted from our prayerbooks till the time of George II. Witches were unanimously believed in and were regularly exorcised, and punished by law, up to the beginning of the last century. Ordeals and duels, most reasonable solutions of complicated difficulties according to the popular theory of religion, were found untrustworthy in practice. The miraculous power of relics and images, still so general in Southern Europe, is scouted in England. The importance attributed to dreams, the barely extinct claims of astrology, and auguries of good and evil lucks, and many other well-known products of superstition which are found to exist in every country, have ceased to be believed in by us. This is the natural course of events, just as the Waters of Jealousy, and the Urim and Thummim of the Mosaic Law had become obsolete in the times of the later Jewish kings. The civilised world has already yielded an enormous amount of honest conviction to the inexorable requirements of solid fact; and it seems to me clear that all belief in the efficacy of prayer, in the sense in which I have been considering it, must be yielded also. The evidence I have been able to collect bears wholly and solely in that direction, and in the face of it the *onus probandi* must henceforth lie on the other side." (pp. 293–4.)

The following section is termed *Enthusiasm* (pp. 294–98), and is concerned not so much with "ardent zeal" in any kind of work, as with the definition of the word in its proper range—"a belief or conceit of private revelation; the vain confidence or opinion of a person that he has special divine communications from the Supreme Being or familiar intercourse with him[1]." Galton remarks that to a large number of the ablest class of mankind the idea of an indwelling divine Spirit, with which they can commune, is so habitual and vivid as to be an axiomatic truth to them. This possibility, he says, has been to him a real and almost lifelong subject of thought, and has been a motive for many of the inquiries in his book, for were this "enthusiasm" a reality and not a vain confidence, it is clear that those races should be encouraged who are characterised by spiritual-mindedness, for they would be far more worthy occupants of the earth than the generality of ourselves (p. 395). Those who have known Francis Galton, and so realised his innate simplicity of mind, will appreciate as no others can that there was no flippancy in his words.

"There is no subject more worthy of reverent but thorough investigation than the objective evidence for or against the existence of inspiration from an unseen world, and none that up to the present time has so tantalised the anxious and honest inquirer with unperformed promise of solution. The arguments scattered or hinted at throughout this book are negative so far as they go, but it must be borne in mind that they would be scattered to the winds by solid objective evidence on the other side, such as could be seriously entertained by scientific men desiring above all things to arrive at truth." (pp. 295–6.)

Galton then cites the points in his *Inquiries* which bear on the axiomatic assumption of inspiration, the visions of apparently objective character, the fluency which is considered automatic unless it deals with devout subjects, the prevalence of extreme forms of religious rapture among the hysterical and insane, the axiomatic necessity, to those that perceive them, of their individual number forms or colour associations. Lastly, Galton claims—and here the dogmatic theologians would not be at one with him—that "it appears to be tacitly recognised by all that the absolute and final court of appeal is not

[1] Galton cites this definition as occurring in a "recent" dictionary.

subjective but objective." We cannot assert that our own instinctive convictions alone are to be trusted, we are forced to grant no less trustworthiness to the convictions and fancies of others. All such convictions should be tested, whenever possible, by appeal to facts which admit of *repetition*, for experience shows that only observations of such facts lead to results which can be universally acknowledged. Galton insists on the duty of suspending our belief and maintaining the freedom of our mental attitude whenever there is strong reason to doubt (p. 298). The section on *Enthusiasm* closes with a fine paragraph, which indicates how far astray the critics went, when they labelled Galton a materialist:

"There is nothing in any hesitation that may be felt as to the possibility of receiving help and inspiration from an unseen world, to discredit the practice that is dearly prized by most of us, of withdrawing from the crowd and entering into quiet communion with our hearts, until the agitations of the moment have calmed down, and the distorted mirage of the worldly atmosphere has subsided, and the greater objects and more enduring affections of our life have reappeared in their due proportions. We may then take comfort and find support in the sense of our forming part of whatever has existed or will exist, and this need be the motive of no idle service, but of an active conviction that we possess an influence which may be small but cannot be inappreciable, in defining the as yet undetermined possibilities of an endless future. It may inspire a vigorous resolve to use all the intelligence and perseverance we can command to fulfil our part as members of one great family that strives as a whole towards a fuller and a higher life." (p. 298.)

It was a great revolution in thought that Galton was proposing and probably few grasped its extent in 1883. He had in mind a new religion, a religion which should not depend on revelation, physical to a few selected men, or psychical to a few individuals. Man was to study the purpose of the universe in its past evolution, and by working to the same end, he was to make its progress less slow and less painful in the future. Darwin had taught evolution as a scientific doctrine; Galton proposed that this new knowledge should be applied to racial and social problems, and that understanding of, sympathy with, and aid in the progress of the general evolution of living forms should be accepted as religious duties. If the purpose of the Deity be manifested in the development of the universe, then the aim of man should be, with such limited powers as he may at present possess, to facilitate the divine purpose. Before Darwin, living forms, indeed the world itself, had no history; there was held to be no serious ethnological difference between the first man and modern civilised man; the reptile and the mammal were coeval. Darwin for the first time gave a real history to living forms, and Galton following him said: Study that history, study the Bible of Life, and you will find your religion in it, and a new and higher morality as well. Thereby he raised Darwinism on to a higher, a spiritual plane. Thus it comes about that the last 40 pages of Galton's *Inquiries into Human Faculty* contain some of the finest passages he ever wrote, for they are devoted to his philosophical or rather religious views, and to their Darwinian basis. Galton saw in his doctrine a new moral freedom for man and a new religion based on scientific knowledge. His theological critics found it pure materialism, a fresh war against Heaven. Who shall determine which party was in the right? These extreme divergences at least confirm Galton's statement that the mental differences of mankind are so great that evolution has ample material to select from!

Galton starts his philosophical creed with a section on *The Observed Order of Events*: his thesis here is that the universe is a single entity and we ourselves are part of a mysterious whole "behind which lies the awful mystery of the origin of all existence," the purpose of the universe. He considers that the conditions which direct the evolution of living forms are on the whole marked by their persistence in improving the birthright of successive generations.

"They determine at much cost of individual comfort, that each plant and animal shall on the general average be endowed at its birth with more suitable natural faculties than those of its representative in the preceding generation. They ensure, in short, that the inborn qualities of the terrestrial tenantry shall become steadily better adapted to their homes and to their mutual needs." (p. 299.)

"If we summon before our imagination in a single mighty host, the whole number of living things from the earliest date at which terrestrial life can be deemed to have probably existed, to the latest future at which we may think it can probably continue, and if we cease to dwell on the mis-carriages of individual lives or single generations, we shall plainly perceive that the actual tenantry of the world progresses in a direction that may in some sense be described as the greatest happiness of the greatest number." (p. 300.)

Galton remarks how, while the motives of individuals in the lowest stages were purely self-regarding, they have broadened out as evolution went on. Subjects of affection and interest other than self become increasingly numerous as intelligence and depth of character develop, and as civilisation extends. He notes that as civilisation has advanced the sacrifice of personal repose to the performance of social duties has become more common.

"Life in general may be looked upon as a republic where the individuals are for the most part unconscious that while they are working for themselves they are also working for the public good." (p. 300.)

This was indeed a refreshingly optimistic opinion! Even the period which the physics of that day fixed for the available heat of the sun, upon which organic life depends, did not daunt Galton. There are countless abortive seeds and germs; among a thousand men selected at random some are crippled, others insane, idiotic or otherwise incurably imperfect in body or mind; what if our "world may rank among other worlds as one of these"? We know that our own life is built up of the separate lives of billions of cells of which our body is composed. They form a vast nation, members of which are always dying, while others grow to take their place. The continual sequence of these little lives—unconscious of the whole—has its outcome in the larger and conscious life of the man as a whole. Even this world of ours and

"our part in the universe may possibly in some distant way be analogous to that of cells in an organised body, and our personalities may be the transient but essential elements of an immortal and cosmic mind." (p. 302.)

Thus Galton, the pantheist, again puts forth as a possibility his beautiful, but unproven and unprovable dream (cf. our p. 114). All he can say of it is that at least it is not inconsistent with observed facts. Yet even while he admits that the slow progress of evolution is due to antecedents and inherent conditions of which we have not yet the slightest conception, he throws out an idea which foreshadows in a startling way Einstein, and in itself predicts how his doctrine may modify man's religious views. I have not seen this strange passage quoted, nor do I know what Galton's readers may have made of it in pre-relativity days. It runs:

"It is difficult to withstand a suspicion that the three dimensions of space and the fourth dimension of time may be four independent variables of a system which is neither space nor time, but something else wholly unconceived by us. Our present enigma as to how a First Cause could itself have been brought into existence—how the tortoise of the fable that bears the elephant, that bears the world, is itself supported,—may be wholly due to our necessary mistranslation of the four or more variables of the universe, limited by inherent conditions, into the three unlimited variables of space and the one of Time." (p. 302).

An obscure passage, indeed, which one of us ought to have asked Galton to interpret, but which now we can only place against Clifford's concept that "matter is a wrinkle in space." Both men might have taught us to think had our minds then been receptive.

Putting these high theories and suggestions on one side, Galton notes two great and indisputable facts:

(i) That the whole of the living world moves steadily and continuously towards the evolution of races that are progressively more and more adapted to their complicated mutual needs and to their external circumstances.

(ii) That this process of evolution has been hitherto carried on with what men from their standpoint must reckon as great waste of opportunity and life, and with little if any consideration for individual mischance.

Measured by man's criterion of intelligence and mercy,

"the process of evolution on this earth has been carried out neither with intelligence nor ruth, but entirely through the routine of various sequences, commonly called 'laws,' established or necessitated we know not how." (p. 303.)

Intelligent man has now been evolved. He has enormously modified the surface of the earth and altered its distribution of plants and animals. This new animal, man, endowed with a little power and some intelligence, ought, Galton holds, to assume a deliberate and conscious part in furthering the great work of evolution.

"He may infer the course it is bound to pursue, from his observation of that which it has already followed, and he might devote his modicum of power, intelligence and kindly feeling to render its future progress less slow and painful. Man has already furthered evolution very considerably, half unconsciously, and for his own personal advantages, but he has risen to the conviction that it is his religious duty to do so deliberately and systematically." (p. 304.)

Thus was the Darwinian doctrine raised by Galton to a religious creed.

The next section of the book, entitled *Selection and Race*, needs, I venture to think, some modification. Galton, having only dealt with the correlation of two variates, misunderstood, as I shall show later, the phenomena of regression. His statement here that the stringent selection of the best specimens of a race to rear and breed from can never lead to any permanent result, is, I feel sure, erroneous, and due to a wrong interpretation of multiple regression. Further, I doubt his assumption of the diminished fertility of highly-bred animals, unless he supposed, as in the case of the race-horse, a selection by *one* character, say, speed, only. There is, I think, no evidence that a selection of man by *both* physique and mentality would lead to an infertile race.

With Galton's statement that a low race, subjected to conditions of life that demand a high level of efficiency, must be submitted to a very rigorous selection involving great pain and misery, we can certainly agree. And we can also do so in the suggestion that the terrible suffering would disappear did we replace it by a higher race.

"The most merciful form," writes Galton, "of what I ventured to call 'eugenics' would consist in watching for the indications of superior strains or races, and in so favouring them that their progeny shall outnumber and gradually replace that of the old one." (p. 307.)

The following section of the *Inquiries* is concerned with the *Influence of Man upon Race* (pp. 308–17). Galton gives in a very few pages an able ethnographical survey of the world; he shows that in almost every known country there are three or four races or sub-races of man competing consciously or unconsciously for dominance. The process of evolution is still going on around us, and we disregard it instead of studying it and facilitating it. He raises a strong protest against that misleading word "aborigines," and points out that it dates from a time when a false cosmogony thought the world young and life to be of very recent appearance. There are to-day practically no original inhabitants of any district; all hold their lands only by the robber-rights of their ancestors. It would be difficult indeed to find a country which being unoccupied was colonised by its *present* inhabitants, and thence to assert their right of occupation[1]. Such reasoning carried to its logical conclusion might demand the complete surrender of Australia to the marsupials or even the monotremes.

"There exists," writes Galton, "a sentiment for the most part quite unreasonable, against the gradual extinction of an inferior race. It rests on some confusion between the race and the individual, as if the destruction of a race was equivalent to the destruction of a large number of men. It is nothing of the kind where the process of extinction works silently and slowly through the earlier marriage of members of the superior race, through their greater vitality under equal stress, through their better chances of getting a livelihood, or through their prepotency in mixed marriages. That the members of an inferior class should dislike being elbowed out of the way is another matter; but it may be somewhat brutally argued that whenever two individuals struggle for a single place, one must yield, and that there will be no more unhappiness on the whole, if the inferior yield to the superior than conversely, whereas the world will be permanently enriched by the success of the superior[2]." (p. 309.)

[1] The preliminary discussion of the recent peace terms at Versailles was accompanied by much futile talk about the 'rights' of small nations and of racial units. No small people, because it at present occupies a certain area, can be said to have a 'right' to mineral resources vastly exceeding its own consumption and essential to the needs of a larger adjacent population. Any allotment of lands based solely on 'aboriginal' or even present occupational 'rights' is certain to be called in question by the pressure of race against race. The peace-makers of Versailles lacked the knowledge that springs from a study of evolution.

[2] A great deal of the missionary argument in favour of the retention of the negro in tropical Africa, as against the Indian, or, as Galton proposed, the Chinese immigrant (see our p. 33), arises from the fact that the negro's emotional nature makes him a more ready convert than the more highly civilised Asiatics. Africa, like Europe of the folk-wandering days, has always been the

Galton notes what enormous influence the men of former generations have exercised unconsciously over the human stock of to-day. How differently world-history would have developed had our forefathers left the 'aborigines' of America, South Africa and Australia to the free occupation of their lands.

"The power in man of varying the future human stock vests a great responsibility in the hands of each fresh generation, which has not yet been recognised at its just importance, nor deliberately employed. It is foolish to fold the hands and to say that nothing can be done, inasmuch as social forces and self-interests are too strong to be resisted. They need not be resisted; they can be guided." (p. 317.)

In the following section, termed *Population*, Galton refers to the increased danger of over-population owing to improved sanitation, lesser mortality and the filling-up of the spare places of the world. He expresses, as Darwin did (see our p. 111), strong disapproval of Malthus' prudential check as prejudicial to the better elements of the race who alone would be prudent and self-denying, while the thriftless and improvident would crowd the vacant space left by the prudent[1]. The 'misery-check'—as Malthus called all influences other than the prudential, such as deaths through lack of food and shelter, overcrowding, war, etc.—does not seem to Galton to cover all the causes which make one race decay in the presence of a second. He thinks that an inferior race becomes listless and apathetic in the presence of a superior one, and loses its virility. He believes that such apathy is less a 'misery' than the prudential restraint where there is a keen desire for marriage (p. 320).

Galton then turns to his own direct proposals for racial betterment. He refers first to Dr J. Mathews Duncan's data for fertility and he uses these as if (i) they represented the survivors at adult ages of those born, (ii) they applied equally to all classes of intelligence and physique; he does not discuss the differential infantile death-rate, nor the differential fertility rate of his selected and rejected classes, nor the important question of the relative number of *survivors* of early and of late marriages. He takes two groups, one of 100 mothers who marry at 20, and another of 100 mothers who marry at 29, and considers their families would be 8·2 and 5·4 respectively. He then says that they would contribute to the next 200 in the ratio of 115 to 85. It seems to me that the ratio should be 121 to 79. He makes the length of a generation on the average 31·5 years, but takes 20 as being 4·5 years earlier and

melting pot of races, and it is safe to say that not a single race is 'aboriginal' in Africa to-day. The Bushman and the Hottentot would have greater 'rights' than the negro to large parts of Africa, but their claim might be worsted by the Rhodesian Man. He having no lineal descendants, would compel us in justice possibly to ascend to a prot-simio human, whence descending to the rightful heirs, we might find the chimpanzees as the true aborigines!

[1] While criticising Malthus' main conclusion Galton pays him a high compliment: "I must take this opportunity of paying my humble tribute of admiration to his great and original work, which seems to me like the rise of a morning star before a day of free social investigation. There is nothing whatever in his book which would be in the least offensive to this generation, but he wrote in advance of his time and consequently roused virulent attacks, notably from his fellow clergymen, whose doctrinaire notions upon the paternal dispensation of the world were rudely shocked." (pp. 318–19.)

29 years as 4·5 years later than the average age at marriage and so takes 27 and 36 years as the lengths of generations in the two classes. Thus in 108 years the early marrying class will have had four and the late marrying class three generations. Galton's numbers on p. 322 should, I think, be replaced by the following:

*Relative contributions to Maternal Populations.*

| Years | Early marriages (20) | Late marriages (29) |
|---|---|---|
| 108 | 214 | 49 |
| 216 | 459 | 24 |
| 324 | 985 | 12 |

The changes emphasise considerably Galton's argument, although exception may well be taken to some of its stages, in particular to the equality of death-rates in large and small families. However, his general principle is probably a correct one: namely that for the physically fit early marriage means more numerous offspring. Galton's next two sections indicate how he proposed to make use of this greater fertility in the case of the early married. His first suggestion is to give marks in competitive examinations for 'family merit.' Thus able youths would be favourably handicapped in civil service examinations if they came of superior breed. A superior breed is one which has been successful in its callings and is physically fit.

"A thriving family may be sufficiently defined or inferred by the successive occupations of its several male members in the previous generation, and of the two grandfathers. These are patent facts ascertainable by almost every youth, which admit of being verified in his neighbourhood and attested in a satisfactory manner. A healthy and long-lived family may be defined by the patent facts of ages at death, and number and ages of living relations, within the degrees mentioned above, all of which can be verified and attested. A knowledge of the existence of longevity in the family would testify to the stamina of the candidate, and be an important addition to the knowledge of his present health in forecasting the probability of his performing a large measure of experienced work." (p. 325.)

Galton would feel his way gradually in these matters, but even a small allowance to family merit would be great in its effect as indicating "that ancestral qualities are of present current value." The second factor is that of 'Endowment' of the able who have 'family merit.' As money has often been left for marriage portions for poor girls, so it might be left for the worthier purpose of marriage portions for able young people of 'family merit.' In the seventies the college statutes of the older universities enforced celibacy on the Fellows.

"The college statutes to which I referred were very recently relaxed at Oxford, and have just been reformed at Cambridge. I am told that numerous marriages have ensued in consequence, or are ensuing. In *Hereditary Genius* I showed that scholastic success runs strongly in families; therefore in all seriousness, I have no doubt, that the number of Englishmen naturally endowed with high scholastic faculties will be sensibly increased in future generations by the repeal of these ancient statutes." (pp. 329–30.)

As Galton very truly states, the wealth of the English race in hereditary gifts has never been properly explored; and when it has been, the natural impulses of man ought to be sufficient to ensure that such wealth should no more be neglected than the existence of any other possession suddenly revealed to man.

In his *Conclusion* Galton sums up the third of the various inquiries in his volume; he points out the vast variety of natural faculty, both advantageous and not, to be found in individuals, and the great differences between human races. Man is variable and has changed widely in the course of hundreds of thousands of years. This idea of growth in man had not been grasped by the early cosmogony makers. Its recognition compels us

"to reconsider what may be the true place and function of man in the order of the world." (p. 332.)

Galton confesses to having examined this question from many points of view, for

"whatever may be the vehemence with which particular opinions are insisted upon, its solution is unquestionably doubtful. There is a wide and growing conviction among truthseeking, earnest, humble-minded and thoughtful men, both in this country and abroad, that our cosmic relations are by no means so clear and simple as they are popularly supposed to be, while the worthy and intelligent teachers of various creeds, who have strong persuasions on the character of these relations, do not concur in their several views." (p. 332.)

He says the results of such inquiries as he has been able to make do not confirm the common doctrines as to our relations with the unseen universe. The one thing that he can see is that man has immensely modified humanity by his action, and if he would only consciously take its betterment in hand, as a freeman shaping the course of future humanity, he might achieve great ends far less ruthlessly and more economically than Nature alone.

"The chief result of these Inquiries has been to elicit the religious significance of the doctrine of evolution. It suggests an alteration in our mental attitude, and imposes a new moral duty." (p. 337.)

Thus Galton finally summarised his labours of more than 12 years. Few of those who, in the following century, have quoted Galton's statement namely that eugenics was a religion, have grasped fully its relation to Darwin and Evolution. Huxley attacked the old orthodox beliefs because he thought they fettered the development of science. Galton attacked them because he thought current religion fettered the development of a higher morality and a more rational religion. Huxley spoke at the ripe moment and produced immediate effect.. Galton spoke—as he himself recognised—before he could be understood. But those who know that even Norway, Switzerland and Roumania have now established their Institutes for the Study of Eugenics, realise that Galton's teaching is likely to affect civilised man—now and in the future—in a manner only comparable with the influence of a new religious faith. He brought the logical application of the doctrine of evolution to the betterment of the human race. That in Galton's judgment was a science, a morality and the only rational religious faith.

D. LATER PSYCHOMETRIC RESEARCHES

After the publication of the *Inquiries into Human Faculty*, Galton's psychical researches ceased to flow with the amplitude of the years 1876–1884, but, as in the case of Geography, he never lost his interest in these matters, and this is shown in a number of minor papers which he continued to issue till at least 1896. These papers may be referred to here, but they mark the transition of his mind to the more definite statistical standpoint of his later years. His anthropometry was largely psychometric, but the statistical basis was growing more developed and more satisfactory.

The first paper of this series which we must consider is entitled "Measurement of Character." It was published in the *Fortnightly* for August, 1884[1], and the material appears to have been largely that of Galton's Rede Lecture in the Senate House at Cambridge[2]. The opening paragraph explains Galton's purpose:

"I do not plead guilty to taking a shallow view of human nature, when I propose to apply, as it were, a footrule to its heights and depths. The powers of man are finite, and if finite they are not too large for measurement. Those persons may justly be accused of shallowness of view, who do not discriminate a wide range of difference, but quickly lose all sense of proportion, and rave about infinite heights and unfathomable depths, and use such like expressions, which are not true and betray their incapacity. Examiners are not I believe much stricken with the sense of awe and infinitude when they apply their footrules to the intellectual performances of the candidates they examine; neither do I see any reason why we should be awed at the thought of examining our fellow creatures as best we may, in respect to other faculties than intellect. On the contrary, I think it anomalous that the art of measuring intellectual faculties should have become highly developed, while that of dealing with other qualities should have been little practised or even considered." (p. 179.)

Galton then emphasises the importance of measuring the emotional characters in man, for only by so doing can the individual know where he stands among his fellow-men, and whether he is getting on or falling back.

"The art of measuring various human faculties now occupies the attention of many inquirers in this and other countries....... New processes of inquiry are yearly invented, and it seems as though there was a general lightening up of the sky in front of the path of the anthropometric experimenter, which betokens the approaching dawn of a new and interesting science. Can we discover landmarks in character to serve as bases of a survey or is it altogether too indefinite and fluctuating to admit of measurement? Is it liable to spontaneous changes, or to be in any way affected by a caprice that renders the future necessarily uncertain? Is man, with his power of choice and freedom of will, so different from a conscious machine, that any proposal to measure his moral qualities is based upon a fallacy? If so it would be ridiculous to waste thought on the matter, but if our temperament and character are durable realities, and persistent factors of our conduct, we have no Proteus to deal with in either case, and our attempts to grasp and measure them are reasonable." (pp. 179–80.)

[1] Vol. XXXVI, N.S. pp. 179–85.

[2] I cannot find that this Rede Lecture was ever independently issued. "He [Frank] was Rede Lecturer at Cambridge on May 27th, and we went to the Vice-Chancellor's and after to Mrs Darwin's, and greatly enjoyed the four days; fortunately I was well at the time." *L. G.'s Record* under 1884.

Galton next gives his reasons for believing that character and temperament are persistent. These are summed up in:

(*a*) Heredity. A son who inherits somewhat exclusively the qualities of his father, "fails with his failures, sins with his sins, surmounts with his virtues." His course of life has been predetermined by his inborn faculties.

(*b*) The life-histories of like twins, who behave like one person. "Whatever spontaneous feeling the one twin may have had, the other twin at the very same moment must have had a spontaneous feeling of exactly the same kind." If we had in our keeping the twin of a man, who was his "double," we could obtain a trustworthy forecast of what the man would do under any new conditions by submitting the twin to those conditions and watching his conduct. (pp. 180–1.)

(*c*) The result of Galton's own inquiry into Free-will (see our p. 245), which indicated how small seems the room left for a possible residuum of free-will.

Galton concluded on the basis of these three researches that the character which shapes our conduct is

"a definite and durable 'something,' and that therefore it is reasonable to attempt to measure it." (p. 181.)

Now-a-days one might think that the statistical material on which Galton based his conclusions was rather meagre, but most of his results have been confirmed by more extensive researches[1]. He appears to have considered that 'character' was in some way a unit entity:

"We must guard ourselves against supposing that the moral faculties which we distinguish by different names, as courage, sociability, niggardness, are separate entities. On the contrary, they are so intermixed that they are never singly in action. I tried to gain an idea of the number of the more conspicuous aspects of the character by counting in an appropriate dictionary the words used to express them. Roget's *Thesaurus* was selected for the purpose, and I examined many pages of its index here and there as samples of the whole, and estimated that it contained fully one thousand words expressive of character, each of which had a separate shade of meaning, while each shares a large part of its meaning with some of the rest." (p. 181.)

From the more modern standpoint it would seem that the direct course would be to measure various factors of character in individuals and to study the extent of the inter-correlations of these factors. At any rate in children it may be doubted whether such factors as shyness, conscientiousness, self-consciousness, temper, etc., are *very highly* correlated together. One would rather anticipate that character was a hotch-potch of factors mixed in different proportions for each individual.

However, Galton starts with 'character' as an entity like intellectual capacity and suggests that as the latter may be sounded by definite tests at individual points, so in character definite acts in definite emergencies may be noted.

[1] For example, that the factors of character are inherited, see *Biometrika*, Vol. III, pp. 131–190; that growth and education have little influence on character, see *Drapers' Research Memoirs*, Biometric Series, No. 4.

"Emergencies need not be waited for, they can be extemporised; traps as it were can be laid....... A sudden excitement, call, touch, gesture, or incident of any kind evokes, in different persons, a response that varies in intensity, celerity and quality." (p. 182.)

Galton suggests that the cardiograph, the sphygmograph, and Mosso's blood-pressure apparatus should be used to test the effect of various small emotional shocks. To those conversant with relatively recent attempts by similar means to measure emotional changes, it will possibly be surprising that Galton suggested such investigations 40 years ago. He even wished to meet the criticism that the presence of the recording instrument might make itself felt and check the expression of the emotion. He accordingly experimented on himself by wearing a Maret's pneumo-cardiograph

"during the formidable ordeal of delivering the Rede Lecture in the Senate House at Cambridge....... I had no connection established between my instrument and any recording apparatus but wore it merely to see whether or no it proved in any way irksome. If I had had a table in front of me, with the recording apparatus out of sight below, and an expert assistant near at hand to turn a stopcock at appropriate moments, he could have obtained samples of my heart's action without causing me any embarrassment whatever. I should have forgotten all about the apparatus while I was speaking." (pp. 183–4.)

Methods of measuring the unmeasured, or trapping unobserved the emotional changes in men and women, delighted Galton above all things. He was particularly pleased with Tennyson's where he tells us that

"Lancelot returning to court after a long illness through which he had been nursed by Elaine, sent to crave an audience of the jealous queen. The messenger utilises the opportunity for observing her in the following ingenious way like a born scientist:

> 'Low drooping till he well nigh kissed her feet
> For loyal awe, saw with a sidelong eye
> The shadow of a piece of pointed lace
> In the Queen's shadow, vibrate on the wall
> And parted, laughing in his courtly heart.'"

And again—with a suggestion of grim possibilities at a social meal in Rutland Gate, which the actuality never to my knowledge fulfilled—

"The poetical metaphors of ordinary language suggest many possibilities of measurement. Thus when two persons have an 'inclination' to one another, they visibly incline or slope together when sitting side by side, or at a dinner table, and they then throw the stress of their weights on the near legs of their chairs. It does not require much ingenuity to arrange a pressure gauge with an index and dial to indicate changes in stress, but it is difficult to devise an arrangement, that shall fulfil the threefold condition of being effective, not attracting notice and being applicable to ordinary furniture. I made some rude experiments, but being busy with other matters, have not carried them on, as I had hoped." (p. 184.)

Other suggestions in the paper dealt with the measurement of temper and fault-finding. Galton concludes:

"The points I have endeavoured to impress are chiefly these. First, that character ought to be measured by carefully recorded acts, representative of conduct. An ordinary generalisation is nothing more than a muddle of vague memories of mixed observations. It is an easy vice to generalise. We want lists of facts, every one of which may be separately verified, valued and revalued, and the whole accurately summed. It is the statistics of each man's conduct in small every day affairs, that will probably be found to give the simplest and most precise measure of his character. The other chief point that I wish to impress is, that a practice of deliberately and methodically testing the character of others and of ourselves is not wholly fanciful but deserves consideration and experiment." (p. 185.)

Galton's Rede Lecture was a very graceful address; it was full of these

Francis Galton in the seventies, from a photograph.

new, if now familiar ideas, but it was a lecture of *suggestion*, and accordingly the reader must not expect to find in it statistics of actual measurements of character. It serves to explain, however, the links in Galton's own mind between his work on Heredity, his paper on Twins and his study of Free-will. At the very time Galton was writing this lecture he was collecting data by aid of his Family Records (see our Chapter XIII) on the distribution of one phase of character, namely, Temper in English Families. A First Report on his results was published in the *Fortnightly Review*, July, 1887[1]. The paper from more than one standpoint is slightly disappointing, and as Galton himself remarks he had to set to work on rough materials with rude tools (p. 29). The criticisms that one may raise are of the following kinds. The descriptions of temper are all verbal, and although many epithets are used, Galton in the main classifies into 'Good Temper' and 'Bad Temper.' His 'Good Temper' contains not only the 'forbearing' and 'self-controlled' but the 'submissive,' 'timid' and 'yielding.' His 'Bad Temper' contains not only the 'quick tempered,' but the 'bickering' and the 'sullen.' My own investigations seem to suggest a fundamental difference in 'Good Temper' between the Self-controlled and the Weak class, and the Sullen cannot profitably be put in the same category with the Choleric. Galton does indeed make a five-group classification, namely: (1) mild; (2) docile; (3) fretful; (4) violent; (5) masterful. The distinction, however, between (1) and (2) is not that of self-controlled and weak good temper, and it is not clear whether such a marked class as the sullen has been put into (3), (4) or (5). Another defect of Galton's material was the large proportion of cases, over 50 °/₀, in which no record of temper was given at all. He calls these neutral and says that approximately

Good Tempered : Neutral Tempered : Bad Tempered :: 1 : 2 : 1,

and he finds in the approximate equality of the Good and Bad Tempered, and their total being equal to the Neutral Tempered, definite evidence of the correctness of the records in this respect. I fail to be convinced by Galton's arguments, for it seems to me that they would have equal application to any classification into alternate categories, e.g. criminal and non-criminal, with a neutral class for those of whom nothing was known, or nothing recorded. On the basis of his classification, omitting the 50 °/₀ of 'neutrals,' Galton deduces that there is no selective mating in human marriage with regard to temper[2], but he concludes that there is emphatic testimony to the heredity of temper. His method of establishing the latter conclusion is somewhat arbitrary and somewhat elementary, but it has undoubtedly been confirmed by later work. He rather weakens his position, however, by introducing a caveat that he does not propose to deal with temper as an unchangeable characteristic. It is difficult to grasp how under such conditions it is possible to assert that temper is

"nevertheless as hereditary as any other quality." (p. 30.)

[1] "Good and Bad Temper in English Families," Vol. XLII, N.S. pp. 21–30. A wrong year and *locus* are assigned to this paper in Galton's *Memories*, p. 328.

[2] This is not confirmed by more recent researches.

The more modern statistician would feel compelled to investigate to what extent temper does change with growth, education and environment before he could assert that it was as hereditary as any other quality. Putting aside these criticisms, Galton undoubtedly indicated in this paper for the first time that statistics of factors of character could be dealt with and inferences drawn as to their distribution and hereditary character.

In the previous year, 1886, Galton had published at least two papers dealing with psychometry. Mr Joseph Jacobs had been interesting himself in "Experiments on Prehension," which he defined as the mind's power of *taking on* certain material, in this case auditory sensations. Nonsense syllables, letters or numerals were delivered at about half-second intervals in a monotonous voice, and the test consisted in the number the subject could repeat[1]. Mr Jacobs found that the 'span,' i.e. number correctly repeated, (*a*) increased with age, (*b*) was greater for those higher in the class than for those lower, and argued that the 'span of prehension' should be an important factor in mental groups, and its determination a test of mental capacity. Galton suggested that the inquiry should be extended to idiots, and visited on June 18, 1886, the Earlswood Asylum with Professor Alexander Bain, and on June 30, 1886, the Asylum for Idiots at Darenth with Professor James Sully. The general conclusion obtained by Galton was that the idiots' 'span of prehension' was only about half that of Mr Jacobs' normal children, three to four figures instead of seven to eight.

In 1886 Romanes published a theory of the origin of varieties, attributing them to peculiarities in the reproductive system of certain individuals which render them more or less sterile to other members of the common stock while they remain fertile among themselves. Galton, who, as we have seen (p. 271), had been working on assortative mating in man, considered that special sexual attractiveness rather than sterility due to peculiarities of the reproductive system was the source of varieties. He writes[2]:

"It has long seemed to me that the primary characteristic of a variety resides in the fact that the individuals who compose it do not, as a rule, *care to mate* with those who are outside their pale, but form through their own sexual inclinations a caste by themselves. Consequently that each incipient variety is probably rounded off from the parent stock by means of *peculiarities of sexual instinct*, which prompt what anthropologists call endogamy (or marriage within the tribe or caste), and which check exogamy (or marriage outside of it). If a variety should arise in the way supposed by Mr Romanes, merely because its members were more or less infertile with others sprung from the same stock, we should find numerous cases in which members of the variety consorted with outsiders. These unions might be sterile, but they would occur all the same, supposing of course the period of mating to have remained unchanged. Again we should find many hybrids in the wild state, between varieties which were capable of producing them when mated artificially. But we hardly ever observe pairings between animals of different varieties when living at large in the same or contiguous districts, and we hardly ever meet with hybrids that testify to the existence of unobserved pairings. Therefore it seems to me that the hypothesis of Mr Romanes would in these cases fail while that which I have submitted would stand." (p. 395.)

Galton then suggests that even in the case of plants insects may exhibit

[1] *Mind*, Vol. XII, pp. 75–9.

[2] "The Origin of Varieties," *Nature*, Vol. XXXIV, pp. 395–6, August 26, 1886.

a selective appetite, and so a variety be preserved from intercrossing with the parent stock. He observes that where we just distinguish two varieties by one or two differences, these may connote a host of differences unknown to us—especially those recognised by the sense of smell, so weak with us and so strong in many of the animals—"whose aggregate would amply suffice to erect a barrier of sexual indifference or even repugnance between their members[1]." Galton, considering the case of man, writes:

"No theme is more trite than that of the sexual instinct. It forms the main topic of each of the many hundred (I believe about 800) novels annually published in England alone, and of most of the still more numerous poems, yet one of its main peculiarities has never, so far as I know, been clearly set forth. It is the relation that exists between different degrees of unlikeness and different degrees of sexual attractiveness. A male is little attracted by a female who closely resembles him. The attraction is rapidly increased as the difference in any given respect between the male and female increases, but only up to a certain point. When this is passed, the attraction again wanes, until the zero of indifference is reached. When the diversity is still greater, the attractiveness becomes negative and passes into repugnance, such as most fair complexioned men appear to feel towards negresses, and *vice versâ*. I have endeavoured to measure the amount of difference that gives rise to the maximum of attractiveness between men and women, both as regards eye-colour and stature, chiefly using the data contained in my collection of 'Family Records,' and have succeeded in doing so roughly and provisionally. To determine it thoroughly and to lay down a curve of attractiveness in which the abscissae shall be proportional to the amounts of difference, and the ordinates to the strength of attraction, would require fresh and special data......." (p. 395.)

I have not succeeded in finding in the *Galtoniana* Galton's rough and provisional results, which, if found, would probably throw light on his method. I have thought over the possibility of his curve and find great difficulties about its determination. Is the man equally attracted by a *plus* or *minus* difference? If so, the curve of attractiveness would have two maxima; but clearly the man's opportunities of mating with both these groups would not be the same, as one would be more frequent than the other, unless the man himself were mediocre. If the man prefers a woman with darker eyes than himself, then the woman's taste—for the system to work—would have to be the reverse of the man's, or she must prefer a man with lighter eyes than herself. I have not found in contingency tables for eye colour in man and wife signs of this double maximum in the arrays[2]. If the preferential difference is in one direction only, say a man prefers a woman relatively shorter than himself, then the tallest class of women and the shortest class of men will have to go without mates or *faute de mieux* marry each other, which will upset badly the curve of attractiveness, if it be based on statistical records. The curve of attractiveness is a fascinating idea, but I do not see how it is to be determined, and Galton's hypothesis that sexual taste and distaste are

[1] The fastidiousness of certain sires is a real trouble to the dog breeder. A bitch may be perfectly willing to mate and other dogs desirous of mating with her, but she may have no attraction for the particular sire chosen until a very late stage of her season, when he becomes excited. Such "last moment" dogs often miss their opportunity.

[2] A little consideration will show that it is very difficult indeed to conceive a surface of frequency of which the arrays both ways are bimodal and yet such that the wife-modes of husband accord with the husband-modes of wife! One is almost driven to the conclusion either that there is no assortative mating, or that after all like must prefer to mate with like.

more important in the creation of varieties than intervarietal sterility is not really affected by it.

In the following year (1887) Galton had a considerable discussion in *Nature*[1] with Professor Max Müller on "Thought without Words." The latter in his *Science of Thought* had propounded theories of the descent of man entirely based on the hypothesis that the most rudimentary processes of true thought cannot be carried on without words. Hence Max Müller asserted that the constitution of the mind of the only truly speaking animal, man, separates him immeasurably from the brutes, and no process of evolution which advanced by small steps could stride over such a gulf. Galton states that if a single instance can be substantiated of man thinking without words the whole of Max Müller's anthropological theory must collapse. Galton then appeals to results he had observed by his own introspection, and holds that he has often thought entirely without words. For example:

"It happens that I take pleasure in mechanical contrivances, and the simpler of these are thought out by me absolutely without the use of any mental words. Suppose something does not fit; I examine it, go to my tools, pick out the right ones, and set to work and repair the defect, often without a single word crossing my mind." (p. 28.)

He then refers to billiards and chess; where the strokes and moves are visualised without words beforehand; also to fencing, where there is no time to think in words, before the counter is given. It seems undoubtedly clear that those who visualise vividly will think in pictures as readily as in words, or even more readily[2]. Galton considers that Max Müller failed in reaching a true hypothesis because he generalised from his own mind, and considered that the mind of every one else was like his own (see our p. 243).

"Before a just knowledge can be attained concerning any faculty of the human race we must inquire into its distribution among all sorts and conditions of men, and on a large scale, and not among those persons alone who belong to a highly specialised literary class. I have inquired myself so far as opportunities admitted, and arrived at a result that contradicts the fundamental proposition in the book before us, having ascertained to my own satisfaction at least, that in a relatively small number of persons true thought is habitually carried on without use of mental or spoken words." (p. 29.)

The reader who wishes to follow the discussion further will find two letters of Max Müller and a further letter from Galton in *Nature*. To the present writer Max Müller's reasoning seems very obscure. Replying to Galton's illustration of chess-playing, he writes:

"You cannot move queen or knight as mere dolls. In chess each one of these figures can be moved according to its name and concept only. Otherwise chess would be a chaotic scramble, not an intelligent game." (p. 101.)

But surely the moves of any piece at chess may be associated with the *form* of the piece and not with its name? Max Müller obscures the matter by adding the words "and concept." A concept may be attached to an

[1] May 12 and June 2, Vol. XXXVI, pp. 28–9 and 100–101.

[2] Some of Galton's correspondents in discussing mental imagery, stated that they depended so much on mental pictures, that if they lost the power of seeing them, they would not be able to think at all.

image or form without any idea of a word. Galton and Max Müller appear to be discussing on wholly different planes. "I add," writes Galton, "nothing about the advantage to modern inquirers due to their possession of Darwinian facts and theories, because we do not rate them in the same way." It was only possible for a pre-Darwinian or at any rate an anti-Darwinian to deny that animals think as well as man. "Dogs, Sir, do a deal of pondering," was a conception which had not and could not reach Max Müller. Galton broke a lance for Darwin, but he might as well have tilted at a windmill as at the Oxford nominalist.

The matter of this controversy remained long in Galton's mind, and seven years later he published a short paper in the *Psychological Review* entitled "Arithmetic by Smell[1]." The purpose of the paper is to show that mental processes may be conducted by the sole medium of imaginary smells, just as well as by visual or auditory images, in other words, to prove that thought does not depend on words. Galton first devised an apparatus by which a whiff of scented air could be sent out as often as required beneath the nostrils. A separate simple apparatus was used for each scent and he worked with the eyes shut. He was thus able to produce at will a whiff of peppermint, camphor, carbolic acid, ammonia, aniseed, etc. He taught himself to associate two whiffs of peppermint with one whiff of camphor, three of peppermint with one of carbolic acid, and so on. He next practised simple addition sums with the scents themselves, and afterwards solely with *the imagination of them.*

"There was not the slightest difficulty in banishing all visual and auditory images from the mind, leaving nothing in the consciousness but real or imaginary scents....... Subtraction succeeded as well as addition. I did not go so far as to associate separate scents with the attitudes of mind severally appropriate to subtraction and addition, but determined by my ordinary mental processes which attitude to assume, before isolating myself in the world of scents."

Galton did not attempt multiplication by smell, because he had convinced himself that arithmetic by scents only, and by imaginary scents, was possible with considerable speed and accuracy. He did, however, try some experiments on taste, using salt, sugar, citric acid, quinine, etc., and found that arithmetic by taste was as feasible as arithmetic by smell. Thus Galton proposed to rout the nominalists[2].

In *Nature* for Nov. 15, 1894 (Vol. LI, pp. 73–4) Galton gave an account of Alfred Binet's book *Psychologie des Grands Calculateurs et Joueurs d'Échecs.* He refers to Inaudi, a Piedmontese, who did his mental sums by the sounds of the numbers, and to Diamandi, a Greek, who worked with

[1] Vol. I, pp. 61–2. New York and London, 1894.

[2] I fear Max Müller might have retorted that without the earlier association of numbers with names arithmetic by smell or taste would be impossible. Such an assertion is like that of the theologian who holds that the agnostic either fails to act morally, or only does so owing to a Christian training or the Christian environment. The one neglects the ages long evolution of morality for which Christianity is a thing of yesterday and the other would neglect the ages long evolution of mind prior to language.

mental images of the figures. Galton had tested Inaudi, in whom he found the visual form of the imagination practically absent. Binet considers that mental 'calculating boys' did not as a rule inherit their gifts, the Bidder family being a conspicuous instance of exception. Galton was not prepared to accept this view; he believed that two mental peculiarities must concur to make a calculating boy, namely (i) special capacity for mental calculation, and (ii) a passion for exercising it. Both are rare and are not necessarily coordinated, so that the chance of their concurrence may be very small indeed. He thought that (i) without (ii) might be commoner than is usually believed, and he cited the case of a lady of remarkable ability, whom he had known, and who did not discover that she possessed (i) until on a long and dull railway journey in middle life. He then gives some account of his own experience in performing arithmetic by imaginary smells and tastes.

In 1888 Galton published a paper on "Mental Fatigue[1]." This was a subject in which from personal experience he felt great interest. Over-fatigue of the brain in schools had been recently discussed and illustrated by experiences which flatly contradicted one another. After the heat of controversy had somewhat cooled Galton was asked to occupy the chair at a meeting of the educational Section of the Teachers' Guild, and he was so struck by the audience on that occasion that he considered that the Guild might be a powerful instrument for the solution of statistical problems, if its intelligent members could be interested in educational inquiries. Galton accordingly issued a schedule of selected questions bearing on mental fatigue. He met with an experience, often repeated in the case of the present writer, namely that circularising societies constituted for definite educational or social work, even on points directly connected with their aims, produces very little by way of useful statistical returns. Galton, although his schedule was accompanied by a covering letter from the Vice-Chairman of the Guild, received only 116 replies to his questionnaire, and all Galton was able to do was to set down in an orderly way the replies received. The questions asked applied not only to the taught, but to the teachers themselves. Of the teachers themselves one-fifth, 23, had at some period in their lives broken down, and no less than 21 of these had never wholly recovered from the effects. The teachers also reported with detail 59 cases known to them of more or less serious prostration from mental overwork. At the same time it is possible that those teachers, who had themselves suffered from or closely observed others suffering from overwork, would be most likely to be interested in Galton's questionnaire, and thus the 116 replies be not a random sample of all teachers. While the answers showed many views on the signs of mental fatigue, and on the studies which could or could not be undertaken when the mind was fatigued, there was little light thrown on the best means of testing mental fatigue, or of *measuring* it in a school-class at large. In fact the only real light on this matter came from Galton himself.

[1] "Remarks on Replies by Teachers to Questions respecting Mental Fatigue," *Journal of the Anthropological Institute,* Vol. XVIII, pp. 157–68.

He refers to his statement in *Nature*, of June 25, 1885[1], entitled "Measure of Fidget." In that paper he described how he had counted in a section of an audience during the reading of a wearisome memoir the varying rate of fidget[2].

"I have since frequently tried this method; it is an amusing way of passing an otherwise dull evening, but in drawing conclusions from the number of movements the average age of the audience and their habits of thought have to be taken into account. The method, however, rather measures the dullness of the performance than the true mental fatigue of the audience."

The second suggestion Galton gives is based upon the experimental fact that the quickness and magnitude of the individual's reaction to a stimulus are greatly affected by fatigue.

"There is an experiment, not so well known as it should be, that after a class has had practice in performing it, can be repeated at any time in a few seconds, which gives an excellent measure of the varying amount of reaction time. The class take hands all round, the teacher being included in the circle, a watch with a seconds hand lies on the table before him. All the pupils shut their eyes. When the seconds hand of the watch comes over a division the teacher gives a squeeze with his left hand to the right hand of the pupil next to him. That pupil forthwith with his left hand squeezes the right hand of the next pupil, and so on. Thus the squeeze travels round the class and is finally received by the right hand of the teacher, who then records the elapsed time since he started it; or he may let it make many circuits before he does so. This interval divided by the number of pupils in the class and by the number of circuits gives the average reaction time of each pupil. The squeeze takes usually about a second of time to pass through each dozen or fifteen persons. We should expect to find

[1] Vol. XXXII, p. 174. In this paper Galton refers first to "the unequal horizontal interspace between head and head" in a bored audience, while in an attentive audience all sit upright with their heads almost equi-distant. In a bored audience the bodies sway from side to side, and the intervals between faces vary greatly. But Galton failed to find any numerical expression for this variability of distance. He was more successful when he counted fidgets as an expression against this "mutiny of constraint." The hall in which the uninteresting paper was read was semicircular and divided by columns into sectors each containing about 50 people. He watched one of these sectors repeatedly and counted the number of distinct movements; this was very uniform, amounting to about 45 per minute or nearly one per person. The audience was elderly, the young would have been more mobile. When occasionally the audience was roused to temporary attention the frequency of fidget was not only reduced to less than one half, but the amplitude as well as the period of the motion were notably reduced. "The swaying of head, trunk and arms had before been wide and sluggish, and when rolling from side to side the individuals seemed to 'yaw,' that is to say they lingered in extreme positions. Whenever they became intent this peculiarity disappeared, and they performed their fidgets smartly. Let this suggest to observant philosophers when the meetings they attend may prove dull to occupy themselves in estimating the frequency, amplitude and duration of the fidgets of their fellow sufferers. They must do so during periods both of intenseness and of indifference, so as to estimate what may be called natural fidget, and then I think they may acquire the new art of giving numerical expression to the amount of boredom expressed by the audience generally during the reading of any particular memoir."

[2] Ignorant of Galton and in a much less scientific manner I can recall practising his method as a child in the sixties. The *locus* was a family pew in the chapel of the Foundling Hospital; there was an old and dull chaplain, the last clergyman of the Church of England that I remember in a Genevan gown. He used to preach for about 35 minutes, and I was accustomed to amuse myself and to measure the dullness of his discourse by counting the number of coughs given in that Sunday's 35 minutes. My brother and sister would remember the text, but I could only say on the basis of my sinful and secret statistics, that the preacher had been rather more or rather less inspiring than usual.

uniformity in successive experiments when the pupils are fresh; irregularity and prevalent delay when they are tired. I wish that teachers would often try this simple, amusing, and attractive experiment, and when they have assured themselves that their class enters into its performance with interest and curiosity, they might begin to make careful records at different periods of the day and see whether it admits of being used as a test of incipient fatigue. I should be exceedingly glad to receive accounts of their experiences. Deception must of course be guarded against." (p. 160.)

I have not so far come across any data in the *Galtoniana*, which suggest that any experiences were communicated to Galton.

From the teachers' replies Galton in the memoir draws two conclusions:

The first suggests the reason why mental fatigue leaves effects so much more serious than bodily fatigue. When a man is fatigued in body he has many of the same symptoms as arise in mental fatigue, but

"as soon as the bodily exertion has closed for the day, the man lies down and his muscles have rest; but when the mentally fatigued man lies down, his enemy continues to harass him during his weary hours of sleeplessness. He cannot quiet his thoughts and he wastes himself in a futile way." (p. 166.)

I am not clear that this diagnosis is of universal truth, especially in the case of men not habitually used to excessive work. Over physical exertion—a fifty-mile walk, or a very strenuous bicycle ride, or a whole day of heavy gardening work—may be followed by muscular fidgets, by unrestrainable fits of shivering, and by actual mental excitement which renders sleep or muscle rest impossible, and the effects may be felt for days afterwards. It says much for Galton's constitution that no experience of this kind seems to have suggested to him that for some individuals bodily and mental fatigue run much the same course[1].

The second conclusion that Galton reaches is that breakdowns usually occur among those who work by themselves, and not among pupils whose teachers keep a reasonable oversight. Too zealous pupils are rare. The chief danger occurs when

"young persons are qualifying themselves for the profession of a teacher, and have also to support themselves and perhaps to endure domestic trials at the same time. Dull persons protect their own health of brain by refusing to overwork. It is among those who are zealous and eager, who have high aims and ideas, who know themselves to be mentally gifted, and are too generous to think much of their own health, that the most frequent victims of overwork are chiefly found." (p. 166.)

There is much in this paper on Mental Fatigue which is of high suggestiveness, and it should certainly be read by any one planning a more elaborate statistical inquiry into a subject still far from completely explored. The recent discovery and discussion of shell-shock show how large a section of a modern population—and not the least intelligent and zealous—bears the terrible load of inherited neuroses. One of the points not touched on in Galton's questionnaire is the family history of those who have suffered serious prostration from mental overwork. We should not be surprised to find a link between this category and that of the shell-shocked. In the

[1] It would be of much interest to inquire into the extent to which nervous breakdowns can be directly traced to over strenuous physical exertion.

present state of our knowledge it should not be impossible to give some warning to those young persons who run a danger if they follow a very strenuous mental occupation, such as that of the school teacher.

Galton, as I have noted, remained interested for many years in psychology although, as in the corresponding case of geography, his main work changed its character. Any mental idiosyncrasy had special attraction for him, and in May, 1896, he published a note on what he considered a very curious mental peculiarity[1]. This occurred in a certain Colonel M. who when in the army had seen flogging, wounds or death without special sensations. But the sight or talk of an injured finger nail at once made him feel sick and faint, and would even bring on a deadly cold perspiration. So much was this the case that at a large dinner party in the prime of life the persistent talk of a guest about a small injury of this kind caused him first to turn faint and then to slide under the table unconscious. His mother apparently attributed the idiosyncrasy to maternal impression, she having pricked her finger (without permanent injury) shortly before his birth. Colonel M. said that his father, brother, three sisters and nephews and nieces had no analogous peculiarity. He had no children; it is not directly stated that his *mother* herself had not the peculiarity. Galton thinks it could not have come by inheritance, and that

"it would be silly to suppose a sickly horror of wounded finger nails or claws to have been so advantageous to ancient man or his brute progenitors as to have formerly become a racial characteristic through selection, and though it fell into disuse under changed conditions and apparently disappeared, it was not utterly lost, the present case showing a sudden reversion to ancestral traits. Such an argument would be nonsense."

He looks upon the idiosyncrasy as a mutation, and fresh evidence of the wide range of possibilities in the further evolution of human faculty. In other words he assumes that it was not inherited, but would have been transmitted had Colonel M. had offspring. The note is interesting as illustrating the working of Galton's mind. It does not seem to me that the evidence for non-inheritance is any more adequate than in the case of Huggins' dog Keppler (see our pp. 66 and 148). But an inquiry into the hereditary character—i.e. the origin and transmissibility of such mental idiosyncrasies—would be well worth making[2].

Another memoir which can best be considered in this chapter is that of the same year, 1896; it deals with the problem of "Intelligible Signals between Neighbouring Stars[3]." Galton tells us that in 1892 Mars made a

[1] *Nature*, Vol. LIV, p. 76, "A curious Idiosyncrasy."

[2] For example there are persons who are made to feel sick by the tearing of a piece of calico in their presence; there are others in whom the mere *imagination* of drawing a knitted glove between their teeth sends a cold shiver through all their limbs; while recently I heard of a workman who was employed to whitewash a room in which there were a few skulls in a glass case throwing up his job, because it made him "ill to work in a charnel house." I think this sort of mental discomfort extends to lower living forms; I have known dogs seriously uneasy when a dressed and cured dog skin was brought to their notice, and seriously distrustful of familiar friends, if they wore gloves made from wool spun from the combings of dogs' coats.

[3] *The Fortnightly Review*, Vol. LX, N.S. pp. 657–64.

near approach to the earth, and that the possibility of exchanging signals with Mars was then discussed in the newspapers; it was considered not impossible, if enormously difficult, to send signals. But there was a general conclusion that if sent, the only thing that could be learnt from them would be that there existed observant, intelligent and mechanical people capable of great enterprises on the other planet. Galton thought that much more might be achieved, and that an *intrinsically* intelligible system of signals could be devised, if the people on the other planet were equally advanced with ourselves in pure and applied science. He amused himself accordingly in thinking out the ground plan of the present paper, but laid it aside for four years during which the craze about Mars died out, "being cooled by copious douches of astronomical common sense." Then, in 1896, came an attack of gastric catarrh, which developed into more serious trouble owing to a visit to Kew—to attend the Observatory—with a temperature of 102. Galton was invalided to Wildbad and its hot baths, and amid their relaxing accompaniments, being able to work only in a desultory fashion, he wrote up his paper on signals from Mars[1]. The main point of this paper is the building up of a system of signals from which ultimately pictures can be constructed. It is half humorous and half serious. It starts with the idea that arithmetical and mathematical notions will be common knowledge of both planet's inhabitants. Signals of $1\frac{1}{4}$, $2\frac{1}{2}$ and 5 seconds are given and termed dot, dash, line. These lead up to a system of numerals. Then comes the ratio of the circumference of a circle to its diameter, the value of the familiar $\pi$. Thence the ratio of the circumference to the radius of the various regular polygons, which introduces signals for the polygons. The 24-sided regular polygon is then indicated as a method of direction, and so angles all round the 360° are gradually learnt in the same way as are the points of a compass, but direction of lines and length of lines being given it becomes possible to give signals indicating a picture by successive "stitches" of definite lengths in definite directions. That is to say, Galton has reached the picture formula of his lecture on the "Just Perceptible Difference" (see the following Chapter, p. 307). But once it is possible to signal pictures, all becomes possible. It becomes possible to indicate motion, and motion will enable one to indicate signals for action, i.e. verbs. Such, very briefly, is the outline of Galton's system of star signals:

> "It would be tedious, and is unnecessary to elaborate further, for it must be already evident to the reader that a small fraction of the care and thought bestowed, say, on the decipherment of hieroglyphics, would suffice to place the inhabitants of neighbouring stars in intelligible communication if they were both as far advanced in science and arts as the civilised nations of the earth are at the present time. In short, that an efficient interstellar language admits of being established under these conditions, between stars that are sufficiently near together for signalling purposes."

[1] Both Galtons were much depressed during this year. Emily Gurney died, and Sir William Grove died on the anniversary of the Galtons' wedding day (August 1st). The season was very wet and Galton suffered much from colds; he complained for the first time (aged 74!) that his brain power was not as vigorous as formerly, that he could not work quickly and that his deafness interfered with his committees.

PLATE XXVI

Galton, aged about 75 years.

Mrs Galton, all unconscious of the near future, after noting the events of 1896, including this paper on star signals, continues in her *Record*:

"So surely do the good things come to us and pass from us, and I try to be thankful for the innumerable blessings we have had even with the pain of feeling them gone. So ends our year, not an eventful one, but a calmly happy one, ending with a merry Xmas at Spencer's [Spencer Butler's], the young folk full of life and ambitions."

Calmly happy sentences—not the depressed or fretful words of some few of the earlier entries of the *Record*—and fitly concluding that brief account of the 43 years of Louisa Butler's married life with Francis Galton.

There is only one more year of entry, 1897, in the *Record*, and something of it may be fitly quoted here, for it will indicate, better than the remarks of some superficial onlookers, the real relationship of the pair. It is hardly necessary to remark that the union could not fail to have been richer had it been blessed with children. Galton's affection for his nieces shows what this would have meant for him.

"1897. It is with painful reluctance that I set down the incidents of this fatal year, and do so on Jan. 6 the anniversary of the day, when I first became acquainted with dear Louisa at the Dean's house, next door to our own at Dover in 1853.

In the early part of the year I was more of an invalid than she was, but we had some pleasant outings together—as to Nansen's great meeting on Feb. 28. Chiefly on account of my persistent asthmatic cough we went to Bournemouth, March 22, partly to be near Dr Chepmell, whose remaining eye was threatened. He told me to go to Cauterets or Royat. Montagu and Agnata [the Master of Trinity and his wife] came to us for a day from Lyndhurst, while we were there. We had had alarming news from time to time of Emma ['Sister Emmy'] from the middle of Feb. onwards. At length she was better, and we went to her April 20–23. Louisa was well enough for some small festivities—a tea party, her last, on May 7th, and the military tournament. We went to Oxford, to Arthur's [Mrs Galton's brother's] June 5–8. June 21, Jubilee day, we went with Mrs Lyell to the Athenaeum and they had excellent places and Louisa was not overtired. Next day Bessie [Galton's sister Mrs Wheler] came to tea and Mrs Lyell. 26th I went to the Naval Review, L. not well enough to go with me. July 14 left for Royat, slept at Boulogne; next day, a weary waiting till 10 p.m. at the Lyons Station, but the night journey comfortable, Louisa not suffering at all. July 24th Puy de Dôme with Mr Livett and a young lady. L. remained in the garden at the *auberge* while we went up, and she had luncheon set out. I never saw her more pleased or nicer as a hostess than when we came down. Aug. 1, M^me^ de Falbe arrived in far from good health. Aug. 3, L. awoke with diarrhoea,—we all had it, but recovered. Very sultry. Arranged for Pont St Laurent in Dauphiné, and wrote to have letters sent to Grenoble. On Sunday 8th she was apparently quite well and half packed for a start next morning. Monday 9th she was ill and sick in the night, not worse than frequently before. Tuesday, Aug. 10 she was worse; I had Dr Petit in, who made light of it, but said he would come the next morning. Wed. 11th she was very ill, but saw M^me^ de Falbe, who was able to leave her bed for the purpose. L. wrote a post-card to Chumley [her maid] in case she was wanted. In afternoon she was very weak indeed....... Thursday 12th worse and in a very serious state. That night, or rather Friday morning early at 2 hrs $\frac{3}{4}$, she quietly passed away. On Saturday she was buried in the cemetery of Clermont-Ferraud in plot 419, which I purchased as a concession *in perpétuité*. So our long married life came to an end. Writing as I do now after nearly 5 months have passed, and I am able to take a fair retrospect, I think that the inevitable blow occurred at a more seasonable period than at any other time. Dear Louisa's vigour was distinctly declining; she was still able to enjoy much, but was I fear rapidly on the way towards permanent invalidism, and she was conscious of a weakening of her mental power, small things fatiguing her much more than formerly. Had I died first, I fear her strength would have been inadequate to carrying on life unaided. She has been in many respects a most valued as well as a loved example to me. May her good influence abide, though she personally is gone. All her friends lay stress on her power of sympathy. How I pulled through the terrible strain and hurried requirements of the occasion I cannot conceive, but I did, thanks largely to the hearty and tactful sympathy received from M^me^ de Falbe and Mr Jennings, who had made Louisa's acquaintance and returned to help. I could not leave Royat on account of letters, till Tuesday night, arriving in London Wed^y^ afternoon, where the sympathy of Spencer, and Mary and of Gertrude [Butler] awaited me. Some few days were spent in sorting her possessions and carrying out Louisa's wishes. Then to dear Emma's at Leamington for a week; thence to the Douglas Galtons' at Himbleton also for nearly a week; thence to Mrs Hills'[1] at Corby, all of which greatly braced me. The general kindness of Louisa's and my relations was extreme. On returning Sept. 13 Frank Butler was ready to live with me, a most valuable help against the sense of isolation....... My own occupations were the inquiries into the Bassett hounds, which led to the "Average Constitution of each Ancestor etc." *Proc. Roy. Soc.*[2], also "Inquiries into Speed of American Trotters," *Proc. Roy. Soc.*[3] and the method of photographic measurement of horses etc., published to-day Jan. 7, '98[4]. The Committee on a Physical National Laboratory has been appointed and is taking evidence. The Evolution Committee has not done much, Kew Observatory prospers; Meteorological Council, the usual routine."

Thus it is when one of our number falls out, the ranks close up; social life as a whole goes on; our intellectual tasks are resumed, and our thoughts are turned again from the immediate environment to the non-personal problems of science. Galton rarely referred to the personal in conversation, or in letters, and it has seemed best to his biographer to maintain his reticence, allowing merely the one entry with which he concludes Louisa Galton's *Record* to tell its own tale.

To sum up the contents of this chapter, I venture to assert that no psychologist, no statistician of energy and imagination can read its pages and not feel that they have provided him with suggestions of many still unsolved problems, for whose solution the world would be not only the wiser but the better. Such is always the outcome of Galton's suggestive mind, and it is on this account—the generosity of ideas—that the reader willingly pardons an occasional conclusion based on apparently scanty data. Beyond those data was always the rich experience of a mind during the whole of a long life perpetually observing and placing in appropriate categories the actions and thoughts of other men as well as of himself.

[1] Wife of Judge Hills of Alexandria, and daughter of Sir William Grove, Galton's close friend. A number of Galton's letters to Mrs Hills have recently been purchased from a bookseller for the Galton Laboratory.

[2] Vol. LXI, pp. 401–43. [3] Vol. LXII, pp. 310–15.

[4] Galton probably wrote the last sentences of this entry on the day following that, Jan. 6, on which he had started to give the account of Mrs Galton's death.

PLATE XXVII

Francis Galton in the 'sixties, from a photograph.

# CHAPTER XII

## PHOTOGRAPHIC RESEARCHES AND PORTRAITURE

"Whatever he touched he was sure to draw from it something that it had never before yielded, and he was wholly free from that familiarity which comes to the professed student in every branch of science, and blinds the mental eye to the significance of things which are overlooked because always in view." *Nature* on CHARLES DARWIN, Vol. XXVI, p. 147.

WE have seen in the preceding chapter how Galton supposed composite portraiture to be connected fundamentally with psychological inquiry. Galton developed composite photography in his search for a method of ascertaining whether physiognomy is an index to mind, i.e. whether facial characteristics are correlated with mental traits. The actual method he employed, however, was curiously enough suggested to him as a result of his attempts to illustrate the multiple geographical features of a country, where he wanted more than could be readily exhibited on the usual type of maps. Galton's own idea of composite portraiture would fully justify our discussing it under the heading of "Psychometric Investigations." But as Galton's contributions to scientific photography are numerous and important, it has seemed to me desirable to devote an entire chapter to the subject, although much that will be contained in this chapter has great psychometric interest.

Galton's contributions to photographic science break up into six sections, namely:

(*A*) Composite Photography.
(*B*) Bi-projections by Photography.
(*C*) Analytical Photography.
(*D*) Measurements by Photography.
(*E*) Indexing and Numeralisation of Portraits.
(*F*) Measurements of Resemblance, chiefly by photographs.

We shall also include in this chapter, as closely related to our present topics, the subjects of the indexing of portraits and the telegraphy of portraits. The matters to be discussed occupied Galton's mind almost continuously from 1878 to 1911, i.e. more than thirty years. They had singular fascination for him not only because they combined fairly simple mathematical investigation with mechanical invention and experiment, but also because they were closely associated with psychological and hereditary inquiries.

### (*A*) *Composite Photography.*

There is a paper on combining various data in maps which explains the origin of Galton's inquiries into composite photography. As he himself says in a paper of 1878:

"It was while endeavouring to elicit the principal criminal types by methods of optical superimposition of the portraits, such as I had frequently employed with maps and meteorological traces, that the idea of composite figures first occurred to me[1]."

This paper by Galton, which I had some difficulty in locating, and of which no copy could be found in his collections, is entitled: "On means of combining various Data in Maps and Diagrams[2]." Galton therein refers again to his stereoscopic map method[3], which plan he regrets had not yet been adopted. He says that it needs good models, but that the number of these increases every year, as the then recent French Geographical Exposition demonstrated. He exhibited in the Loan Exhibition stereoscopic views of models taken by the Royal Engineers, but I do not know what subjects they represented. He suggested the stereoscope, not only as a means of showing things in the solid, but of superposing plans and maps for comparative purposes.

He next turns to superposition by means of a telescope; he remarks that if one half of the object-glass be covered up, the sole effect on the image of a distant object will be to reduce its brilliancy by a half. Now let us suppose two lenses placed one in front of one half of the object-glass and the other in front of the other half, each with its own object at its focal distance; the two objects will then be combined superposed in the field of view of the telescope. Instead of two lenses four might be distributed over the area of the object-glass to combine four objects, etc. Actually Galton placed his telescope vertically and ran a horizontal tramway under the objective; in the blackened roof of a 'tramcar a number of lenses were inserted and opposite each on the floor of the carriage its own object at the focal distance of the lens. If now lens 1 be brought under the objective we see only object 1, as we push the carriage farther lens 1 tends to pass out of the field and lens 2 to share the field; thus different intensities of objects 1 and 2 can be combined. A further push of the carriage causes 2 alone to be seen, and the process continued combines it with 3, and so on. Galton's model had six lenses of the same focal distance and size in the roof of his carriage. By this means a series of geographical data which would overcrowd any single map can be combined in sets of two.

"It affords a peculiarly suitable method for picturing changes whether in physical or political geography. I will not describe the mechanism by which complex and powerful instruments of this kind might be constructed; where the images should be thrown by a lime light on a screen, and a string of perhaps only three large achromatic collimators would serve for an indefinite number of pictures." (p. 315.)

That Galton should have spoken of the old 'wheel of life' in connection with his apparatus, shows that he had a foreshadowing of the modern cinematograph. By using lenses of different focal length objects at different distances could be combined, and by using inclined mirrors facing definite parts of the object-glass the objects need not be placed in parallel planes.

[1] *Journal of the Anthropological Institute*, Vol. VIII, p. 135, 1878. In this paper (ftn. p. 135) Galton gives 1878, instead of 1876, for the year of the Map paper.

[2] South Kensington Museum Conferences held in connection with the Special Loan Collection of Scientific Apparatus, 1876; Chemistry, Biology, Physical Geography, Geology, Mineralogy and Meteorology (pp. 312–15). London, Chapman and Hall.

[3] See our p. 33.

Galton very soon discovered that the methods of optically combining images are very various indeed. Thus in a paper of 1878 he writes[1]:

"I have tried many other plans; indeed the possible methods of optically superimposing two or more images are very numerous. Thus I have used a sextant (with its telescope attached); also strips of mirrors placed at different angles, their several reflections being simultaneously viewed through a telescope. I have also used a divided lens, like two stereoscopic lenses brought close together, in front of the object-glass of a telescope.

I have not yet had an opportunity of superimposing images by placing glass negatives in separate magic lanterns, all converging upon the same screen; but this or even a simple dioramic apparatus would be very suitable for exhibiting composite effects to an audience, and if the electric light were used for illumination, the effect on the screen could be photographed at once. It would also be possible to construct a camera with a long focus, and many slightly divergent object-glasses, each throwing an image of a separate glass negative upon the same sensitised plate." (p. 140.)

Among Galton's instruments in the Galton Laboratory is a piece of apparatus for compounding six objects. It is of the following nature. Six different photographs, arranged symmetrically round a blackened screen, face six different object-glasses which, set round the base of a conical tube, form a composite image of all six on a small focusing screen towards the vertex of the cone. This image is examined by an eye-piece passing through the centre of the vertical screen and entering centrally the base of the cone. The focusing screen is only about 2 inches in diameter, but the image is magnified by the eye-piece. Six components can be superposed and examined visually, but there does not seem any special provision in the

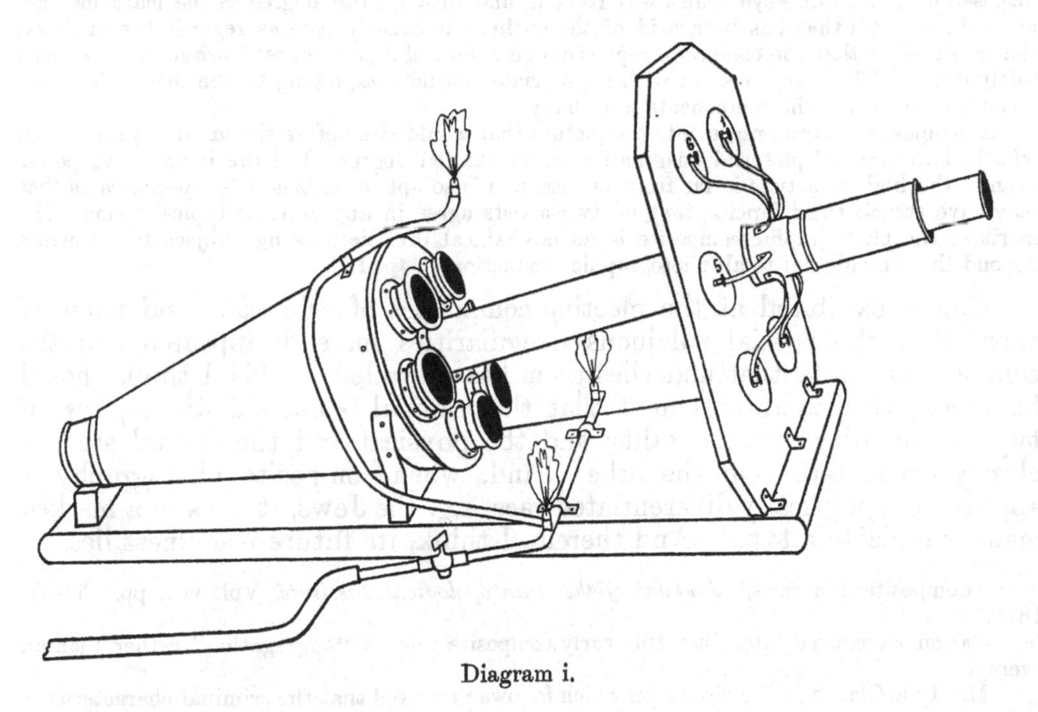


Diagram i.

[1] "Composite Portraits," made by combining those of many different persons into a single resultant figure. *Journal of the Anthropological Institute*, Vol. VIII, pp. 132–42, 1878.

apparatus for photographing the composite image either from below the screen or from the eye-piece side.

The paper of 1878 to which reference has more than once been made[1] describes for the first time the very simple arrangement—a window with two cross-hairs or wires and two pinholes in the frame—by which Galton at first registered the series of photographs to be compounded. For full face one hair was taken horizontally bisecting the pupils and the other, the vertical, bisecting the distance between the pupils. A prick made in each pinhole then registered the photograph. It is requisite that the whole series of photographs should be practically of the same size, or if not, reduced to the same size. All that is needful is, if $n$ seconds be the correct exposure and there be $m$ photographs, to give $n/m$ seconds to each. If we suppose $n = 50$ and $m = 8$, we combine eight portraits. If we wish to combine more, it is better to combine composites of 8 to 10 each to obtain the full composite. Of this Galton writes:

"Those of its outlines are sharpest and darkest that are common to the largest number of the components; the purely individual peculiarities leave little or no visible trace. The latter being necessarily disposed on both sides of the average, the outline of the composite is the average of all the components[2]. It is a band and not a fine line, because the outlines of the components are seldom exactly superimposed. The band will be darkest in its middle whenever the component portraits have the same general type of features, and its breadth, or amount of the blur, will measure the tendency of the components to deviate from the common type. This is so, for the very same reason that the shot-marks on a target are more thickly disposed near the bull's-eye than away from it, and in a greater degree as the marksmen are more skilful. All that has been said of the outlines is equally true as regards the shadows; the result being that the composite represents an averaged figure, whose lineaments have been softly drawn. The eyes come out with appropriate distinctness, owing to the mechanical conditions under which the components were hung.

A composite portrait represents the picture that would rise before the mind's eye of a man who had the gift of pictorial imagination in an exalted degree. But the imaginative power even of the highest artists is far from precise, and is so apt to be biased by special cases that may have struck their fancies, that no two artists agree in any of their typical forms. The merit of the photographic composite is its mechanical precision, being subject to no errors beyond those incidental to all photographic productions." (p. 134.)

Galton exhibited at the meeting composites of criminals, and notes of them that the special villainous irregularities have disappeared and the common humanity that underlies them has prevailed[3]. This I think should have been used as an argument that the criminal is not a distinct *physical* type, criminality is a mentality and the physical and the mental are not closely correlated. On the other hand, when composite photography is applied to a physically differentiated race, e.g. the Jews, it does in a marked manner indicate a type[4]. And therein, I think, its future usefulness lies.

[1] "Composite Portraits," *Journal of the Anthropological Institute*, Vol. VIII, pp. 132–42, 1878.

[2] Galton recognised later that this early composite was an "aggregation" rather than an average.

[3] Mr Hyde Clarke in the discussion which followed asserted that the criminal characteristics were eliminated, and they had a natural type of man instead, and attributed the result to the process producing merely an 'average,' instead of arguing that there was not a distinct physical criminal type.

[4] See our pp. 293–4 and Plates XXVIII and XXIX.

Composites, made from Portraits of Criminals convicted of Murder, Manslaughter or Crimes of Violence.

PLATE XXIX

Comparison of Criminal and Normal Populations.

In the next section Galton records various methods he had hit upon for superposing images. Thus

(*a*) He had used a sextant with its telescope attached.

(*b*) Strips of mirror at various angles, their several reflections being simultaneously viewed through a telescope.

(*c*) A piece of glass inclined at a very acute angle to the line of sight and a mirror beyond it also inclined but in the opposite direction to the piece of glass; the latter must be thin, a selected piece of the best glass used for covering microscopic specimens. Several such pieces inclined at different angles may be used for multiple compounding.

(*d*) A divided lens like two stereoscopic lenses brought close together in front of the object-glass of a telescope (see our p. 285).

(*e*) Glass negatives in separate magic-lanterns all converging on the same screen.

(*f*) A camera with a long focus and many slightly divergent object-glasses, each throwing an image of a separate glass negative upon a screen (cf. our p. 285).

(*g*) A double image prism of Iceland spar.

Of this Galton says (p. 138):

"The best instrument I have as yet contrived and used for optical superimposition is a 'double-image prism' of Iceland spar. The latest I have had......has a clear aperture of a square half an inch in the side, and when held at right angles to the line of sight will separate ordinary and extraordinary images to the amount of two inches, when the object viewed is held at seventeen inches from the eye. This is quite sufficient for working with carte-de-visite portraits. One image is quite achromatic, the other shows a little colour. The divergence may be varied and adjusted by inclining the prism to the line of sight. By its means the ordinary image of one component is thrown upon the extraordinary image of the other, and the composite may be viewed by the naked eye, or through a lens of long focus, or through an opera-glass (a telescope is not so good) fitted with a sufficiently long draw tube to see an object at that short distance with distinctness. Portraits of somewhat different sizes may be combined by placing the larger one farther from the eye, and a long face may be fitted to a short one by inclining and foreshortening the former. The slight fault of focus thereby occasioned produces little or no sensible ill-effect on the appearance of the composite. The front and profile faces of two living persons sitting side by side, or one behind the other, can be easily superimposed by a double-image prism."

The apparatus itself with accessories is figured and described in Galton's paper, and remains after more than thirty years intact to this day in the Galton Laboratory[1].

Galton remarks that the truth of the composite photograph can be assured by the substantial agreement between the results from different batches of components—"a perfect test of truth in all statistical conclusions." He tried changing the order of exposure of the components and found substantial identity[2] (p. 135).

[1] There are in fact two such Iceland spar compounders.

[2] It might seem that this point wanted greater experimental demonstration than the short series Galton speaks of. I have, however, Galton's evidence before me; he took three portraits, $A$, $B$, $C$, and compounded $A+B$ and $B+A$, the composites are practically identical; then he took $A+B+C$, $A+C+B$, $B+A+C$, $B+C+A$, $C+A+B$ and $C+B+A$, and again the composites are practically identical. He had thus good evidence that order of exposure did not

The paper finally discusses the uses to which composite portraiture may be put. Galton refers to:

(i) Typical pictures of different races of men (see our pp. 290, 293–4, and Plate XXXIV).

(ii) Selection of some strongly marked type within a race, e.g. Criminal or Phthisical subjects (see our pp. 286, 291–2 and Plates XXVIII, XXIX and XXXIV).

(iii) Composite portrait of the same individual to obtain more than a single momentary expression. Galton considered that such a composite would have 'varied suggestiveness.'

(iv) Composites from independent portraits of historical personages. It may be from coins, medals or busts. Thus Galton later did Alexander the Great, Napoleon, Cleopatra, etc. (see our p. 295 and Plates XXXVI—XLIV).

(v) Composite portraits of ancestry and collaterals, each individual being given his or her relative 'weight' in terms of exposure. Galton thus hoped to produce 'family types,' and to forecast the physical appearance of the offspring of proposed marriages (see our Plates XXXI and XXXII).

(vi) In the same manner as (v) composite portraits might be produced to aid breeders of pedigree stock to judge the result of any proposed union better than they are able to do at present (see our Plate XXX). Galton took the opportunity of appealing for family portraits taken in the same attitude, $\frac{4}{10}$ inch or say 10 mm. between the interpupillary line and the line that separates the lips, in *right* profile, full face, and three-quarters always showing the left side, "in this the outer edge of the right eyelid will be only just in sight." I repeat these suggestions of Galton in case any of my readers wish to make experiments in this somewhat difficult art for themselves.

At the York meeting of the British Association, *Trans. B. A.* 1881, p. 3, Galton read a paper "On the Application of Composite Portraiture to Anthropological purposes." He exhibited the first cranial composite, the profile of the Andamanese skull based on eight components (see our Plate XXXIII). The large original composite is still in the *Galtoniana*, but although it is distinctly better than much later work—which has tended to discredit composite portraiture in craniology—I venture to think that from the standpoint of the profile better fiduciary lines might be selected.

The next paper with which we have to deal is that of 1881, in which year Galton gave an account of "Composite Portraiture" to the Photographic Society[1]. Four years had produced great changes and improvements not only in Galton's methods, but in his apparatus. He now figures and describes a much more elaborate instrument not only for compounding, but at the same time for reducing individual transparencies to a standard size. He not only makes the interpupillary line and its vertical bisector the same

weight the different components. He also demonstrated that the effect of an exposure for $n$ seconds was sensibly the same whether it was continuous or given in $p$ equal doses; the experimental prints giving intensity of tint due to exposure for a variety of values of $p$ have survived to this day.

[1] *Journal and Transactions of the Photographic Society of Great Britain*, Vol. v. pp. 140–46, June 14, 1881.

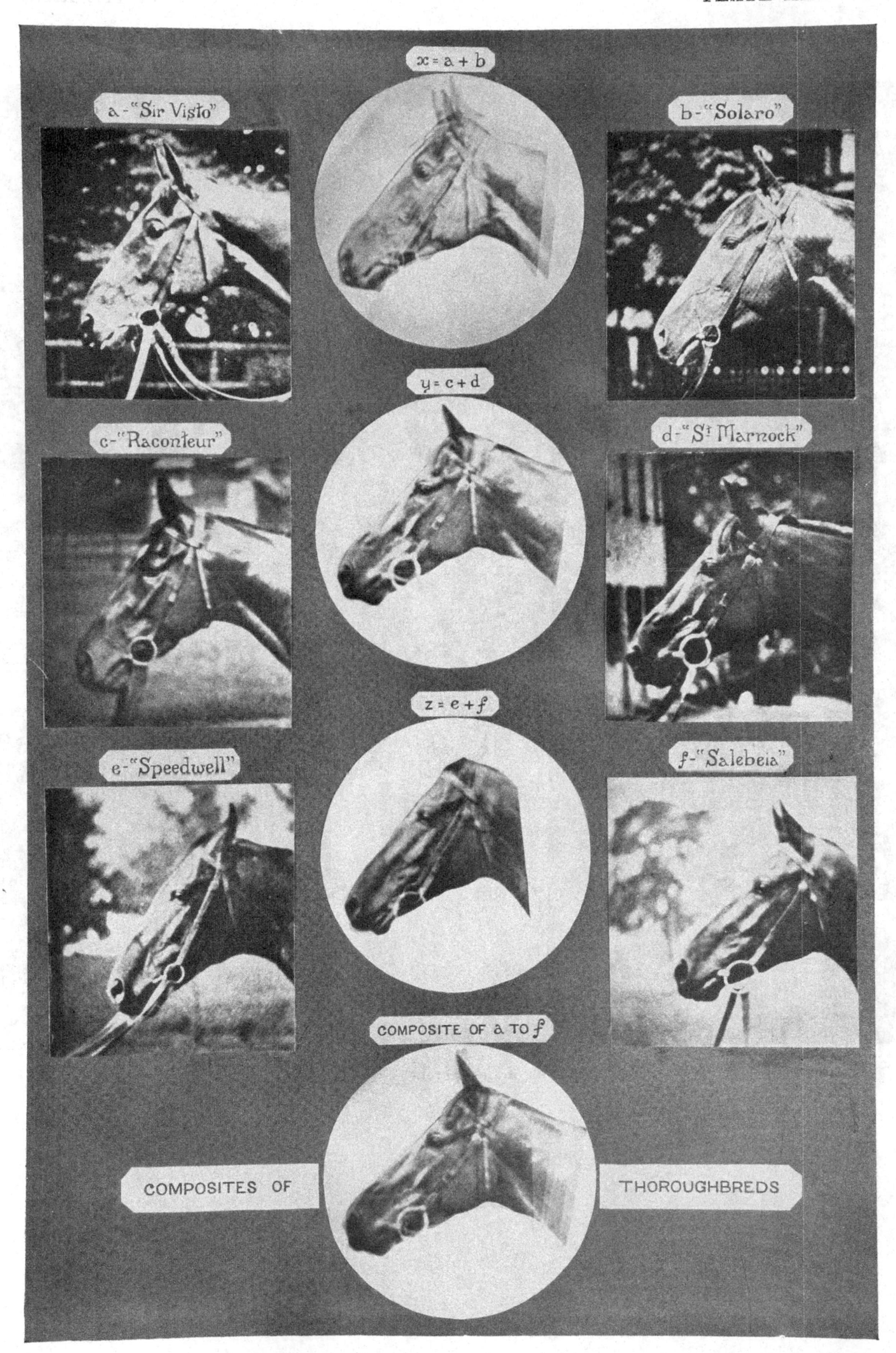
x = a + b
a - "Sir Visto"
b - "Solaro"
y = c + d
c - "Raconteur"
d - "St Marnock"
z = e + f
e - "Speedwell"
f - "Salebeia"
COMPOSITE OF a TO f
COMPOSITES OF
THOROUGHBREDS

PLATE XXXI

Composites of the Members of a Family.

Portraits of three Sisters, full face and profile, with the corresponding Composites.

PLATE XXXIII

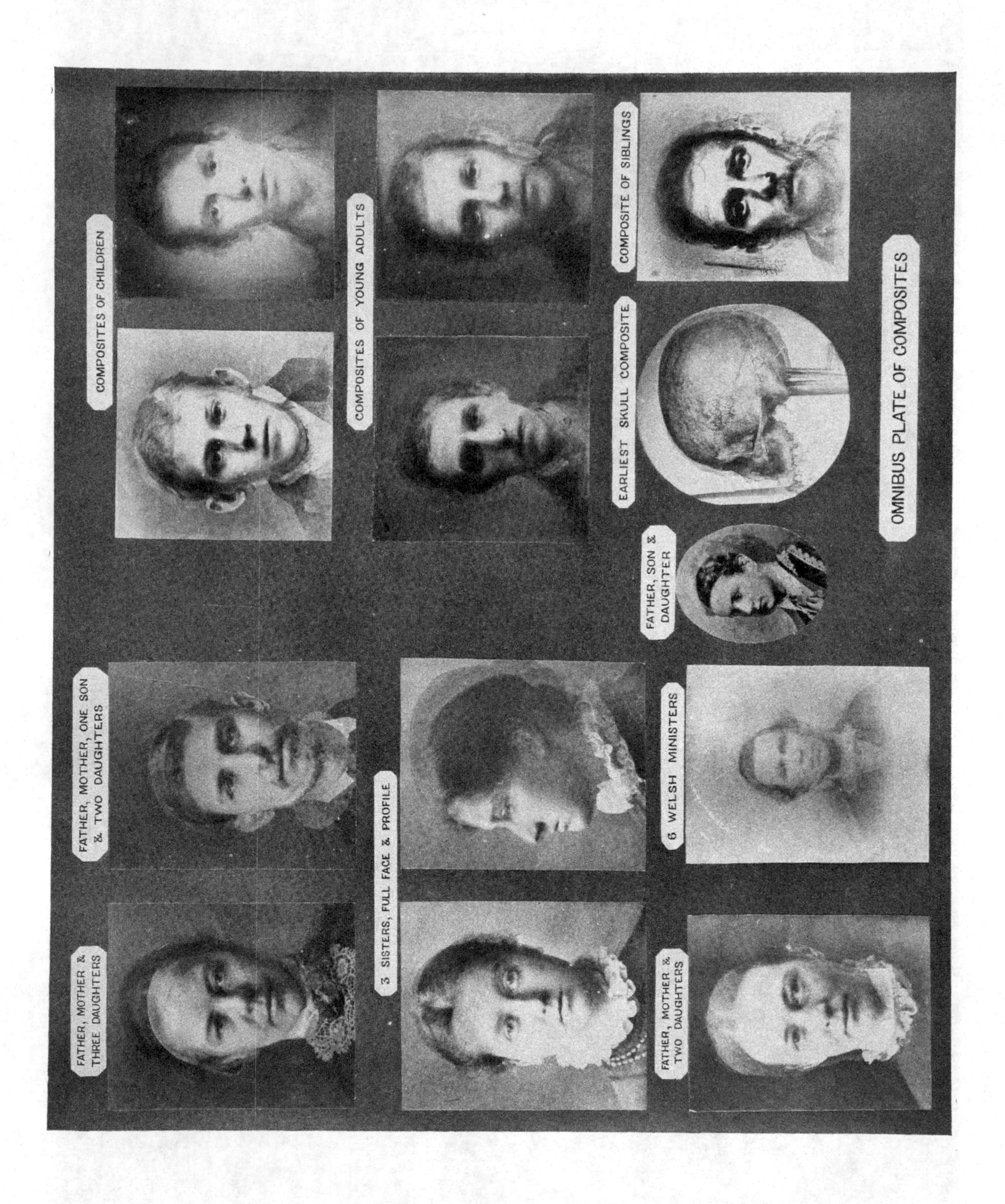

for all subjects, but the distance from interpupillary line to the line of the lips is also made constant[1]. The general principle of the apparatus is that of a modified copying-camera, only the alteration in scale and the adjustment in position are not done on the focusing-screen, because it is desirable that the negative which is to be several times exposed should remain fixed in position. By the simple artifice of a mirror let down at 45° across the camera, Galton gets an image on a horizontal screen in the roof of his camera, and upon this screen also are thrown the three fiduciary lines which serve as register-marks for his adjustments. The details of the apparatus will be sufficiently indicated by the accompanying line engraving and Galton's description of it; we may merely remark that gas would now-a-days be probably replaced by electric light. The apparatus is still preserved in the Galton Laboratory.

DIAGRAM SHOWING THE ESSENTIAL PARTS.

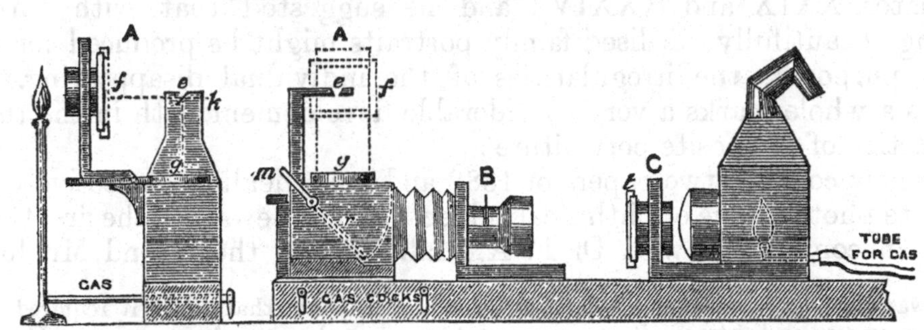


A The body of the camera, which is fixed.
B Lens on a carriage, which can be moved to and fro.
C Frame for the transparency, on a carriage that also supports the lantern; the whole can be moved to and fro.
*r* The reflector inside the camera.
*m* The arm outside the camera attached to the axis of the reflector; by moving it, the reflector can be moved up or down.
*g* A ground-glass screen on the roof, which receives the image when the reflector is turned down, as in the diagram.
*e* The eye-hole through which the image is viewed on *g*; a thin piece of glass immediately below *e* reflects the illuminated fiducial lines in the transparency at *f*, and gives them the appearance of lying upon *g*—the distances *fk* and *gk* being made equal, the angle *fkg* being made a right angle, and the plane of the thin piece of glass being made to bisect *fkg*.
*f* Framework, adjustable, holding the transparency with the fiducial lines on it.
*t* Framework, adjustable, holding the transparency of the portrait.

Diagram ii.

"For success and speed in making composites, the apparatus should be solidly made, chiefly of metal, and all the adjustments ought to work smoothly and accurately. Good composites cannot be made without very careful adjustment in scale and position. An offhand way of working produced nothing but failures." (p. 143.)

Galton exhibited certain results of very considerable interest tending to meet criticisms which had been raised. He drew on a square card a circle of about 2·5 inches diameter with a vertical and a horizontal diameter. Where these diameters met the circle he placed four circular discs of different tints, and in one quadrant he placed a black dot. He then made a composite

[1] It would probably be possible by a slight rotation of the frame for the transparency about a central vertical axis to make the interpupillary distance (or the external ocular distance) constant without in any way injuring the result.

of the axial four positions of this card and found: (i) a sharply defined cross, showing the accuracy of adjustment, (ii) four very faint dots one in each quadrant, indicating that any single irregularity hardly survives, (iii) the equal tint of the four dots showing the equality of the exposure, (iv) the uniformity of the resulting tint of the four 'wafers' arising from the exposure of the tinted discs, $A, B, C, D$, in the four orders, $ADCB$, $BADC$, $CBAD$, and $DCBA$, demonstrating that order of exposure is not material.

He also showed composites in one of which a portrait $X$ was exposed $\frac{2}{3}$ and a portrait $Y$ $\frac{1}{3}$ of the total exposure, and in the other $X$ was exposed $\frac{1}{3}$ and $Y$ $\frac{2}{3}$ of the total. The individuality of $X$ predominated in the first and that of $Y$ in the second; thus justifying 'weighting' by length of exposure.

He further exhibited composite portraits of male and female phthisical subjects, and of men and of officers of the Royal Engineers (see our pp. 291–3 and Plates XXIX and XXXIV), and he suggested that with 'artistic touching' beautifully idealised family portraits might be produced for commercial purposes; the irregularities of the individual disappearing. The paper as a whole marks a very considerable development both in the theory and practice of composite portraiture[1].

We now come to two papers of 1882 and 1885 dealing respectively with composite photographs of phthisical subjects and of Jews. In the first Galton worked in conjunction with Dr F. A. Mahomed, in the second Mr Joseph

[1] It was, perhaps, a misfortune for composite photography, that while it required really extraordinary care and patience, it was very easy to compound in an inferior manner. It became popular, especially in America, and a good deal was published which is of small scientific value and in which no attempt was made at real analysis of the results. Thus *Science* published May 8, 1885, composite portraits of American (*a*) Mathematicians, (*b*) Naturalists, (*c*) Academicians and (*d*) Field Geologists, which lead us hardly further than the conclusions that all American scientists of those days were hairy, and that mathematicians while being least so had more frown. Composites of Washington in three aspects (*Science*, Dec. 11, 1885) are somewhat more successful: *Science* also published (on May 7, 1886) composites of American Indians, but the components were few in number. Further composite photographs were made of undergraduates and graduates of various American Colleges (Jastrow, 1887, did 21 Johns Hopkins doctors of philosophy for the years 1886–7: *Century*, March, 1887. The latter also contains a fairly successful family composite of father, mother, five sons and one daughter). A possibly more scientific use of composite photography was that by Persifor Frazer (*American Philosophical Society*, Vol. XXIII, pp. 433–41, 1888 and *Franklin Institute Journal*, Feb. 1886, p. 123) to obtain an average signature. He illustrates it by one of George Washington, and thinks the process could be used to determine the maximum deviation compatible with a non-forged result. In our own country Arthur Thomson in 1884 (*Journal of Anatomy*, Vol. XIX, p. 109) tried to apply it to obtain type Australian and European crania; the components being too few, and the superposition not very satisfactory, the results are not to be taken as condemning the application of the method in craniometry. The possibilities of composite photography in this matter had been referred to by Galton at the 1881 York meeting of the British Association: see *Transactions*, p. 690. Whipple adopted the process for the reduction of meteorological observations (*Quarterly Journal of Meteorological Society*, Vol. IX, p. 189), and it can clearly be employed in harmonic analysis. There is no doubt that Galton's idea was taken up by many, but it may be doubted whether any one but the originator appreciated the amount of care and patience required to produce a good composite. At the same time I cannot pass over the fact that in the *Galtoniana* there exists a splendid series (not by Galton) of racial composites, Wends and Saxons; they are probably of German origin. I am unaware if they have been published.

PLATE XXXIV

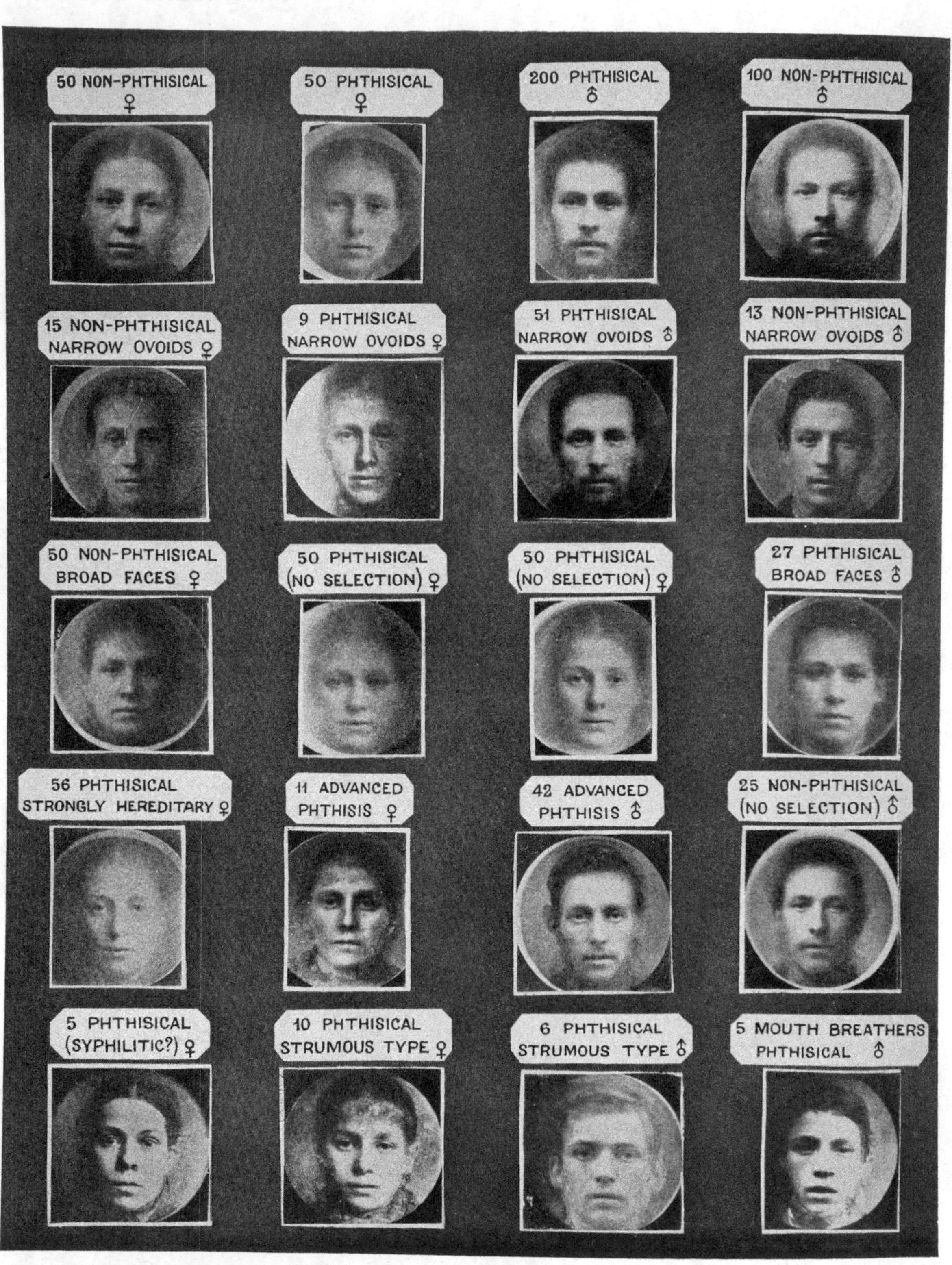


Composites of Phthisical and Non-phthisical Hospital Populations.

Jacobs gave him help. The first paper is entitled: "An Inquiry into the Physiognomy of Phthisis by the Method of Composite Portraiture[1]." It contains illustrations of 47 composites and of 113 individual portraits. There is thus a great wealth of material to judge by. Unfortunately, and probably unavoidably, the portraits are all small, the individual smaller than the composite portraits, and this, I venture to think, lessens the accuracy of judgments based on comparisons of this illustrative material.

The question raised by Mahomed and Galton was whether there was any justification for a belief in a phthisical diathesis; it is of course clear that such may exist without involving a phthisical physiognomy. It is further possible that if such a physiognomy exists, it might be produced by the action of the disease itself. The material consisted of 261 male and 181 female photographs of phthisical subjects between 15 and 45 years of age taken partly at Guy's Hospital and partly at the Brompton and the Victoria Park Hospitals for Diseases of the Chest. A schedule for each subject dealt with: Age—extent of disease (advanced, moderate, slight)—duration of disease (chronic, over 3 years; medium, 1 to 3 years; brief, under 1 year)—hereditary taint (strong, some, none)—onset of disease (insidious, or preceded by severe haemoptysis, bronchitis, pneumonia, pleurisy, syphilis, gout, alcoholism). These classifications enabled composites[2] of various groups to be made. As control two series of female patients, each fifty in number, and a series of male patients 100 in number, all suffering from diseases *other than phthisis*, were taken without selection.

When all individual phthisical patients were compounded without selection, a composite was obtained strikingly like the composite portrait of the non-phthisical: see our Plate XXXIV. If there be made selections of the narrow, ovoid or 'tubercular' faces from the phthisical and non-phthisical patients, or again of broad faces with coarser features from the two groups, we again reach composites closely resembling each other. In other words both phthisical and non-phthisical patients contained representatives of the same two types. Further, of the non-phthisical women 15 °/ₒ gave the narrow ovoid face while only 11·6 °/ₒ of patients with phthisis presented it. Among males the proportion of narrow ovoids was only 13 °/ₒ in the non-phthisical patients and 19·0 °/ₒ among the phthisical on the best estimation. Taken altogether the phthisical cases showed 14·3 °/ₒ and the other than phthisical 14·0 °/ₒ of the narrow ovoid or so-called 'tubercular' physiognomy. The 'tubercular' physiognomy is therefore not more common among the phthisical than the non-phthisical diseased population. Mahomed and Galton write:

"Let us here emphasise the fact that we are now comparing phthisis with *other diseases*, and not with the healthy population, and these observations would seem to show that a delicate person may fail in many ways besides being phthisical, and that a delicate narrow ovoid face may mean liability to other diseases not necessarily tubercular." (p. 13.)

[1] *Guy's Hospital Reports*, Vol. xxv, February, 1882.

[2] In this paper a compound of composites is termed a *co-composite*, and if several co-composites are compounded the result is termed a *co-co-composite*. Composites and co-co-composites are positives, and require to be reversed before printing from them.

When the individuals with a markedly hereditary taint were taken the resultant face had distinctly more delicate features, but the composite seems to the present writer too faint to provide much information; further, these cases may well have shown on the average more emphasised emaciation. On the other hand, if a composite be formed of the far-advanced cases, where the emaciation is shown in the deeply sunken eye, the hollow cheeks, and thinly covered lower jaw, the face was not by any means of the narrow ovoid type. The authors do not state whether the chronic cases were more frequent in the hereditary and the rapid cases in the latter group. Possibly they might have gone further in compounding the material on the basis of the schedule data, but we must remember that the composite photographer has not only a temptation to compound the well-fitting faces, but that to do so is almost a mechanical necessity. As our authors put it:

"We would also draw attention to the fact that this is the first attempt at applying the new process of composite portraiture on a large scale[1], and that many technical difficulties, mechanical and others, could only gradually be overcome." (p. 18).

Mahomed and Galton conclude finally that their results

"lend no countenance to the belief that any special type of face predominates among phthisical patients, nor to the generally entertained opinion that the narrow ovoid or 'tubercular' face is more common in phthisis than *among other diseases.* Whether it is more common than among the rest of the *healthy* population we cannot at present say.

It is true that taking both sexes together we find 14·3 per cent. of faces that may be classed as 'narrow ovoids[2],' and 9·3 per cent. that come under the head of 'broad faces with coarse features[3],' making in all 23·6 per cent. of our cases which may be grouped under one or other extreme departure in either direction from the normal average; but we doubt if this is more than would be found among the general population. Our results are therefore negative, but it may be they are no less valuable; although we commenced our investigations with the expectation of establishing a 'type' on a firm foundation, we shall be little less satisfied with them if they have succeeded in refuting an error.

Although these conclusions would seem to indicate that there is no foundation for the belief that persons possessing certain physical characters are especially liable to tubercular disease, yet it may hereafter be proved that some explanation of the doctrine may be found in the course of the disease when it attacks such persons." (p. 18.)

In the last paragraph our authors seem to have made an unallowable extension of their result. Were it true, we must totally deny the existence of any hereditary tendency to phthisis. Such, in my opinion, cannot be accepted in view of existing statistical data. Yet any hereditary tendency must depend upon a differentiation in physical structure, for that ultimately is what determines the efficient working of the various bodily organs. But it is idle in the present state of our knowledge to assume that there is a high correlation between the dynamical efficiency of the bodily organs and the physiognomy in particular. It is possible that nasal shape and carriage of the mouth might have some—probably not very intense—correlation with a tubercular diathesis. But no special study of mouth and nose was

[1] When we note that composites of 50 to 200 components were made for the first time, we can appreciate the magnitude of the task.

[2] What our authors term the 'tubercular' type.

[3] What they term the 'strumous' type.

made in this paper, and it may be doubted whether it could be made on portraits of so small size and all reduced to the same standard length from interpupillary line to lip line.

The main achievement of the paper lies undoubtedly in its demonstration (i) that mechanical difficulties of compounding large series of portraits had been overcome, and (ii) that no marked association exists between a phthisical tendency and physiognomy. The belief that it does exist probably arises from the more emphatic impression on the observer produced by emaciation in the narrow ovoid face.

The second paper to which I have referred is entitled "Photographic Composites[1]," and is remarkable for the two plates of the Jewish type in profile and full face. While many will criticise, and I think rightly criticise, the analysis Mr Jacobs gives of the 'Jewishness' in these portraits, they must agree with him in appreciating the extraordinary fidelity with which they portray Jewish physiognomy, or rather youthful Jewish physiognomy, for we are dealing with young Jews. Mr Jacobs writes:

> "But words fail one most grievously in trying to split up into its elements that most living of all things, human expression[2]; and Mr Galton's composites say in a glance more than the most skilful physiognomist could express in many pages. 'The best definition,' said the old logicians, 'is pointing with a finger' (*demonstratio optima definitio*); and the composites here given will doubtless form for a long time the best available definition of the Jewish expression and the Jewish type." (p. 268.)

There is little doubt that Galton's Jewish type formed a landmark in composite photography, and its success was, I think, almost entirely due to (*a*) increased facility in the process, and (*b*) to the fact that his composites were based on physiognomically like constituents. In the case of criminality and phthisis he had based his composites on mentally and pathologically differentiated components, and had expected to find mental and pathological characters highly correlated with the facial. His negative results were undoubtedly of value, but they cannot appeal to the man in the street like his positive success with the Jewish type. We all know the Jewish boy, and Galton's portraiture brings him before us in a way that only a great work of art could equal—scarcely excel, for the artist would only idealise from *one* model.

Plate XXXV (described over page) reproduces Galton's Jewish composites. The original photographs are in the Galton Laboratory.

[1] *The Photographic News*, Vol. XXIX, April 17 and 24, 1885. The Jewish profile occurs with the earlier, the full face with the later issue. Galton's paper occupied pp. 243–45 of the earlier issue. In the later issue is a paper by W. E. Debenham on Galton's "Composite Portraits" (pp. 259–60), which does not seem to do more than repeat, probably unconsciously, certain methods already referred to by Galton (see our pp. 285, 287), except in the one matter of acquiring the stereoscopic power without any instrument. There is also a paper entitled: "The Jewish Type, and Galton's Composite Photographs," by Joseph Jacobs (pp. 268–9).

[2] Mr Jacobs here uses "expression" not like Darwin in the kinematic sense, but in the statical sense of physiognomy.

PLATE XXXV, Left (Profile).

$b$ is the composite of five portraits of young Jewish boys; and similarly $c$ is the composite of five others. $d$ is a co-composite of $b$ and $c$ reversed in position, and thus represents all the ten components. $a$ is a composite of five other older faces; the components of $b$ and $c$ are given in Galton's original plates.

PLATE XXXV, Right (Full Face).

$f$ is the composite of the five full-faced portraits corresponding to $b$, while $g$ is the composite of those corresponding to $c$. $h$ is a co-composite of $f$ and $g$, and represents therefore all ten components. $e$ is a composite of the five older faces. The influence of a black curl on the forehead of $f_1$ can be traced in $f$, and even in $h$, where it is reversed (or as in $f_1$).

The great bulk of Galton's own paper in the earlier issue deals with modifications and improvements of his technique, and should be consulted even to-day by any would-be compounder. His final advice with regard to composite photography may be cited:

"It must be borne in mind by those who attempt it, that offhand methods will not avail. The adjustments must be made with judgment and extreme care to produce good effects. The difference between a very carefully-made composite and one that has been combined with only moderate care is great." (p. 245.)

In the paper Galton also gives for the first time his fiducial system for profiles; it consists of a sloping straight line with two horizontal straight lines proceeding from it to the right. The portrait is adjusted so that this sloping straight line touches the forehead, and passes through what the photographer estimates to be the alveolar point, i.e. the point of the gum between the middle incisors of the upper jaw[1]. The horizontal lines are then taken to bisect the pupil and to coincide as far as possible with the lip line respectively. Galton further notices that if he brings one of the fundamental points $A$ of his fiducial system on to the marked optical axis of his instrument, and makes the corresponding point $A'$ of his image agree with $A$, then throughout all further adjustments $A$ will coincide with $A'$, and this will much simplify the complete adjusting. Beautiful as Galton considered the adjustments of his own compounding camera to be, he believed great improvements might be made in it, especially in the direction of automatically setting the component in position after taking a series of measurements upon it. He further emphasised the need of a simple optical method of combining a considerable number of photographs to test what the compound would be like before actually photographing a composite.

The success of the 'Jewish type' convinced Galton that the future of composite photography lay largely in ethnological and genetic work. He refers in this matter to the typical crania of different races prepared by Dr Billings, Surgeon-General of the U.S. War Department[2], Mr A. Thomson of the Edinburgh Medical School (see our p. 290 ftn.), and earlier by himself, using composite photography. But he clearly placed less stress on this than on purely ethnographic portraiture of the living.

In 1879 Galton gave a Friday evening lecture at the Royal Institution

[1] The alveolar point is a well-recognised craniometric point, and it seems slightly better in this respect to use it than to make with Galton the sloping fiducial line touch the upper gum between the mid-incisors. It might even avoid the difficulties of the superciliary ridges in adult males to take the fiducial line from nasion to alveolar point.

[2] Copies of these are to be found in the Galton Laboratory.

PLATE XXXV

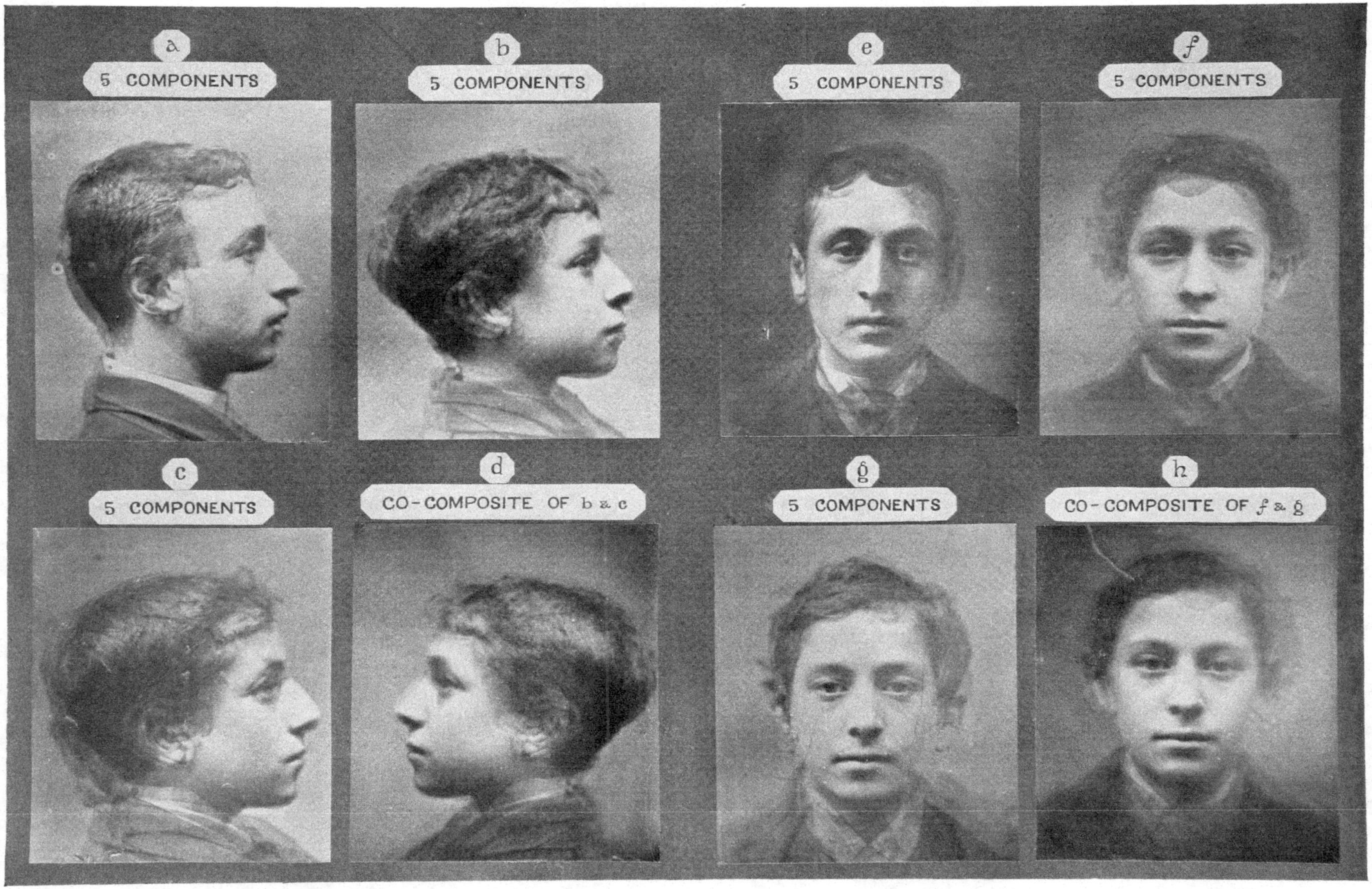

The Jewish Type.

Profile. Full Face.

on composite portraiture[1]. The lecture is called "Generic Images," according to what Galton terms "the happy phrase of Professor Huxley."

"The word generic presupposes a genus, that is to say, a collection of individuals who have much in common, and among whom medium characteristics are very much more frequent than extreme ones The same idea is sometimes expressed by the word typical, which was much used by Quetelet, who was the first to give it a rigorous interpretation, and whose idea of a type lies at the basis of his statistical views. No statistician dreams of combining objects into the same generic group that do not cluster towards a common centre, no more can we compose generic portraits out of heterogeneous elements, for if the attempt be made to do so the result is monstrous and meaningless." (p. 162.)

We thus see that Galton demands a clustering round Quetelet's 'mean man' as a success for a composite portrait; in such a case the mediocre characteristics prevail over extreme ones; the common traits reinforce each other and the extreme ones tend to disappear. In the course of the lecture Galton showed the following composites:

(*a*) A family portrait of two brothers and a sister. He built this up by the aid of three converging magic lanterns carefully adjusted, and showed that he obtained the same effect as a composite photograph of the three components.

(*b*) Alexander the Great [6] (reproduced in printed lecture: see our Plates XXXVI and XXXVII).

(*c*) Antiochus, King of Syria [6] (not hitherto published: see our Plate XXXVIII).

(*d*) Demetrius Poliorcetes [6] (not hitherto published: see our Plate XXXIX).

(*e*) Cleopatra [5]. The composite was here as usual better than the components, "none of which gave any indication of her reputed beauty; in fact, her features are not only plain, but to an ordinary English taste are simply hideous" (not hitherto published: see our Plate XL).

(*f*) Nero [11] (not hitherto published: see our Plate XLI).

(*g*) Greek female face [5] (not hitherto published: see our Plate XLII).

(*h*) Roman female face [6] (reproduced in printed lecture: see our Plate XLIII).

(*i*) Napoleon I [5] (reproduced in printed lecture: see our Plate XLIV).

(*j*) The English criminal [18] (reproduced in printed lecture). Galton here recognises two types of criminals, one with broad and massive features like Henry VIII, but with a much smaller brain; the other with a weak and certainly not a common English face[2] (see our Plates XXVIII and XXIX).

While Galton exhibited in this lecture more composite portraits than I think he showed on any other occasion, his main object was to compare

[1] *Proceedings of the Royal Institution*, April 25, 1879, Vol. IX, pp. 161–70, with an autotype reproduction of the Roman Lady, Alexander the Great, Napoleon the Great, and the English criminal.

[2] The material on which all these composites were based is still in the Galton Laboratory, although many of the photographs are sadly faded and some of the negatives have perished (owing to the use of poor chemicals, or to inadequate washing). With the exception of the phthisical plate all our reproductions are from the original material.

general impressions of the mind founded upon blended memories with blended portraits. He writes:

"In the pre-scientific stage of every branch of knowledge, the prevalent notions of phenomena are founded upon general impressions; but when that stage is passed and the phenomena are measured and numbered, many of those notions are found to be wrong, even absurdly so. This is the case not only in professional matters, but in those with which everyone has some opportunity of becoming acquainted. Think of the nonsense spoken every day about the signs of coming weather, in connection, for example, with the phases of the moon. Think of the ideas about chance, held by those who are unacquainted with the theory of probabilities; think of the notions on heredity before the days of Darwin. It is unnecessary to multiply instances; the frequent incorrectness of notions derived from general impressions may be assumed, and the object of the following discourse is to point out a principal cause of it.

Attention will be called to a source of error that is inherent in our minds, that vitiates the truth of all our general impressions, and which we can never wholly eliminate except by separating the confused facts upon which our general impressions are founded and treating them numerically by the regular methods of statistics. It is not sufficient to learn that an opinion has been long established or held by many, but we must collect a large number of instances to test that opinion, and numerically compare the successes and failures." (p. 161.)

Galton assumes the physiological basis of memory to be of the following character:

"Whenever any group of brain-elements has been excited by a sense-impression, it becomes, so to speak, tender, and liable to be easily thrown again into a similar state of excitement. If the new cause of excitement differs from the original one, a memory is the result. Whenever a single cause throws different groups of brain-elements simultaneously into excitement, the result must be a blended memory. We are familiar with the fact that faint memories are very apt to become confused. Thus some picture of mountain and lake in a country which we have never visited often recalls a vague sense of identity with much we have seen elsewhere. Our recollections cannot be disentangled, though general resemblances are recognised. It is also a fact that the memories of persons who have great powers of visualising, that is of seeing well-defined images in the mind's eye, are no less capable of being blended together. Artists are, as a class, possessed of the visualising power in a high degree, and they are at the same time pre-eminently distinguished by their gifts of generalisation. They are of all men the most capable of producing forms that are not copies of any individual, but represent the characteristic features of classes." (p. 162.)

Galton holds that the brain has the capacity for blending memories together, and that general impressions are faint and perhaps faulty editions of blended memories. Thus there is some analogy between general impressions and composite photographs, both are generic images.

"A generic mental image may be considered to be nothing more than a generic portrait stamped on the brain by the successive impressions made by its component images[1]."

But while the photographic generic image gives each component a weight proportional to its exposure, Galton says that the mental composite does not give weight in the same manner.

"The physiological effect of prolonged action, or of reiteration, is by no means in direct proportion to the length of the one or to the frequency of the other."

He then cites the Weber-Fechner Law of the geometrical mean[2] as one at least of the sources of error in general mental impressions.

"Exceptional occurrences leave an impression on the brain of far greater strength, and habitual occurrences of far less strength, than their numbers warrant."

[1] Galton here cites Huxley (*Hume*, p. 95) as independently reaching the same conclusion.

[2] Illustrated in the lecture itself by spinning discs painted black and white in concentric rings, one giving an arithmetical the other a geometrical series of tints; the eye repudiates the former as a graduated scale.

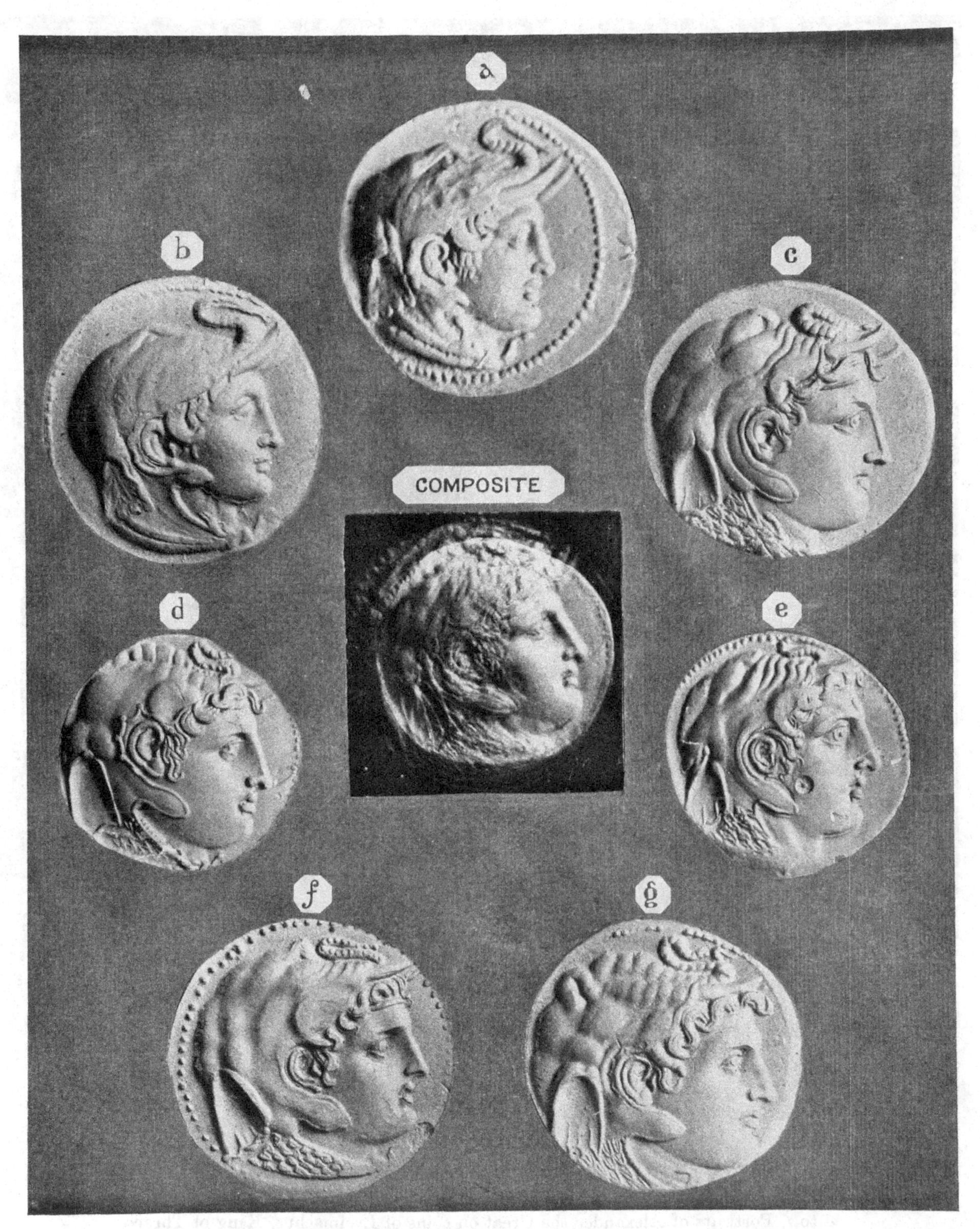


Indian Portraits of Alexander the Great with Composite in centre.

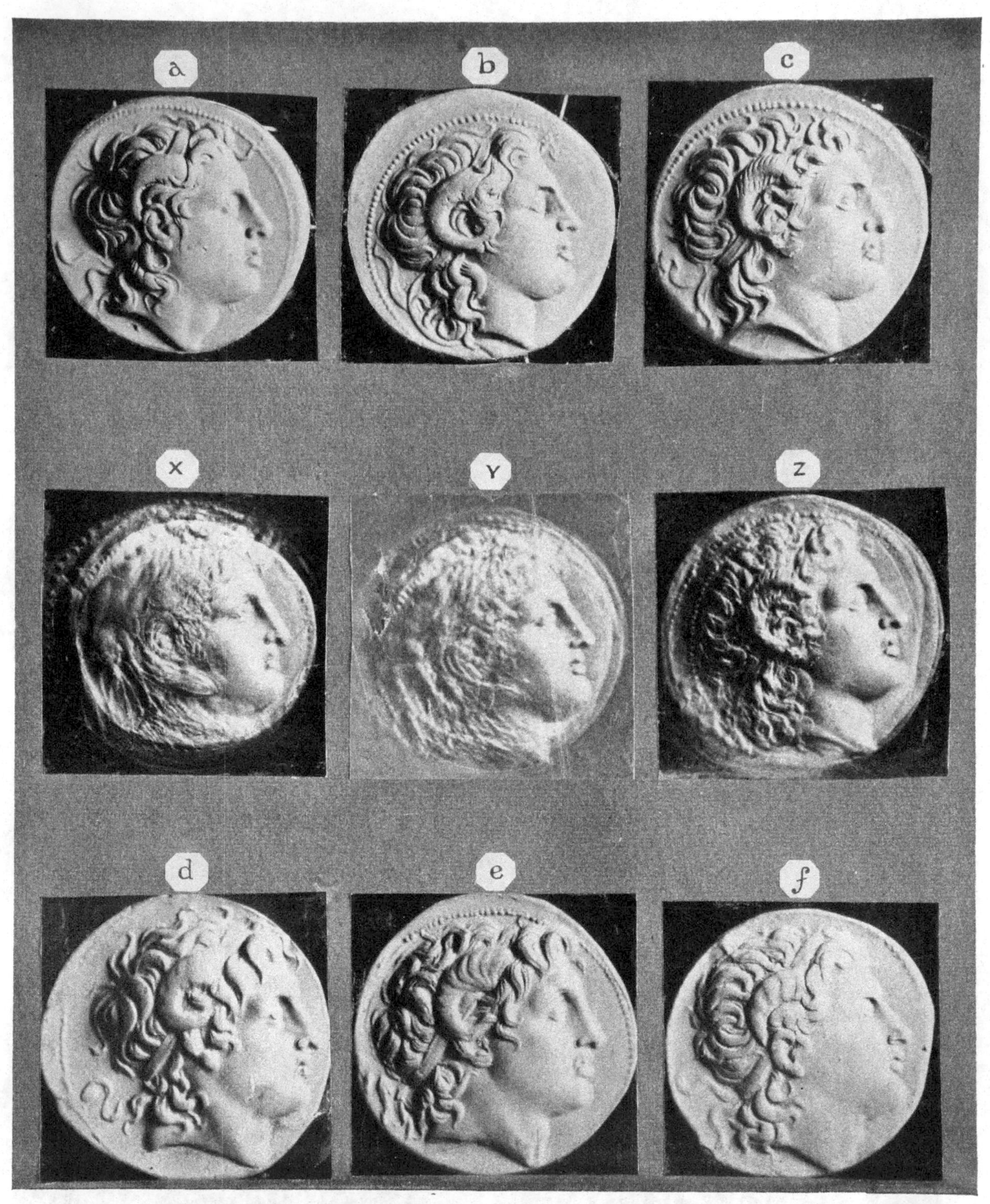


*a* to *f*, Portraits of Alexander the Great on coins of Lysimachus, King of Thrace.
*X* = Composite of Indian Alexander (see Plate XXXVI).
*Z* = Composite of *a* to *f*.
*Y* = Co-composite of Indian and Greek Portraits.

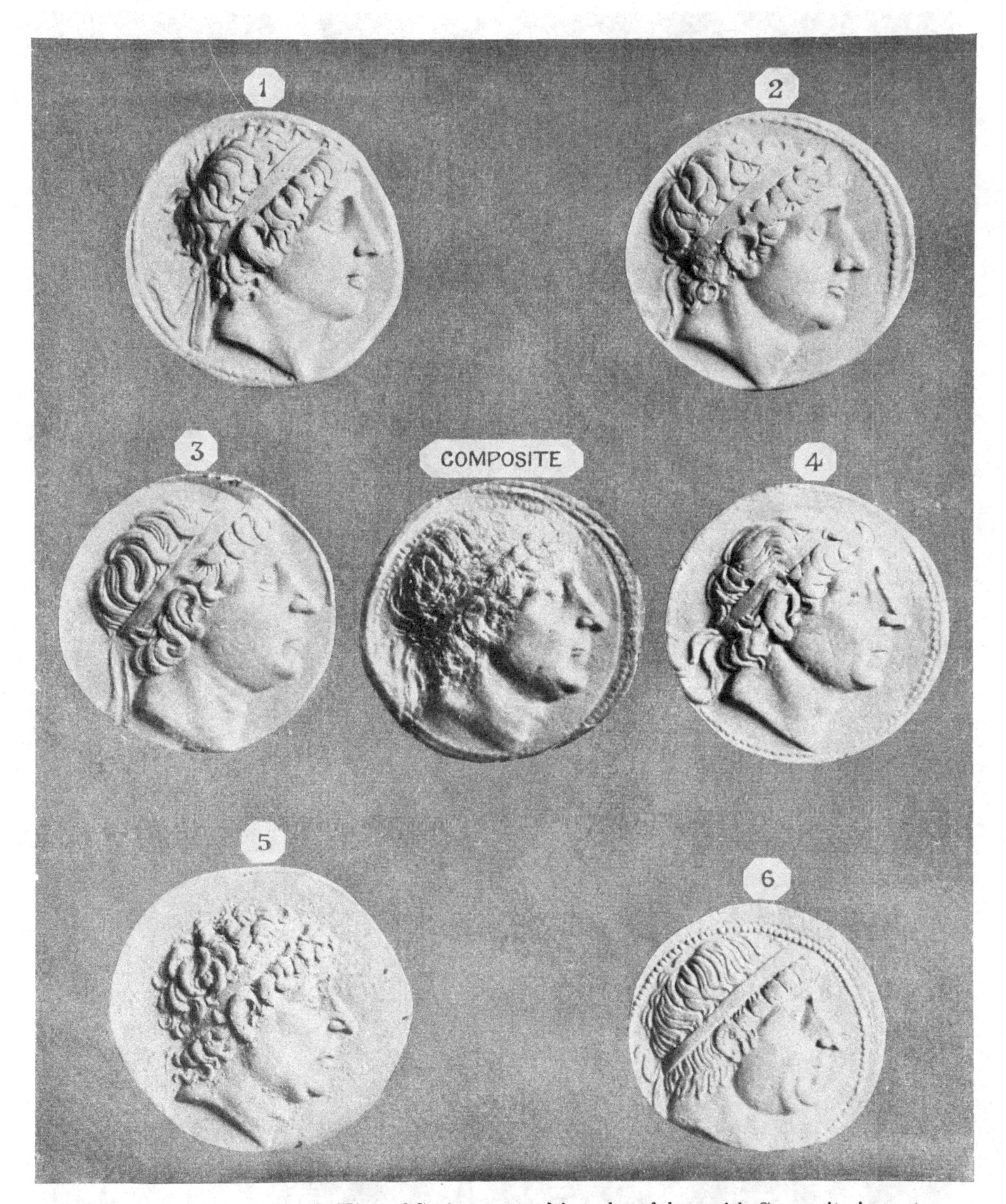


Six Portraits of Antiochus I, King of Syria, arranged in order of date with Composite in centre.

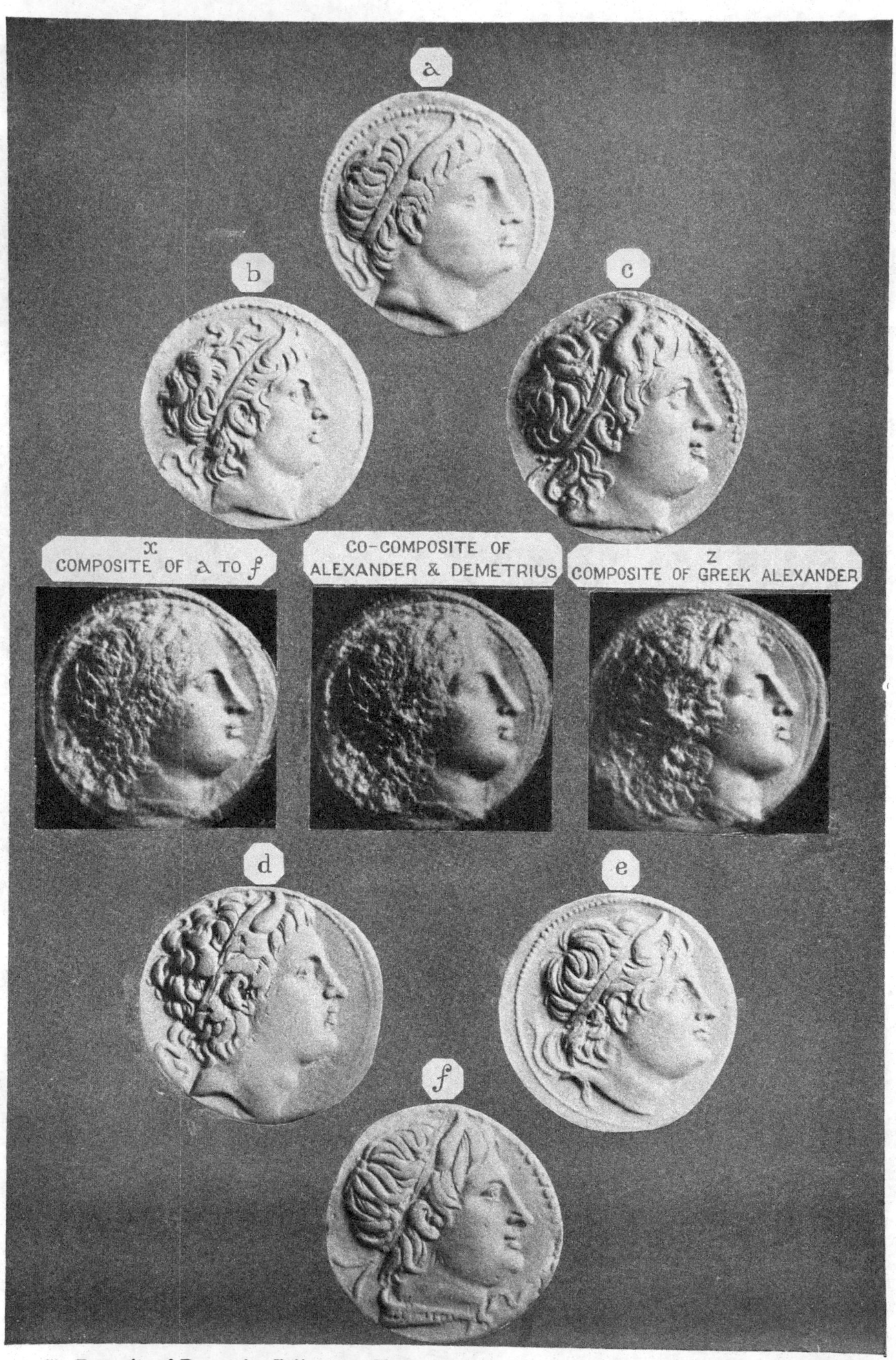


**Six Portraits of Demetrius Poliorcetes, King of Macedonia, *a* to *f*, giving typical Greek Head.**

Five Portraits of Cleopatra, Queen of Egypt, *a* to *e*, with Composite *x*.

PLATE XLI

COMPOSITE
12 PORTRAITS OF NERO, EMPEROR OF ROME.
WITH COMPOSITE.

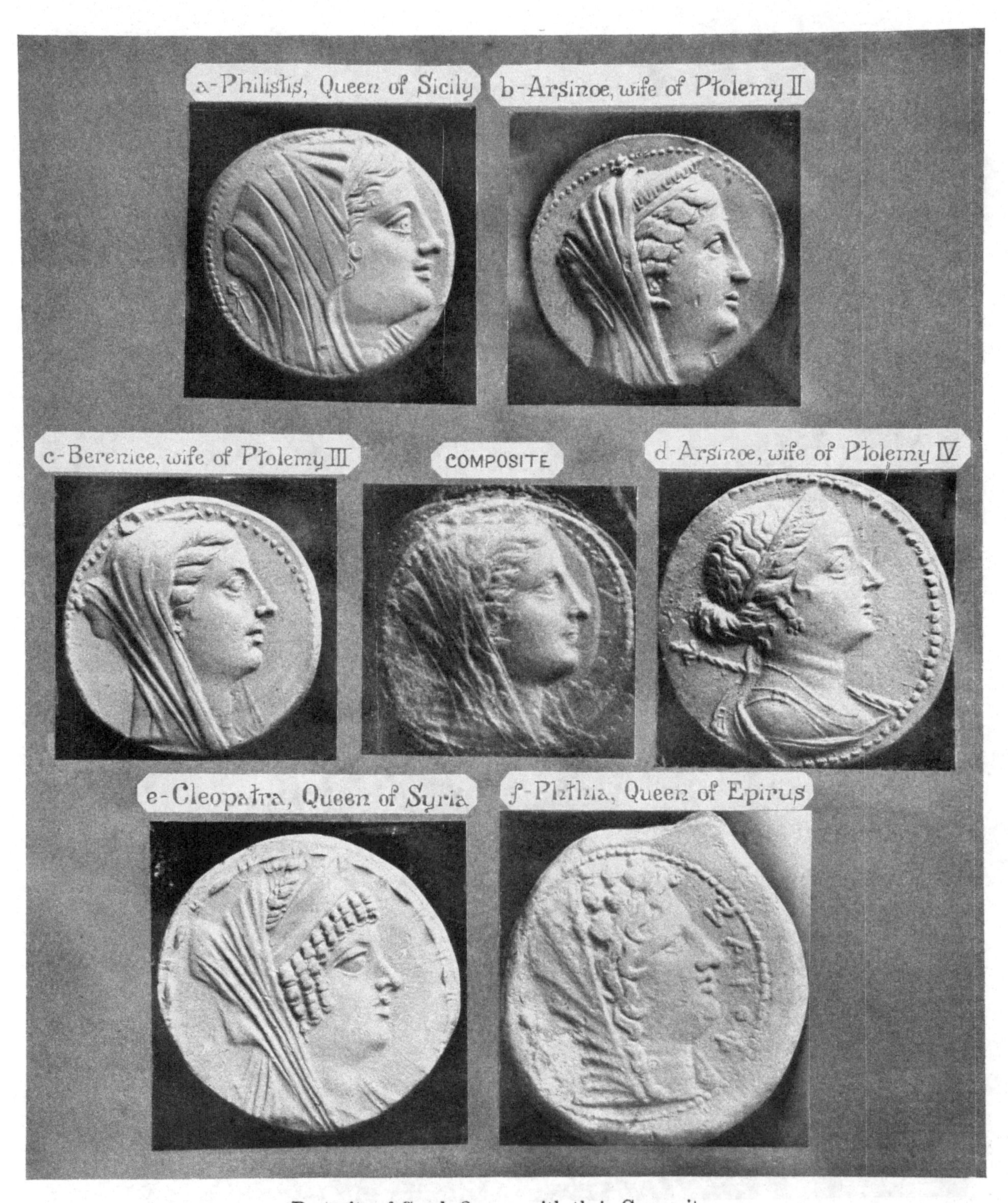


Portraits of Greek Queens with their Composite.

Roman Ladies with Composite.

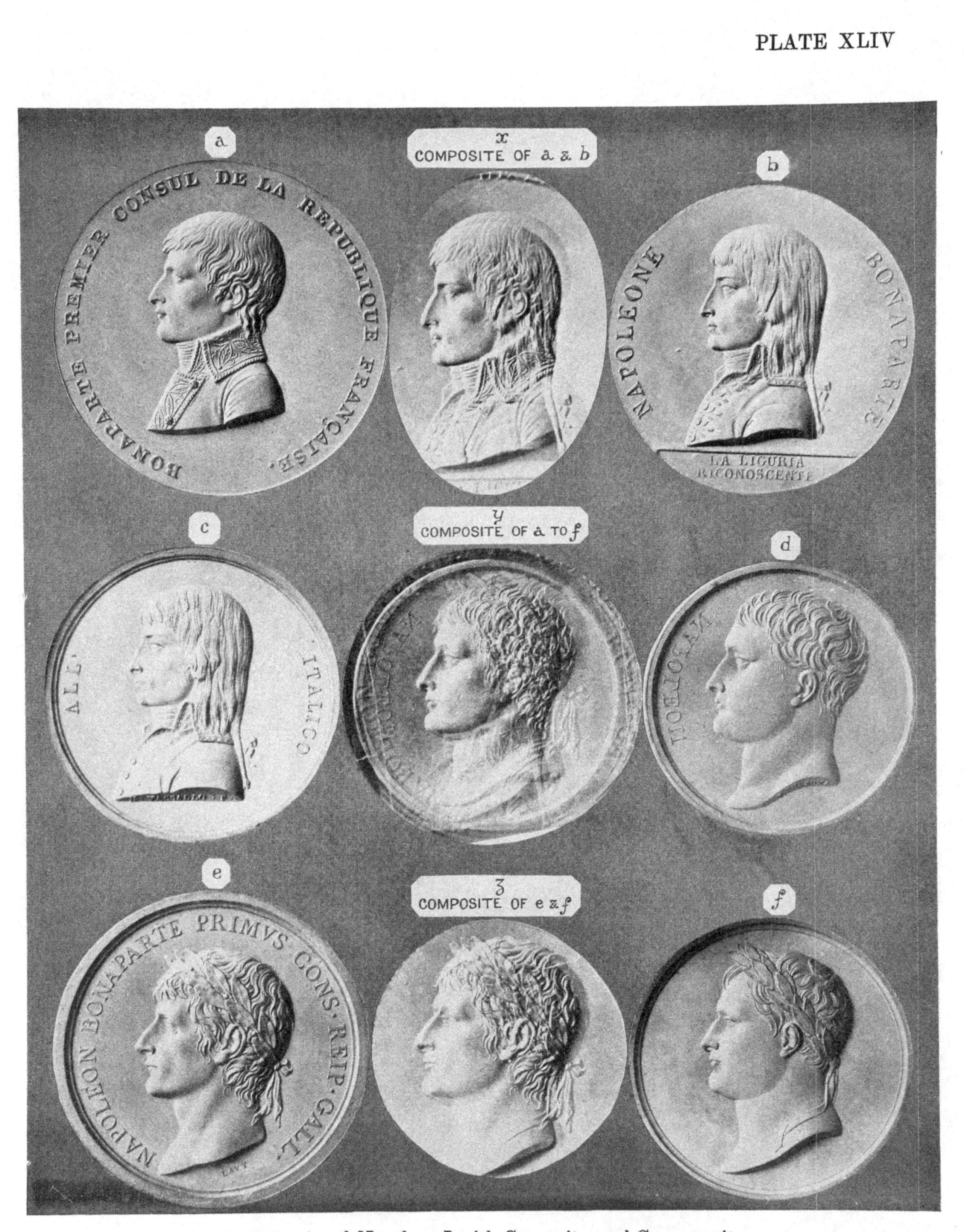


Six Portraits of Napoleon I with Composites and Co-composite.

Just as in the composite photograph some images may be alien to the genus, so in the case of the mind superficial and fallacious resemblances may be associated.

"Seeing as we easily may, what monstrous composites result from ill-sorted combinations of portraits, and how much nicety of adjustment is required to produce the truest possible generic image, we cannot wonder at the absurd and frequent fallacies in our mental conceptions and general impressions."

Galton continues:

"Our mental generic composites are rarely defined; they have that blur in excess which photographic composites have in a small degree, and their background is crowded with faint and incongruous imagery. The exceptional effects are not overmastered, as they are in the photographic composites, by the large bulk of ordinary effects. Hence in our general impressions far too great weight is attached to what is strange and marvellous, and experience shows that the minds of children, savages and uneducated persons have always had that tendency. Experience warns us against it, and the scientific man takes care to base his conclusions upon actual numbers.

The human mind is therefore a most imperfect apparatus for the elaboration of general ideas. Compared with those of brutes its powers are marvellous, but for all that they fall vastly short of perfection. The criterion of a perfect mind would lie in its capacity of always creating images of a truly generic kind, deduced from the whole range of its past experiences. General impressions are never to be trusted. Unfortunately when they are of long standing they become fixed rules of life, and assume a prescriptive right not to be questioned. Consequently, those who are not accustomed to original inquiry entertain a hatred and a horror of statistics. They cannot endure the idea of submitting their sacred impressions to cold-blooded verification. But it is the triumph of scientific men to rise superior to such superstitions, to devise tests by which the value of beliefs may be ascertained, and to feel sufficiently masters of themselves to discard contemptuously whatever may be found untrue." (pp. 168–70.)

The words just cited—almost lost in their manner of publication—are among the finest Galton ever wrote in the service of science.

In reply to the recent inquiry of a friend as to what point I had reached in my account of Galton's labours, I said: To the discussion of composite portraits in his researches on psychology. It seemed to him an inappropriate association. Yet almost all Galton's photographic work in his own mind had relation to psychology, and up to the end of his life he continued to develop photographic methods for statistically studying mental characters. From composite portraiture the stage was for him an easy one to generic mental images, thence he passed to the Weber-Fechner Law, and this turned his thoughts to the statistical bearings of the latter as we have already seen. The relation Galton held to exist between generic mental images and composite photographs is well illustrated by Galton's paper entitled "Generic Images" published in *The Nineteenth Century* for July, 1879[1]. This is in some respects an enlargement of the Royal Institution Lecture, with less technical description and no plate. The point he emphasises in this paper is the bearing of composite portraiture on then current metaphysical conceptions. He writes:

"Composite portraits are, therefore, much more than averages, because they include the features of every individual of whom they are composed. They are pictorial equivalents of those elaborate statistical tables out of which averages are deduced. There cannot be a more perfect example than they afford, of what the metaphysicians mean by generalisations, when the objects generalised are objects of vision, and when they belong to the same typical group, one important characteristic of which is that medium characteristics should be far more

[1] Vol. VI, pp. 157–69. In this paper Galton compares the composite portrait to Quetelet's "mean man." (p. 162.)

frequent than divergent ones. It is strange to notice how commonly this conception has been overlooked by metaphysicians, and how positive are their statements that generalisations are impossible, and that the very idea of them is absurd. I will quote the lucid writing of Sir W. Hamilton to this effect, where he epitomises the opinions of other leading metaphysicians. I do so the more readily because I fully concede that there is perfect truth in what he says, where the objects to be generalised are not what a cautious statistician would understand by the word generic. Sir W. Hamilton says (*Lectures*, II, p. 297):

'Take for example, the term *man*. Here we can call up no notion, no idea corresponding to the universality of the class, or term. This is manifestly impossible. For as *man* involves contradictory attributes and as contradictions cannot exist in one representation, an idea or notion adequate to *man* cannot be realised in thought. The class *man* includes individuals, male and female, white and black and copper coloured, tall and short, fat and thin, straight and crooked, whole and mutilated, etc., and the notion of the class must therefore at once represent all and none of these. It is therefore evident, though the absurdity was maintained by Locke, that we cannot accomplish this; and this being impossible, we cannot represent to ourselves the class *man* by any equivalent notion, or idea.... This opinion, which after Hobbes, has been in this country maintained among others by Berkeley, Hume, Adam Smith, Campbell, and Stewart, appears to me not only true, but self-evident.'"

To this Galton replies, demolishing by a concrete representation an imposing philosophical dogma:

"If Sir W. Hamilton could have seen and examined these composite portraits, and had borne in mind the well-known elements of statistical science, he would certainly have written very differently. No doubt, if what we are supposed to mean by the word *man* is to include women and children and to relate only to their external features and measurements, then the subject is not suitable for a generic picture, other than of a very blurred kind, such as a child might daub with a paint-brush. If, however, we take any one of the principal races of man and confine our portraiture to adult males, or adult females, or to children whose ages lie between moderate limits, we ought to produce a good generic representation."

Bold indeed to face the metaphysician in his own cave, and assert that his generic pictures were as those of a child daubing with a paint-brush, solely because he had not adequately, i.e. statistically, defined what was to be understood by a genus, and a generalisation! It is the old tale of the scientist, analysing phenomena, coming up against the metaphysician bandying words[1]. No wonder that Galton's psychology was of small influence with philosophising dialecticians!

### (*B*) *Photographic Bi-projection, Indexing of Profiles, etc.*

As late as April, 1888, Galton was still thinking over composite photography. In his earlier work he had made the vertical distance between the

[1] My friend Professor W. P. Ker warns me to avoid an *ignoratio elenchi*. It seems to me that any argument would turn on how far the "general idea" is that of a limited class. I feel sure that Berkeley and I think Hamilton would have argued that the abstract idea of a Jewish Boy was impossible. Yet Galton shows that we can form a concrete image of such a Boy, and he sees no reason why we should not, if so constituted, visualise him. Berkeley (*Works*, Vol. I, pp. 76–7, 1843) confesses that what other minds can do, he knows not, but he himself cannot abstract the qualities from a number of individuals and compound them to a general notion. Both Berkeley and Hamilton surveyed their own minds, and they do not appear to have experimented on the visualising faculty of other minds.

What Galton asserts is that it is possible to reach a general idea or a generic image provided the individuals generalised form a limited class or *genus*, and he holds that the metaphysicians, proceeding purely by introspection, had overlooked the statistical criteria for the homogeneity of a group or *genus*.

pupillary line and the line of the lips the same for all his components. But the result of this vertical distance only being the same was a great diversity in breadths, leading to an absence of sharpness in the outline; thus, as Galton expressed it[1], the result was an *aggregate* rather than a *mean*. He now considered that the value of composites would be much increased, if they were at the same time reduced to a common breadth as well as a common vertical standard. Galton chose as his breadth the interpupillary distance, but for some purposes it might be more useful to take, say, the external ocular distance or even the breadth of face at ear level. The average value of the selected vertical and horizontal lengths was to be determined, and each photograph reduced to these dimensions. Thus the problem becomes precisely that referred to on our p. 45. A photographic arrangement which would act as a bi-projector was needful, and one must be devised in which foreshortening would not be accompanied by any sensible blur. Galton first considered that this result could be obtained by a form of pin-hole camera he had seen discussed in the *Photographic News*; namely one in which the pin-hole was replaced by two adjustable diaphragms. The first of these diaphragms would contain a vertical slit, and have a motion horizontally but perpendicular to the optical axis, and another motion along the optical axis; the second diaphragm would contain a horizontal slit, and have one motion vertical and perpendicular to the optical axis, and another motion along the optical axis. By proper adjustment of the two diaphragms, interchanging them if needful, any desired modifications in height and breadth of the object could be made. Theoretically the scheme is admirable; it is precisely that of the bi-projector referred to on p. 46, first ftn., except that the beam of light is replaced in the latter by a "mechanical straight line." The practical difficulty lies in the need of a very intense light on the object, not only to reduce the long exposure, but to enable the operator to adjust the image on the focal plane. When I discussed with Galton in 1903 the possibility of double photographic reduction, he did not refer to this pin-hole scheme, possibly he had discarded it after trial[2].

Galton's notebooks and papers show that he spent in that year much time over this problem of reducing photographically a circle to an arbitrary ellipse. He proceeded, however, by an entirely different method. He proposed to rotate his object plane round a vertical axis until it made an angle $\theta_1$ with the vertical plane, and then photograph it, trusting to stopping down to

[1] *Photographic News*, April 27, 1888.

[2] Trials have recently been made in the Galton Laboratory of this method of bi-projectional photography. To get rid of blurring the slits had to be extremely narrow, and thus a four hours' exposure might be necessary for the reduction of a black and white drawing. It was then found that the negative of the drawing had a series of light and dark bands across it. I am not certain whether these are due to some diffraction effect, or to slight inequalities in the breadth of the slits. I have found that a precisely similar system of bands, of course in one direction only, may arise in photographing the sun with a focal plane shutter, when owing to clearness of atmosphere it is needful to reduce the breadth of the slit in the shutter to a minimum. Another objection to the method is that the circular dots used by draughtsmen for points become ellipses, and vertical and horizontal lines do not remain of the same thickness, but this objection applies to all methods of photographic bi-projection.

cure the blurring. A print was then taken of the result, and this print placed in the objective plane rotated through an angle $\theta_2$ in the *opposite* direction to $\theta_1$ is rephotographed. Then Galton discovered that with a certain relation[1] between $\theta_1$ and $\theta_2$ the second photographing can be made to restore linearity, or an original circle be converted into an ellipse. The accompanying diagrams indicate the three stages of Galton's process. On the last diagram he had written the words "Show this to Pearson." He never did so and only

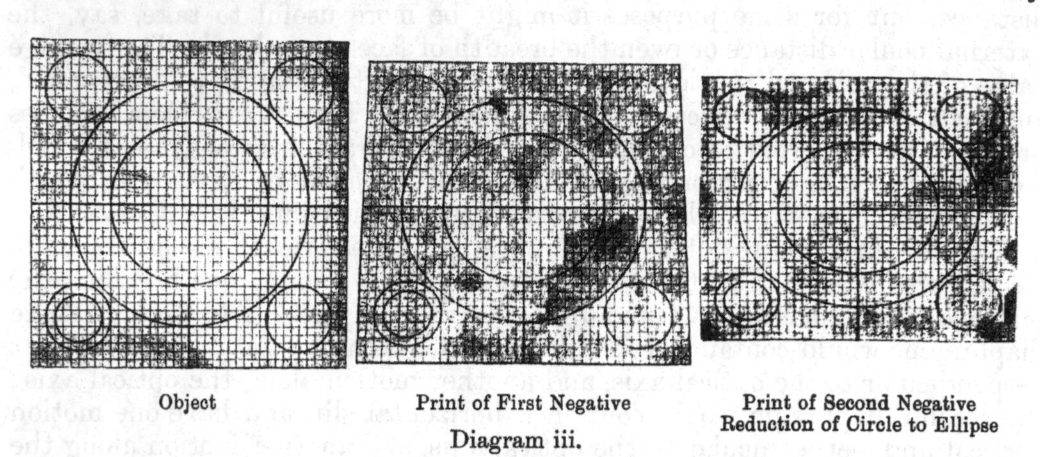

**Object** **Print of First Negative** **Print of Second Negative Reduction of Circle to Ellipse**

Diagram iii.

after his death has his biographer discovered the large amount of time and energy Galton spent over this matter; there are elaborate tables of $\theta_1$ and $\theta_2$ with the corresponding vertical and horizontal reductions. There is also the first draft of a paper intended for publication, but I cannot find that the paper and the tables[2] were ever completed, still less published. Like so many of Galton's ideas it was simple and suggestive, but Galton was too old in

[1] We have found that a single negative will suffice, if a second camera be employed to photograph the image on the focal plane screen of the first, this camera being adjusted so that its focal plane makes an angle $\theta_2$ with the focal plane of the first camera. The difficulty lies in the length of exposure requisite if the objectives of both cameras are cut down so as to reduce blurring to insignificance. Nothing is gained theoretically or practically by tilting as well as rotating the object plane of the first and the optical axis of the second camera. The theory is as follows: Let $d_1$ and $b_1$ be respectively the distances from the optical centre of the first camera to the object plane and to the focal plane, and $d_2$ and $b_2$ the distances from the optical centre of the second camera to the focal plane of the first camera and to its own focal plane. Then for rectilinearity in the photograph we must have

$$d_2 \tan \theta_1 = b_1 \sin \theta_2;$$

and if $R_v$, $R_h$ be the vertical and horizontal scales of reduction, then

$$R_v = d_1 d_2/(b_1 b_2), \quad R_h = R_v \cos \theta_1 \cos \theta_2.$$

Thus $R_h$ must be less than $R_v$, but this is no limitation as the object can always be turned through a right angle. Actually the chief difficulty lies in the suitable choice of $d_1$ and $d_2$. The apparatus takes a simpler form if we keep the two optic axes in the same straight line, and the object perpendicular to them, but rotate the focal planes in opposite directions.

[2] There is also a bundle entitled "Photographic Reduction in breadth" with models in both wood and card of the proposed camera apparatus. As far as I can see Galton always proposed making an auxiliary intermediate print.

1903 to spend the necessary time in working out the practical details of camera dimensions, or spend the hours required in dark-room experimental work. As in the similar case of analytical photography, what is needed is a young and enthusiastic photographer.

To grasp fully Galton's photographic activities at this time we must bear in mind two important facts. He was still searching for some physical features which should have high association with the mental characters. This attitude was perfectly reasonable at that date, because not only no correlations between such characters had been determined, but the methods of measuring correlations were of the crudest kind. Further Galton was a traveller, and every traveller is accustomed as he passes along to notice that the racial mentality changes with the change of the physical characters. The conception therefore naturally arises that physique and mentality are highly correlated. The American Indian, the Negro, or the Arab has each his individual physique, and each also his individual mentality. But this appearance of high correlation may be most deceptive; it does not follow that there is any organic linkage between the physical and psychical characters. If a race be started from a pair of individuals both possessing a physique of type $A$ and a mentality of type $A'$, we may find in later generations an apparent linkage of $A$ and $A'$ in all the members; but this is not a true correlation, and a cross-breeding may show that $A$ and $A'$ have no organic relation, and can be at once separated. In the second place Galton did, like most men of his generation and probably like most of us to-day, consciously or unconsciously, give weight to physiognomy. So impressed by physiognomy is mankind in ordinary every-day life, that we hardly realise how much confidence we place in it. We say a person is good or bad, is intelligent or stupid, is slack or energetic, on what is too often only a rapid physiognomic judgment. The custom is so universal as a rough guide to conduct, that we are almost compelled to believe that there is in human beings an intuitive or instinctive appreciation of mental character from facial expression. Galton differed only from the mass of us in desiring to ascertain on what physiognomic appreciation is based. He belonged to a generation in which the influence of Lavater and the belief in some form of phrenology were still appreciable. He accordingly sought to isolate types and to measure deviations from facial type, in order to determine whether facial variations were correlated with mental variations. He was really attempting to make a true science out of the study of physiognomy. The anthropologist up to Galton's date had employed portraiture to distinguish racial types physically. Galton employed portraiture to distinguish if possible between mental types. He may have been pursuing a will-o'-the-wisp, but this psychical investigation was really at the basis of all his photographic work, and he was interested in my desire for a photographic 'bi-projector,' not in the first place because it would relieve the difficulties of an editor, but chiefly because it would be of great service in composite and analytical photography. It may be that it is rather on the *play* of features than on their static form that our intuitive judgment as to mental and moral character is based[1]. In this case a static photograph would only lead to a negative, albeit important conclusion[2].

From Galton's outlook on mankind the mentality and physique of its stirps were of first-class importance to the child, and he emphasised the value of a family record made on a standardised plan to the child as early as 1882 (*Fortnightly Review*, January, 1882, pp. 26–31), and of such a record Galton held an essential feature to be a series of photographs.

"Obtain photographs periodically of yourselves and of your children, making it a family custom to do so, because unless driven by some custom the act will be postponed until the opportunity is lost. Let these periodical photographs be full and side views of the face on an adequate scale, and add any others you like, but do not omit these. As the portraits accumulate have collections of them autotyped. Take possession of the original negatives, or have them stored in safe keeping, labelled and easy to get at. They will not fade[3], and the time may come when they will be valuable for obtaining fresh prints or for enlargement. Keep the prints methodically in a family register, writing by their side all such chronicles as those that used to find a place on the fly-leaf of the family Bibles of past generations, and much more besides. Into the full scope of that additional matter I do not propose now to enter. It is an interesting and important topic that requires detailed explanation, and it is better for the moment not to touch upon it."

Here we see Galton's thoughts turning in the direction whence afterwards arose his *Record of Family Faculties* and his *Life-History Album*.

"This, however, may be said, that those who care to initiate and carry on a family chronicle, illustrated by abundant photographic portraiture, will produce a work that they and their children and their descendants in more remote generations, will assuredly be grateful for. The family tie has a real as well as a sentimental significance. The world is beginning to perceive that the life of each individual is in some real sense a prolongation of those of his ancestry. His character, his vigour and his diseases are principally theirs; sometimes his faculties are blends of ancestral qualities, more frequently they are aggregates, veins of resemblance to one or other of them showing now here and now there. The life-histories of our relatives are, therefore, more instructive to us than those of strangers; they are especially able to forewarn and to encourage us, for they are prophetic of our own futures. If there is such a thing as a natural birthright, I can conceive of none greater than the right of each child to be informed, at first by proxy through his guardians, and afterwards personally, of the life-history, medical and other, of his ancestry. The child is brought into the world without his having any voice at

[1] I think Charles Darwin realised this fully in 1873, and indicates it in the opening sentences of his *Expression of the Emotions*; for him "Expression" itself means kinetic facial changes. "Many works have been written on Expression, but a greater number on Physiognomy,—that is, on the recognition of character through the study of the permanent form of the features. With this latter subject I am not here concerned." (p. 1.)

[2] There might still be a chance for the film. It would need a super-Galton to organise the technique of a composite film!

[3] This is alas! not the fact. Galton had a large collection of prints and negatives of individuals and of composites. A very large proportion of the prints are now so faded as to be useless; of many the subject is indistinguishable. When I turned to the negatives to replace the prints, I found many negatives had perished also, gone as yellow and faded as the prints, and others were in process of decay. Unless immediate steps be taken to reproduce it in a permanent way, Galton's unpublished photographic work will have practically perished entirely within 50 years of its preparation. Failing some form of ink reproduction—and then it must not be on paper laden with china clay—there is no real security for permanency in photographic negatives and prints. The patchy preservation of Galton's negatives—some faded, some excellent—shows that there is no security that negatives will survive fifty years, the source may lie in varied technique, or in varied quality of chemicals used.

all in the matter, and the smallest amend that those who introduced him there can make, is to furnish him with the most serviceable of all information to him, the complete life-histories of all his near progenitors." (p. 31.)

The idea of portraiture as expressing mental character and that of individuality as measured by deviation from type fascinated Galton throughout the whole of his long life, and he returned to these subjects with great energy even in his last years. He sought to measure the degree of resemblance or of difference in portraits. The amount of labour he put into this research was immense; there is a great mass of manuscript matter, there are endless profiles drawn by his assistants, there are models of apparatus and there is apparatus itself. Without a more definite key than we possess it is often very difficult to trace what line of thought he was following up, although not infrequently one lights on most suggestive ideas in side tracks from the main problem.

That the work in this direction arose from the composite photograph investigations is clear from a lecture Galton gave on May 25, 1888 at the Royal Institution, entitled "Personal Identification and Description[1]". It opens with the following words:

"It is strange that we should not have acquired more power of describing form and personal features than we actually possess. For my own part I have frequently chafed under the sense of inability to verbally explain hereditary resemblances and types of features, and to describe irregular outlines of many different kinds, which I will not now particularise. At last I tried to relieve myself as far as might be from this embarrassment, and took considerable trouble, and made some experiments. The net result is that while there appear to be many ways of approximately effecting what is wanted, it is difficult as yet to select the best of them with enough assurance to justify a plunge into a rather serious undertaking. According to the French proverb, the better has thus far proved an enemy to the passably good, so I cannot go much into detail at present, but will chiefly dwell on general principles." (*Nature*, Vol. XXXVIII, p. 173.)

Galton then states that while recognising different degrees of likeness and unlikeness we have not so far as he knows made any attempt to measure them. He now proposes to take for his unit of measurement the *least-discernible difference*.

"The measurement of resemblance by units of least-discernible difference is applicable to shades, colours, sounds, tastes, and to sense-indications generally."

Galton illustrates his method on sight differences; he takes two superposed oval contours (see Fig. *a*, Diagram iv, p. 304), intersecting one another, and then halves the distance between their boundaries for a new contour, and then halves again until he reaches—in his case in the fourth stage—a contour indistinguishable from one of the original contours. He then says there are 16 grades of least-discernible difference between $A$ and $B$. The method is suggestive, but obviously liable to difficulties, for it is clear that its measurement is largely *subjective*. It depends on the fineness of drawing of the original contours and of the subdividing contours. It depends also on the scale upon which they are drawn. It is modified by the subjective conditions of the observer, whether his sight is good, and whether he uses or does not

[1] *Nature*, Vol. XXXVIII, pp. 173–77, 201–2, 1888; *Proc. Royal Institution*, Vol. XII, pp. 346–60, 1889; my references are to the pages of *Nature*.

use glasses. Also it is clear that the least-discernible difference may be reached at some points long before it is reached at others, or the measure of resemblance would vary from part to part, and ultimately be a measure

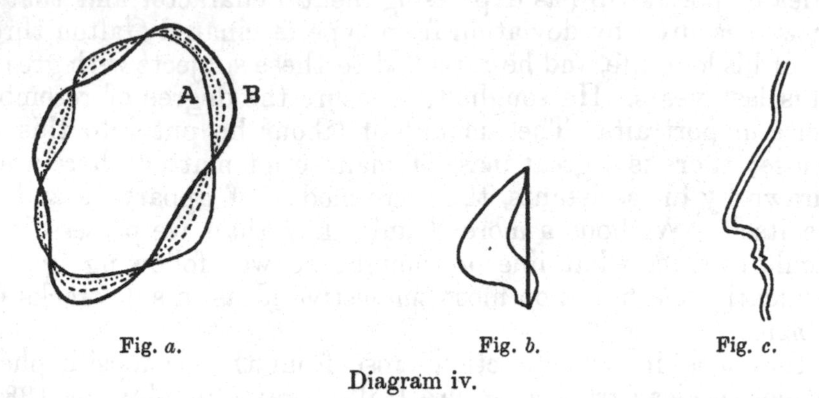


Fig. *a*. Fig. *b*. Fig. *c*.

Diagram iv.

of only the most unlike parts. If we agree to an average fineness of line, and an average keenness of sight, we shall still be left with the question of scale[1].

Dealing with the silhouette, Galton remarks that:

"All human profiles of this kind, when they have been reduced to a uniform vertical scale, fall within a small space. I have taken those given by Lavater, which are in many cases of extreme shapes, and have added others of English faces, and they all fall within the space shown in Fig. *b*. [Galton is working with the distance from the notch that separates the brow from the nose (nasion) to the parting of the lips as standard length.] The outer and inner limits of the space are of course not the profiles of any real faces, but the limits to many profiles, some of which are exceptional at one point and others at another. We can classify the great majority of profiles so that each of them shall be included between the double borders of one, two, or some small number of standard portraits, such as Fig. *c*. I am as yet unprepared to say how near together the double borders of such standard portraits should be drawn; in other words, what is the smallest number of grades of unlikeness that we can satisfactorily deal with. The process of sorting profiles into their proper classes, and of gradually building up a well-selected standard collection, is a laborious undertaking if attempted in any obvious way, but I believe it can be effected with comparative ease on the basis of measurements as will be explained later on, and by an apparatus that will be described." (p. 174.)

The reader will now be able to perceive better what Galton was really attempting to do by this special illustration: he was aiming at identifying individuals by their profiles, and in order to do this it was needful to index profiles. This leads Galton to the topics of indexing and of entering indices. He first refers to Bertillon's system of identifying criminals, and states that the actual method by which it is done is not all that theoretically could be desired. He notes a fundamental difficulty that arises:

"The fault of all hard-and-fast lines of classification when variability is continuous, is the doubt where to place and where to look for values that are near the limits between two adjacent classes." (p. 175.)

[1] For example, suppose it be required to find the degree of resemblance between two maps, *A* and *B*, of the same district on different scales; shall we reduce *A* to *B*, or *B* to *A*, for that will clearly affect our judgment? Or, shall we look at them both placed on the table before us, or both hung at some little distance on a wall?

Bertillon divides each of his four fundamental characters into three groups —large, medium, small—and Galton points out that the difference between the men at the extremes of the medium group is, for stature, say 2·3 inches while the possible error of determining stature may be $\pm$ 0·5 inch; that is to say, that there is a total doubtful range of 2 inches, while the medium range itself is only 2·3 inches. He further points out that nearly all Bertillon's characters, we may anticipate, will be highly correlated together and accordingly his 81 ($=3^4$) groups will contain very unequal numbers.

"No attempt has yet been made to estimate the degree of their interdependence. I am therefore having the above measurements (with slight necessary variations) recorded at my anthropometric laboratory for the purpose of doing so." (p. 175.)

I do not think these measurements were ever taken in adequate numbers or that Galton ever determined actually their correlations. This was, I believe, first done by the late Dr Macdonell, on actual criminal data, and he pointed out how, by the use of proper "independent variates," the trouble of correlation in the characters could be eliminated[1].

The first difficulty, however, of the border-line cases, which involve such a large proportion of the population and therefore the multiplication of cards in several groups, Galton got over by what he termed a "mechanical selector." I have not found any 'selector' described before 1888, but many since, all involving Galton's principle, some patented, without any recognition of Galton's priority. The idea is indeed a very simple one; each individual has a card 8 to 9 inches long. If there are 4 or 6 indexing characters each is allotted something less than a quarter to a sixth of the card. This portion of the card represents the range of the corresponding variate and a notch is cut into the card at the value of the variate within this range[2]. The breadth of this notch represents twice the possible error of measurement, once in excess and once in defect, for that variate. The cards are placed vertically and loosely in a box divided into batches by partitions so that there is not sufficient friction to interfere with their independent motion. The bottom of the box, except sufficient at the ends for the cards to rest on, is replaced by a "keyboard" as Galton termed it; this keyboard is of the breadth of the variate portion of the cards, and can be elevated by a lever. Adjustable wires can be arranged across a gap in the keyboard of the size of the series of cards, and these wires are adjusted to give the measurements of an individual to be selected, just as the notches are cut in the cards. When the keyboard is elevated its wires pass into the notches of those cards which are within possible errors of the individual set on the keyboard—all the other cards but these are raised and thus discriminated from those which require examination. It is clear that the cards do not require classification by size of organs, but may be placed by age or alphabetically. Galton considered that this mechanical

[1] *Biometrika*, Vol. I, pp. 177–227. The Bertillon system of indexing by physical measurements has now been replaced by direct indexing of finger-prints.

[2] Actually the notch would not be cut at the exact value of the variate except when near the boundary of the sub-range; in other positions it would suffice to cut it at the middle of the sub-range. For Bertillon's index it would suffice practically to have pin points marked for each variate on the card, where notches should be cut.

selector of which he gives ample drawings could deal with 500 cards at a time. Of his 'selector' Galton writes:

"Its object is to find which set, out of a standard collection of many sets of measures, resembles any one given set within any degree of unlikeness. No one measure in any of the sets selected by the instrument can differ from the corresponding measure in the given set by more than a specific value. The apparatus is very simple; it applies to sets of measures of every description, and ought to act on a large scale as well as it does on a small one, with great rapidity, and be able to test several hundred sets by each movement. It relieves the eye and brain from the intolerable strain of tediously comparing a set of many measures with each of a large number of successive sets, in doing which a mental allowance has to be made for a *plus* or *minus* deviation of a specified amount in every entry. It is not my business to look after prisoners, and I do not fully know what need may really exist for new methods of quickly identifying suspected persons. If there be any real need, I should think that this apparatus, which is contrived for other purposes, might after obvious modifications supply it."

Galton then returns to the measurements of the profile and indicates those he would propose to take. These measurements he then suggests should be used with a "mechanical selector." He considers that measurements on the profile would be nearly as trustworthy as those on the limbs for approximate identifications, and he states that their values are less highly correlated than those on the limbs[1].

This paper shows that Galton at this time had not fully made up his mind as to the best characters by which to measure or index individuality. He considered that personal characteristics existed in much more minute particulars than in the profile:

"The markings of the iris of the eye are of the above kind. They have never been adequately studied except by the makers of artificial eyes, who recognise thousands of varieties of them. These markings well deserve being photographed from life on an enlarged scale."

Besides the handwriting, Galton refers to the bifurcations and interlacements of the superficial veins, and the shape and convolutions of the external ear, and then turns for the first time, I believe, in published work to the small furrows and intervening ridges on the palmar surfaces of hands and feet. To this matter I shall return when dealing with Galton's work on finger-prints. In the concluding paragraph of his lecture Galton tells us that he was induced to make these researches into individuality and personal identification in order to discover independent features which might be suitable for inquiries into heredity.

"It has long been my hope, though utterly without direct experimental corroboration thus far, that if a considerable number of variable and independent features could be catalogued, it might be possible to trace kinship with considerable certainty. It does not at all follow because a man inherits his main features from some one ancestor, that he may not also inherit a large number of minor and commonly overlooked features from many ancestors. Therefore it is not improbable and worth taking pains to inquire whether each person may not carry visibly about his body undeniable evidence of his parentage and near kinships." (p. 202.)

[1] It seems unnecessary to specify Galton's profile measurements here, for opinions will differ as to the suitability of his axes and choice of points. In the Galton Laboratory, by means of special apparatus we mark the auricular point and 'Frankfurt horizontal plane' on the silhouette. The nasion to the auricular point is then taken as a fundamental axis and as the standard length. We have obtained on this basis mean silhouettes for men and women. I should be inclined to measure certain deviations of the individual profile from the mean profile, when the nasio-auricular lines of both coincide in direction and magnitude, as the indexing characters.

While finger-prints are now an accepted form of evidence in our courts of law—only a few newly-appointed and yet uninstructed magistrates questioning their validity—it is singular that no use has hitherto been made of them in cases of doubtful paternity, only vague impressions as to family likeness being given in evidence or apparently thought of importance.

Another lecture closely allied to that just discussed was the Royal Institution Friday evening discourse on January 27, 1893. It is entitled "The Just-Perceptible Difference"[1]. In this lecture Galton starts with a definition:

"We seem to ourselves to belong to two worlds, which are governed by entirely different laws; the world of feeling and the world of matter—the psychical and the physical—whose mutual relations are the subject of the science of Psycho-physics, in which the just-perceptible difference plays a large part.

It will be explained in the first of the two principal divisions of this lecture that the study of just-perceptible differences leads us not only up to, but beyond, the frontier of the mysterious region of mental operations which are not vivid enough to rise above the threshold of consciousness. It will there be shown how important a part is commonly played by the imagination in producing faint sensations, and how its power on those occasions admits of actual measurement." (p. 13.)

Galton started by referring to Weber's Law and illustrating its action by an ingenious mechanical model. He placed on an axle a wheel, a logarithmic or equiangular spiral (perpendicular to the axis and with its pole at the centre of the axle) and an index-hand marking on a scale the angle turned through by the axle. All these were accurately balanced, so that they could rest in any position of the axle; round the wheel was taken a cord carrying a scale-pan at one end and a counterbalance weight to the pan at the other. Round the spiral was taken a second cord fixed at one end to the axle and carrying a ball at the other. If now a weight be put into the scale-pan, the axle will rotate until the increasing ray of the spiral provides leverage enough to balance the weight in the scale-pan. The weight in the scale-pan measuring the 'stimulus, the angle turned through by the index-hand measures the sensation[2]. Galton demonstrated on the model that as the stimulus grew large the increases of sensation were very small.

"The progressive increase in the effective length of the logarithmic arm is small at first, but is seen soon to augment rapidly, and then to become extravagant. We thus gain a vivid insight through this piece of mechanism into the enormous increase of stimulus, when it is already large, that is required to produce a fresh increment of sensation, and how soon the time must arrive when the organ of sense, like the machine, will break down under the strain rather than admit of being goaded farther.

The result of all this is, that although the senses may perceive very small stimuli, and can endure very large ones without suffering damage, the number of units in the scale of sensation is comparatively small. The hugest increase of good fortune will not make a man who was already well off many degrees happier than before; the utmost torture that can be applied to him will not give much greater pain than he has already suffered. The experience of a life that

[1] *Proc. Royal Institution*, Vol. XIV, pp. 13–26; *Nature*, Vol. XLVII, pp. 319–21, 342–45.

[2] If $b$ be the effective radius of the wheel, $w$ the weight in scale-pan, $W$ the weight of the ball, $\phi$ the angle of the spiral, $\theta$ the angle of the scale and $a$ the linear constant of the spiral, so that its equation is $r = a\,e^{\theta \tan\phi}$, then the principle of moments gives us $w \times b = W \times r \sin\phi$, whence $\theta = \cot\phi \log\frac{r}{a} = \cot\phi \log\frac{b \times w}{aW\sin\phi}$, which is Weber's Law if $\theta$ be read as sensation and $w$ as stimulus. It is easy to modify the mechanism to take account of the 'threshold.'

we call uneventful usually includes a large share of the utmost possible range of human pleasures and human pains. Thus the physiological law which is expressed by Weber's formula is a great leveller, by preventing the diversities of fortune from creating by any means so great a diversity in human happiness." (pp. 14–15.)

Galton notes how the threshold of sensation differs in different persons and how delicacy of perception is a criterion of a superior nature. It may be modified in the same person by health and disease, by drugs and hypnotism.

He notes, however, that external causes of stimulation may be reinforced by internal causes, and that external stimuli which would fail to exceed the threshold may by aid of the imagination be magnified to the production of a just-perceptible sensation. As illustration of this Galton quoted a personal experience which certainly deserves record in a biography for it indicates how Galton worked "habitually searching for the causes and meaning of everything that occurred to him[1]." After citing Wordsworth and Tennyson as cases in which the force of imagination could master their sense of the present real, Galton notes that his own deafness prevented him when seated in the middle rows at a scientific meeting from following memoirs read in tones suitable to the audience at large. He could, however, distinguish the words of the speaker if he had the unrevised proof of the memoir before his eyes. If the speaker used words not in the proof, he failed to catch them, and if he raised his eyes from the proof nothing whatever of the reading could be understood, the overtones by which words are distinguished being too faint to be understood. He found that he had to approach the speaker by one quarter of his distance from him to follow him without the aid of seeing the words. The loudness of the overtones at the two distances would be as 9 to 16, and Galton concluded that his auditory imagination is to that of a just perceptible sound as 16 minus 9 or 7 is to 16.

"So the effect of the imagination in this case reaches nearly half-way to the level of consciousness. If it were a little more than twice as strong it would be able by itself to produce an effect indistinguishable from a real sound." (p. 19.)

He suggests that experiments as to this might be easily made with two copies of the same newspaper, a few words being altered here and there in the copy to be read from.

People growing deaf, although they cannot lip-read, appear to interpret sounds better when they watch the lips of the speaker. Spectators at the theatre, e.g. at the French plays, *hear* better if they follow with a "Book of Words."

Whatever may be thought of Galton's explanation—the internal stimulus due to the imagination—we must recognise that he discovered a most interesting psychological problem in an experience which the bulk of men would never have thought of analysing[2].

The next part of Galton's paper deals with optical continuity and the just-perceptible distance between two dots. The ordinary eye is just able or just unable to see two dots about a minute of angle asunder. Taking

[1] Charles Darwin: see our p. 1.

[2] A somewhat similar experience occurs in deciphering very bad handwriting; we find it impossible to read the words, until we take to imagining what the writer is likely to be talking about, and with this assistance the eye can often realise what the hieroglyphics stand for.

ordinary reading distance as 12 inches, a row of five dots each $\frac{1}{300}$th of an inch in diameter arranged on the page of a book would be like an almost invisible fine and continuous line. A row of 300 dots to the inch will look at a foot like a continuous line, but far fewer dots are interpreted by the imagination as a line. The ordinary cyclostyle works by dotting and has about 140 dots to the inch; the usual half-tone engraving is produced also by dots, but without a lens the illustration appears continuous in its tones and shading. Galton found that with only 50 dots to the inch he could reproduce a profile which many persons to whom it was shown failed to discriminate from an ordinary woodcut. 250 to 350 points gave exceedingly well the profile of a Greek girl copied from a gem[1].

Taking his points at equal distances Galton found that the direction from one point to the next could be in most cases adequately given by the points of the compass, the top of the paper being treated as north. He takes the letter *a* to represent north, *b* for north-north-east, *c* for north-east and so on in order up to *p*. This presumed, it is possible to represent any profile by a formula. Letters beyond *a* to *p* give points of reference or mark by a sort of bracket points not to be drawn in as when we pass from brow to eye. For convenience Galton breaks up his directional letters into words of five letters each. Thus the profile of the Greek girl involved about 400 letters or 80 words, and might have been sent by telegram. In 1893 it would have cost about £8 to cable it across the Atlantic. Galton illustrated by examples the accuracy with which such portraits, maps or plans could be reproduced. In a postscript added to the printed lecture he gives a coordinate system which allows of somewhat greater exactness, but it requires two numbers to each direction; at the same time it allows variety in the length of the steps.

The whole paper is very characteristic of its author; it leads us from psychological theory to a practical end, the sending of portraits by telegraph; but beneath the whole we find Galton really working at the idea of inherited resemblance as measured by the degree of likeness in the formulae for the profiles of relatives.

We have noted in our first volume that the Galton family was portrayed in a considerable number of very characteristic silhouettes. When Francis Galton turned to the problems of quantitatively measuring resemblance and of indexing portraits, he was compelled by the nature of his subject to deal chiefly with profiles, and from this standpoint he recognised the great value of the silhouette. No doubt thinking of his own family portraits, he addressed two letters to the editor of *The Photographic News*[2]. Silhouettes, he tells the readers of that journal,

"were very familiar to those who lived in the pre-photographic period. They were quickly cut out of paper by a deft hand with a small keen pair of scissors, and at least one of the many operators in this way ranked as an artist capable of making excellent likenesses[3]. The paper

[1] This profile, about 12 inches high, was in the *Galtoniana*, and probably still is, but could not be found recently for reproduction here.

[2] July 15th and July 22nd, 1887.

[3] No doubt Edouard, who did the Galton and Darwin families. See our Vol. I, Plates IV, XVII, and XXXIV.

was black on one side, and the silhouette that had been cut out was pasted then and there, with the black side upwards, upon a white card, and framed. A perfectly durable and often a good likeness was thereby produced in a very short time. This art was superseded by photography, and is now temporarily extinct; but I want to show that it might with great facility—and I think with some profit in a humble way—be advantageously re-introduced by the help of the very agency that extinguished it."

Galton next suggests photographing the profile of a sitter, either in a strong light against a dark background, or *vice versâ*, and then taking a print of this result, cutting out the profile and blackening it[1]. In his second letter Galton gives an example of a silhouette prepared in this way. Such silhouettes are, he says,

"particularly useful in studying family characteristics which, I think, are on the average far better observed in profiles than in any other single view of the features. The truth of this statement may be verified in church, where whole families, each occupying a pew, can often be seen sideways, and each family can be taken in and its members compared at a single glance.

Galton's photographic silhouettes of himself, aged 65.

The instances will be found numerous in which the profiles of a family are curiously similar, especially those of the mother and her daughters. This is most noticeable where their ages and bodily shapes differ greatly, as when the daughters are partly children and partly slim girls, and the mother is not slim at all."

It must be admitted that Galton went to church rather for scientific than religious purposes; but the reader of this passage will hardly be inclined to accept Dr Beddoes' statement that Galton was wanting in a sense of humour! See Vol. I, p. 59.

Another photographic problem which occupied a good deal of Galton's thoughts at one time was the problem of keeping the object and the focal plane at the conjugate foci of the optical centre of the object-glass. This

[1] If the sitter be placed in front of a window, a half plate will give a silhouette of about four inches high, which is often a very characteristic portrait. The chief need is the 'deft hand' in cutting out the print and avoiding angles. In the Galton Laboratory we have a silhouetting arrangement in which the sitter's head is adjusted to the 'Frankfurt horizontal plane,' and the shadow is cast by an arc light some fifty yards away. The shadow is traced by an artist's hand and the resulting silhouette preserved with the anthropometric records of the subject. It is used for measurements and by compounding series of subjects to obtain type profiles.

is a very important point in the case of reduction to an exact size. Galton's papers show a large number of attempts at its solution. He ultimately sought the aid of Mr (now Sir) Horace Darwin, who in 1878 published in *Nature*[1] a satisfactory theoretical solution by aid of a double Peaucellier's cell. Galton found, however, that the cells would have to be of unwieldy size, if these arrangements were used. I am not aware that the problem has even yet been solved practically, although for scientific photography its solution is very important.

## (*C*) *Analytical Photography.*

At the same time that Galton was working out his idea of composite portraiture a new problem occurred to him, that of creating what he termed a "transformer" which would transform the type into any individual component. The transformer would thus be a measure of the difference between individual and type, or indeed between any two individuals. He proposed by this method to analyse the differences between types (or races), between individuals (or between an individual and his family type), or between an individual on different occasions. Galton termed the production and study of transformers *Analytical Photography.* The idea appears first to have occurred to him in 1881; but not till 1900 did he write a letter, which appeared in *Nature*[2], August 2, on the subject, stating the outlines of the process, and speaking somewhat doubtfully of his own power of carrying it out. In this letter, after describing the theory[3] and something of the technique, he writes:

"I photographed two faces, each in two expressions, the one glum and the other smiling broadly. I could turn the glum face into the smiling one, or *vice versâ*, by means of the suitable transformer; but the transformers were ghastly to look at, and did not at all give the impression of a detached smile or of a detached glumness."

Later Galton realised that transformers were hieroglyphics which required a key to their interpretation; the photograph of a "smile" is really the photograph of facial modifications which failing the stable basis of the face we do not recognise as a smile at all. I owe to Mr Egon S. Pearson the photographs on p. 312. *A* is the normal, *B* the smiling subject. *C* and *D* are the transformers. *D* is the "glumness" and *C* the "photograph of a smile." All that can be said of the latter is that it does not closely correspond with John Tenniel's conception of the grin which remained some time after the rest of the Cheshire cat had vanished[4].

[1] Vol. XVIII, p. 383. [2] Vol. LXII, p. 320.

[3] If $x$ be the transformer, Galton lays down two equations

$$\text{(i) } pos.\ a + neg.\ a = \text{grey}, \quad \text{and} \quad \text{(ii) } pos.\ a + x = pos.\ b,$$

whence he deduces

$$\text{(iii) } pos.\ a + \{pos.\ b + neg.\ a\} = pos.\ b + \text{grey}, \qquad \text{(iv) } pos.\ b + \{pos.\ a + neg.\ b\} = pos.\ a + \text{grey}.$$

Thus the quantities in curled brackets are the transformers, one the negative of the other (the "smile" and the "glumness").

[4] *Alice's Adventures in Wonderland*, Edn. 1872, p. 93.

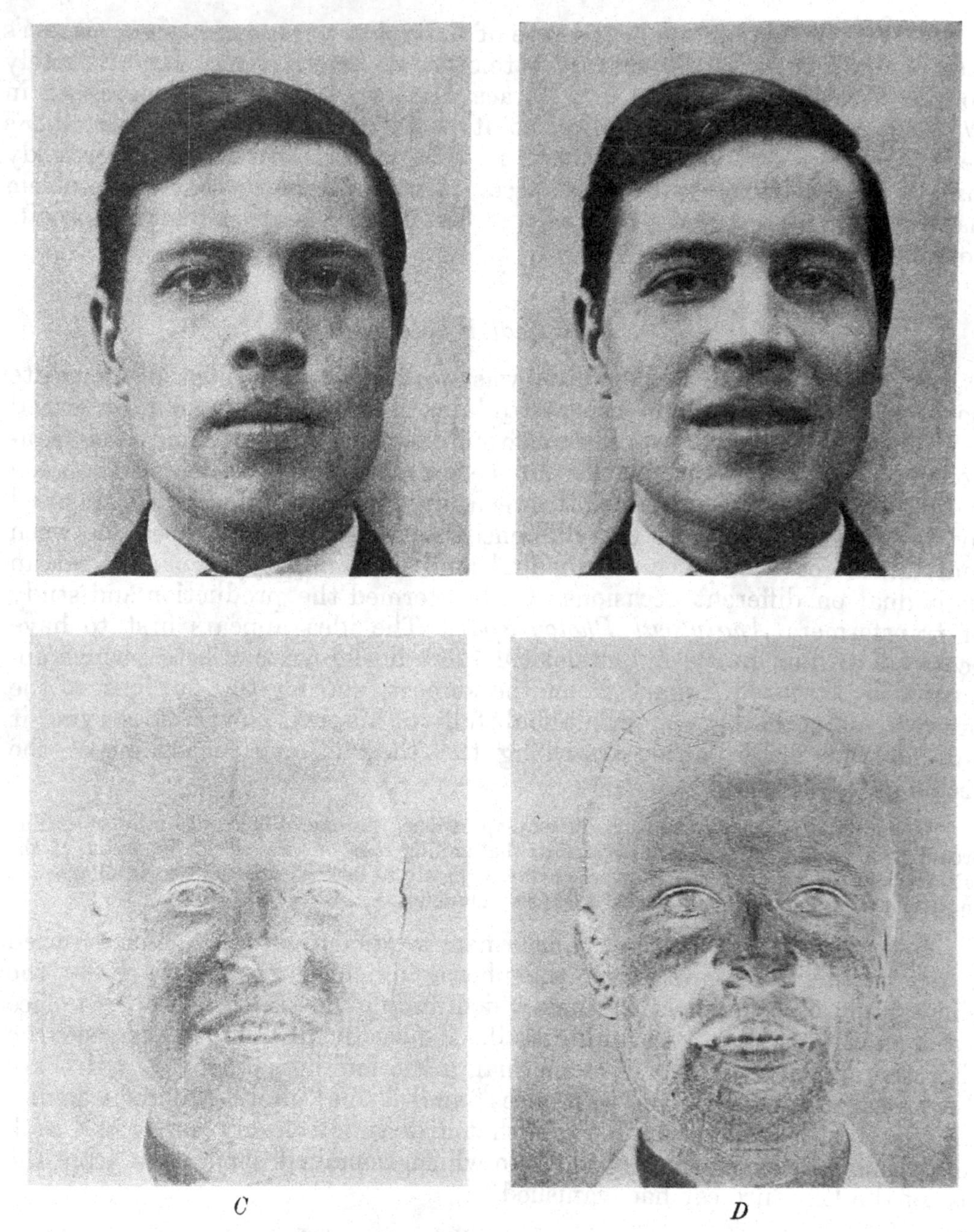

*C* *D*

By November 27, 1900, Galton had devised a simple small apparatus[1]

[1] This is now in the Galton Laboratory. It consists essentially of a triple camera; one object is thrown directly on to the focal plane; the other two by aid of two reflecting prisms are also thrown on to the same plane. The three objects thrown on to the plane are positive *a*, negative *a* and positive *b*. By throwing out one or two of the three we can throw on to the screen (i) positive *a*, (ii) negative *a*, (iii) positive *b*, (iv) positive *a* + negative *a* to show the uniform grey, (v) negative *a* + positive *b*, the transformer, or (vi) positive *a* + the transformer, giving the darkened *b*.

with the aid of Mr T. R. Dallmeyer for carrying out his project, and this was exhibited on the date just mentioned to the Royal Photographic Society; Galton's paper is printed in *The Photographic Journal*, Vol. xxv, pp. 135–38. The idea at the bottom of Analytical Photography is extremely simple, as most of Galton's methods. A subject $A$ and a subject $B$, taken in similar positions and of similar size, have faint transparent positives and faint transparent negatives taken of each. If now positive $A$ and negative $A$ be thrown accurately adjusted on the same screen, they will antagonise each other and give a uniform grey background. If further positive $A$ and negative $B$ be thrown on the same screen, they will only antagonise one another where the originals are identical; where they are different, they will only in part antagonise each other. Thus the combination of positive $A$ and negative $B$ gives a representation of their difference on a grey ground. This Galton calls the "transformer." If the transformer be thrown on the screen with positive $B$, it converts positive $B$ into positive $A$. Similarly negative $A$ and positive $B$ is the transformer, which superposed on positive $A$, converts it into $B$. The two transformers are in fact positive and negative of the same difference. In both cases the transformed portrait is that of a darkened subject. The fact that our combination of faint positive and faint negative gives a uniform *grey* or half tone is very important; because it follows that where our transformer adds nothing in the way of difference to $A$ to make $B$, it will still add everywhere this grey or half tone. The transformed $B$ will therefore be a *darkened* picture of $A$.

Galton illustrated this point by obtaining a 'real' scale of tints. He took nine teetotums: the first had a white surface, the second a sector of 45° painted black, the third two sectors of 45° black, the fourth three sectors and

Diagram v. Galton's photograph of a spinning wheel of tints.

so on up to the ninth which was all black. On spinning these nine teetotums, he obtained a 'real' scale of tones from white to black[1]. Having thus obtained a scale of nine tones from white to black, Galton terms the fifth of these (180° painted black) the medium tone. Pictures painted with tones less than, but up to and including, the medium tone he calls 'faint'; pictures painted in tones from the medium to the black, he calls 'dark'. He then caused three portraits of a lady to be painted with these tones:

(*a*) is the normal painting, using all the tones from 0 to 8, i.e. white to black;

(*b*) is the faint painting, using all the tones from 0 to 4, i.e. white to medium grey;

(*c*) is the dark painting, using all the tones from 4 to 8, i.e. medium grey to black.

On Plate XLV will be seen Galton's scale and the portraits. Of these he writes:

"I exhibit three sketches of the same portrait to show the differences of effect under these conditions, and how very little the mere question of more or less likeness is affected by them. All the tones from 0 to 8 were used in painting the first picture. Then a grey mixture that matched the medium grey was made in one corner of a palette and pure white squeezed out in another. The artist by using mixtures of this grey and white, and nothing else, made the second picture as a copy of the first. It is evident that its resemblance is not affected by the limitation of the range of tones. The third picture was made on the same principle as the second, except that black and medium grey were employed instead of white and medium grey, and here again the resemblance to the original is perfect. It follows that the value of the analytical process is not much affected by the fact that it is unable to transform, in other words that it cannot produce a transformer, or in still other words that it cannot isolate the differences between *any* two portraits, but only those between a light half-toned copy of the one and a dark half-toned copy of the other. It should be remarked that although the light-toned *a* and the dark-toned *b* severally contain one-half of the complete scale of tones, yet the transformer of the light-toned *a* into the dark-toned *b* contains the complete scale." (pp. 136–7).

Galton illustrates the whole process well by showing the steps taken to convert any mosaic of four tones into another mosaic of four different tones. He takes tones 6, 4, 2, 2 for *A* and 4, 6, 2, 6 for *B*: see our scale Plate.

| | | | | | |
|---|---|---|---|---|---|
| (1) | *A* (Original full-toned portrait) ... ... | 6 | 4 | 2 | 2 |
| (2) | *B* " " " ... ... | 4 | 6 | 2 | 6 |
| (3) | *pos. a* (faint half-tone), i.e. ($\frac{1}{2}A$) ... ... | 3 | 2 | 1 | 1 |
| (4) | *neg. a* (faint half-tone), i.e. ($4 - \frac{1}{2}A$) ... | 1 | 2 | 3 | 3 |
| (5) | *pos. b* (faint half-tone) ... ... | 2 | 3 | 1 | 3 |
| (6) | darkened *pos. b* (i.e. *pos. b* + 4) ... ... | 6 | 7 | 5 | 7 |
| (7) | *pos. a* + *neg. a* (uniform grey) ... ... | 4 | 4 | 4 | 4 |
| (8) | *neg. a* + *pos. b* (the "transformer") ... ... | 3 | 5 | 4 | 6 |
| (9) | *pos. a* + (*neg. a* + *pos. b*), i.e. (3) + (8) ... (which is the same as the darkened *b*, or (6)) | 6 | 7 | 5 | 7 |

The greatest difficulty in the above process is to ensure that positive *b*

[1] Galton carefully distinguishes between the scales of 'real' tones perceived or sense tones and actinic tones. (p. 137.)

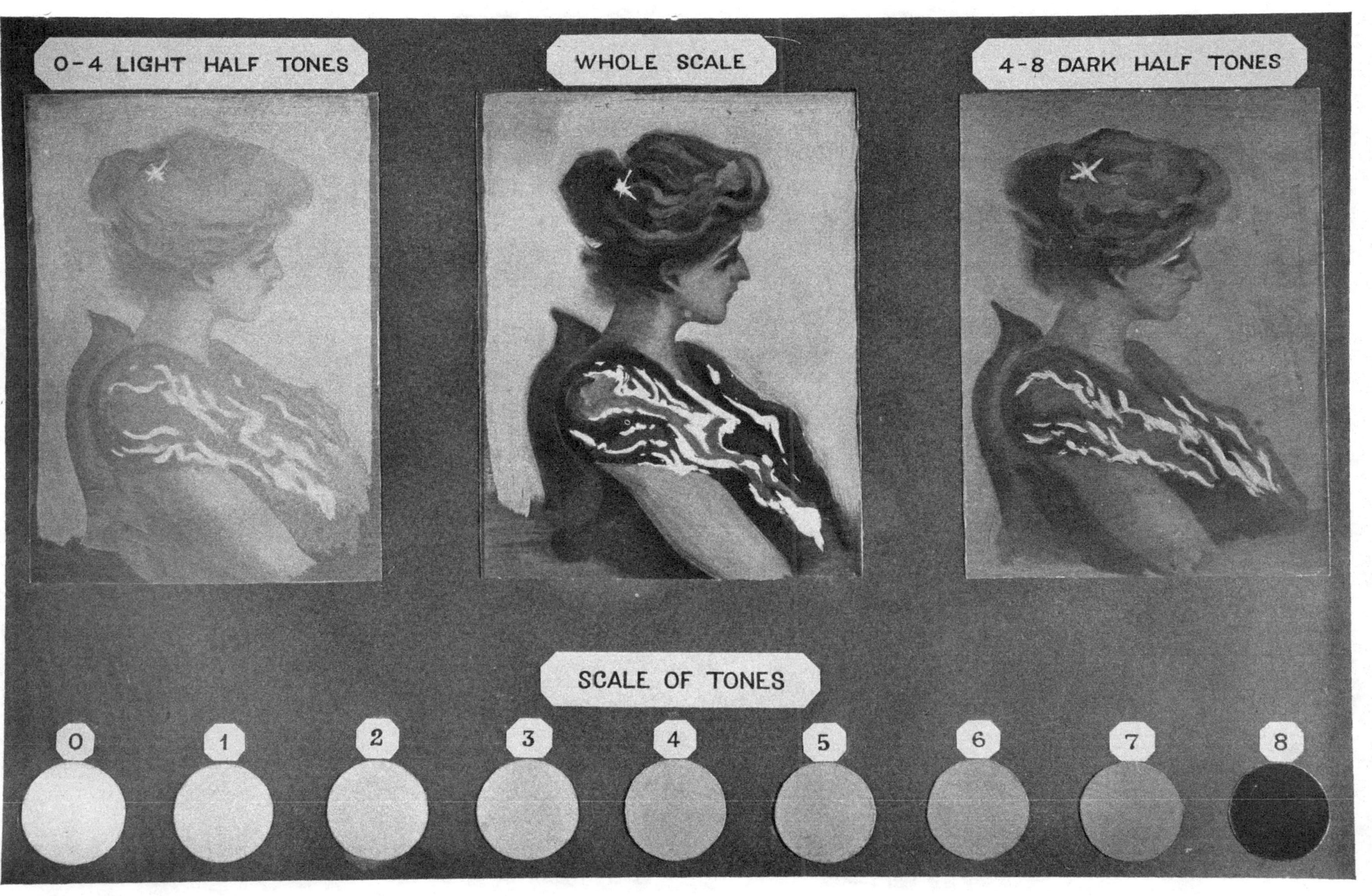


Analytical Photography.
Oil paintings of a portrait in various tones.

and negative *a* have the same tone, so that if *b* were *a*, the composite would give merely medium grey. It is needful also that *pos. a* and *neg. a* should accurately antagonise one another. Our figure shows one of the illustrations Galton gave at the lecture, namely the "transformer" which will change an **F** into a **G**.

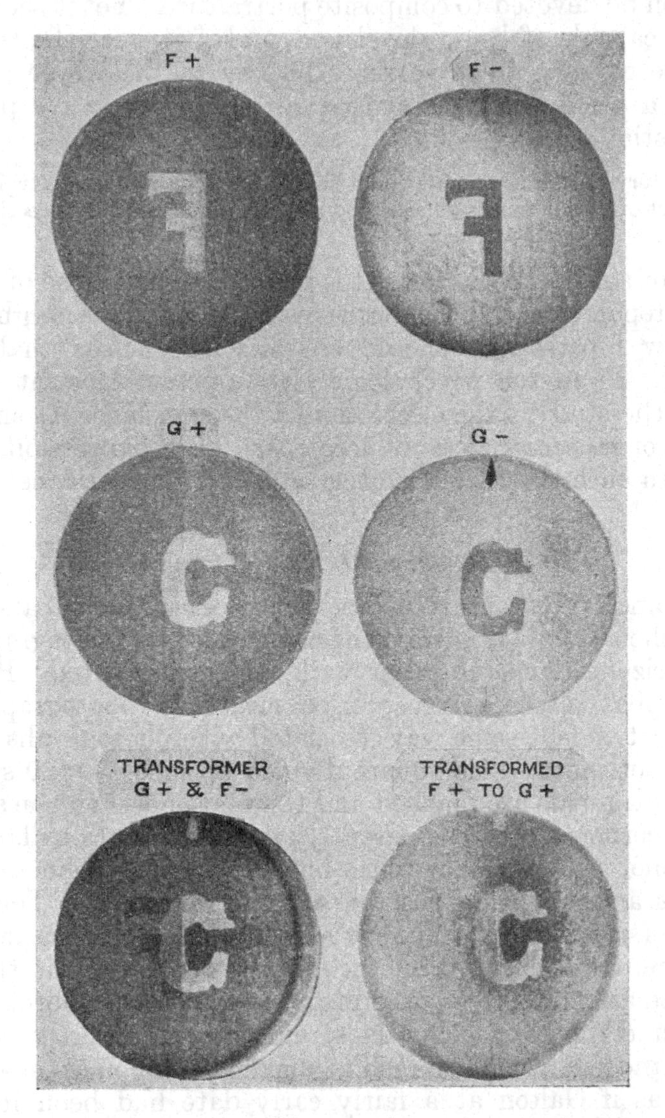


**Diagram vi.**

Galton also transformed a mosaic St George's Cross into a St Andrew's Cross (slides in *Galtoniana*). It may be said of these results that they were only partially successful—this is the view taken by Sir George Darwin in his account of Galton in the *Dictionary of National Biography* (*Second Suppl.* Vol. II, p. 72). But Galton recognised the main difficulty himself:

"But negatives and positives do not wholly obliterate one another. They do so to all intents and purposes when the tones are not very far from the middle of the scale; an extreme white is not obliterated by its negative." (p. 137.)

Galton was 78 years of age when the paper was published, and it was hurriedly written (p. 138). He never worked out the technique with the care and elaboration he devoted to composite portraiture. Yet it seems to me that the method is capable of being developed, and if it were the results reached by it would be of very great value. The key to the whole position is the production of a perfectly obliterating positive and negative pair. Galton's original suggestion that

"the only satisfactory experiments now would be those made by two converging lanterns on a screen, one at least of which admits of easy and delicate adjustment in direction and intensity of its illumination"[1]

might still be of aid in the matter. It is possible that the use of homogeneous light in the preparation of both positives and negatives might be of some value. Anyhow I personally should be sorry to dismiss analytical photography as idle. From the psychological standpoint it ought to be of first class value in the study of the expression of the emotions. It should indicate what physical or muscular changes accompany such expression. The subject needs to-day an enthusiastic cultivator, who has the patience to develop its technique.

### (*D*) *Measurements by Photography.*

In 1896 Francis Galton started another inquiry. He appears first to have developed a scheme for taking from the same spot two photographs, one with the camera horizontal and the other with the camera tilted. By aid of two such photographs distances were to be measured photographically. The method of reduction is in some way associated with photographs (positive and negative) of a horizontal ruled square divided up into 20 × 20 small squares. The two chief diagonals are marked and there are posts on the square at the centre and the corners. It was apparently capable of rotation about the centre and there are photographs of it with the optical axis of the camera in one of the diagonal planes, and in a plane bisecting the diagonal planes. There are photographs of streets, roofs and chimneys seen from a high parapet, possibly of a bridge, apparently to be used in testing the method. But the manner in which the reticulation was to be used in these cases is very obscure, and so far I have only found the box of negatives, and no explanatory notes or papers, in the *Galtoniana*. It would be rash to make any dogmatic assertion, but it would appear as if Galton at a fairly early date had been independently working at photographic triangulation.

*Photographic Measurement of Distances and Lengths.* No less than eleven notebooks in the *Galtoniana* deal with this topic. They appear to have been started in 1894–1895. They contain not only experimental measurements, but a succession of drafts of papers, changing sometimes by

[1] *Nature*, Vol. LXII, p. 320. Attempts in this direction would have been made before this in the Galton Laboratory, but for inadequacy of funds.

very little the method of procedure, or changing sometimes its applications. There are a good many negatives or prints also referring to the matter, some intelligible, some needing an interpretation, which I have been unable to supply. To one series of such, clearly involving distant objects, I have already drawn attention (see p. 316); a good deal of the matter refers to fairly near objects, and the first experiments—to judge by the photographs—seem to have been made on a series of shelves or racks at the Kew Observatory. Then Galton photographed a bronze horse (Carousel) and determined the three coordinates of eight to eleven points of it in a variety of ways. It is really wonderful thirty years later to see the amount of labour he put into work of this kind. A reader of *Nature* might conclude that his communications to it were brilliant suggestions written in a few hours. This, I believe, was never the case; he rarely refers to a tithe of his experimental work, the calculations, trials and failures he had made before preparing his paper; in many cases a paper was written over and over again before it assumed its final form, and if a reader of the latter thinks the result could have been more easily reached by another method, it is extremely probable that that method could be found, experimentally tested and silently rejected, in one or other of Galton's preliminary notebook records. If he tried and condemned a method, he scarcely ever stated that he had done so. He assumed that his readers would suppose him to have surveyed the country before plotting the selected path to his goal.

Galton started with the general problem of studying the perspective of a photograph; he did this by the simple method of photographing with his subject some horizontal reticulation, or if needful both a horizontal and a vertical reticulation, and this served as the basis for analysing the perspective properties of the photograph. Galton shows that photographic measurements of objects may be divided into two classes; those in which we measure lengths parallel to the focal plane of the camera, and those in which we measure other lines, and in this case we may require two photographs of the same object taken simultaneously from different aspects. The mathematics of the latter are by no means complicated, and are provided by Galton, but his dominant passion for the study of heredity soon led him to the measurement of animals, and by proper orientation of the animal the principal measurements Galton was seeking can be obtained from a single standardised photograph, provided it is accompanied by suitable reticulations or fiducial lines.

In one of his many notebooks I find the draft of a paper which starts thus:

"Architectural draughtsmen are familiar with the art of translating objects into their perspective representations, but the converse process of translating perspectives into their objective equivalents has never, I believe, been yet brought into practice[1]. So long as pictures had to be

[1] It was not an uncommon problem before even 1890 to ask engineering students to draw a model in perspective and then take the measurements of parts of the original model from the perspective drawing. It must be confessed, however, that it was done with the view of testing drawing accuracy and possibly suggesting the superiority of plan and elevation drawings. It would certainly have been good experience to have obtained measurements of the parts of machines by double photographs of them accompanied by suitable reticulations.

drawn by hand and therefore inexactly, there was no inducement to consider the possibilities of this converse process, for which exactitude in the picture is essential to success, but now that photography has become common the old difficulty has disappeared, and the possibilities of the neglected process well deserve consideration. The applications would be numerous and especially valuable in determining and measuring restless animals in their momentary attitudes, or even when in rapid motion, which could not otherwise be measured without difficulty, or when in rapid motion be otherwise measured at all. The object I have especially in view is to establish a system of measuring a large number of domestic animals of various pedigree stocks, whether horses, cattle, sheep, dogs, poultry etc., in order to provide material to advance our knowledge of heredity of a kind that is greatly needed. It is not qualitative facts and exceptional instances that are now wanted by students of heredity, but a large collection of quantitative facts in the form of trustworthy measurements. They are needed to determine with far greater precision than they are at present known the statistical laws and coefficients of heredity. Among these are the conditions and rate of 'regression' of the offspring of exceptional parents; the gradual or sudden alterations of position of the point towards which regression tends, as the breed becomes more pure; the relative influence of the male and female parent in respect to various measurable peculiarities; the intensities of prepotencies; the frequencies and magnitudes of sudden sports, and the degrees of their subsequent stability through successive generations....I should add that the direct measurement of creatures so sensitive, timid and sudden in their actions as thoroughbred horses, who at the same time are often vicious, is difficult and dangerous; similarly as regards bulls and some of the breeds of dogs. Photography is a simpler, more exact and far safer method of measurement in these cases than the direct application of rod, tape and callipers."

We shall consider later Galton's method of determining lengths parallel to or nearly parallel to the focal plane of the camera. This he has published (see our p. 320). His two-camera method of determining the three coordinates in space of any point of a subject has not, as far as I know, been published and deserves a paragraph here.

Diagram vii, figs. 1–5, is taken from Galton's manuscript. Fig. 3 represents the plan on a working scale; $M_1$, $M_2$ are the plans of the optical centres of the two camera lenses; bak, fcd are fiducial lines drawn on the base plane upon which the object stands; b, a and c, d are fiducial points, $M_1$a and $M_2$c being the traces of the vertical planes through the optical axes of the two cameras, and these are so arranged that $M_1$a and $M_2$c are accurately at right angles to bk and fd. Figs. 1 and 2 represent the two photographs, and $p_1$ and $p_2$ the point P in them whose coordinates are to be determined. From $p_1$, $p_2$ perpendiculars $p_1g$ and $p_2f$ are dropped on the images of the fiducial lines *ba* and *cd* in the photographs. But clearly $ag/ba =$ ag/ba and $fc/cd =$ fc/cd. Hence g and f on the plan drawing can be scaled. Produce $M_1$g and $M_2$f to meet in Q, then Q is the plan of the given point P. If S be the perpendicular from Q on $M_1$a, we may take SQ and aS for our coordinates $x$ and $y$. Now draw to the same scale an elevation (fig. 4) of the system on the vertical plan through the optical axis; $N_1M_1$ is the height of the optical centre, ST is the elevation of P. Since Sa is known, by joining S to $N_1$ we obtain s. If the elevation of P, or $z$, be TS, we require to determine ts, for knowing it we have TS/ts = $M_1$S/$M_1$a. But ta is $p_1g$ the apparent height in the first photograph altered in the ratio of ab to *ab*. Fig. 5 illustrates this clearly. Of course we must settle the scale for the drawing-board ab by the value of the fiduciary distance ab in the reticulation on the base plane of the object photographed. Such is Galton's very simple process of taking measurements on photographs of an object. It can indeed be achieved if the object be inanimate by placing it on a reticulated turntable and rotating this turntable through 45° or 90° to obtain the second photograph with the same camera. Of course if we can find the coordinates of one point we can find those of any number, and the

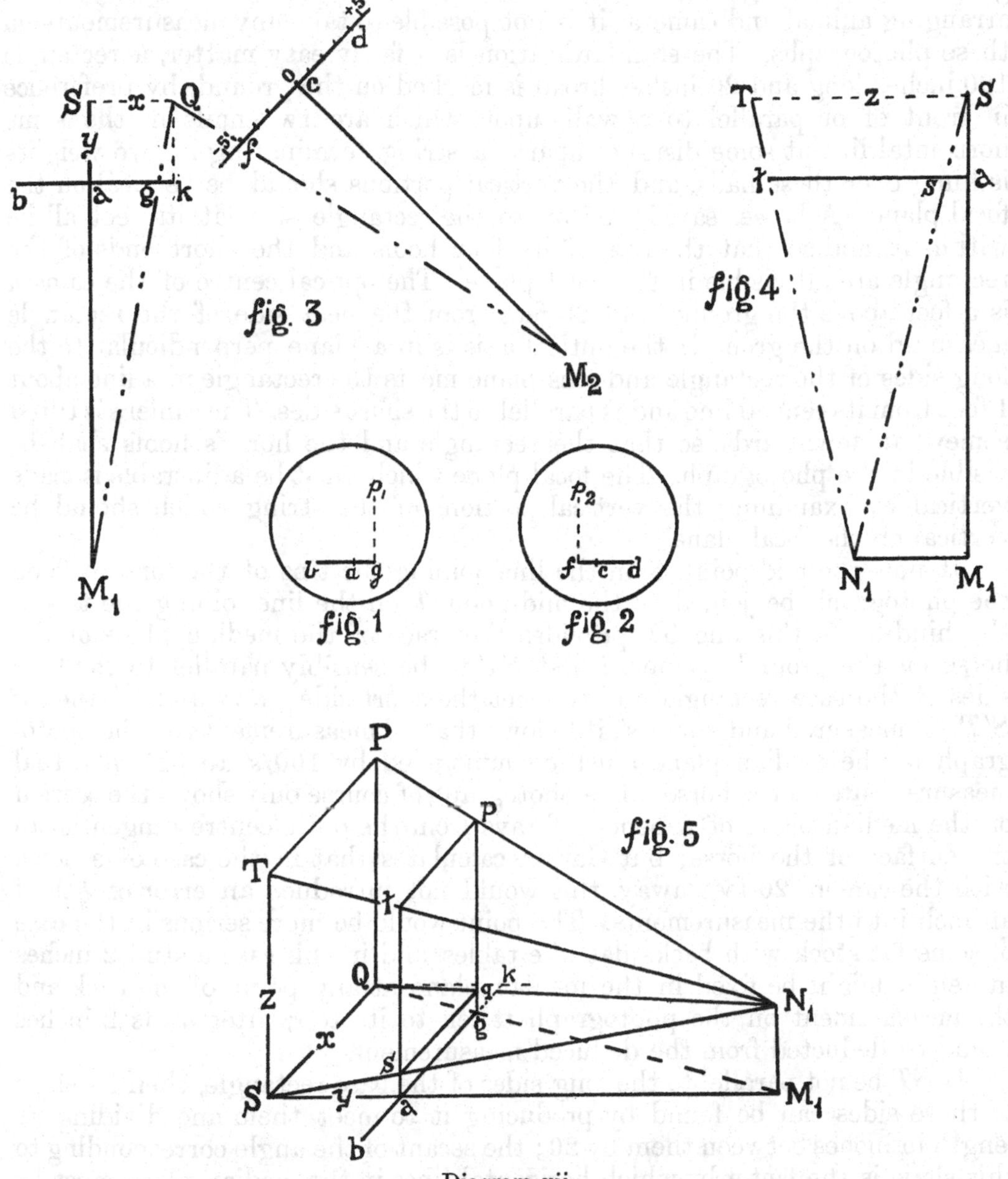


Diagram vii.

distances between them can then be found in the usual way. Another method is to find a whole series of points T and S or construct plan and elevation drawings of the object from the two photographs.

A paper by Galton entitled "Photographic measurement of Horses and other animals" was published in *Nature* on January 6, 1898[1]. It belongs perhaps more closely to our chapters on biometry and heredity, but I have included it here as concerned chiefly with photographic technique. Galton points out how frequently valuable horses and other show animals are photographed, but owing to the fact that there is no standardised method of arranging animal and camera, it is not possible to take any measurements on these photographs. The standardisation is a fairly easy matter, a rectangle 100 inches long and 20 inches broad is marked on the ground, by preference in front of or parallel to a wall, upon which are two nails in the same horizontal line at some distance apart; a string terminating in two weights is hung over these nails, and the vertical portions should be vertical on the focal plane. A horse, say, is led on to the rectangle so that its feet all lie within it, and so that the tips of its four hoofs and the short ends of the rectangle are all visible in the focal plane. The optical centre of the camera is 5 feet above the ground and 20 feet from the near side of the rectangle measured on the ground; the optical axis is in a plane perpendicular to the long sides of the rectangle and this plane meets the rectangle in a line about 1 foot from its central line and is parallel to the short sides. The camera is tilted somewhat downwards, so that the rectangle and the horse's hoofs shall be visible in the photograph. The focal plane which must be adjustable is made vertical by examining the vertical portions of the string which should be vertical on the focal plane.

If now the mid-point $S$ on the line joining the tips of the fore-hoofs on the photograph be joined to the mid-point $T$ on the line joining the tips of the hind-hoofs, this line $ST$ provides the trace of the median plane of the horse on the ground. Suppose first $ST$ to be sensibly parallel to the long sides of the base rectangle and to meet the short sides in $S'$ and $T'$, then if $S'T'$ be measured and equal $s'$, it follows that all measurements on the photograph in the median plane must be multiplied by $100/s'$ to obtain actual measurements on the horse. The photograph of course only shows the section on the median plane of a pencil of rays from the optic centre tangential to the surface of the horse; but Galton calculates that in the case of a horse with the camera 20 feet away, this would not introduce an error of $\frac{1}{8}$th of an inch into the measurements. The point would be more serious in the case of some fat stock with backs flat like tables, and in this case a stud 2 inches in height might be fixed in the median plane at any point of the back and the measurement on the photograph taken to its top; afterwards 2 inches would be deducted from the deduced measurement.

If $ST$ be not parallel to the long sides of the base rectangle, then its slope to those sides can be found by producing it to meet them and dividing its length in inches between them by 20; the secant of the angle corresponding to this slope is the factor by which horizontal lines in the median plane must be multiplied in order to obtain their true value. Galton also indicates how lines not parallel to the median plane can be obtained as from shoulder to haunch bone (p. 232), but a discussion would carry us into too great detail.

[1] Vol. LVII, pp. 230–32.

At the British Association in 1898 Galton applied for and obtained the appointment of a committee consisting of Professor E. B. Poulton, Professor W. F. R. Weldon and himself "to promote the systematic collection of Photographic and other Records of Pedigree Stock." This Committee made a Report to the Association meeting at Dover in 1899 and it was published in the *B. A. Report* for that year (pp. 424–29). The report emphasises the fact that while it is possible in the various Stud-books and Herd-books to trace the ancestry of pedigree stock, these works

"afford scant means for obtaining that distinct presentment of each of the nearer ancestry which is needed for an exact study of the Art of Breeding." (p. 424.)

Information is almost entirely confined to colour, or in the case of horses to height at the withers. While photographs exist it is very difficult indeed to obtain those of sire, dam and produce as adult—what Galton terms a genealogical *triad*—and groups including the grandparents, even in the case of pure-bred shorthorn cattle, are practically unattainable[1]. The reason Galton finds is not far to seek:

"Heredity is a comparatively new science and few people are as yet acquainted with the character of the records most suitable for its study, or are sufficiently impressed with the need for their exactness and persistence. The most important of these records which it seems feasible to obtain are photographs, not merely pretty and well worked-up productions satisfactory to an artistic eye, but rather such as are analogous to the portraits made of criminals, for storage at the central police office, to serve as future means of identification. The desired photographs need to be taken under such conditions as shall ensure their being comparable under equal terms and shall admit of the accurate translation of measurements made upon them into corresponding measurements made on the animals themselves." (p. 424.)

The report then describes the *Standard Conditions*. These are modified considerably from those given in the *Nature* paper. There is to be a solid wall or screen painted blue, a solid pathway in front of it of 6 feet width of light-coloured bricks to show the horse's hoofs up in the photograph. Two lines are drawn on the pathway, one two feet from the wall, and the other two feet from the first; the edge of the path towards the camera is to show in the photograph as a sharp line. On the wall are to be small marks or studs each about the size of a sixpence, arranged in three vertical and three horizontal lines each at an exact distance of 3 feet apart, the bottom row being at a foot above the path level; the camera is to be 30 feet from the wall, and its optical centre at a height of 5 feet.

"The equivalent focus of the lens should not be less than 9 inches, otherwise the photograph will be too small for convenient measurement; the lens used in the experiment [at the Royal Agricultural Hall in March 1899] was of 13 inches focus, with plates of $6\frac{1}{2} \times 4\frac{3}{4}$ inches, and proved exactly suitable." (p. 426.)

The verticality of the focal plane and its parallelism to the wall are ascertained by the squareness of the stud-network of the wall on the focal plane screen. The camera once adjusted is to remain undisturbed during a whole series of operations. Prominent points on the horse or on cattle may be marked by

[1] What Galton wrote in 1899 remains equally true a quarter of a century later.

white wafers attached by paste[1]. Many detailed suggestions for taking the photographs are given. Galton indicates as in his earlier paper that if all the animal's hoofs fall in the middle two feet of the path, a slight obliquity to the wall will not introduce an error of any importance into the photographic measurements (see our p. 320). Galton's procedure is now somewhat simpler than that which he gave in *Nature*. He projects from the optical centre on to the wall behind the horse, takes the measurement there and reduces in the ratio of distance from that centre to median plane of horse to distance of that centre from the wall. He now considers it adequate instead of determining the median plane by bisecting lines between tips of fore- and hind-hoofs, to take the median plane 6 inches behind the line joining corresponding fore- and hind-hoofs. He gives reasons for believing that the errors of measurements made on the photographs are less than those made by different persons on the same animal.

In the course of the paper he refers to standard photographs made of 28 premium stallions at the Royal Agricultural Hall and to photographs of 31 triads made on pure-bred shorthorn cattle, chiefly at Alnwick Park. The Royal Commission on Horse-Breeding was asked to permit a trial installation at their show at the Royal Agricultural Hall in 1898, and permission was cordially granted. The installation was made and 35 horses were photographed in 3½ hours. The total cost including that of two veterinary measurers was under £25, and Galton believed that if the operations became customary, the cost would be paid by the sales of copies of the standard photographs.

Galton's Report occupies pp. 12–16 (with plate of figures) of the *7th Annual Report of the Royal Commission on Horse-Breeding* (C. 9487, 1899). He gives some strong arguments for standardised photography and describes again in detail the standardised methods, and the divergences between the photographic and veterinary measures. He considers the latter as far more reliable than the former. In the matter of

"length of body, the photographic method is the only one to be depended on, and it seems to be as trustworthy as that of height." (p. 15.)

Putting aside the length[2] we have for the horse "Maroni" for example:

Height in inches.

| | At Withers | At Back | At Croup |
|---|---|---|---|
| From Photograph | 65·1 | 60·7 | 64·6 |
| Veterinary Measurement | 65·5 | 61·0 | 64·5 |

as an illustration of the extent of agreement.

[1] Associated with this section of the material in the *Galtoniana* is a careful drawing of a horse by W. F. R. Weldon with anatomical sketches of the skeleton indicating the points which could be approximated to on the living animal, and suggesting suitable measurements.

[2] "A direct and trustworthy measurement of the length of a vicious or timid horse is extremely difficult, perhaps impracticable." (p. 15.)

"The main result of this experiment has been to prove the feasibility of taking photographs of horses at a Show, that shall be acceptable as ordinary portraits, and will at the same time be of sterling scientific value. I beg in consequence to express a hope that the Royal Commissioners may think fit to arrange that photography under standard conditions shall become a permanent feature of their annual Shows, it being impossible to ensure that those conditions shall be strictly attended to when animals are photographed at their homes, though easy to do so at a public exhibition. The experience gained by this trial...proves how inexpensive, and at the same time how necessary it is to have an appropriate installation, one that might be removed and replaced when desired....If the Royal Commissioners led the way, other societies who exhibit at the Royal Agricultural Hall would doubtless be glad to follow their example and to avail themselves of the installation. The managers of local exhibitions would in time pursue a like practice, until the custom of utilising exhibitions for the purpose of photographing prize winners under standard conditions became general, and probably more or less self-supporting, and the principal object of the Committee of the British Association...would be attained. Horses and other pedigree animals are usually exhibited more than once, so occasional failures due to bad weather admit of being subsequently rectified." (p. 13.)

The general idea of standard photographs of pedigree stock was a splendid one; the results of the trials seem to have been quite satisfactory, but I can find no trace after 1899 either of further work by the British Association Committee[1] or of further photography at the Agricultural Hall. Galton was already 77 years of age, and this was only one of many inquiries he had in hand. There was need probably of an active and younger man to push the matter to a complete success. This want was not satisfied, and Galton's suggestion did not at that time bring forth fruit. Possibly after another quarter of a century we shall find it successfully carried out in America or in some continental country with a keener appreciation than our own of the value of scientific breeding.

### (*E*) *Indexing and Numeralisation of Portraits.*

For 13 years after 1893 (see our p. 307), Galton published nothing further about methods of indexing and numeralising portraits, but he worked most energetically at them. He divided the profile into parts—forehead, nose, lips, chin. He had discovered—what is soon forced on the craniologist—that the component bones of the skulls of two individuals may be extremely alike, but that great differences may be produced by change in angle at the sutures or joints. Now Galton's collection of profiles was most extensive; it ran to many hundreds. He obtained them from drawings—he took 68 from Dance alone—from photographs, and from engravings of all sorts, and again directly by silhouettes; he proceeded to break up his profiles into component parts. From hundreds of noses or chins, he constructed a mean nose or chin. Then he proceeded to measure deviations from these mean noses or chins, and constructed standard patterns of noses or chins. A new profile might be described as having Forehead No. 3, Nose No. 31, Lips No. 26 and Chin No. 8. The individuality of the profile was thus determined and a means given for indexing it. All that was needed in order to get something of a likeness was to add the angles between the joints and fix the magnification to be given to each component.

[1] No further report was made and the Committee lapsed in 1901. I do not know the reason why; nor have I been able to discover what became of the material collected by the Committee. Professor Poulton knows nothing about it; it is not in the *Galtoniana*, nor among Weldon's papers.

A profile therefore would be specified by four numbers, each of which might be two figures, four magnifications say each of two figures, and three angles say also given by two figures to nearest degree. In other words 22, say 30, symbols in a telegram would suffice; thus *six* words in code would convey a creditable likeness of a man. Galton even supposed in his later days that wireless could be used to communicate to the captain of a liner the profile of a person suspected to be on board, whom it might be desirable to keep a watch upon.

Drawings of profiles of men of different races, Copts, Arabs, Negroes, etc. were taken as well as men of every grade of distinction in our own country; at least three or four artists were employed at different times in the preparation of these profiles. There are endless notebooks, measurements, materials of all kinds, and drafts of papers, I believe, never completed and published. I can trace no sign of discontent with the methods adopted, but it would appear as if Galton was always seeking for something better. He had collected data for a work which would certainly have eclipsed Lavater's, being based on much more accurate methods; there is material and suggestions enough for a scientific treatise on physiognomy. Let us remember what Galton had in view, for there is more than one strand in his researches:

(i) He wanted to numeralise physiognomies; he dealt chiefly with profiles, but not wholly. For each profile he wanted a formula from which it could be satisfactorily reproduced. Thus an individual could be identified by 80 words of 5 letters or figures each. This enabled a very sufficient likeness to be telephoned, telegraphed or 'wirelessed.'

(ii) He wanted to index portraits, in particular, profiles. This needed a simplification of the individual formula, and in 1907 he reduced his formula for the purpose of indexing to 4 or at most 6 standard points.

(iii) He wanted to obtain a quantitative measure of the degree of resemblance with three special aims:

(*a*) for the purpose of measuring hereditary likenesses or differences,
(*b*) for the purpose of measuring racial likenesses or differences,
(*c*) for the purpose of ascertaining whether special types of physiognomy were correlated with definite mental or moral characters.

He may be said to have solved (i) in a fairly satisfactory manner before his lectures of 1888 and 1893. In 1907 he was satisfied with his method of "lexiconising" or indexing profiles by standard points. In 1906 he was busy with (iii), and he then apparently threw over any idea of measuring resemblance by likeness of formulae[1] and turned to optical methods, at first that of distance and ultimately that of "blurring," to get a measure of "mistakability." I have a set of "blurrers" he presented to me shortly before his death, and the method was at least ingenious, if not reduced to a final scientific statement. Not having completed his solution of (iii), he never lived to apply his methods to the mass of material he had collected for the discussion of (iii) (*a*), (*b*), (*c*).

[1] The problem presents exactly the same difficulties as the discovery of a single coefficient to measure racial differences when 30 or 40 measurements have been made on two series of crania.

To conclude our consideration of this matter we must give some account of the published papers of 1906 and 1907, and of Galton's unpublished ideas as to "blurrers."

I deal first with the paper of 1907[1]. Galton writes:

"It will be shown that it is easy to 'lexiconise' portraits by arranging the measurements between a few pairs of these points [standardised or cardinal points] in numerical order, on the same principle that words are lexiconised in dictionaries in alphabetical order, and to define facial peculiarities with greater exactness than might have been expected." (p. 617.)

The cardinal points selected by Galton are (*c*) the tip of the chin (*pogonion*), (*n*) the tip of the nose, (*f*) the hollow between nose and brow (*nasion*), (*m*) the hollow between upper lip and nose, (*l*) tip of lower and (*u*) tip of upper lip. None of these are really points but vaguely limited regions, and Galton proceeds to define them more closely. A tangent to chin and nasal hollow $YY$ is drawn, a line $Y'Y'$ parallel to this to touch the nose is then drawn, and finally a tangent to nose and chin intersects $YY$ and $Y'Y'$ in points $C$ and $N$, which give the first two cardinal points. A line drawn from $N$ tangential to the nasal hollow gives $F$ by intersection with $YY$, and tangents

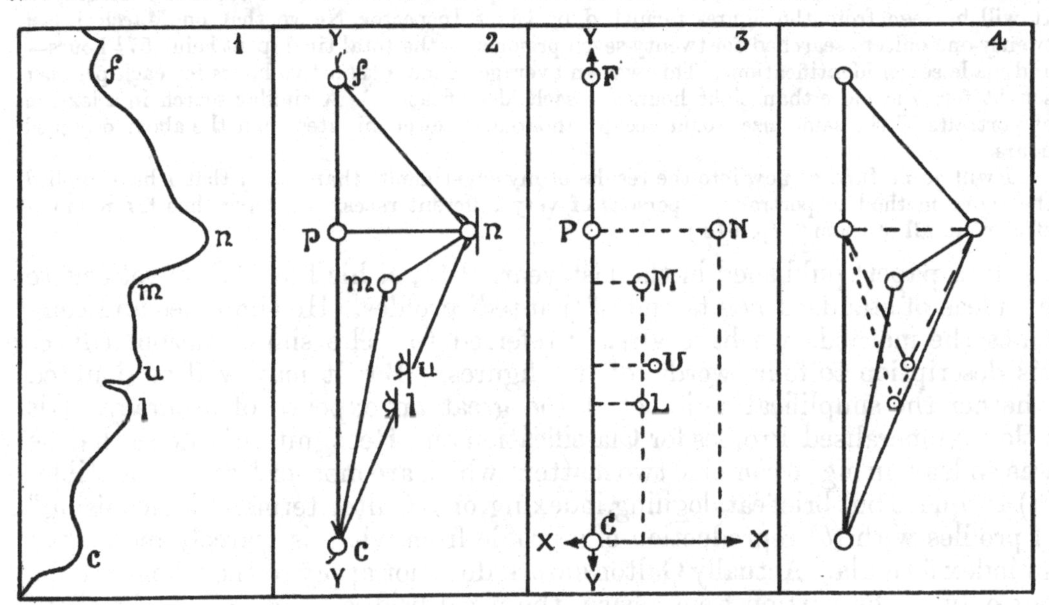


Diagram viii.

to the region $m$ from $N$ and $C$ intersect in the cardinal point $M$. To obtain the upper lip point $U$ we draw a tangent parallel to $CF$ to touch $u$ and a tangent to touch $u$ from $N$, their intersection is $U$. Similarly we draw a tangent to $l$ parallel to $CF$ and to $l$ through $C$ and their point of intersection is $L$. Galton found that the position of the six cardinal points $F$, $C$, $N$, $M$, $U$, $L$, when reduced to a common scale in which $CF$ represented 100 units or "cents[2]," was sufficient to "lexiconise" profiles. The processes might

[1] "Classification of Portraits," *Nature*, Vol. LXXVI, pp. 617–18, October 17, 1907.

[2] A "cent" on the mean profile for a life-sized adult portrait is about 1·25 mm. or $\frac{1}{20}$ inch.

be by distances between the points or by coordinates taking *YCFY* and its perpendicular *XCX* as axes. Galton preferred on the whole the indexing by coordinates. Working merely with the four coordinates of *M* and *N* read only to the nearest cent, Galton was able to index Dance's 68 profiles so that no two of the numerical formulae agreed. In two-thirds of the series the smallest difference between the most resembling pairs was 3 cents in one or more measures.

"This conspicuous difference, equivalent to between $\frac{1}{8}$th and $\frac{1}{7}$th of an inch in a portrait of the natural size, could never be due to the inherent imperfection of the art of measurement, but to some gross blunder." (p. 618.)

Galton thinks that in 1000 profiles indexed on the basis of the coordinates of *N* and *M* *only* there would be some duplicates and perhaps some triplicates. Even these would be reduced by indexing *U* or *L*, or possibly both of them. Galton concludes as follows:

"In the report of a Committee appointed by the Secretary of State in 1894 (C. 7263, price 10*d.*) to inquire into the best means available for identifying habitual criminals, the following remark appears on p. 18: 'An enormous amount of time is spent in examining the books of photographs. It will be seen from the figures furnished by Chief Inspector Neave that on March 1 last twenty-one officers searched for twenty-seven prisoners—the total time spent being $57\frac{1}{2}$ hours—and made seven identifications. This was an average of more than two hours for each prisoner sought for, and more than eight hours for each identification.' A similar search in a lexicon of portraits of the same size would occupy apparently fewer minutes than the above occupied hours.

I will go no further now into the results of my experiments than to say that I have applied the above method to portraits of persons of very different races, and have thus far found it efficient in all of them." (p. 618.)

In a paper[1] published in the last year, 1910, of his life, Galton returns to his ideas of standard points and of 'jointed' profiles. He simplifies and combines the methods we have already referred to. The simplification reduces his description to four 'words' of five figures. But it may well be doubted whether the simplification is not at too great an expense of accuracy. His title "Numeralised Profiles for Classification and Recognition" shows that he was endeavouring to combine two matters, which are more or less incompatible: (*a*) adequate but brief cataloguing, indexing, or as Galton terms it "lexiconising" of profiles with (*b*) reproduction of a profile from what is scarcely more than an index formula. Actually Galton's work does not apply to the whole profile, but only to the portion from nasion, the nasal bridge to pogonion, the tip of the chin. He takes as his five standard points[2]: the nasion *F*, the nose tip, or say, the rhinion *N*, the notch between the upper lip and the nose, the nasolabial point or hypercheilon *M*, the parting of the lips or syncheilion *S*—no longer the two lip-tips (see our p. 325)—and the tip of the chin or pogonion *C*.

[1] *Nature*, Vol. LXXXIII, pp. 127–30, March 31, 1910.

[2] These points are clearly to be determined by tangents to the profile as described on our p. 325. Although Galton does not here state this, yet his standard types and profiles indicate it. It is doubtful how he defines his syncheilion, marked as a point in his standard types and profiles; probably it was taken to bisect the line joining the theoretical lip-tips of the previous paper. The nasion and pogonion are accepted names in anthropometry; I have ventured to supply names to the other standard points.

Two of these five standard points, nasion $F$ and pogonion $C$, are used to get an absolute base. $CF$ is taken vertical, treated as axis of $y$, and made 50 units in length, the unit being with Galton a millimetre; no fractions being given. The axis $CX$ of $x$ is taken perpendicular to $CF$, and the coordinates of $N$, $M$ and $S$ require two double-figure numbers each for plotting. Thus far we have reached a lexicon in which naso-pogonial length and the coordinates of rhinion, hypercheilon and syncheilion would enable us to identify a profile—the errors of measurement being as Galton says small as compared to the variations due to individuality.

Galton now proceeds to the specification by nine types of each (ten in the case of the nose) of the seven parts of the profile from nasion to pogonion. These are (i) shape of nasion and slope of brow to be superposed at $F$,

| | 1 | 2 | 3 | 4 | 5 | 6 | 7 | 8 | 9 | 0 | |
|---|---|---|---|---|---|---|---|---|---|---|---|
| (i) | RAD: 2 BACK | RAD: 2 VERT'L | RAD: 2 FOR'DS | RAD: 6 BACK | RAD: 6 VERT'L | RAD: 6 FOR'DS | RAD: 10 BACK | RAD: 10 VERT'L | RAD: 10 FOR'DS | STR'T | Dot = $F$ |
| (ii) | SLIGHT SIN'S | SLIGHT C.CAVE | SLIGHT C.VEX | MEDIUM SIN'S | MARKED C.CAVE | MARKED C.VEX | UPPR 3RD C.VEX | LOWR 3RD C.VEX | | | |
| (iii) | RAD: 2 STR'T | RAD: 2 SIN'S | RAD: 2 C.VEX | RAD: 6 STR'T | RAD: 6 SIN'S | RAD: 6 C'VEX | RAD: 10 STR'T | RAD: 10 SIN'S | RAD: 10 C'VEX | | Dot = $N$ |
| (iv) | ANGLE STR'T | ANGLE C.CAVE | ANGLE C.VEX | RAD: 2 STR'T | RAD: 2 C.CAVE | RAD: 2 C.VEX | RAD: 6 STR'T | RAD: 6 C.CAVE | RAD: 6 C.VEX | | Dot = $M$ |
| (v) | EVEN SHUT | EVEN PARTED | EVEN V. OPEN | UPPR PR SHUT | UPPR PR PARTED | UPPR PR V.OPEN | UNDR PR SHUT | UNDR PR PARTED | UNDR PR V.OPEN | | Dot = $S$ |
| (vi) | SMALL SMALL | SMALL MEDIUM | SMALL LARGE | MEDIUM SMALL | MEDIUM MEDIUM | MEDIUM LARGE | LARGE SMALL | LARGE MEDIUM | LARGE LARGE | | |
| (vii) | RAD: 2 UPPR 3RD | RAD: 2 MIDDLE | RAD: 2 LOWR 3RD | RAD: 6 UPPR 3RD | RAD: 6 MIDDLE | RAD: 6 LOWR 3RD | RAD: 10 UPPR 3RD | RAD: 10 MIDDLE | RAD: 10 LOWR 3RD | | |

**Diagram ix.**

**The dots represent the position of standard points.**

(ii) nose from nasion to rhinion, (iii) nostril from rhinion to hypercheilon, (iv) upper lip from hypercheilon to lip-parting, (v) nature of the lip-parting with reference to syncheilion as origin, (vi) size of upper and lower lips respectively, (vii) outline of chin between border of lower lip and pogonion.

The type of each portion is here given by a single number. We have thus twelve figures for coordinates and seven for types and we can communicate this in four 'words' of five figures with one figure to spare. After the standard points have been put in on tracing paper, Galton suggests that tracings should be taken of the seven selected standard forms on to this paper very faintly; next

"to harmonise the whole tentatively with faint and brushlike strokes; lastly, with a free and firm hand to draw the outline through them." (p. 128.)

Now there is little doubt that Galton's original method of numeralising profiles allowed their reproduction with astonishing accuracy, and that his original six standard points permit of their accurate lexiconising. Only experience could determine whether the loss of exactness in this his final four-word method would not be at the cost of a considerable part of the certainty of recognition. Galton in his paper in *Nature* gives eight illustrations and says—with which any one examining the results would agree—that they are by no means deficient in resemblance to their originals.

"I think they are considerably more like to them than the sketches, usually printed in the illustrated newspapers, are to the public characters whom they profess to represent. They are, to say the least, of considerable negative value sufficing to eliminate at the rate of about nineteen out of every twenty individuals as *not* being the person referred to." (p. 129.)

It must be remembered that the resemblance provided is between a profile and a profile, not between the actual person and the four-worded reproduction of his profile. Dance almost in the manner of a caricaturist emphasised individual characters especially the nasal, and this I fancy renders in the illustrations given in *Nature* identification of the accurate profiles, and their rough reproductions, relatively easy; it would be a harder matter with the living subjects of the profiles. Only some experience could test the utility, but it would be worth testing as the police value would undoubtedly be large.

Galton fully recognised the limitations of these rougher methods, and noted that the next step to an accurate profile is a large one[1], requiring our four-word formula to be replaced by one of fifty or more words. Galton had numeralised many profiles in this more elaborate way and found that normal sighted persons who examined them at a distance of 12 inches in a somewhat careless way did not distinguish them from the originals. By such profiles it would be possible to recognise the living. I am far less certain that the rough profiles suggested in the 1910 paper would be adequate, they certainly would not preserve anything in the nature of an artistic characterisation, which 50 to 80 word formulae undoubtedly achieve. Here we must leave the matter as Galton left it, until another scientific worker feels able to spend a like number of years and an equal enthusiasm on the analysis of portraiture.

[1] "I do not find that a general resemblance can be much increased by using one or a *few* more quintets or words." (p. 130.)

## (*F*) *Measurements of Resemblance.*

I have already referred to Galton's long-continued researches on the measurement of resemblance. He gave in *Nature*, October 4, 1906[1], some account of his method and of his apparatus[2] for measuring his "index of mistakability." He opens his account with the following words:

"At the distance of a few scores of paces the human face appears to be a uniform reddish blur, with no separate features. On a nearer approach specks begin to be seen corresponding to the eyes and mouth. These gradually increase in distinctness, until—at about thirty paces—the features become so clear that a hitherto unknown person could thereafter be recognised with some assurance. There is no better opportunity of observing the effects of distance in confounding human faces than by watching soldiers at a review. Their dress is alike, their pose is the same, the light falls upon them from the same direction, and they are often immoveable for a considerable time. It is then noticeable how some faces are indistinguishable at distances where great diversity is apparent in others, and the rudely-defined idea will be justified that the distance at which two faces are just mistakable for one another might serve as a trustworthy basis for the measurement of resemblance. The same may be said of obscurity, of confused refractions, and of turbid media."

In the apparatus described in this paper in *Nature*, Galton used distance. But he also looked at two portraits through a graded series of "blurrers[3]," or glasses with different thicknesses of Canada Balsam placed upon them. Finally he adopted a neutral coloured wedge (like the wedge photometer used for star magnitudes), looking at the portraits through thicker and thicker parts of it until they were "mistakable." The apparatus is fairly simple for the distance observations. There is a six-foot base board upon which are two sledges carried along its length by endless cords each going round their own pair of wheels, one at either end of the board. At the summit of the base board, which slopes slightly downwards from the observer, is a screen with an eye-slit to carry spectacle lenses for examining the photographs; it can be replaced by a bracket upon which optical combinations can be mounted for throwing the photographs to a considerable distance, i.e. greater than that of the base board, in the manner of an inverted telescope. The sledges each carry a standard to which the portraits to be examined can be attached, and when attached they can be rotated in azimuth to compensate for differences of degree in the photographs of "half" face. The position of the photographs with regard to the observer's eye can be read on a scale which runs down the centre of the base board.

Galton's procedure is as follows: He first measures in millimetres the distances $u$ and $u'$ from the pupil line to the lip line of each portrait. He then takes from the base board scale the distances $d$ and $d'$ of the two portraits from the eye screen in centimetres. If now the indices $n = 100u/d$ and $n' = 100u'/d'$ be formed, then when they are equal, the two portraits subtend the same angle at the eye, and this allows for any difference in size.

[1] Vol. LXXIV, pp. 562–3.

[2] Now with a good many additions, devised by Galton himself, in the Galton Laboratory.

[3] He presented me in 1907 with a series of "blurrers" and there are a good many sets in the *Galtoniana*, but I have not come across any account of their preparation and standardisation. The photometric wedge is a much more permanent measurer.

If we vary $n=n'$ taking a whole series of values for them, we reach a value $N$ at which the two portraits can be mistaken for portraits of the *same* individual. This value of $N$ is Galton's "index of mistakability." Two persons will have little resemblance to each other, if they must be put at a great distance off to be mistaken for each other; when they are very like each other the distance will be small or the index of mistakability will be large. The index if it can be determined is therefore a measure of general resemblance. "Faces," Galton writes,

"that are alike are certainly [in-]distinguishable at shorter distances than unlike ones, and I notice no excessive clustering of values closely round particular values of $N$ in my results, which there would be if mistakability always occurred near a particular stage, such as that at which the whites of the eyes cease to be visible, or at twice or three times that distance. A strong likeness in small details may so dominate the perception that a want of likeness in larger features is overlooked. Here the distance of maximum mistakability will be small, the portraits appearing more unlike when removed farther off, and the small details ceasing to be visible. Extreme cases of partial likeness, whether in contour or in detail, would of course be noted and allowed for. With these exceptions the index of mistakability appears to be a fair, even, as I think, a close, approximation to an index of resemblance when the quality of the observed likeness is recorded by appropriate letters as will be described later.

The observational value of mistakability lies in its asking a simple question which different persons would answer in the same way, when they had become familiar with the method." (p. 562.)

The difficulty about the distance measure of mistakability lies in the fact that the comparison of two portraits of different sizes involves continued resetting of the portraits at different distances. To expedite matters Galton tabled $d$ for given $u$ and $n$, so that the operator knowing $u$ and $u'$ could quickly send the two portraits to their proper distances for a given $n$. Nevertheless the continual shifting for each new judgment is laborious. Galton then proceeded to prepare test types and noted the $d$ at which each row of figures was just unreadable. If now a test line be put against the portraits themselves when they are just mistakable in a clear light, we can interchange $d$ and readability of a certain type. By marking the types by bold values of $d$ we replace our distance scale by a type scale. Now if the hindrance to vision increases the portraits with the test card must be brought nearer to the eye, and they will increase simultaneously in legibility. The written $d$ will always show what the true $d$ would be in a clear light. We now see how the "blurrers," wedge or inverted telescope[1] are to be used; we can keep the actual $d$ constant, and measure the apparent $d$ on the card of test types placed alongside the portraits.

Galton's reduction to test types seems to emphasise an obvious criticism —the judgment of the index of mistakability will be dependent on the keenness of vision of the operator. Hence different operators would need differently marked test cards, and there would be a need to correct the index for personal equation, if the results of operators with marked dif-

[1] At one stage of Galton's experiments he made "blurrers" of gauze of different meshes. I think it likely he discarded these because the visibility so largely depends on the position of the network between eye and object. Type absolutely illegible if the gauze be midway between eye and object becomes legible if the gauze screen be quite close to object or to eye.

ferences of keenness of vision had to be pooled. Galton considers that mutual mistakability may occur under. any one or more of the following conditions, which he thinks should be noted alongside the index:

"*aa.* The portraits are apparently exact copies or reductions on different scales.

*a.* They appear to be portraits of the same person at about the same age, though differing in pose and dress.

*b.* They would be mistaken for portraits of the same person, even though they differ in sex and considerably in age, if the hair had been cut and dyed alike, and the dress arranged in the same way.

*c.* As above, if much disguised, as for theatrical impersonations.

*b–c.* Applies to cases intermediate between *b* and *c.*

*p.* Their resemblance is partial only, being confined to specified features.

. . . . . . . . . . . . . . . . . .

The applications of the process are numerous, as must always be the case when a hitherto vague perception is brought within the grip of numerical precision. To myself it has the special interest of enabling the departure of individual features from a standard type to be expressed numerically. The departure may be from a composite of their race, or from a particular individual. The shortcomings of a pedigree animal from a highly distinguished ancestor could be measured in this way. Many other examples might be given." (p. 563.)

As in his profile work Galton used a very large number of pairs of photographs of relatives to test his index of mistakability upon. He asked in the newspapers for photographs of families, and they appear to have been rained down upon him; some material was suitable, some quite unsuitable! It seems to me that to get reliable measures of resemblance special photographs should be taken—full face and profile, the hair being screened under a tight fitting elastic cap. Further if bearded individuals are to be one of the "comparates," then the comparison must further be made with the chin and lips screened; the eye is very apt to be misled in its judgments by extraneous characters such as hair and pose.

A manuscript typed and prepared for press in February 1906, entitled "The Measurement of Visual Resemblance," seems never to have been published. It adopts a somewhat different index to that finally chosen by Galton in October of the same year. He begins by saying that visual resemblance between any two objects may be measured in units whose value is strictly defined.

"Resemblance is independent of actual magnitude and has therefore to be expressed in angular units. It is curious that no popular terms exist to express them in the language of any civilised country, for not only would they be useful, but the diameter of the sun when paled by an intervening screen affords an excellent and practically constant standard for rough measurement. It would often be well to indicate objects in a distant landscape by describing them as so many sun-breadths to the right or left of some conspicuous feature, or to speak of a mountain seen from a specified place as towering so many sun-breadths in height, or as bulking so many [square] sun-breadths in area. But as sun-breadths are not terms in popular use, and as they are not the best unit for the purpose of this memoir, I will employ another that is. The sun's diameter may be taken as subtending an angle of 31·0 minutes of a degree. I will employ for my unit the diameter of an imaginary mock sun that subtends 34·4 minutes, and is therefore wider than that of the real sun in the proportion of 10 to 9. Its merit lies in the fact that the tangent as also the arc of 34·4 minutes differ insensibly from 0·01; in other words the angular unit is that which is subtended by 1 measure of any kind, at the distance of 100 measures of the same kind. I will call the arc subtended by this angle at any specified distance a 'sol'."

Galton now gives a paragraph of considerable importance which shows that he had anticipated and met the criticism which naturally arises on reading his second paper.

"The portions of objects to be compared and between which Resemblance is to be measured must be strictly defined. Non-essentials may be either marked out or be simply ignored, but there must be no vagueness as to the limits of the portions selected for comparison. If the objects be portraits the selected portion may be any specified part of the whole of it. It may be a single feature, it may be the face irrespectively of hair, and of beard if any, it may be the whole head, or it may be the entire person. But, whatever it may be, it must be defined."

After defining the objects for comparison as comparates, Galton continues:

"The comparate is limited to the portion under comparison, the two comparates are supposed to be reduced to similar scales, to be mounted side by side on the same moveable screen squarely to the line of sight, and to be viewed in a good light through a perfectly transparent atmosphere."

Now in order to conduct the experiments successfully the experimenter requires to have the power of adapting the focus of his eye sharply to the various distances of the screen or to use an optical contrivance to supply this faculty if he should be deficient in it—at the time of writing this paper Galton was 84 years old, and the following words are very characteristic and indicate at least the nature of one of his 'toys,' which I had puzzled over:

"The range of adaptability of my own eye, as in that of most elderly persons, has become very narrow, and during a long time was the cause of serious embarrassment in my various experiments on Resemblance. But all this difficulty was happily removed by a small inverting telescope of very low power, that I made abroad in a very makeshift way, out of two small magnifying glasses that I had by me, with pasted paper tubes and corks. It acted so well that I was loath to replace it by a better. Its field of view was ample and enabled me to focus my eye sharply on 'comparates' at any distance from a few inches upwards. I will call telescopes that neither magnify nor minify by the name of *Isoscopes*; their use is simply to secure a sharp focus for the eye at any distance. Two convex lenses of 2 inches focal length seem to be on the whole most suitable for an isoscope. The tubes must admit of a wide range of adjustment. Either lens may serve as the eye-piece, but when used as such it should be covered by a cap with an eyehole. Distances must be measured from the object-glass. An isoscope should be fitted with two eye-pieces one of them furnished with a micrometer of crossed lines [i.e. an areal micrometer]. If the eye-piece be of 2 inches focus, and the distances between the lines one 50th of an inch, the intervals between them will subtend 1 sol and each small square will subtend one square-sol."

Galton proposes to take as his index of mistakability the number of square sols covered by either comparate when they are at the 'critical distance,' and the corresponding angle is the critical angle. The measure of Resemblance between two comparates, he says, is the angular area of either of them at the critical distance when the comparates as a whole are mutually mistakable. The angular area as a whole is proportional to the number of just-distinguishable plots (i.e. for the normal eye plots of about 1 minute diameter) which they contain, the possibility of mistaking one comparate for another being due to apparent identity in every one of the just-distinguishable plots. The more numerous the plots, the more minute is the coincidence, and consequently the closer the resemblance. It will be seen that the entire difference between this earlier and the later paper of Galton is the measurement by a solid instead of a plane angle.

PLATE XLVI

Elizabeth Anne Galton (1808–1906), Mrs Wheler, February 21, 1904, aged 96.
From the last photograph taken.

Francis Galton, aged 83, from a photograph taken by W. F. R. Weldon in July, 1905.
"I enclose the best I can do with one of the negatives you were kind enough to let me make. Please forgive my caricaturing you in this way. You know enough about the lower forms of man to know that respect and affection show themselves in strange ways—look upon this as one of them, and pardon it." W. F. R. W.

Apart from the question, however, of whether the eye does pass under review and ascertain the apparent identity of all these just-distinguishable plots, we may ask what would happen to the relative measurement in the following cases:

*A*. All the plots are absolutely alike except a few which are extremely different, say two identical twins one only with a mole on the face.

*B*. All the plots are different but not widely different.

The comparates in case *B* would present a higher index of mistakability than those in *A*, for we should have in the case of *A* to proceed with distancing the comparates until the isolated anomalous feature disappeared. It might be nothing of course so easily observed and allowed for as a mole, but the measure seems largely to depend upon items of extreme divergence rather than on an average of *all* comparable plots.

Nevertheless the whole investigation is of great suggestiveness, and its originality striking for a man of Galton's then age. He saw a great need, and he did his best to supply it, spending a large part of many years over the problem. If he did not fully solve it, no one has done so much towards its solution, and no one to this day has tested his work, his apparatus or his method and ascertained how far they would carry us.

"The measurement of Resemblance is of wider importance than may appear at first sight. It covers a field of research that escapes the ordinary measurements by foot rule, scales and watch. It is particularly applicable to a variety of biological studies in which hereditary likenesses and family or racial peculiarities are inquired into, and seems eminently suitable for comparing composite photographs. The account of the method I propose has been given merely in outline. It presents many side-issues of interest, and deserves a large amount of photographic illustration such as I am now unable to give."

Thus wrote the veteran of 84! What is needed is that some one should take up the subject where Galton was forced to leave it, starting possibly with his material and apparatus. What are the average degrees of resemblance of parent and child, of brothers and sisters, of first cousins, etc.? And would the results obtained from Galton's Index of Mistakability correspond with those found by the principle of correlation from a single character in kinsmen of various degrees?

The number of years over which Galton's photographic researches are stretched is in itself remarkable, but more remarkable still is the amount of time in those years he devoted to them. I have spent weeks in going through his manuscripts, his published papers, his photographic apparatus, his negatives and his prints, with the view of writing the account in this chapter, but it is more than possible that I have missed points of interest in the overwhelming mass of his material. The suggestiveness of his contributions to portraiture seems to me great, but long as he lived he had only time to blaze the trail. In heredity and statistics a younger generation has been found to take up his work; in photography and portraiture his pioneer steps have not yet been trodden into a well-marked track by enthusiastic disciples.

# CHAPTER XIII

## STATISTICAL INVESTIGATIONS, ESPECIALLY WITH REGARD TO ANTHROPOMETRY

"Until the phenomena of any branch of Knowledge have been submitted to measurement and number it cannot assume the status and dignity of a science." FRANCIS GALTON.

### A. STATISTICS IN THE SERVICE OF ANTHROPOLOGY

THERE is no branch of knowledge to which Galton's remark applies more closely than anthropology; and there is certainly no field of research which owes more to Galton than that of anthropometry and in particular that branch of it which deals with craniometry. Here again as we have so often had occasion to remark Galton's contribution was essentially one of method, and lay in his insistence that the only way to permanent and safe deductions was the path of measurement and number. The reader has only to examine craniological papers of the 'sixties or 'seventies, even by such authorities of those days as Dr George Busk or Sir William Flower, to grasp how indefinite and inconclusive craniometry was before it became permeated with Galton's ideas of measurement and number. Half-a-dozen measurements on half-a-dozen skulls screened by a smoke-fog of vague remarks were considered an adequate basis for attack on the most elusive problems of racial differentiation. There was no conception of the number of individuals or of the number of characters which require to be measured before we can reach definite conclusions. Anthropology was considered as a field to be left for a recreation ground almost entirely to men busy in other matters, for it had developed no academic discipline of its own, until Galton's methods gave it the status and dignity of a real science.

What troubled Galton, when travel and geography in the wider sense had led him to anthropology, was not only the lack of quantitative method but the lack also of ample material[1]. He at once set about supplying both in his own original way. Yet having reached some certainty himself, he proceeded, owing to the weakness of his brethren, in administering it only in homoeopathic doses. At the Brighton British Association of 1872, a recommendation was made by the General Committee, probably on Galton's suggestion, that brief forms of instruction should be prepared for travellers. Two years later the *Notes and Queries on Anthropology, for the Use of Travellers and Residents in uncivilised Lands*, drawn up by the Committee of the British Association (which included Lane Fox, Beddoe, Lubbock, Tylor, Galton and others), was issued. To the first edition of this handbook

[1] Neither lack was fully recognised even to Galton's death in many of the papers published by the representative English Anthropological Society; and I remember on more than one occasion his saying with a sigh: "Poor dear old Anthropological." All his efforts had produced little if any impression upon its members.

Francis Galton in the late 'sixties.

Galton contributed the hundredth section entitled "Statistics." He opens with the characteristic sentence:

"The topics suitable to statistics are too numerous to specify; they include everything to which such phrases as 'usually,' 'seldom,' 'very often' and the like are applicable, which vex the intelligent reader by their vagueness and make him impatient at the absence of more precise data." (p. 143.)

He then refers to the necessity of homogeneity, the breaking up even of homogeneous groups when there is a variation largely governed by a dominant influence, e.g. age, and the need for a truly random selection. He says that precision varies as the square root of the number of observations, but that number must not be reached at the expense of accurate reporting. He then turns to the "law of deviations" and suggests the "ranking" of characters in individuals, and the measurement of the mid (500th), the 250th and the 750th individuals in ranks of a thousand, or what we now term the median and quartiles[1]. The ranking gives him his so-called ogive curve, and his whole appeal to theory consists in the statement that when individual differences in a homogeneous population are due to many small and independent variable influences then the excess of the $(m+t)$th individual if $m$ be the mid number will equal the defect of the $(m-t)$th individual from the mid individual. Galton does not enter into the mathematics of the matter. He says this

"law of deviations holds for the stature of men and animals, and apparently in a useful degree for every homogeneous group of qualities or compound qualities, mental or bodily, that can be named."

Galton gives no proof of the "normal curve of deviations," but suggests that it is mathematically deducible on making certain rather forced suppositions to render calculation feasible. Comparing fact, however, with theory

"wherever comparison is possible, it is found that they agree very fairly and in many cases surprisingly well." (p. 144.)

He concludes with the statement that a good book on these matters has yet to be written.

"Quetelet's Letters on the Theory of Probabilities is perhaps the most suitable to the non-mathematical reader." (p. 146.)

It will be clear that Galton was proceeding gradually, and the dose was a very small and simple one[2].

In the second edition we find Galton contributing some further sections.

[1] Galton then termed the 500th individual in a thousand the "average." The middle man is practically the 500th, but not so theoretically. The diagram in later editions disappeared.

[2] Other contributions by Galton to the first edition were No. XCIV on "Population," which begins characteristically with "Count wherever you can," No. LVIII on "Communications," reminiscent of the *Art of Travel*, No. LIV "Causes that limit Population," No. XXI "Astronomy," with special reference to the seasons, and to steering by sun and stars. There is also (pp. 21–2) a note on heredity, giving a list of hereditary characters which admit of precise testing; "those who confuse the effects of nature and nurture give information that is of very little use." The first edition also contains a section (No. IX) on "Physiognomy," by Charles Darwin, who was collecting material for his work on Expression of the Emotions, a section which Dr Garson had the temerity to revise in later editions.

In a note on the "Physical Powers" he suggests taking three weights $A$, $B$, $C$ of definite and increasing heaviness and noting how many fail to lift each, and so ranking the community. He suggests similar tests for running the same distance in specified times and for running three distances in a specified time, etc., etc. He says that such methods

"afford a complete and approximately correct picture of the distribution of the qualities tested, and not merely general averages." (p. 41.)

There is also in this edition a note on testing sight (pp. 43–4). The distance at which the tested can distinguish between two white squares on a black ground, one with the side and the other with the diagonal vertical, is recorded and used as a measure of acuity of vision.

"The testing must be performed when the light is perfectly good, but not dazzling. Always test yourself when you are testing others, because if your own efficiency comes up to its normal standard, it is fair evidence that the conditions of light etc. are normal also, otherwise very probably they are not."

Like Darwin's contribution, Galton's were "revised" in later editions, and ultimately they disappeared, perhaps desirably as they had been deprived of any characteristic value.

Galton's anthropometric projects were, however, far from being confined to travellers; he had much more comprehensive schemes in view. One of his earliest proposals was the establishment of anthropometric laboratories in schools[1], but here again he exhibited at first only the thinnest end of the wedge. He had realised that statistical material for such fundamental characters as height and weight did not exist for the British people.

"We do not know whether the general physique of the nation remains year after year at the same level, or whether it is distinctly deteriorating or advancing in any respects. Still less are we able to ascertain how we stand at this moment in comparison with other nations, because the necessary statistical facts are, speaking generally, as deficient with them as with ourselves." (p. 308.)

Galton's proposal was to take samples of reasonably homogeneous classes, and then by aid of the census to combine the returns in the proper proportions. He considered that homogeneous groups of boys, girls and youths already existed in several large schools, under conditions which offered extraordinary facilities for obtaining anthropometric data. He proposed to measure children in the great public schools, middle class schools and others down to those for pauper children (p. 311). Galton held that the masters in such schools were "trustworthy and intelligent in no common degree," that they knew their pupils well, and that the general organisation and discipline of the school was favourable to collecting full and accurate statistics. He believed that the school authorities might be induced in not a few instances to cooperate heartily and with great intelligence. Once the system of anthropometric measurement was established in schools it would spread elsewhere.

"The boys when they grow up into men would retain favourable recollections of the whole procedure, and application might then be made to Universities, Factories, and other large bodies of adults, with greater probability than at present of obtaining the required information." (p. 309.)

[1] "Proposal to apply for Anthropological Statistics from Schools," *Journal of the Anthropological Institute*, Vol. III, pp. 308–11, 1874. Paper read in 1873, proposals drafted 1872.

Galton confines his attention to the data for age, height and weight, and remarks:

"It seems to me better not to speak at present of the attractive and numerous problems that might be solved by a wider range of inquiry; because if we confine the attention of those we ask to few and simple questions, we are far more likely to have them well and thoroughly answered, than if we had issued a more ambitious programme." (p. 310.)

Anthropometric measurements were soon taken at a number of schools and in some schools anthropometric laboratories established. From the schools they spread to the Universities (Cambridge, 1884; Oxford, 1908; London, Galton Laboratory, 1920). But on the whole there has been a tendency to take in routine fashion a few superficial measurements, and not use the anthropometric laboratory as a means of solving definite problems, physical or mental. They might still be of value if a little inspiration were thrown into their work and psychic or dynamic qualities measured rather than superficial static characteristics. One result of the proposal was those returns from the public schools, upon which Galton based his paper on the weight and height of boys in town and country schools discussed on our p. 125.

Another somewhat slender paper of this period is entitled: "Excess of Females in the West Indian Islands from documents communicated to the Anthropological Institute by the Colonial Office[1]." This paper gives statistics showing the excess of females in most of the West Indian colonies, although there is an excess of male births. The anomaly is partly due to mortality following dissipation in the young of the male sex, but more extensively to adult male emigration. The whole topic might now be rediscussed with fifty years additional statistics, and would not be without interest. As Galton remarked in 1874 each of the West Indian Islands is an individual social experiment, and each therefore deserves the pains of a separate and thorough statistical investigation.

The collection of statistical data was, however, not the only point that Galton had in view; he sought to make statistical theory simple and of easy application, and he risked the possibility that loss of refinement might involve decreased accuracy and a drawing of over hasty conclusions. His "Proposed Statistical Scale" was first given at a Royal Institution Friday evening discourse on February 27, 1874. He followed the lecture up by a letter to *Nature* on March 5, 1874[2]. His communication embraced the idea of "ranks," and the whole theory of ranks has been developed from this origin. It is easy to recognise that it is often less difficult to place two persons in order as to the intensities they possess of any physical or psychical character than actually to measure those intensities. A trained schoolmaster can "rank" his class for intelligence with very considerable accuracy. If a number of individuals be placed in order of their intensity for any character, they are arranged according to Galton on a "statistical scale" (S.S.). The grade of any individual is then determined by the percentage of the whole population who stand above that individual on the statistical scale. The middle man—or the man who would stand half-way between the two middle men if there were two—was later said

[1] *Jour. Anthrop. Inst.* Vol. IV, pp. 136–7, 1874. [2] Vol. IX, p. 342 (abstract of lecture, p. 344).

by Galton to have the *median* value of the character ($m$). The two men with 75 °/ₒ and 25 °/ₒ of the population above them are said to have the lower and upper quartile values ($q_1$ and $q_2$). If the distribution be symmetrical about the median then $m - q_1$ and $q_2 - m$ will be equal; if the distribution obeys the so-called normal curve of deviations, then all the constants of the distribution can be found by measuring the intensity of the character in the median and in the quartile individuals. Thus Galton would place a hundred and one savages in a row, the curves formed by the apices of their heads would be his "ogive[1]" for their stature, and by measuring only the 25th, the 51st and 76th men he would obtain a reasonable distribution for the stature of adult men in that tribe.

*Theoretically* there are difficulties about Galton's "ogive," if we suppose it to correspond to a normal curve of deviations, in particular at the terminals. Galton endeavoured to get over these difficulties by replacing the normal curve by a symmetrical binomial, which has a finite range. He treats of this matter in a paper on "Statistics by Intercomparison with Remarks on the Law of Frequency of Error[2]." In this paper after mentioning that Quetelet had shown that a binomial to the 999th power was practically a normal curve of deviations, Galton goes on to indicate that the same holds very closely for symmetrical binomials of quite low powers. Thus he plots (p. 39) the Binomial Ogive of 17 elements against the Binomial Ogive of 999 equal elements, which is practically identical with the Exponential Ogive, and argues therefrom to the binomial of the 17th power being very close, indeed (which is a fact), to the normal curve[3]. Galton then passes to some suggestive remarks on the origin of the distribution of deviations according to the normal law. He rejects any idea of its source in a very large number of small and independent contributory causes. He supposes the exponential curve to arise because it nearly resembles the curve based upon a binomial of moderate power, i.e. he supposes that in nature the contributory cause-groups are relatively few; but he has to suppose in this case that nature works all her processes by equal additions or subtractions, i.e. prefers the mathematics of coin-tossing to those of the dice-box.

"I shall show," he writes, "by quite a different line of argument that the exponential view contains inherent contradictions when nature is appealed to, that the binomial of a moderate power is the truer one and that we have means of ascertaining a limit which the number of elements [independent cause-groups as the individual coins of a combined toss] cannot exceed." (p. 40.)

Galton takes the mean ($m$) and divides it by the quartile deviation ($q_2 - m$ or $m - q_1$) and computes the ratio $m/(q_2 - m)$. In the case of the binomial $(a+b)^n$ this will be

$$nb/{\cdot}67449\sqrt{nab} = 1{\cdot}48257\sqrt{\frac{nb}{a}},$$

[1] *Philosophical Magazine*, January, 1875, Vol. XLIX, pp. 33–46.

[2] If $X$ be the abscissa, i.e. the rank, and $Y$ the ordinate or value of the variate, $m$ the mean, $\sigma$ the standard deviation of the population $= 1{\cdot}48257\,(q_2 - m)$, and $N$ the total population, then the equation to Galton's "ogive" will be $X = \frac{1}{2} + \int_{-\infty}^{Y} \frac{N}{\sqrt{2\pi}\sigma} e^{-\frac{1}{2}(Y-m)^2/\sigma^2}\,dY.$

[3] I have not been able to agree with the values given in the Table on p. 42.

and it will be clear that we could not determine $n$—the number of independent cause-groups in general—without a knowledge of $a$ and $b$, for which we require higher moments than the first two. If we suppose with Galton that "nature tosses," i.e. $a = b = \frac{1}{2}$, then clearly a knowledge of $m/(q_2 - m)$ will give $\sqrt{n}$ and so determine the number of elements, or contributory cause-groups.

Galton obtains (p. 43) the following results:

| Number of Elements or Contributory Groups | Value of $\frac{m}{q_2 - m}$ | |
|---|---|---|
| | Galton | Above formula |
| 17 | 5 | 6 |
| 32 | 10 | 8 |
| 65 | 15 | 12 |
| 107 | 20 | 15 |
| 145 | 25 | 18 |
| 186 | 30 | 20 |
| 999 | 48 | 47 |

The source of the divergence is two-fold. First, there is no theoretical means of discovering the quartiles in a binomial of discrete terms. Galton determined them by drawing a freehand curve through the tops of the plotted binomial blocks in order to reach a continuous ogive. This method is not capable of great accuracy. Secondly, although the standard deviation of a binomial is well known, the probable deviation (or quartile) will not be equal to 67449 multiplied by the standard deviation unless $n$ is fairly large. Still the deviations seem too large to be wholly attributable to this cause.

I have enlarged on this matter because it provides an illustration of the cases in which a standard deviation can be determined and a quartile cannot. On the other hand there is more than an assumption of $a = b$ in Galton's method. If we are given any data, for example statures of a definite group, there appears to be no reason why the zero of stature should correspond to the start of the binomial; nature is more likely to take its additions and subtractions from some definite value, and zero stature to be not only a great improbability, but an impossibility. Further, nature's unit of addition or subtraction will not be that of the measurement of stature and this introduces another unknown. When we approach the problem with all these quantities—$a$, $n$ and both the unit of addition (or subtraction) and the centre about which nature works—unknown, we can still solve the problem of fitting a binomial to our data. Experience shows, however, that in the great majority of cases the equations lead to imaginary values for our constants, or to such as are uninterpretable (e.g. $n$ negative) on the basis of a simple binomial. In other words nature does not work on the basis of a finite number of *independent* cause-groups, such as are assumed in the binomial frequency;

it is more probable that the products of the contributory cause-groups are correlated. That is to say, that the first contribution influences later ones[1].

As far as I am aware, however, Galton was the first to endeavour to unriddle something of nature's method of working from the frequency distribution of a given variate. We may see now-a-days that his solution of 1874 is not valid, but we have to confess that we have not got much farther than he did[2].

In the remainder of the memoir Galton discusses

"how a medley of small and minute causes may, as a first approximation to the truth, be looked upon as an aggregate of a moderate number of 'small' and equal influences." (p. 42.)

He considers that small disturbing influences would weld the binomial blocks into a continuous ogive. He concludes by showing that the sum of three symmetrical binomials taken in certain proportions may lead to a result indistinguishable from a single binomial. He justifies the exponential law, or normal curve, on the ground that it is very close in the results it gives to any binomial ogive, and would propose to use it for intercomparison in cases where no scale of equal parts has been or can be applied. As we have endeavoured to show the paper is extremely suggestive, but not every reader will be induced by the arguments to accept its conclusions.

Galton, influenced by his own motto: "Whenever you can, count," seldom went for a walk or attended a meeting or lecture without counting something. If it was not yawns or fidgets, it was the colour of hair, of eyes or of skins. But the record of several characters involves a considerable effort of memory, and using a pencil invites attention to the work of the recorder. The Galton Laboratory possesses no less than five implements of a type which Galton later termed "registrators." One consists of a pair of cotton gloves; on the palmar face of one glove across the fingers is a pocket capable of containing a card, about the size of a gentleman's visiting card; just below the tip of the thumb is a thin piece of wood or metal sewn into the inside of the glove and carrying a needle point projecting very slightly through the material of the glove. If the thumb be pressed against the palmar surface of any one of the four fingers a fine hole is recorded on the card. "A great many holes may be pricked at haphazard close together without their running into one another or otherwise making it difficult to count them afterwards." Another registrator consists of a thimble which being pressed against a card or even a newspaper makes a pinhole by aid of a needle point which projects on the thimble being pressed. A third registrator is a single dotter and contains a guarded needle-point which on a slight squeeze stabs a strip of paper, the action of the instrument being such that a 'stab' slightly pulls the strip of paper forward, so that a line of dots is made; the instrument can be held in the palm of the hand in the pocket of an overcoat. Another simple pocket

[1] Nature prefers a hypergeometrical series to a binomial series!

[2] No stress whatever, in my opinion, can be laid on the results of those writers who believe that the direction of evolution of a character can be determined from the asymmetry or skewness of its distribution, or of those who assert that certain forms of distribution connote "instability" in the character.

recorder consists of a brass disk sliding with a range of about 3″ vertically, and rather more than $\frac{1}{3}$″ horizontally, so that a needle which projects from the disk on pressing a spring is capable of holing about one square inch of visiting card supported on chamois leather. The range is adequate for the record of two, possibly three characters.

The most complete registrator was one made for Galton by Hawksley; the needle point is done away with, and the instrument records on five dials the number of separate pressures on five pins. These pins or stops communicate by a ratchet with a separate index-arm that moves round its own dial. The dials are covered by a plate which can be removed to read off the results. The instrument is $\frac{1}{4}$″ thick, 4″ long and $1\frac{3}{4}$″ wide and it can be held unseen in either hand with a separate finger and thumb on each stop. When any finger is pressed on the stop below it the corresponding index-arm records a unit. Guides are placed to keep the fingers in their proper positions. The instrument may be used in the pocket or under a loose glove or other cover. "It is possible by its means to take anthropological statistics of any kind among crowds of people without exciting observation, which it is otherwise exceedingly difficult to do[1]." I may remark that it requires some little training to press with the correct finger. With an instrument of this kind Galton recorded the percentage of attractive, indifferent and repellent looking women he met in his walks through the streets of various towns with the object of forming a "Beauty-map" of the British Isles—a project he never completed, although he held London to have most and Aberdeen fewest beautiful women of the towns he had observed. He once also remarked to me that he had found Salonika to be the centre of gravity of lying, though I have no direct evidence that he used a registrator to tick off liars and truth-speakers in his travels in Greece.

While busy with his *Hereditary Genius*, 1869, Galton had noticed how apt are the families of great men to die out and that genius has been asserted to be related to sterility. He endeavoured to explain the matter in the case of the judges and in the case of peers by special causes (see our pp. 93–96). De Candolle also referred to this topic in his *Histoire des Sciences*, four years later, and suggested without mathematical investigation that families in the *male* line must always tend to die out, the name becoming extinguished when a son failed to be born. He suggested that a mathematician ought to be able to solve this problem of the extinction of surnames. Galton saw the importance of the determination of the rate of extermination of surnames as a preliminary investigation to the inquiry as to the dying out of the families of men of ability, in whose cases heredity had been too often traced in the male line only—e.g. the extinction of peerages granted for great achievements—and this extinction of the line attributed to some unexplained sterility in able men. Galton accordingly propounded the problem in the *Educational Times*, and there it met with poor success at first—one erroneous solution. Ultimately the late H. W. Watson, a personal friend of Galton's,

[1] See the paper: "Pocket Registrator for Anthropological Purposes," *British Association Report*, Swansea, 1880, p. 625.

was persuaded to take it up and sent his discussion of it to the above Journal[1]. His discussion with certain preliminary remarks by Galton was also published in the *Journal of the Anthropological Institute*[2]. The kernel of Watson's paper is as follows: The symbols $t_0, t_1, \ldots t_s, \ldots t_q$ denote the chances of a man having no children or one, ... $s$, ... $q$ children. Then the chance of a surname having $s$ representatives in the next succeeding generation, if it has $p$ in any generation, will be the coefficient of $x^s$ in the multinomial

$$(t_0 + t_1 x + t_2 x^2 + \ldots + t_q x^q)^p = T^p, \text{ say.}$$

Let ${}_u m_v$ be the fraction of $N$, the original number of distinct surnames, which in the $u$th generation have $v$ representatives, then the number of surnames with $s$ representatives in the $v$ generation must be the coefficient of $x^s$ in

$$\{{}_{r-1}m_0 + {}_{r-1}m_1 T + {}_{r-1}m_2 T^2 + \ldots + {}_{r-1}m_{q^r-1} T^{q^r-1}\} N = f_r(x) N, \text{ say.}$$

It follows that ${}_{r-1}m_1$, ${}_{r-1}m_2$, etc. are the coefficients of $x$, $x^2$, etc. in the expression $f_{r-1}(x)$. As soon as the $t$'s are known, it should be possible, although laborious, to find the succession of functions given by

$$f_r(x) = f_{r-1}(t_0 + t_1 x + \ldots + t_q x^q).$$

As the numerical values of the $t$'s are not known, Watson takes two hypothetical systems. In the first he takes $q = 2$ and $t_0 = t_1 = t_2 = \frac{1}{3}$. He finds by a brute-force expansion that out of a million distinct surnames 333,333 will disappear in the first, 148,147 in the second, 89,620 in the third, 70,030 in the fourth, and only 34,150 in the fifth generation. In this case the total male population is clearly constant and two-thirds of the surnames have disappeared in five generations. Watson's second hypothesis is that the $t$'s are the successive terms in the binomial

$$(\lambda_1 + \lambda_2)^q,$$

where $\lambda_1 + \lambda_2 = 1$. In this case

$$f_1(x) = (\lambda_1 + \lambda_2 x)^q, \quad \text{and} \quad {}_1m_0 = \lambda_1^q,$$

$$f_2(x) = \{\lambda_1 + \lambda_2(\lambda_1 + \lambda_2 x)^q\}^q,$$

and

$${}_2m_0 = (\lambda_1 + \lambda_2 \lambda_1^q)^q$$

$$= (\lambda_1 + \lambda_2\, {}_1m_0)^q = \lambda_2^q \left(\frac{\lambda_1}{\lambda_2} + {}_1m_0\right)^q,$$

and generally

$${}_rm_0 = \lambda_2^q \left(\frac{\lambda_1}{\lambda_2} + {}_{r-1}m_0\right)^q.$$

The extinctions in each generation can then be easily calculated. Watson takes the case of $\lambda_1 = \frac{3}{4}$, $\lambda_2 = \frac{1}{4}$ and $q = 5$. In this case the $t$'s are

$$t_0 = {\cdot}237, \quad t_1 = {\cdot}396, \quad t_2 = {\cdot}264, \quad t_3 = {\cdot}088, \quad t_4 = {\cdot}014, \quad t_5 = {\cdot}001,$$

and the extinctions in the first ten generations of 1000 original distinct surnames:

237, 109, 65, 40, 27, 18, 14, 10, 7, 6,

[1] *Educational Times*, Vol. XXVI, 1873, p. 17 and p. 115.
[2] Vol. IV, pp. 138–44, 1874.

or a total loss of 533 surnames. Here the population increases since

$$t_1 + 2t_2 + 3t_3 + 4t_4 + 5t_5$$

is greater than unity. As before the extinction rate is quick to begin with but soon slackens down, as the number of persons holding each surname increases, while the number of surnames diminishes. On the above hypothesis nearly a quarter of newly-created peerages would become extinct in the first generation and half of them by the sixth generation. With any such hypothesis there is no need to appeal to sterility as rendering rapidly extinct a large proportion of the peerages created for ability. It will be clear that if we take not the number of sons, but the number of children, in computing the $t$'s, the problem becomes that of the extinction of definite stirps; it is highly probable that families die out in approximately the same manner as they die out in the male line. If mankind has not sprung from a single pair, it seems possible that even the most numerous nation may tend with the ages to be the product of a very few stirps, if not of a single pair. The fable of Adam and Eve may be somewhat truer for an old world than for a young one!

Beside the data noted in the paper on the stature of boys from urban and rural schools[1], several schools provided material of a more extended kind, notably Marlborough School, which had established something like an anthropometric laboratory[2]. The school medical officer and the natural science master took the measurements: namely weight, stature, horizontal circumference of the head, chest girth, girth of the flexed arm over the biceps muscle, girth of the leg over the calf, both the last two being the maximum measurements. The ages of the boys ranged from 10 to 19 and there were 550 of them. The authors of the paper give three correlation tables for age with stature, weight and head circumference, but make no reductions, citing merely in the case of the extreme boys in each measurement the other measurements of those boys. One remark deserves citing. The authors state that they

"are unable to trace any distinct connection between intellectual vigour and head measurement; for although many of those who possess the higher girths of head are intelligent boys of considerable ability, it must be confessed that many boys whose heads measure less than 22 inches, are in ability, perseverance, and general culture, quite equal to those who possess the higher measurements." (p. 129.)

This remark bears on a point already referred to in this *Life* (p. 94).

Galton's short accompanying paper confines itself to one character, stature, and he tells us that he proposes to illustrate the statistical methods which will be adopted, when sufficient material of a homogeneous nature is available. He takes the boys for each year of age and finds their means, which give for the central ages $12\frac{1}{2}$, $13\frac{1}{2}$, etc. the law of growth. He thus obtains what we should now term the regression line. But here he strikes a new point: he finds that the arithmetical means of the arrays are not identical with

[1] See our p. 125.

[2] "On a Series of Measurements for Statistical Purposes, recently made at Marlborough College." By Walter Fergus, M.D., and G. F. Rodwell, F.R.A.S. *Journal of the Anthropological Institute*, Vol. IV, pp. 126–30. "Notes on the Marlborough School Statistics" by Francis Galton himself follow this paper, pp. 130–5.

what he terms the "typical means." The normal curves for the age-arrays are not fitted to the arithmetical means, but to the "typical means." Galton does not describe how he obtained either "typical means" or normal curves; probably but not certainly his "typical means" are not "modes," but what he later termed "medians," and his normal curves were then found from the quartiles. The median or typical mean in all cases corresponds to a less stature than the arithmetic mean; and there is thus some little evidence that the arrays are asymmetrical and are not normal curves. Galton emphasises the point that we have really not a mere system of arrays, but a continuous frequency surface. Further he points out that the variation of these arrays widens as the age increases, a condition we now know to be incompatible with a normal frequency surface. It is a striking fact that in this first anthropometric surface Galton should have actually run up against the line of medians generally diverging from the line of means, when the variability of the arrays is not constant, i.e. that he should have come across the asymmetrical frequency surface, which is still proving a hard nut to crack.

We have seen how Galton urged anthropologists to turn from the sole discussion of external physical characters to the mental characters in man, such as personal equation and rapidity of judgment[1]. He early perceived the importance of the school not only for anthropometric physical but for anthropometric psychical measurements, and he endeavoured to enlist the schoolmaster in the service of psychical anthropometry. He rightly looked upon the school as not only an institution for educating the young, but as a laboratory for studying their mentality, and so by increased knowledge of psychical character improving education. In 1880 Galton wrote a strong letter on this point to *Nature*[2] as a result of his receiving from Mr W. H. Poole, then science master of Charterhouse, very valuable material on visual images (see our p. 237). It seemed to Galton—as it has often seemed to some of us who do not fully realise the pressure of school routine work—that other schoolmasters might emulate the exceptional Mr Poole.

"The observation I desire"—writes Galton—"to make is that as every hospital fulfils two purposes, the primary one of relieving the sick, and the secondary one of advancing pathology, so every school might be made not only to fulfil the primary purpose of educating boys, but also that of advancing many branches of anthropology. The object of schools should be not only to educate, but also to promote directly and indirectly the science of education. It is astonishing how little has been done by the schoolmasters of our great public schools[3] in this direction, notwithstanding their enviable opportunities. I know absolutely of no work written by one of them in which his experiences are classified in the same scientific spirit as hospital cases are by a physician, or as other facts are by the scientific man in whose special line of inquiry they lie. Yet the routine of school work is a daily course of examination. There, if anywhere, the art of putting questions and the practice of answering them is developed to its highest known perfection. In no other place are persons so incessantly and for so long a time under close inspection. Nowhere else are the conditions of antecedents, age, and present occupation so alike as in the boys of the same form. Schools are almost ideally perfect places for statistical inquiries....If a schoolmaster were now and then found capable and

[1] Address to Section H, British Association, 1877: see our p. 228.

[2] May 6, 1880, Vol. XXII, p. 9.

[3] The stress laid on the appointment of classical and clerical heads, to the neglect of scientific candidates, largely accounts for the matter.

willing to codify in a scientific manner his large experiences of boys, to compare their various moral and intellectual qualities, to classify their natural temperaments, and generally to describe them as a naturalist would describe the fauna of some new land, what excellent psychological work might be accomplished? But all these great opportunities lie neglected. The masters come and go, their experiences are lost, or almost so, and the incidents on which they were founded are forgotten, instead of being stored and rendered accessible to their successors; thus our great schools are like mediaeval hospitals, where case-taking was unknown, where pathological collections were never dreamt of, and where in consequence the art of healing made slow and uncertain advance.

Some schoolmaster may put the inquiry: What are the subjects fitted for investigation in schools? I can only reply: Take any book that bears on psychology, select any subject concerning the intellect, emotions, or senses in which you may feel an interest; think how a knowledge of it might best be advanced either by statistical questioning or by any other kind of observation, consult with others, plan carefully a mode of procedure that shall be as simple as the case admits, then take the inquiry in hand and carry it through."

I have cited Galton at length because in 1924 his words remain as true as in 1880, but I have faint hope that they will by repetition here reach a new generation of teachers more responsive than the old. In this country we have exceptional men who promulgate new ideas, but the average mind is an inert one. The school as laboratory, the factory as laboratory, the prison as laboratory, and the asylum as laboratory, these are essentially true conceptions, but their truth and their profit will be seen in America, in Germany—even in France—before they are grasped here! Galton scarcely realised that it required greater *ingenium* to discover a solvable problem than to carry it through when propounded, and that the average schoolmaster finds it easier to take prescribed measures of his boys—even to fill folios with them—than to discover an important problem and design new measurements to solve it. The school anthropometric laboratory must be futile if it be only a laboratory of record and not one of discovery. The fault lies rather with our current academic training than with the schoolmaster—for it lays greater stress on the average man solving set problems, than on finding novel problems himself.

The boy is never discouraged, and Galton retained his boyhood to the end. He could put on one side his teaching as to eugenics because the time was not ripe for it, and propound it with all his youthful enthusiasm nearly forty years later; the relative barrenness of the harvest resulting from his school anthropometric proposals did not cause him to despair of profits resulting from anthropometric inquiry in schools. In the eighty-third year of his life, thirty-three years after his first attempt, he returns to the charge, and with additional proposals, which would immensely increase the work—while needless to say they would enormously increase the utility—of school anthropometric laboratories.

In 1905, at the London Congress of the Royal Institute of Public Health, Galton gave an address on "Anthropometry at Schools[1]."

"**Anthropometry**, or the art of measuring the physical and mental faculties of human beings, enables a shorthand description of any individual to be given by recording the measurements of a small sample of his dimensions and qualities. These will sufficiently define his bodily proportions, his massiveness, strength, agility, keenness of sense, energy, health, intel-

[1] *Journal of Preventive Medicine*, Vol. XIV, pp. 93–98. London, 1906.

lectual capacity, and mental character, and will substitute concise and exact numerical values for verbose and disputable estimates[1]. Its methods necessarily differ for different faculties; some measurements are made by the foot-rule, others by scales, others by the watch; health is measured by the frequency and character of illness; the remainder by performances in the school or on the playground. Anthropometry furnishes the readiest method of ascertaining whether a boy is developing normally or otherwise, and how far the average conditions of pupils at one institution differ from those at others. Though partially practised at every school—for example in all examinations—its powers are far from being generally understood, and its range is much too restricted. But as an interest in anthropometry has arisen and progressed during recent years, it is to be expected that the good sense of school authorities, assisted by the expert knowledge of medical men, anthropologists, and statisticians, will gradually introduce improvements in its methods and enlargement of its scope."

This passage is noteworthy as it indicates how fully Galton had come to realise that the complete anthropometric laboratory must take measurements not only of statical physique and psychical characters, but also of the dynamic workings of the body, and generally of its physiological and medical fitness. What a stage onwards from that thin end of the wedge which suggested a measurement of stature, and obtained some half-dozen statical characters! But when we have got all this information, what is its value? Galton was not bent on describing what the school anthropometric laboratory should do for the boy, but what it should do for the man into which he developed. He regretted the deplorable and widespread lack of knowledge of the true value of anthropometric forecasts. Who can answer the questions:

"How far does success or failure in youth foretell success or failure in later years? What is the prophetic value of anthropometry at school in respect to health, strength and energy in after-life?"

Indeed these matters are only yet on their trial: Will the data collected in a fully equipped anthropometric laboratory recording the physical, mental, medical or other characters be able to make a forecast of the best career for a young man, or the probable success or failure in after life of its examinees? It will take twenty to thirty years to correlate well-selected measurements with experience in after careers. Galton realised this and wished to prepare the way for obtaining a life-history of the boys who had been measured in the school anthropometric laboratories.

"The first conclusion to be emphasised is that no programme for anthropometry in any school can be considered complete unless it provides for the collection of data during the after-lives of their pupils."

Every fourth year, Galton suggests, the "old boy" should receive a schedule and return it with an account of his doings in life, his health, vigour, his profession and achievements, his marriage and children. These four-yearly reports would be combined in one dossier with his school anthropometric measurements record. The schedule of these records would leave a space for one sheet of family history to be obtained from the boy's parents when he was about to leave school, which he himself would verify later, and there would be space for a few photographs.

Such was Galton's scheme in brief abstract. It will be seen to approach closely the eugenic record proposed many years previously, but now asso-

[1] It would be difficult to excel this passage as a description of anthropometry.

ciated with more detailed anthropometric measurements, and with the school as record office.

"The school authorities would rejoice in the possession of the whole history of those over whose early development they exercised large control. Anthropologists would know where to lay hands on a mass of material suitable for comparing the health, bodily qualities, and scholastic achievements in early life with the health, vigour and achievements afterwards. Statisticians would possess a four-yearly census, out of which unexpected conclusions would probably be derived. Lastly some few of the records would be invaluable to future biographers.... The effect of the present proposals would be...to prolong and intensify the kindly fellowship between past and present pupils and their school, and to make it serve more than sentimental purposes. The addition of a scientific motive could not fail to invest that relationship with a more durable and businesslike character, and to open a way to fields of research of no small importance that have hitherto been neglected."

Galton suggests that the return of the four-yearly schedule should always be made on February 29th, thus associating the return with Leap Year. On this day there should be school regatherings[1] and thoughts of the old school and former friends should predominate.

"The celebration of the day in schools would be much concerned with the works of living men, who were formerly pupils, but then engaged in the battle of life. Their doings would be spoken of, and hearty sympathy evoked. Affection and duty should co-operate in maintaining the bands of fellowship between school and former scholars; in short, its maintenance should be considered a 'pious' object."

"The child is father to the man,
And I would wish my days to be
Bound each to each by natural piety."

Galton's dream was a noble one, if the time for its fulfilment be not yet. Possibly it may one day be realised in ways the dreamer thought not of. I cite it here to show how rarely he let fall, rather more often amplified in his old age, the ideas of his younger days.

It must not be thought that Galton's principle: "Count whenever you can," led him to a slavish admiration of all types of statistics. There is a very striking illustration of the contrary. In 1877 the Council of the British Association had been much troubled by the proceedings of Section F (Economic Science and Statistics), and appointed a committee to report on

"the possibility of excluding unscientific or otherwise unsuitable Papers and Discussions from the Sectional Proceedings of the Association."

While Galton reserved a final judgment the remarks he put before the committee were adverse to the maintenance of Section F. He analysed all the papers of the years 1873–75 and remarked that

"not a single memoir treats of the mathematical theory of Statistics, and it can hardly be doubted that if any such paper should be communicated to the Association, the proper place for it would be Section A."

Galton admitted that Section F dealt with numerous and important matters of human knowledge, but such as are akin, for example, to History, not to Science, and are therefore inappropriate subjects for the British Association.

"Usage has drawn a strong distinction between knowledge in its generality and science, confining the latter in its strictest sense to precise measurements and definite laws, which lead

[1] Revisits of the old pupil to his school after a long interval might provide opportunities for recording a few simple measurements such as weight, stature, eyesight, strength.

by such exact processes of reasoning to their results, that all minds are obliged to accept the latter as true. It is not to be expected that these stringent conditions should be rigorously observed in every memoir submitted to a scientific meeting, but they must not be too largely violated; and we have to consider whether the subjects actually discussed in Section F do not depart so widely from the scientific ideal as to make them unsuitable to the British Association."

Galton's test of what constitutes science is clear—it is that of a mathematical physicist—and rigidly applied it would exclude large regions of biology, including possibly the doctrine of evolution.

But it emphasises exactly Galton's feeling with regard to much of what passed for statistics in 1877, that old type of statistics which had no theoretical basis, while Galton was working for a new type; he would willingly have transferred Section F to the Social Science Congress. But what could be said against Section F applied equally to and remained true till at least the end of the nineteenth century of the Royal Statistical Society itself. The opposite point of view was taken by Dr W. Farr; he cited a long list of *mathematical* statisticians from Halley to Poisson, who were undoubtedly men of science. But this was no real reply to Galton, for these men would have frequented Section A, and the atmosphere of Section F, or indeed of the Statistical Society, would have been as distasteful to them as to Galton. Probably the right procedure would have been to permeate Section F with the newer type of statisticians. This process has been more or less successful in the course of the last twenty years in the case of the Statistical Society. There is still opportunity for the modern school of statisticians to adopt a similar course with regard to Section F.

### B. STATISTICS BY SCHEDULE-ISSUES.

We have seen that Galton had great hopes from the schoolmaster as a collector of statistical data, but he by no means confined himself to this source of information. We have also noted how he appealed to English men of science, and to his many personal friends and others, by issuing schedules in preparing his books. The *Galtoniana* contain numerous instances in which he issued inquiry schedules, and in some cases we possess considerable numbers of these filled in. As a rule, however, I cannot find that he published anything bearing on the subject of the proposed inquiry. Either he never issued the schedule after printing it, or having issued it he was discontented with the quantity or the quality of the returns, and so made no use of them. Yet several of these schedules are so suggestive of the workings of Galton's mind, that they deserve a brief notice here.

Before 1876 Galton was much interested not only in the inheritance of longevity, but also in the influence of the age of parents on the vigour of their offspring. The schedule he issued is entitled "Inquiry into the Relation between Vigour in the Offspring and Age in the Parents," and it is prefaced by the remark:

"Instances are sought of old persons of both sexes, who have retained their bodily vigour and activity in very advanced life. It is desired to know the ages of their fathers and mothers at the time of their birth."

The schedule contains spaces not only for the facts illustrating the special vigour in old age of the subject, but for the ages of his parents at his birth, the size of his co-fraternity and his position in it, and further for other instances of exceptional longevity in the kinship. Among the somewhat meagre data collected are several instances of marked hereditary longevity, and one of a man who above eighty became the father of healthy children. Galton was undoubtedly interested in this inquiry owing to the hereditary longevity in his own family[1], but the knowledge of this fact did not relieve him from having at times considerable anxiety as to his own health, and in the sense of the proverb "that cranky doors hang longest on their hinges," he was interested to know whether "a considerable proportion of aged persons have been more or less ailing through a great portion of their lives."

I am not sure whether a printed document I have found with the longevity dossier was issued with the schedule or prepared for some later inquiry; it bears evident traces of Galton's complete or co-operative production. It is so suggestive for an inquiry which apparently has never been made, and still might be made with great profit, that I have reproduced it bodily here.

Those who knew Galton personally will trace some of his beliefs and some of his doubts seeking statistical confirmation in this document.

*An Inquiry concerning Persons who have attained or passed the Age of Eighty Years.*

This inquiry, as will be seen by the card, is intended to be general, the object being to obtain by Collective Investigation on a large scale, information respecting the present and past condition, habits, and maladies, as well as the family history and other circumstances, of those who have attained to advanced periods of life, in order that we may be able to ascertain, with greater certainty than we now can, what are the circumstances which favour longevity, the means by which it may be promoted, and the maladies which are most, and those which are least incidental to it.

The following are some of the questions which arise in connection with this subject, and for answers to which we may look.

What bodily conformation, temperament, and habits, are most associated with, or conducive to, longevity?

Do women more frequently attain to great age than men, and have women somewhat below the ordinary stature the advantage in this respect?

Are the married or the unmarried, the stout or the spare, the active or the sedentary, the industrious or the idle, the indoor student or the outdoor workers, the well-to-do or the poor, the town dwellers or the country dwellers, the more likely to become octogenarians?

It is said that "small eaters and short sleepers are long livers." Is this so? Will the "early to bed and early to rise" maxim receive confirmation? What is the influence of alcohol?

It has been remarked that a considerable proportion of aged persons have been more or less ailing during a great part, or the whole, of their lives. Is that the case? It has also been remarked that many of them have been troubled with constipation, and that many have long been in the habit of resorting to aperient medicine.

The cartilages of the ribs and the trachea have been found soft and elastic in some very aged people, old Parr forming no exception in this. Should this be shown to be generally the case, the inference would follow that persons in whom they are not so are not destined to attain to great age.

Do octogenarians often suffer, or do they enjoy a comparative immunity from affections of

[1] See *Memories of my Life*, p. 7—a paragraph which contains the only reference I have seen to the age of Elizabeth Collier's mother at death, if it be not a slip for the age at death of Erasmus Darwin's mother.

the urinary and genital organs, and of the abdominal organs, also from malignant disease and scrofula?

Are they on the whole comparatively exempt from disease?

To what affections are they most liable, and to what morbid influences are they most susceptible? Do any maladies seem to have an influence in promoting longevity? What influence upon the longevity of an individual has the age of the parents at his birth? Do twins or the children of twins often attain great age?

Information, though not positive, yet of much interest and importance, upon these and other points, will accrue from the replies to the questions on the cards. It need hardly be said that the questions are by no means exhaustive, and that information upon other points which are judged to be of interest and importance by those who fill up the cards, will be valuable, as also information on any special points in particular cases which seem worthy of note.

Though the questions are such that they may for the most part be answered by the persons themselves, or by their friends, it is hoped that in most instances the observations will be made and the information given by medical men; and the person who fills up the card is in each instance requested to state whether he is a medical man or not.

It will be an additional advantage if some information can be gleaned respecting the succession of maladies in the same person, and in different individuals of the same family, or respecting the preservative influence upon the system of certain maladies against the inroads of others.

Something in the Hereditary Problem may be also learned respecting the cross-action and modifying influences of certain diseases. For instance, is there any foundation for the view that chronic gouty affections retard the development of other diseases.

The strength and enduring quality of the body, like that of a chain, must be measured at its weakest point; and though in it, more than in a chain, the strength and quality of some parts may compensate for deficiency in others, yet the very opposite may be the result. The stronger organs may relieve, but they may also oppress, the weaker members. A strong digestive system may overload a weakly circulation, and prove injurious to the liver, lungs, or kidneys, in fact a disturbing agent to the general nutrition. The requisite for longevity, therefore, we may expect to be not so much strength of organs as their enduring quality, a good mutual adjustment, in other words, their good balance. The replies relating to "plethora" and other features of general condition will have an important bearing on this view.

Another matter about which Galton's mind was much exercised was that of "social stability." He was anxious to know whether and to what extent individual stirps move up and down in the social scale, or whether our society is in the main made up of "castes," which stand fast by their grade of occupation and to their social position. To throw light on this matter he sought a comparison between the average social position of householders of all classes in the present day and in that of their fathers. He prepared accordingly a schedule—which appears never to have been issued—of inquiries concerning householders and their fathers; the age of the householder, his profession, trade, or employment, his position among men of the same occupation, and the corresponding data for the father of the householder were to be recorded, and Galton hoped to get material ample enough to provide a measure of social stability—the frequency with which sons advanced on, remained equal to or regressed from the father in social status. While this schedule deals with social stability in its narrower economic sense, the same dossier contains notes by Galton giving the term social stability a much wider significance; he notes that many view with alarm the progressive disappearance of those ancient landmarks—such as theological beliefs—by which conduct has been traditionally regulated and fear that mankind must sink into brutality. The motives which lead to social stability ought therefore to be measured and analysed, the main fact being that it is discoverable among the most divergent populations and the most varied surroundings. The inquiry, Galton tells us, must be *statistical* in its character, referring to the acts of a population as a whole, and not regarding the units of which it is composed, for it is only in this way that we can neutralise and eliminate the effects of individual character and circumstances, which are far too numerous to be severally allowed for. Galton's notes then turn to the "valuation of motives," and he asserts that while they are of the most varied kind, they are yet *commensurable*; they may be equally efficacious in producing a particular result. "There are an indefinite variety of bribes, and experience shows the amount of bribe of each several sort that is necessary to produce a given average result[1]." The attractive forces of each of many shows at a fair, appealing to many diverse tastes, are comparable in a statistical sense without any other reservation by the money they take. I will not venture to cite more of Galton's rough notes; he was thoroughly convinced that "motives" like other psychical characters are capable of statistical evaluation. To press the matter would be to call forth from some readers a protest similar to that which the editor of the *Spectator* made after Galton's Royal Institution Lecture of 1874, wherein he applied the method of "ranking" to psychical characters.

"We can only express our wonder, and repeat our belief that what Mr Galton has succeeded in doing, is in exposing the utter inapplicability of physico-scientific methods to intellectual and moral subjects....We can imagine no more profitless or idle task than the attempt to draw out a Statistical Scale (say) of Candour or of Power of Repartee, and to arrange the public men of this generation in it, except indeed doing the same thing for a considerable number of qualities, and giving the reasons for the place assigned in the biographies, which would be rendered unreadable by the process[2]."

There might be difficulty in "ranking" Gladstone and Disraeli for "Candour," but few would question John Morley's position relative to both of them in this quality. It would require an intellect their equal to rank truly in the quality of scholarship Henry Bradshaw, Robertson Smith and Lord Acton, but most judges would place all three above Sir John Seeley, as they would place Seeley above Oscar Browning. After all there are such things as brackets, which only make the statistical theory of ranking slightly less simple in the handling.

Drafted much about the same time was Galton's first circular on "Fatigue," by which he sought to measure any permanent ill effects of mental work. This again was a topic on which Galton felt strongly, having his own experience always in mind. The proposed circular was to be addressed apparently to the fellows and scholars in Cambridge (and possibly Oxford) Colleges, and related not only to mental overwork, but to its possible association with physical overstrain, in both school and college periods of life. He probably

[1] "The gingling of the guinea soothes the hurt that honour feels"—which is not exactly Tennyson. Galton was wont to say, on seeing a hilarious party of middle-aged persons, that it struck him as strange that notwithstanding their glee they were all of them orphans.

[2] See the *Spectator*, May 23, 1874, and Galton's letter with the editorial rejoinder May 30, 1874. "It is about time we drew the *Spectator* again," W. Kingdon Clifford would say, and Galton was only too apt to do so without malice prepense!

refrained from circulating his questionnaire, as so many of the recipients might reasonably associate "mental overstrain" and "mental breakdown" with a form of mental illness they would be unwilling to admit having suffered from. As we have seen[1], Galton took up the topic again in 1888, endeavouring to obtain the requisite data from school teachers.

The next circular I pick up is entitled: *Ethnological Inquiries on the Innate Character and Intelligence of Different Races.* By Francis Galton. The object of these inquiries is clear, they were intended to obtain statistical data upon which a judgment might be made as to how far racial character or training influences the mental characters. The "subjects" dealt with are to be those "who have been reared since childhood in European or American schools, families, asylums or missionary establishments. By this restriction, it is hoped to eliminate all peculiarities that are due to the abiding influence of early education, and to the manners and customs of their own people." The *standard* to be kept in mind in answering these questions is the *average* Anglo-Saxon character; paying strict regard to the influence of sex, age, education and social position. Where there is no *decided* divergence from this standard, it will be best to reply—'ordinary.'

The *Galtoniana* contain no replies to this circular; I do not know whether it was ever issued in mass, nor have I anywhere seen a reference to it, nor to data obtained by its circulation. The origin of it may be connected with the idea conveyed by Galton's treatment of unlike twins under like environment (see our p. 126 *et seq.*). As we might suppose the questions are well chosen, and bear closely on Galton's own experience with uncivilised races. As the questionnaire would be distinctly helpful to anyone embarking on an inquiry of like kind—and one might be well worth pushing with more vigour than Galton seems to have given to the matter—I reproduce the questionnaire here:

1. Signature, title and full address of the sender of the information.
2. Name or initials, sex and age of individual whose character is described.
3. His (or her) country and race. State specifically if his race is known to be pure, if not describe the admixture.
4. Age at which he was removed from his parents and people, also particulars showing the extent to which he has since been separated from their influence.
5. What language, or languages, does he commonly speak? Does he retain the use of his native tongue?
6. State any circumstances that may or may not justify his being considered a good typical specimen of his race.
7. Is he capable of steady and sustained hard labour; or, is he restless and irregular in his habits?
8. Is he capable of filling responsible situations? Does he show coolness of temper when in difficulty? (It is said that Hindoos are incapable of steering large ships, that is, of acting as quartermasters; while in British vessels that duty is commonly performed by native Christians of the Philippines.)
9. Is he docile or obstinate?
10. Children of many races are fully as quick, and even more precocious than European children, but they mostly cease to make progress after the season of manhood. Their moral character changes for the worse at the same time. State if this has been observed in the present instance.

[1] Cf. our p. 276.

11. Has he any special aptitudes, or the reverse, such as in mimicry, sense of the ludicrous, taste in colours, music, poetry, dancing, calculating power, keenness of sight or hearing, quickness of observation, manual dexterity, horsemanship, ability to tend cattle?

12. Is he naturally polished and self-composed in manner or rude and awkward?

13. Is he modest and self-reliant, or servile and cringing? Is he vain?

14. Is he solitary or sociable; morose or cheerful?

15. Is the passion of sexual affection strongly developed in him, or the reverse?

16. Is he fond of children, and are children fond of him?

17. Does he cherish malice for long periods, or does he forgive frankly?

18. Is he liable to outbursts of rage?

19. Did he for long show uneasiness at the restrictions of civilised life, or did he readily accept them; such as keeping regular hours, acting on a steady system, wearing shoes and other clothing?

20. Children of savages, who have been reared in missionary families, have been known to throw off their clothes, and quit the house in a momentary rage, and to go back to their people, among whom they were afterwards found in apparently contented barbarism. State authentic instances of this, if you know of any, with full particulars.

21. Has he a strong natural sense of right and wrong, and a sensitive conscience?

22. Does he exhibit to his religious teachers any strong conviction of an original sinfulness in his nature, or the reverse?

23. Is he much influenced by ceremonial observances, such as those of the Roman Catholic Church?

24. Is he a willing keeper of the Sabbath?

25. Has he any strong religious instinct; is he inclined to quiet devotion?

26. Is he ascetic, self-mortifying and self-denying, or the contrary?

27. Is he inclined to be unduly credulous or unduly sceptical?

28. Is he active or impassive in social duties?

29. Is he much governed by superstitious feelings, such as [are indicated by the use of] charms or omens of good or ill luck?

30. Has he any tendency to be sanctimonious and hypocritical?

31. Is he honest, truthful and open, or cunning and intriguing?

32. Is he grateful or ungrateful?

33. Does he, in conversation, make frequent use of abstract terms? Does he adequately understand their meaning when he employs them?

34. Are there any other marked peculiarities in his character or intellect?

*Please address copies to*
FRANCIS GALTON,
42 Rutland Gate,
London.

This is a schedule which—if the employers of native labour could be induced to fill it up accurately in large numbers—would still be certainly of much value.

Francis Galton's next venture was entitled:

*Inquiry into the alleged Darkening of the Hair of the English in the Present and Recent Generations.*

The explanation of the reason for the inquiry is given on the back of the schedule. It had been alleged that on the whole the hair of English children was darker than that of their parents, and it was asserted that the English race was gradually but surely becoming dark-haired. The object of the inquiry was to test the truth of this statement. Galton remarks that it is probable that the recent and rapid changes in English habits may have caused certain sub-types, that were previously repressed, to prevail in the struggle for existence, and that it is of interest to know what these sub-types are. The colour of animals is often found to be intimately correlated with their power or incapacity to thrive under certain conditions, and it may well be the same in the case of man. Galton cites Baxter[1] to prove that in America, where the pressure of life peculiar to modern civilisation is even greater than with us, the black-haired persons are less liable to nearly every form of disease than the fair-haired. He observes, however, that it is needful at the same time to determine the relative fertility of the light and dark haired, and that it would be very important to distinguish between the children of a dark-haired man who had sprung from a light-haired stirp, and those of a similar man from a dark-haired stirp. The schedule is fairly straightforward and contains the first statement of Galton's system of numerals for relationship, i.e. child 1, parents 2, 3, grandparents 4, 5, 6, 7, etc., the even numbers standing for males and the odd for females (No. 1 excepted, which may have either sex); the number of any individual when doubled gives that of his father, and his mother's number is obtained by the addition of one to the number of his father[2]. The characteristic Galtonian statement is made incidentally that:

"The inquiry will have the merit of being accompanied by incidental pleasures; it will be an excuse for corresponding with distant friends and relations on topics of common interest, and it is probable that not a few facts of family history much prized by its members will in many cases be incidentally brought to light by its means."

Galton himself was so interested in family history that he quite naively overlooked the fact that nine-tenths of humanity either fear to examine it or are frankly bored by it. Against that dead-weight of inertia Galton could effect little, and there is no evidence that these circulars were ever returned in sufficiently adequate numbers to serve as a basis for an answer to his inquiry.

[1] *Medical and Anthropological Statistics of the Provost-Marshal-General's Bureau*, Washington, 1875.

[2] Galton published a letter on this numerical system of relationship in *Nature*, Sept. 6, 1883, under the title: "Arithmetic Notation of Kinship." Taking $f$ = father of, $m$ = mother of, he gives the following equivalent systems of notation:

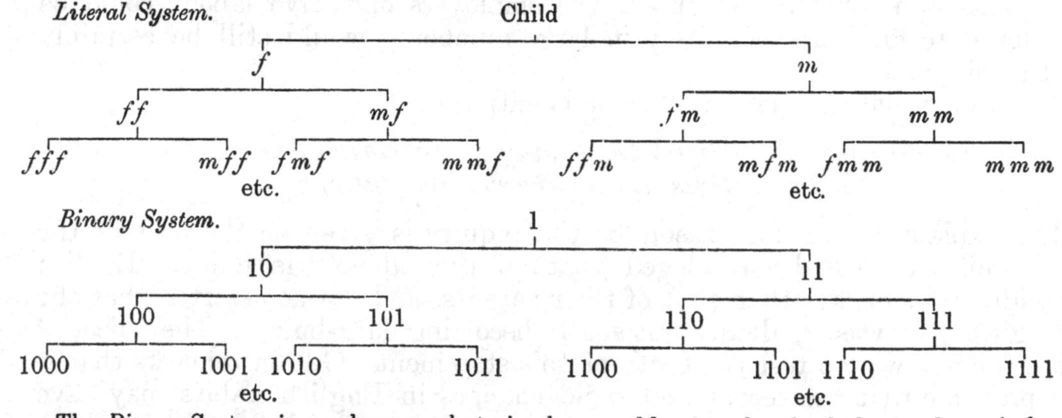


The Binary System is cumbersome but simple, we add a zero for the father and a unit for the mother of any individual to that individual's number. The decimal system is as follows:

The next schedule I have come across is termed a "Biographical Register." It starts with a genealogy of the subject as far as the grandparents and their descendants with a space for more distant relatives. Then follows the biographical register proper with a column for each age period of seven years, with spaces for education (class lists), amusements (tastes and pursuits), accidents and bad illnesses, anthropometric tests at various ages, and other characteristics. The "Notes" show that personal appearance, pigmentation, height, weight, etc. were to be included, and eventually marriage and children. There is not a doubt but that this was the original scheme from which the *Life-History Album* sprung. The interesting point is that this biographical register was designed for undergraduates. The returns were apparently to be preserved in the archives of the colleges for future statistical purposes and for the compilation of college histories.

"It is believed that a large collection of personal and family records such as these, would furnish important data for investigating the social and hereditary antecedents that are most favourable to success in college and after life. They will certainly protect from oblivion many facts that may hereafter prove of considerable biographical interest to the undergraduates themselves and to their families; possibly to a much wider circle."

Again there appears to have been no result from this schedule, even if it were ever issued to an undergraduate population. The author of this biography knows only too well—having collected with the aid of colleagues two long series of schedules from undergraduates—how hard is the task; each series took four to six years to collect even by those who were actually working and teaching among the population; and Galton had none of these advantages. The very wealth which enabled him to carry out effectively his experimental ideas, prevented him from seeking and holding a teaching post, whereby he could have created more quickly a school, and been able to collect adequate material. It would be hard to say whether the balance was one of gain or loss to the world. There were factors in Galton's character—his invariable courtesy and kindliness, his love of simple methods, his sympathy with younger minds, and his suggestive enthusiasm—which would have made

*Decimal System.* We translate the binary into the ordinary scale. Thus:

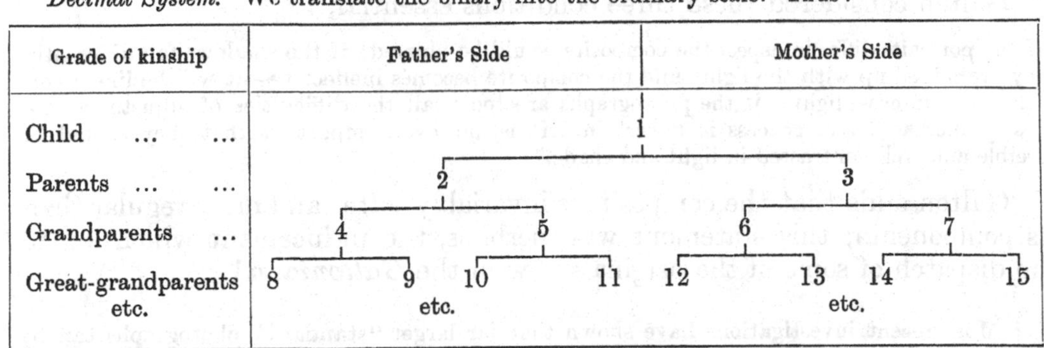

Grade of kinship
Father's Side
Mother's Side
Child ... ...
Parents ... ...
Grandparents ...
Great-grandparents
etc.
1
2
3
4
5
6
7
8
9
10
11
12
13
14
15
etc.
etc.


A want of these systems is an expression for the sibship of any individual, his or her brothers and sisters, or again for his or her nephews and nieces, uncles and aunts. Perhaps decimal figures might be added.

him a great teacher, but a teaching post would probably have cost him that travel-experience and that leisure to ruminate on which so much of his scientific success depended. He loved to work and to play with absolute freedom, and fixed duties would probably have been irksome to him, even if his health could have stood the *opus cathedrae strepitusque*.

The "Biographical Register" was followed by a "Genealogical Table of the family of brothers and sisters that includes ——." We need not linger over this, it was the immediate forerunner of the "Family Records," which when Galton hit upon the idea of offering money prizes for filling in schedules became at once a great success—the material source whence sprung his two books the *Life-History Album* and *Natural Inheritance*. The latter will be duly considered in our chapter on Galton's contributions to Heredity.

One remaining schedule may be noticed here—Galton's circular letter of March, 1882 entitled: "Application of Composite Photographic Portraiture to the Production of Ideal Family Likenesses." This circular is remarkable for its artistic printing and "get up." It is an appeal to amateur photographers to provide full-face and profile portraits of members of families, and contains a characteristic family composite. The "bribe" in this case was a print of the family composite together with the negative if they desired it. Galton also stated that he should await with great interest the family's opinion on the family likeness. The response to this circular was very considerable, and the ruins of the material—for most of the photographs have perished or are perishing—are still in the *Galtoniana*. The conditions Galton demanded for the composite are worthy of preservation:

1. The set of portraits must be all *absolutely* in full-face, looking straight at the camera just above the lens, or they must be all in profile, with the eyes directed straightforward along their own level.
2. The light must fall from the same direction in every case; it is best that the sitters should occupy successively the same seat.
3. The portraits of which the head alone is used, must be of about the size of the sketches on the previous page, that is, a little more than an inch from the chin to the top of the head[1].

Galton considered these three conditions essential,

"if the portraits differ in aspect the composite would be blurred; if the shadows fall differently they are mixed up with the lights and the composite becomes ineffective—it will be like a portrait taken in cross-lights; if the photographs are too small the difficulties of adjustment are greatly increased and success is uncertain....It is, however, important that they should be forcible and well contrasted in light and shade."

Galton adds that the composite is invariably softer and more regular than its components; this statement was, perhaps, the inducement which led to the dispatch of some of the originals now in the *Galtoniana*!

[1] More recent investigations have shown that far larger "standard" photographs can be advantageously employed.

PLATE XLIX

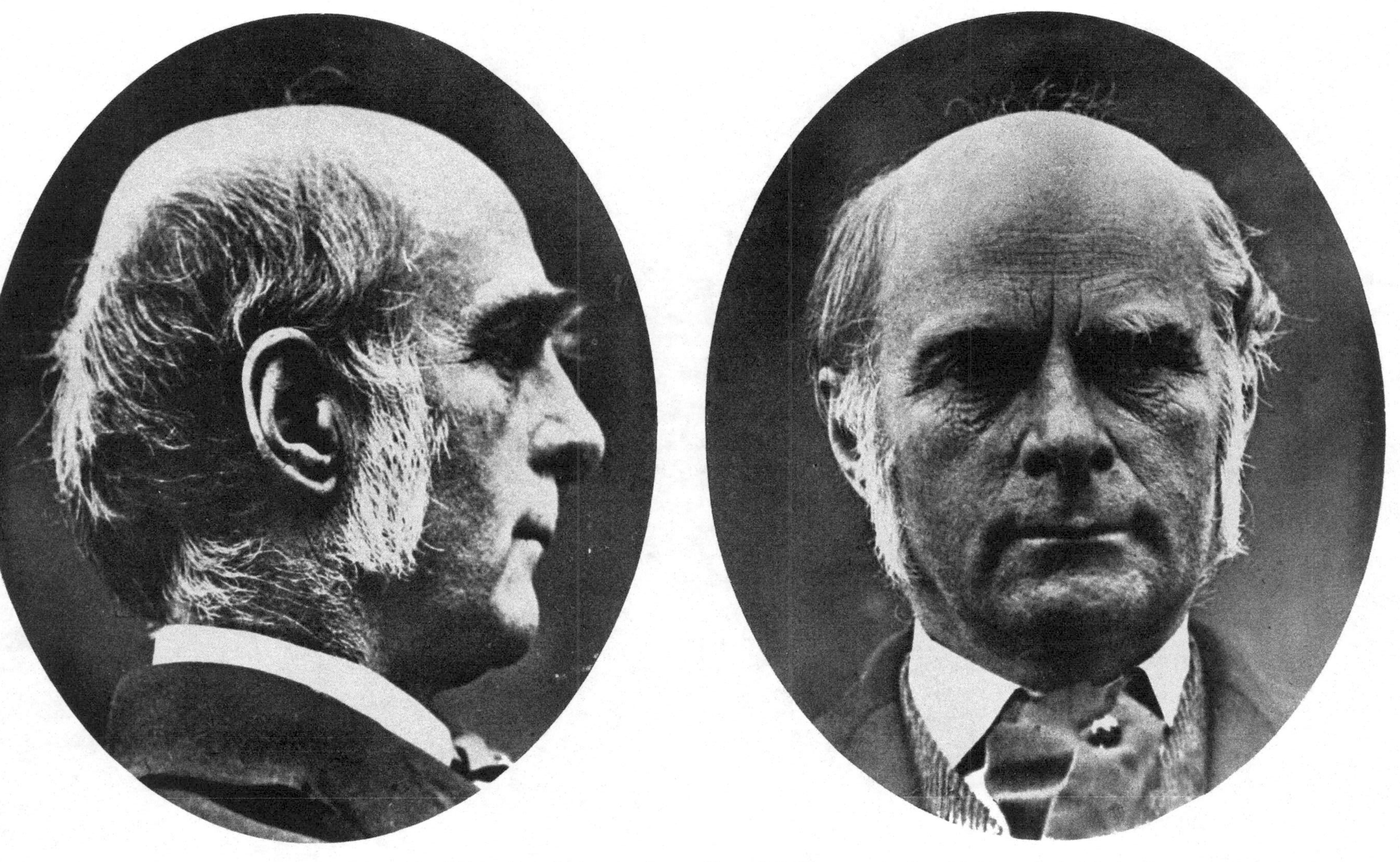

Francis Galton's "Standard Photograph" of himself to illustrate the profile and full-face portraits which are desirable in the case of Family Records and Life-History Albums and are suitable for composite photography.

### C. GALTON'S ANTHROPOMETRIC LABORATORIES

The above series of schedules will show how fertile were Galton's plans for collecting statistical data during the decade 1874–84. It was, however, only in the course of this schedule experience that he learnt how reluctant most people are to fill up a schedule. As a result of this experience Galton changed his method of action. Failing the establishment of school anthropometric laboratories Galton determined to set one up at his own cost, and catch the world when on its leisurely and inquisitive peregrinations. He called into existence the first Anthropometric Laboratory at the International Health Exhibition in London, 1884. On the closing of this exhibition the laboratory was removed to the Science Museum, South Kensington, and the total number of visitors measured before it was closed was well over 9000. These included both sexes and all ages from five to eighty years. This splendid material, which is only at the present time being fully reduced and utilised[1], together with Galton's "Family Records" embracing between three and four hundred families, some 150 'stirps,' provided him at last with the material he had so long sought. The discussion of this material furnished Galton with occupation for at least ten years; and the need for novel statistical methods, which its problems demanded, led him to the correlational calculus, the *fons et origo* of that far-reaching ramification—the modern mathematical theory of statistics. One quakes to think of what might have happened had Galton not obtained through that first anthropometric laboratory and his family records the data he needed! The latter led him at once to the quantitative measure of heredity—the correlation of kinsmen for any faculty—and the former showed him that the same problems repeat themselves in all statistical material, and that the conception of correlation is not peculiar to heredity, but embraces all recordable qualities which without being causally linked together yet vary more or less stringently one with the other. From that conception arose a new view of the universe, both organic and inorganic, which provides all branches of science with a *novum organum*, far wider-reaching in its effects than that of Bacon, and as characteristic of the last quarter of the nineteenth century as the fluxional calculus was of that of the seventeenth. I have sought in vain for any forerunner of Galton in this matter[2], and feel convinced that he was the first to grasp not only the need of measuring associated variations, but the first to provide any real measure of them. Galton wrote to Darwin on December 24, 1869 that the appearance of the *Origin of Species* had formed a real crisis in his life and freed him from his old superstition as if he had been roused from a nightmare (see Vol. I, Plate II). For some of us Galton's new calculus acted in precisely the same manner; it enabled us to reach real knowledge—"to submit phenomena to measurement and number"—in many branches of inquiry where

[1] See for example Koga and Morant, "On the Degree of Association between Reaction Times in the case of Different Senses," *Biometrika*, Vol. XV, pp. 346–72, 1923.

[2] See a paper by the present writer entitled "Notes on the History of Correlation," *Biometrika*, Vol. XIII, pp. 25–45, 1920.

opinion only had hitherto held sway. It relieved us from the old superstition that where causal relationships could not be traced, there exact or mathematical inquiry was impossible. We saw the field of scientific, of quantitative, study carried into organic phenomena and embracing all the things of the mind. It was for us the dawn of a new day, and we smiled indeed over the attempts of the *Spectator* to obscure such a daybreak by looking westward and asserting it was and must remain night.

To those who realise what Galton's work meant for some of us in the 'eighties, when fresh from Cambridge we encountered his papers, there is something of supreme interest in the path by which he reached his conceptions, his long failure to collect data and its final solution in the Anthropometric Laboratory.

The growth of Galton's plan for creating an Anthropometric Laboratory is fairly well exhibited in his papers. We have first the idea of very simple statistics being collected in schools, then the plan of a somewhat more extended school laboratory and in 1882 a paper in the *Fortnightly Review*[1], "The Anthropometric Laboratory." The points of present day importance in this paper are the following:

(*a*) Galton propounds the need of an institute where a man may from time to time get his family and himself measured physically and mentally and photographed according to a standardised method.

(*b*) He reasserts his conclusion that circumstances and education have very little to do with an individual's capacities. These are provided by his heredity, they form his stock-in-trade, the amount of which admits of definition, and by means of which he has to earn his living and play his part as a citizen. Just as far as we succeed in measuring them, so far we shall be able to forecast what a man is fit for, and what he may undertake with the least risk of disappointment. In other words we have the first foreshadowing of industrial or occupational anthropometry.

(*c*) He then proceeds to speak very briefly of the old type of anthropometric records (chiefly statical), height, weight, vital capacity, grip[2], pigmentation, etc.

He next turns to Energy and Endurance. He considers that the true tests would be physiological and very delicate, measuring excess of waste over repair. Just as a clothdealer tests a piece of cloth by moderate tension without tearing it, so the balance of the living system might be artificially disturbed by a definite small force and its stability under the influence of greater forces be thereby inferred. He admits that at present tests of a person's endurance under sustained bodily or mental work have not been adequately developed. But he recognises that dynamic tests—the functioning of the body—are far more important than static tests. He would have *agility* tested by gymnasium or athletic sports tests. Co-ordination of muscles and eye by measured skill in well-known games from racquets to billiards[3].

[1] N.S. Vol. xxxi, pp. 332–8.

[2] Vital capacity and sustained grip belong rather to the dynamic characters.

[3] Since 1884 balancing (a slender column on a flat board raised from hip to shoulder), maze

(*d*) Keenness and discrimination of the senses are next emphasised as indispensable tests.

(*e*) Reaction times and judgment times follow; memory of form and memory of number. These points are sufficient to indicate that Galton from the earliest time laid as much stress on the psychical as on the physical tests of an anthropometric laboratory. Nay, he went further; he asserted that:

(*f*) There is need for a medico-metric section in an anthropometric laboratory. This section would make as exact and complete a report of the physiological and medical status of an individual as is feasible in the present state of science by the help of the microscope, chemical tests and physiological apparatus.

Such a "medico-metric" laboratory Galton holds would be useful to the general practitioner who could send his patients to be examined in the same manner as physicists send their delicate instruments to Kew Observatory to have their errors ascertained. Great stress is laid on the physician writing case notes of the successive illnesses of private patients even as he takes clinical notes at the bedsides of his hospital patients. These notes should be preserved by the patient and accumulating with the years would form his medical life-history, and be a unit-contribution to the medical history of the family. Galton emphasises the value they would be as an heirloom to the children of the subject and to their medical attendants in future years by throwing light on hereditary peculiarities. In short Galton saw in the anthropometric laboratory a centre for standardised family records of biographical interest to all members of the family, of value from the medical point of view to each individual during his life, and to his descendants as suggesting hereditary dangers and vital probabilities. Lastly and perhaps for Galton himself the most important advantage was the material they would ultimately provide for much needed statistical research into human genetics.

For the race the value of such records is incontestible, but all men have not Galton's power of calm self-introspection, and the effect of studying his family medical history in the case of a neurotic subject might well be disastrous for the individual.

The idea of medical family histories was further developed by Galton in a paper entitled "Medical Family Registers" in the *Fortnightly* for August, 1883[1].

In this paper Galton defines more closely what he means by medical histories and states that he has consulted a number of eminent medical men (Simon, Beddoe, Duncan, Gull, Ogle, Ord, Richardson and Wilkes) who have approved the scheme. In this article he suggests for the first time—as far as I am aware—a system of monetary prizes.

"I have made arrangements to initiate the practice of compiling them [Medical Family Registers] through the offer of substantial prizes, open to competition among all members of the medical profession. The prizes will be awarded to those candidates who shall best succeed

(a pencil carried in measured time round the convolutions of a maze without touching the sides), needle (a fine knitting needle put through a series of holes of decreasing diameter without contact) and other similar tests have been introduced replacing skill in games.

[1] N.S. Vol. xxxiv, pp. 244–50.

in defining vividly, completely, and concisely the characteristics (medical and other) of the various members of their respective families, and in illustrating the presence or absence of hereditary influences."

We have seen how Galton grew from traveller to geographer, from geographer to ethnologist, from ethnologist to anthropologist, and now the last stage appears: he is chiefly interested in anthropometry because of the contributions he expects from it to heredity; the anthropologist becomes a geneticist. Looked at superficially Galton's work seems like a comprehensive but confused mosaic of many branches of science. Studied in relation to his life we see a definite pattern, a picture of a long-continued mental development; each branch of knowledge he acquired fell into its fitting place, and formed a stepping stone to a further advance.

His own interest in Medical Family Registers arises, he tells us, from

"all that can throw light on the physiological causes of the rise and decay of families, and consequently on that of races. Some diseases are persistently hereditary, and others are not; they are variously found in different varieties or subraces of men, and these have various other attributes including various degrees of fertility. We cannot as yet foretell, but we may hope hereafter to do so in a general way, which are the families naturally fated to decay and which to thrive, which are those who will die out and which will be prolific and fill the vacant space." (p. 245.)

In this paper Galton shows that he has realised more fully the difficulty about medical registers:

"Most men and women shrink from having their hereditary worth recorded. There may be family diseases of which they hardly dare to speak, except on rare occasions, and then in whispered hints or obscure phrases, as though timidity of utterance could hush thoughts and as though what they fondly suppose to be locked-up domestic secrets may not be bruited about with exaggeration among the surrounding gossips. It seems to me ignoble that a man should be such a coward as to hesitate to inform himself fully of his hereditary liabilities, and unfair that a parent should deliberately refuse to register such family hereditary facts as may serve to direct the future of his children, and which they may hereafter be very desirous of knowing. Parents may refrain from doing so through kind motives; but there is no real kindness in the end." (pp. 245–6.)

Still Galton recognised that the difficulty remains, that the majority of men do fall into his category of ignoble cowards and will not record their family secrets as to disease. Accordingly he proposed to get over the difficulty by inducing medical men, under the bribe of £500 in prizes, to give confidential records of their own families. He hoped that the custom of medical family records having been introduced in this way, doctors would thereafter be not infrequently called upon to draw them up for the satisfaction of the patients themselves, and—Galton adds naively as a lure—"at their expense"! The particulars Galton proposed should be included in the registers to be provided deal not only with medical details, but with race, conditions of life, marriage, fertility, vigour, keenness of sense, artistic capacity, intelligence, character, etc. He was clearly working up to much of the material he finally asked for from laymen in his "Family Records." A great part of the paper is taken up with the conditions under which the prizes he proposed would be given.

There is an interesting paragraph as to fraud:

"As regards the probable trustworthiness of the information received, I am perfectly aware that a modern De Foe or Swift might write an interesting romance, and make a register apparently true to life, wholly out of his own head; but De Foes and Swifts are not common, and such persons would be sure to find better occupation than that. Moreover they could not gain a prize without committing a downright fraud. Able men are generally above petty tricks, and there will be abundant internal evidence in every register to show whether the writer be able or not. It is almost needless to remark, that every statistician worthy of the name is wary and slow to accept startling conclusions without much indirect confirmation.... What I fear most is that the registers sent by many of the candidates will afford internal evidence of being little trustworthy, not through deliberate intent, but owing to the incapacity of the writers to state their cases clearly, and to support their statements with judiciously selected data." (pp. 248–9.)

Perhaps this latter remark was rather severe on the profession to whom Galton was appealing for aid; it failed to give due weight to what should be the effect of the clinical training in a hospital on a man's powers of observation, record and deduction. Yet our poor hero saw with yearning those 23,000 qualified medical men and thought here at last was a source of the material essential to his work!

"I should hope that the examination would be complete after some three months' labour of myself and the examiners. The prizes being allotted and done with, it will remain to work up the results....The statistical meal will be a large one; I gloat over it in anticipation, and know that it will take long to digest. I cannot doubt that new ideas will be derived from a careful study of so unique a collection, enough I hope to justify to myself the cost and time spent on it. When I shall have done with this collection, its ultimate destination will probably be as a gift to some appropriate medical or anthropological institution. It will then be in the form of anonymous documents bearing mottoes, but with no mark by which any one of them could be distinguished....Considering that prizes for essays usually attract numerous competitors, although the pains taken in working for them are rather barren of result except to the winners, I conclude that similar prizes leading to inquiries beneficial in every case, and from many points of view, ought to attract yet more numerous candidates, and to result in producing shelves full of family histories of unprecedented completeness and concentration, and of extreme value for a long time to come to medical and anthropological investigators." (p. 250.)

What killed this scheme of prizes to the doctors for medical registers of their families? I can find but one further reference to it (see our p. 367). All letters for the year 1883 seem to be wanting; so that we cannot trace the causes which led Galton to drop the emphasis on the medical register and offer his prizes for family records to laymen as well as medical men. Three extracts from *L. G.'s Record* may be given here as conveying about all the knowledge we have of the matter:

"1882. Frank began his book on *Human Faculty* early in the year and it has gone on through the year, and was a great pastime to him during our summer ramble on the Rhine, in the Black Forest, Constance, and lastly Axenfels. Bad weather haunted us, but we were happy and I kept well and began sketching again. It was such a boon not to be kept by a British Association Meeting this summer. Mr Darwin's death in May had cast a deep gloom over us....Besides what I have mentioned as to Frank's work during the year he gave a Lecture at Eton on Anthropometric Registers and Life Histories, and wrote a paper in the *Fortnightly* on the same, and gave a lecture to the Committee of the Medical Association[1]. He was invited to lecture at the Lowell Institution in America, but refused. In Meteorology he designed a clock for cumulative temperature. He is elected on to the Council of the Royal Society and was begged to accept the presidency of the Anthropological, but refused.

[1] Probably the Committee of the B.M.A. for Collective Investigation.

1883. Frank went to the British Association at Southport in September. In the early part of the year he corrected proofs of his *Inquiries into Human Faculty*, which was published by Easter. He also worked, helped by Croom-Robertson, at means of measuring the sensations. In the August *Fortnightly* he wrote an article on Medical Life Histories, offering Prizes up to £500. He spent much time on the details. His *Record of Family Faculties* has just come out, also the *Life-History Album*, which he edited. He was Chairman of the Anthropometric Committee of the British Association, which published this year its fourth and final Report, and also Chairman of the Local Scientific Societies Committee.

1884. More and more home seems the most fitting place for me and for Frank as he is always full of occupation. This last year he has been chiefly occupied in abstracting and collating from the Family Records for which he offered prizes to the amount of £500. About 150 of these Records were sent in by May 15th, and the statistics they afford will be food for many a month and possibly years. Then Frank took great interest in establishing an Anthropometric Laboratory at the International Health Exhibition, South Kensington, and gave no end of time and money to its prosecution; he gained, however, full recognition from Sir James Paget and others, whose opinion he cared about....In the summer my great desire to go abroad was stopped by cholera in the south of France and after long debate Frank settled on the English Lakes, and I was quite willing in default of my pet scheme, which I especially had cherished, as there was no English British Association to spoil our holiday, it being held at Montreal."

The *Record* tells us indeed as much as we know of the whole matter, namely, that in some way the medical profession disapproved of the prizes, and that they were then offered to laymen, who to a considerable extent responded[1].

The interval, however, between the *Fortnightly* paper of 1883 and the appeal to laymen for family records was well employed, and Galton having got in touch with several leaders of medical opinion over the medical registers obtained their aid in preparing his *Life-History Album*. It may be well to deal with these matters before we return to the Anthropometric Laboratory.

[1] In the announcement of the prizes of December, 1883, this note occurs after the statement of conditions, etc.: "The above conditions are in lieu of those provisionally sketched out by Mr Galton in the *Fortnightly Review* of August, 1883, for the purpose of eliciting suggestions, and which were subsequently submitted in a more elaborate form to many members of the medical profession. Their present shape is fixed in accordance with the balance of opinions elicited by these preliminaries, which was in favour of throwing them open to general competition and not to medical men only, as at first intended." This is obviously only a formal explanation, and does not indicate the nature of the opinions against confining the scheme to men who were in a much better position than the layman to record family ailments and the causes of death. Possibly the article in the *Fortnightly* rather impeded than aided his plans; possibly the medical profession of those days resented the intrusion of an outsider into what they might consider to be their own domain, even if they had not cultivated this portion of it, and showed no haste to do so. There is also another point which has much weight with me: Galton even to the time of his death had a great belief in working his projects through committees. I think it arose by reason of an innate modesty which was always seeking advice from others and accepting their opinions as worth more than his own. These committees often became unwieldy, were composed of incongruous and irresponsible elements, and on more than one occasion perverted or destroyed Galton's original scheme. It is conceivable that the Life-History Sub-Committee of the Collective Investigation Committee of the British Medical Association, of which Galton was Chairman, contained elements of this character. Anyone who has endeavoured like the writer to pick out from Galton's *Record of Family Faculties* a definite disease like phthisis by aid of its numerous lay synonyms or rather intentional pseudonyms will be rapidly convinced that the widening of the field of candidates was not an unmixed advantage. I had already written this note when I chanced to turn up Galton's preface to the second edition of the *Life-History Album* (p. ix). He there admits to the full the evil of working through a committee, but alas! it did not cure him of the habit.

D. THE *RECORD OF FAMILY FACULTIES* AND THE *LIFE-HISTORY ALBUM*

In the advertisement of the prize competition Galton suggested that information should be collected with regard to the child, its parents, its grandparents and great-grandparents, Numbers 1 to 15 of his scale of relationship, and as far as possible of all the collaterals of these, i.e. members of the same sibships as these, or about 70 to 90 individuals. Cousins Galton omitted, although we now know that, both on Mendelian theory and by actual observation, they exhibit as much of the constitution of the common stirp as aunts, uncles or grandparents and more than great-grandparents[1]. The prizes were to be given to those who filled in the blank spaces of the *Record of Family Faculties* most completely and perspicuously. The *Record* was published by Macmillans in 1884 and has been long out of print. A new and somewhat modified edition of it is certainly needed. It records, for the family of an individual—his stirp—what the *Life-History Album* of the same year does more copiously for the individual himself. In other words, the *Record of Family Faculties* could be extracted from the separate Life-History Albums of its units, but the inverse process is not possible. The one gives—except for the medical section—a brief account of the *adult* characteristics of each unit of the stirp, the other traces the unit through all phases of growth.

In the *Record* the following questions are asked: (1) Date of Birth; (2) Occupation, Birthplace, Residences; (3) Age at Marriage of individual, number of sons and daughters alive and their ages, and the same for those deceased with age at death; (4) Age at Marriage of spouse; (5) Mode of life so far as affecting growth or health; (6) Was early life laborious? why and how? (7) Adult height, adult colour of hair—colour of eyes; (8) General appearance; (9) Bodily strength and energy, if much above or below the average; (10) Keenness or imperfection of sight and other senses; (11) Mental powers and energy, if above or below the average; (12) Character and temperament; (13) Favourite pursuits and interests, artistic aptitudes. Then comes the medical history: (14) Minor ailments to which there was special liability (*a*) in youth, (*b*) in middle age; (15) Graver illnesses, (*a*) in youth, (*b*) in middle age; (16) Cause and date of death, and age at death. There are pages for male and female relatives of whom little is known but the age at and cause of death. There are pages for summaries of the anthropometric and medical characteristics of the stirp, and two Appendices to be devoted respectively to the Biological History of the Father's and of the Mother's Family. By 'biological history' Galton understood the constitutional history and hereditary peculiarities of mind and body on either father or mother's side. A third appendix deals with an examination of the way in which the faculties of the father and mother are blended or otherwise combined in the child.

[1] This result of theory and observation always troubled Galton, but I do not think there is any doubt of its accuracy.

The whole work is prefaced with an account by Galton of how the *Record* should be filled in. It contains many characteristic statements. A few of these may be cited here, as the book is very scarce.

"This book is designed for those who care to forecast the mental and bodily faculties of their children, and to further the science of heredity. The natural gifts of each individual being inherited from his ancestry, it is possible to foresee much of the latent capacities of a child in mind and body, of the probabilities of his future health and longevity, and of his tendencies to special forms of disease, by a knowledge of his ancestral precedents. When the science of heredity shall have become more advanced, the accuracy of such forecasts will doubtless improve; in the meantime we may rest assured that fewer blunders will be made in rearing and educating children, under the guidance of a knowledge of their family antecedents, than without it."

In the third paragraph Galton rightly points out that it is needful to study as many ancestral lines as possible, and that the book gives no countenance to the vanity that prompts most family historians to trace their pedigree to some notable ancestor and to pass over the rest in silence. Galton remarks that

"one ancestor who lived at the time of the Norman Conquest, twenty-four generations back, contributes (on the supposition of no intermarriage of kinsfolk) less than one part in 16,000,000 to the constitution of a man of the present day." (p. 1.)

This is rather a theoretical than an observational result. It is true a man may have 20 to 30 generations back 16,000,000 direct line ancestors if so many were available, but it is equally true that a distinguished man of that day might have several million descendants, and, if any system of alternative or factorial inheritance be true, the *distinguished* individuals among those descendants may owe their nature to that distinguished ancestor. It does mean something to trace even in one line—and there are four or five—a link between Darwin or Galton and Alfred the Great. It signifies nothing to trace the same link between a mediocrity and Alfred the Great[1]. Galton suggests that we need not go back beyond our great-grandparents, and this is absolutely true of characters which blend. But when he tells us that if an alien element of race or disease has been introduced into a family—a touch of Hebrew, of Huguenot (or even negroid) blood—it may be traced far further, he seems to me to be contradicting his previous statement. Albinism for example may remain latent through far more than three generations. But Galton recognises fully this latency at other points.

"Brothers and sisters are alike in blood, but it commonly happens that one of them exhibits some faculty in a conspicuous degree, which exists only in a latent form in another, and which the latter is, perhaps, equally capable of transmitting to his children. Therefore records of the faculties of the brothers and sisters of direct ancestors are of great value in disclosing hidden characteristics." (p. 2.)

I think it would be more just to say that the limitation to great-grandparents is only a question of the limitation of knowledge in the case of most families; and without being conclusive a great deal may still be learnt, if we

[1] The illustration can be given in another form: Some 15 years ago piebaldism appeared in my stirp of dogs and soon disappeared. After 7 or 8 generations it has reappeared. The piebald ancestor means little to the average dog of my stock; he means everything to the isolated piebald puppies of to-day.

cannot get beyond the three generations and their collaterals. We must make a start somewhere and if we record three generations now our great-grand-children will be able to consider *six*. What we need is to store data for the future.

"The advance of the science of heredity is seriously delayed through the want of such data. We do not yet know whether any given group of different characteristics which may converge by inheritance upon the same family will blend, neutralise or intensify one another, nor whether they will be metamorphosed and issue in some new form. Our ignorance is especially great in hereditary maladies, where much alarm undoubtedly exists which inquiry will dispel. It is possible that the different disease tendencies of different ancestors may in some cases neutralise one another, and on the other hand, that some ancestral combinations may be more hurtful to the descendants than we have at present any idea of....Our present ignorance of the conditions by which the level of humanity may be raised is so gross, that I believe if we had some dictator of the Spartan type, who exercised absolute power over marriages, assigning *A* to be the wife of *B*, and *C* to be the wife of *D*, and who acted with the best intentions, he might possibly do even more harm than good to the race."

Which remark I commend to the good Mr Chesterton who believes that he is better able than the Founder of Eugenics to appreciate what Eugenists propose.

Galton discusses the questionnaire of the *Record* showing how the answers will bear on such vital problems as the relative influences of nature and nurture, as the effect of overstrain in the parents on the robustness of their offspring, as the possibility of secular changes in the English race, and as to the influence of various racial types on fertility.

Considering the senses Galton remarks:

"Keenness of sensation in each of its forms is a valuable natural gift; unfortunately no means are as yet easily accessible for testing it in different persons; there are no anthropometric laboratories as yet in existence to which any one may go, and on payment of a small fee have all his faculties measured and registered by the various ingenious appliances known to modern science. We must therefore be content for the present with such definite facts bearing on the keenness or imperfection of the various senses as may have been incidentally observed." (p. 9.)

Galton demonstrated even in the same year as he wrote these lines what an anthropometric laboratory could achieve in measuring the keenness of the senses. Fitly attached to the Laboratory in the University of London, which now bears his name, is an Anthropometric Laboratory with complete equipment for testing not only the keenness of the senses but for measuring many other physical, psychical and physiological characters. The difficulty does not now lie in the absence of means of testing, but in the discovery of suitable ways of reaching those who are desirous of being tested, or who ought to be tested.

When Galton comes to the medical record he gives a list of "minor ailments" to which an individual may be subject and also one of the "graver illnesses." These lists may be repeated here as they are still of value in considering hereditary transmission of disease.

"*Minor Ailments.* Colds in the head or throat, sick headaches, sleeplessness, boils, quinsy, enlarged glands in the neck, bleeding at the nose, indigestion, bilious attacks (state whether accompanied by jaundice, vomiting or headache), constipation, skin eruptions (their nature should be stated if known), varicose veins, etc.

*Graver Illnesses.* Gout, rheumatism, consumption, spitting of blood, struma (scrofula), cancer (and other forms of tumour), bronchitis, asthma, paralysis (state whether of both legs or of one side), epilepsy, insanity, heart disease, dropsy of abdomen, general dropsy (Bright's disease), diabetes, stone, goître, fistula, the peculiar liability to bleed seriously from slight cuts, etc."

It will be noted that Galton omits such things as the tendencies to zymotic diseases which undoubtedly run in families, or those to hereditary eye diseases such as cataract, *retinitis pigmentosa*, etc. He does state that malformations which are extremely hereditary should be included.

Then follows a brief but useful list of sources from which family information may be obtained and he concludes as follows:

"Whatever may be the value of these results, the facts incidentally obtained during the course of the inquiry will form a separate document much prized by the family. The scientific importance of each investigation will, however, be soon appreciated by the author of it, for his researches will lay bare many far-reaching biological bonds that tie his family into a connected whole, whose existence was previously little suspected. Few, if any, have seriously studied the facts of heredity without being impressed with the conviction that no man stands on an isolated basis, but that he is a prolongation of his ancestry in no metaphorical sense, and I shall be surprised if the compilation of these registers does not extend this conviction very widely." (p. 13.)

We now turn to the *Life-History Album*. The first edition appeared in 1884 and bears on the title-page the words: "Prepared by direction of the Collective Investigation Committee of the British Medical Association. Edited by Francis Galton, F.R.S., Chairman of the Life-History Sub-Committee."

The original proposal seems to have been the return at fixed periods of the records in the Album to the Collective Investigation Committee. I think this proposal was not carried out, for I know of no publication of results from Life-History Album data ever being issued. The foreword to the owner of the book concludes with Galton's words from the *Fortnightly Review* article of January, 1882 (see our p. 358).

"The life-histories of our relatives are, therefore, more instructive to us than those of strangers; they are especially able to forewarn and to encourage us, for they are prophetic of our own futures."

The second edition of this work appeared in 1902 and there has recently been a reissue by the Cambridge University Press for the Galton Laboratory Publications. The second edition bears the sub-title "Tables and Charts for recording the Development of Body and Mind from Childhood upwards with Introductory Remarks. Rearranged by Francis Galton, D.C.L., F.R.S."

In the preface to the second edition Galton refers to the enthusiasm in the production of the work of the late Dr Mahomed, who had firmly persuaded himself that a work of the kind would be favoured and promoted by medical men throughout the country; and this idea led to its being produced under the auspices of the Collective Investigation Committee of the British Medical Association. Dr Mahomed

"made it a further condition that my name should appear as editor, I being known at that time to be much occupied in such matters. To this I agreed with some reluctance, for I wished to bear the entire responsibility or none at all. So a small committee of medical men was formed who met frequently at my house, where the book was mostly composed. But the result did not at all satisfy myself, neither do I think it satisfied the others. It was too bulky and ill-arranged. In fact it was emphatically a case of too many cooks. Each had his own views, and I believe that any one of us acting alone would have produced a better balanced book than we did working together."

Soon after the issue of the first edition Dr Mahomed died and none of the remaining medical members of the committee concerned himself much further with the work. Accordingly when Galton in 1902 was pressed for a second edition he largely rearranged and rewrote it. Thus the second edition and not the first really represents the book as Galton had originally planned it nearly twenty years earlier. It alone will be considered here[1].

While the author in his *Introductory Remarks* states that the *Album* may be started at any age in life and if then filled in as far as records are available will be of considerable interest, he notes that the book is especially suitable as a present in readiness for infants expected to be born, or for very young children. It should be handed over by the parent to the child only when the latter is old enough to appreciate its importance, to take charge of it and to continue the record for himself.

"The future of each man is mainly a direct consequence of the past—of his own biological history, and of those of his ancestors. It is therefore of high importance when planning for the future to keep the past under frequent review, all in its just proportion, and this is exactly what this album is intended to help him to do." (p. 2.)

Much of what Galton writes about filling in the details of the individual in the *Album* need not detain us; it is more or less a development of the information given in the *Record of Family Faculties.* Ample space is left for photographs in the standard positions and even for pictures of the homes and for finger-prints. The actual scale of coloured wools is omitted in this edition, although the heap of bits of variously coloured wools is still suggested as a test for colourblindness. An appendix gives tests for keenness of vision. Two paragraphs to illustrate the wise inferences that our sage of eighty could draw from his experiences may be reproduced here. The first relates to memory:

"Memory alone is an imperfect and deceitful guide; it preserves only a trifling part of the events of early life, and that part far from correctly. The extreme vividness with which a few childish incidents are usually recalled gives a very exaggerated view of the power of its grasp. Anybody who attempts to compile a sustained history of his early years will soon be persuaded of the truth of this remark, for he will surely become aware of huge gaps of time that he is totally unable to give account of. Every autobiography that I have seen testifies to these lapses of memory. Again when one happens to meet a friend not seen since early life, and to compare recollections, it is astonishing to find how differently the two memories have behaved. The one man fails to recall a multitude of events that have strongly impressed the other. Even as to those they alike remember as wholes, it will often occur that their memories disagree in essential details. In fact, the experience gained by such interviews is commonly humiliating to both." (p. 2.)

The second quotation I shall make relates to veracity in recording the nature of disease. Galton had come up against that hesitation to be literally

[1] The second edition differs not only in quality but quantity, for instead of carrying on the record only to the seventy-fifth year, it continues it to the hundredth. Galton was himself eighty when it was issued.

truthful which at once meets all of us when we attempt to deal with the heredity of pathological conditions. In the humbler classes of society, who form the bulk of hospital patients, there is sometimes almost a pride in the possession of malformations and of pathological conditions. Many of the women are only too ready to talk about them, and to exhibit any children who may possess them. It is possibly the great interest which they observe is excited among medical men by pathological states, which leads them to be pleased at being noteworthy even for a deformity[1]. Unfortunately the family records and traditions in hospital cases are often woefully scanty. The great-grand-parents are rarely to be traced, and only too often little if anything is known of the families of uncles or aunts. On the other hand, when we turn to social classes where the knowledge of ancestry and collaterals is considerable, we are too often met by an indignant refusal to give information, even if it be needed for scientific purposes. No retreat for the insane or sanatorium for the tuberculous designed for patients of middle or upper class status is able to provide adequate material for heredity. The inquirer is solemnly informed that the grandfather died of inflammation following a chill, that the sister pined away after the death of her fiancé, or that the father was at times eccentric, the aunt had attacks of the nerves, or the cousin brain-fever due to overwork. It does not profit to give way to discouragement, or feel sore before rebuffs, if you want to study human heredity. The only resource is to try and educate the so-called educated classes and produce if possible a more reasonable attitude towards hereditary matters amongst them. Galton achieved much in this direction and if it is still difficult to make rapid progress, it is certainly easier than it was before he started his campaign against the folly, not to say crime, which would screen family history not only from the future wife, but from adult children. Galton writes on this second point in 1902 with the sagacity of old age; he has learnt that to brand such action as "ignoble cowardice[1]" may be absolute truth, but it will hardly obtain what we need from the cowards. He says:

"It is too much to expect that even the most scrupulously kept records will be written throughout with perfect veracity. Healthy minded persons are seldom disposed to lay themselves wholly bare in written words. There will be omissions in every Album, sometimes of matters of fact and at other times of the real inwardness of events, that are of high importance to the right understanding of a life-history. The writer of the Album will mentally supply the omissions and interpret the misleading euphemisms when he refers to its pages; other persons who read his records must be prepared for their existence. Thus in matters of disease, an unsurmountable prejudice exists in many sensitive persons against ascribing cancer and insanity to their ancestors in direct terms. They shrink from the thought of recording hereditary possibilities that might destroy the peace of mind of their descendants, and perhaps work their own fulfilment. The duty of parents to be truthful histographers seems overborne by what they consider to be a still more pressing duty to their children. It is almost useless to attempt to calm hypersensitive feelings by pointing to the fact that healthful tendencies are just as heritable as morbid ones, and that every child is sure to be endowed with both. So I will confine myself to the mention of an instructive

[1] The psychology of these cases is similar to that of many persons, who being connected with some appalling criminal trial seem pleased with the momentary notoriety which carries their portraits into the cheap illustrated daily papers, possibly the one chance of publicity in their lives. I am by no means certain that it is not the same human frailty, under a thin covering of veneer, which causes the bride of another class to send her photograph to the 'society' papers.

experience of my own, which was gained while working at the family histories of a multitude of individuals. They were so tabulated that the medical history of any individual could be concealed by the hand or by a sheet of paper, while those of all his immediate ancestors, namely, parents, grandparents, uncles, and aunts, were exposed. I experimented frequently in guessing at the cause of death of a deceased individual from the knowledge of those of his ancestry, and I found my guesses to be on the whole grossly incorrect[1]. But though the stated cause of death could not be predicted with any approach to a useful degree of accuracy, the inheritance of minor ailments was conspicuously manifest. For this reason some stress is laid in the Album upon recording them. For my own part, I find no reason why the diseases of ancestors should be described otherwise than with perfect honesty, especially as a knowledge of them may induce their descendants to take reasonable precautions against inborn tendencies, instead of taking no precautions at all and doing themselves irreparable injury out of pure ignorance." (p. 3.)

The publication of the *Record of Family Faculties* and the *Life-History Album* brought Galton into touch with many families and individuals who were deeply interested in heredity. These books also brought him material, even if it was not quite as adequate as he had hoped for; further, they gave him an acknowledged position with regard to human heredity, which enabled him to prosecute more widely his inquiries, and to focus a great deal of the somewhat scattered attempts which were then being made in this field. Nor did his works fail to find competitors. Mr Jonathan Hutchinson, who had been a member of the British Medical Association Committee, published in 1888 *The Life Register*, which we might suppose to have derived some ideas from Galton's work, did not the author tell us that it was in no sense a copy, "having been in existence long before the Album was published." It is less elaborate than Galton's and fails in any adequate description of how and what entries are to be made. Quite recently much more ambitious *Albums* have been published in America demanding entries far more numerous than Galton's book. On the whole that work takes the *juste milieu*, and combines anthropometric with medical details in fair proportions. Galton was undoubtedly first in the field with a published volume; that volume shows his characteristic talents, and still remains the best *Life-History Album* available[2]. The insignificant annual sale of the work is, however, fairly conclusive evidence of how little the many persons who talk about eugenics and enrol themselves in societies for the advance of eugenics do to carry out Galton's fundamental idea that every individual should keep, in a standardised form, as perfect a register as possible of his own and his stirp's genetic possibilities. That is still a want for the future to supply; and as without its fulfilment we must ourselves fail in eugenic conduct, few of us have a right to preach eugenic morality to others.

[1] One may hope shortly to see this result tested on actual tabulated material. One of the chief difficulties lies in the fact that the lives of relatives are not exposed to equal risk, the environments and occupations may be so different; probably the risk in the case of sisters will be more nearly similar and give more conclusive results. Undoubtedly there are many pedigrees in which it would be safe to predict from the ancestral generations that at least 20 °/ₒ of the new generation will be affected with a given deformity or pathological condition. Would it be reasonable by predicting for isolated individuals to say that one's prediction was grossly incorrect if four individuals out of five were free of the defect? I think prediction must proceed on an average or probability basis, and that in Galton's test where apparently only *one* individual's cause of death was considered the *probability* of death from a particular form of disease was not really incorrect because that individual failed to die of it.

[2] It is still published by the Cambridge University Press, Fetter Lane, E.C. 4.

The *Record of Family Faculties* and the offer of prizes supplied Galton with the material upon which to work for the next five years, and the resulting memoirs and his book *Natural Inheritance* of 1889 will be discussed in a later chapter. When Galton came to distributing the prize money he had offered, he was in some difficulty as to measuring relative values and finally assigned forty £7 and forty-five £5 prizes, so that eighty-five competitors shared the £505. The awards were made on June 24, 1884, and were published in the *Times, Nature, British Medical Journal* and *Guardian.* Of the returns it is of interest to note that 70 were made by men, 80 by women; (i) 20, or rather fewer, concerned titled persons and the landed gentry; from (ii) Army and Navy, (iii) Church (various denominations), (iv) Law, (v) Medicine, (vi) Commerce (higher), (vii) Commerce (lower) there were 110 returns in all. The remaining twenty were from land agents, farmers, artisans, literary men, schoolmasters, clerks, students, and one domestic servant—a fairly representative collection. The original records were returned by Galton to their authors after he had made statistical abstracts.

### E. THE ANTHROPOMETRIC LABORATORY (*continued*)

We have already indicated the labour Galton gave to the construction of instruments for this laboratory and described some of the psychometric and other measurements he took. We have further discussed the paper on the "Outfit for an Anthropometric Laboratory" which he drew up in conjunction with the late Professor Croom-Robertson and privately circulated (see our p. 212). Other papers actually deal with the instruments and measurements made at the first Anthropometric Laboratory (International Health Exhibition). The first is a penny pamphlet of some fourteen pages, "Issued by Authority," and intended to be sold to visitors to the Exhibition. It is entitled: "Anthropometric Laboratory arranged by Francis Galton, F.R.S., for the Determination of Height, Weight, Span, Breathing Power, Strength of Pull and Squeeze, Quickness of Blow, Hearing, Seeing, Colour Sense, and other Personal Data. The Laboratory is situated in the East Corridor Annexe, Entrance from South Gallery. Admission to the Laboratory 3*d*., for which a schedule filled up with the above details will be furnished. London, William Clowes and Sons. International Health Exhibition, 1884." The pamphlet describes the purpose of the Laboratory, namely to show to the public the great simplicity of the instruments and methods by which anthropometric measurements are taken. It then states what measurements are taken and how they are taken. It further indicates two uses of such measurements, namely (i) the personal use, to ascertain whether the growth of child or youth is proceeding normally, and to draw attention to defects with a view to their being remedied; and (ii) the statistical use, to discover the efficiency of a nation as a whole and of its several parts, and the direction in which it is changing, whether for better or worse. Galton then notes the need there is for a more systematic registration of physical measurements.

PLATE L

Francis Galton's First Anthropometric Laboratory at the International Health Exhibition, South Kensington, 1884–5.

"Their value is indisputable, the cost of making them is trifling, and the facility of registration in any permanent institution is obvious. It seems strange that they should be neglected at any school or university." (p. 4.)

Dealing with the question of hair and eye colour, Galton remarks that:

"The British nation is partly a blend and partly a mosaic of very distinct types. The short black-haired ancient British race unites imperfectly with the tall fair-haired Danish or Scandinavian. Their union resembles what druggists call an emulsion, that is a mixture of oil and water, so well shaken together that they form an apparently homogeneous substance; but the compound is not durable. Leave the emulsion alone and after a longer or shorter time it will separate into its component elements. Types are stable, but the forms of their mongrel offspring are not; and whenever the external features of the old types are found in something of their original purity, it is reasonable to suppose that their inward characteristics are present also." (p. 5[1].)

Galton notes that Baxter has shown from an analysis of 330,000 to 340,000 reported cases of invalidism in the medical examinations for the American army during the Civil War, that the light-haired men suffered more than the dark-haired from every form of disease except chronic rheumatism[2]. It is to solve problems of this kind that a record of pigmentation in individuals and families as well as a record of disease is desirable.

The actual floor space of the Laboratory was only 6 feet wide by 36 feet long. It was fenced off from the side of a gallery by open lattice-work, through which the public could see what was going on without impeding the examinees. These entered by a door at one end and left by a door at the other; and some 90 individuals were passed through in the course of a day. The pamphlet gives an account of what the visitor is expected to do at each station and the nature of the instrument used for the test. A more elaborate account of the Laboratory was published in the *Journal of the Anthropological Institute* for 1885[3]. It is entitled: "On the Anthropometric Laboratory at the late International Health Exhibition." The memoir begins by stating that the exhibition being over it is desirable that some account should be given of the methods and experiences of the Anthropometric Laboratory, and that the author wishes to invite criticism and suggestion. He states that the Laboratory aroused considerable interest, 9337 persons were measured, each in 17 different ways[4]. Duplicates of the instruments had been ordered by executive officers in foreign countries, and much interest had been expressed in the apparatus by many places of education. As it was most desirable that for comparative purposes there should be a standard set of instruments Galton brought his apparatus and the attendants who had supervised the measuring before the Institute[5]. Galton remarks that the total expenditure having been covered by a charge of 3*d*. per head, he

[1] More recent research indicates little association between external and psychical characters after hybridisation.

[2] Baxter, J. H. *Medical and Anthropological Statistics of the Provost-Marshal-General's Bureau*, Washington, 1875.

[3] Vol. XIV, pp. 205–21.

[4] The schedules containing these measurements are now in the Galton Laboratory and work on their complete reduction is in progress.

[5] All the work was done by Serjeant Williams, Mr Gammage, an optical instrument maker, and a doorkeeper, of course under Galton's supervision.

did not see why a similar periodic anthropometric examination should not be made in all the great schools of the country; and he makes the excellent suggestion of an itinerant anthropometric laboratory, which should circulate from school to school. The idea is a very good one, and would keep the school laboratory, where such existed, up to modern standards of methods, tests and apparatus.

Discussing what should be measured, Galton observes:

"One object is to ascertain what may be called the personal constants of mature life. This phrase must not be taken in too strict a sense, because there is nothing absolutely constant in a living body. Life is a condition of perpetual change. Men are about half an inch shorter when they go to bed than when they rise in the morning. Their weight is affected by diet and habit of life. All our so-called personal constants are really variables, though a large proportion of their actual variations may lie between narrow limits. Our first rule then is, that the trouble of measurement is best repaid when it is directed upon the least variable faculties." (p. 207.)

He then touches on a point which still troubles the anthropometrician, especially in the case of mental tests:

"There are many faculties that may be said to be potentially constant in adults though they are not developed, owing to want of exercise. After adequate practice, a limit of efficiency would in each case be attained and this would be the personal constant; but it is obviously impossible to guess what that constant would be from the results of a single trial. No test professes to do more than show the efficiency of the faculty at the time it was applied, and many tests do even less than this, being so novel to the person experimented on that he is maladroit, and fails to do himself justice[1]; consequently the results of earlier trials with ill-devised tests may differ considerably from those of later ones. The second rule then is that the actions required by the tests should be as familiar as possible[2]."

Galton makes a suggestion as to practice which might be worth following up, although great care would have to be taken in experimenting between sessional and secular variations. The sessional variation, or variation in a sitting (or in one closely contracted series of tests), may be largely a result of chance, partly of practice, and partly of fatigue, while the secular variation may show the marked effect of continual practice[3]. The suggestion runs:

"There is some hope that we may in time learn to eliminate the effect of an unknown amount of previous practice by three or more distinct sets of trials. There exists a rough relation between practice and proficiency which ought to be apparent wherever progress is not due to acquiring a succession of new knacks, but proceeds regularly. When no practice has previously taken place, the progressive improvement will be very rapid; then its rate will smoothly decrease until it comes to an entire stop. I suspect that a curve might be drawn between proficiency and practice, and that the data afforded by at least three successive series of tests would roughly

[1] I think Galton has conceivably overlooked a point here. One test of fitness to environment is the readiness with which an individual can adapt himself to new conditions and respond promptly to the everchanging experiences of his life. In many cases therefore a *novel* test is a truer test of mental agility, than one which has become a familiar routine.

[2] This rule appears to be stated too generally; when familiarity comes in, then there will be a correlation between length of practice and efficiency, of which we do not know the intensity. When I was a boy I was taught to dovetail and learnt to do so creditably; any mediocre carpenter would excel my work now both in speed and neatness. On the other hand I should probably copy out far more clearly and rapidly than he would a page of letterpress. Any test to determine our relative powers of mentally controlling the rapidly moving hand ought, I think, to be made in a manner which should be as unfamiliar to us both as possible.

[3] The reader may consult a paper by E. S. Pearson on the "Variations in Personal Equation, etc.," *Biometrika*, Vol. XIV, pp. 23–102.

determine the position in the curve of the person who was being tested. They would show what he was capable of at the time, and approximately how much conscious or unconscious practice he had already gone through, and the maximum efficiency to which the faculty under test admitted of being educated." (p. 208.)

On the same page Galton raises a protest against pursuing, in anthropometric measurements, the strict methods appropriate to psycho-physical investigations.

"We do not want to analyse how much of our power of discriminating between two objects is due to this, that or the other of the many elementary perceptions called into action. It is the total result that chiefly interests us."

Galton is content if the judgment finally made rests upon a blend of many different factors.

A good deal of the paper is taken up with a discussion of the means of passing people rapidly through a long series of tests with a minimum of supervision. This was on account of the finite period for which the laboratory was open and the cost of attendants, a necessity in Galton's case, but it is, I think, open to a certain amount of reasonable criticism. Our own plan in the Galton Laboratory is to have as far as possible one supervisor for each test or at least kindred group of tests. He is able to explain what is needed, assist the examinee when in doubt and save delicate instruments from rough handling. If such supervision is feasible I feel certain that examinees pass through at the maximum speed, and there is greater accuracy in the records.

In Galton's first laboratory no head measurements were made, and he discusses the omission on p. 210. They were purposely omitted under the peculiar circumstances of a mixed crowd.

"I feared it would be troublesome to perform on most women on account of their bonnets, and the bulk of their hair, and that it would lead to objections and difficulties[1]."

However, Galton actually designed a spanner to take the height of the head from the auricular passages, and it came into use in Cambridge in 1885. In the case of this instrument he measures the maximum height

"above the plane that passes through the upper edges of the orbits and the orifices of the ears." (p. 210.)

Presumably he means the central line of the orifices and the *upper* border of the orbits. It is clear that Galton's standard plane differs on both counts from the Frankfurt horizontal plane[2].

The reader will find on pp. 211 and 217 descriptions of Galton's first instrument for measuring the swiftness of a blow (see our p. 220) and of what

[1] It may be worth noting that we feared the same trouble in our anthropometric laboratory at University College, but it has proved an idle fear. It is true bonnets no longer exist and hair is often 'bobbed.' Even if bulky it lies now-a-days low on the head and the head length is generally taken without disturbing it. Occasionally the tape measurements are a difficulty, but if help is asked for, I have found it most readily given as in passing the tape under the hair knot, or releasing the knot if necessary. I have been told occasionally that the hair will be let down if required, but there is usually a little justifiable pride about this offer, and a practised operator can take all the usual measurements fairly accurately without accepting it!

[2] The change to the *lower* border of the orbits was made at Cambridge later, but I do not know when.

appears to be an ingenious instrument for measuring delicacy of touch on the principle of the Roberval balance on p. 212. "The action of the instrument seems perfect, but it exists as yet only as a working model." I do not know whether any further account was published of it, nor how it exactly functioned physiologically.

There are two Appendices to the paper; the first contains chiefly extracts from the Exhibition pamphlet already discussed. The second indicates the methods Galton was using for the reduction of his material, and gives a certain number of his results[1]. In this Appendix he says that he is prepared to admit that the persons who applied to be measured were not possibly a random sample of those who attended the Exhibition, nor the crowd who visited the Exhibition a random sample of the British population, but he considers that such a criticism must not be pushed unreasonably far. Probably the data afford materials for testing the relations between various bodily faculties, and the influence of occupation and birthplace. Although in this paper he is dealing chiefly with statical anthropometric characters—i.e. those of least interest—such treatment was a necessary preliminary to further discussion and served to exemplify Galton's method of the statistical scale, i.e. the use of percentiles. He considered—and at that time he was justified in considering—that he was

"presenting in a compact and methodical form a great deal more concerning the distribution of the measurements of man than has hitherto been attempted in a numerical form." (p. 275.)

Galton deals only with adult males and adult females, of whom there were 4726 of the former and 1657 of the latter.

His first table gives the maximum or highest records among these adult cases for seven characters.

| Character | Highest Record | |
|---|---|---|
| | 4726 Adult Males | 1657 Adult Females |
| Stature without shoes ... | 6 ft. 7·5 in. | 5 ft. 10·3 in. |
| Weight ... ... ... | 22 st. 0 lbs. | 15 st. 8 lbs. |
| Vital Capacity ... ... | 354 in.[3] | 270 in.[3] |
| Strength of Pull ... ... | 148 lbs. | 89 lbs. |
| Strength of Squeeze ... | 112 lbs. | 86 lbs. |
| Swiftness of Blow ... | 29 ft. per sec. | 20 ft. per sec. |
| Keenness of Sight, distance of reading diamond type | 39 in. | 40 in. |

In all characters but sight the male had a higher record, and if we deal with the median values even there the male is higher than the female. Galton concludes that

"the female differs from the male more conspicuously in strength than in any other particular, and therefore that the commonly used epithet of 'the weaker sex' is appropriate." (p. 278.)

The lady, however, who could give a squeeze of 86 lbs.[2] is not to be despised,

[1] See "Some Results of the Anthropometric Laboratory," *J. A. I.* Vol. XIV, pp. 275–87.

[2] This is slightly in excess of the squeeze of the *median* adult male. Galton remarks that

and the idea of her tickled the fancy of *Mr Punch*, who in the issue that followed the publication of Galton's paper thus apostrophises her:

*The Squeeze of* 86.

Maiden of the mighty muscles,
  There recorded, you would be
Famous in all manly tussles,
  And its very clear to me,
That if in the dim hereafter
  Any husband should play tricks,
You would with derisive laughter,
  Give a "Squeeze of 86."

Husbands be it sadly stated,
  Have been known their wives to whack,
You, unless you're over-rated,
  Could give such endearments back.
Yours the task to try correction,
  Till your husband and your "chicks,"
Had a lively recollection
  Of your "Squeeze of 86."

*Punch*, April 15, 1884.

Galton's second table exhibits the full advantages of his method of percentiles, but presents also its disadvantages, as it does not allow us to deduce the usual frequency constants with any reasonable accuracy. At the same time it is easily understood by the non-statistical, who can readily determine from it their rank for any character. Thus suppose a man between 23 and 26 years of age to have a vital or breathing capacity of 190 cubic inches, he sees at once from the table that he is surpassed by 70 °/₀ of men (199) and himself surpasses 20 °/₀ (187). A simple rule of three sum then indicates to him that he lies $\frac{3}{12}$ or $\frac{1}{4}$ of the way from 80 to 70 or at 77·5. In other words he would rank between the 77th and 78th men in a population of 100 young adults similar to those who visited the anthropometric laboratory.

The ages and numbers in the table[1] on the following page possibly require some justification. All Galton says is that he had

"groups of appropriate cases extracted from the duplicate records by Mr J. Henry Young of the General Registry Office. I did not care to have the records exhausted, but requested him to take as many as seemed in each case to be sufficient to give a trustworthy result for these and certain other purposes to which I desired to apply them. The precise number was determined by accidental matters of detail that in no way implied selection of the measurements." (p. 278.)

Now-a-days we should consider it needful to keep the probable errors in view, which are occasionally somewhat large for small series treated by the method of percentiles. Galton deduced his percentiles from the frequency distributions by summing, plotting, drawing a curved line through the plotted points and then interpolating for the actual percentiles by graphical interpolation[2].

"the population of England hardly contains enough material to form even a few regiments of efficient Amazons."

[1] This table with the description of its method of preparation was also published in *Nature*, January 8, 1885 (Vol. XXXI, pp. 223–5), a month or two before its appearance in the *Journal of the Anthropological Institute*.

[2] The method with more detail was discussed by Galton under the title "The Application of a Graphic Method to Fallible Measures" in a paper communicated to the Jubilee Congress of the Royal Statistical Society and published in the *Jubilee Volume of the Statistical Society*, 1885, pp. 262–5. The method consists in drawing an ogive curve from the data and interpolating for the percentiles. The same end could be reached by integrating the frequency histogram and dividing its final ordinate into equal parts corresponding to the percentiles. This may, of course, be rapidly done by the integraph.

*Anthropometric Percentiles*[1].

*Values Surpassed and Values Unreached, by various percentages of the persons measured at the Anthropometric Laboratory at the late International Health Exhibition.*

| Character measured | Age | Unit of measurement | Sex | No. of Persons | Values surpassed by percents. below: 95 % | 90 % | 80 % | 70 % | 60 % | 50 % | 40 % | 30 % | 20 % | 10 % | 5 % |
|---|---|---|---|---|---|---|---|---|---|---|---|---|---|---|---|
| | | | | | Values unreached by percents. below: 5 % | 10 % | 20 % | 30 % | 40 % | 50 % | 60 % | 70 % | 80 % | 90 % | 95 % |
| Height standing | 23— | Inches | M. | 811 | 63·2 | 64·5 | 65·8 | 66·5 | 67·3 | 67·9 | 68·5 | 69·2 | 70·0 | 71·3 | 72·4 |
| without shoes | 51 | ,, | F. | 770 | 58·9 | 59·9 | 61·3 | 62·1 | 62·7 | 63·3 | 63·9 | 64·6 | 65·3 | 66·4 | 67·3 |
| Height sitting from | 23— | Inches | M. | 1013 | 33·6 | 34·2 | 34·9 | 35·3 | 35·4 | 36·0 | 36·3 | 36·7 | 37·1 | 37·7 | 38·2 |
| seat of chair | 51 | ,, | F. | 775 | 31·8 | 32·3 | 32·9 | 33·3 | 33·6 | 33·9 | 34·2 | 34·6 | 34·9 | 35·6 | 36·0 |
| Span of Arms ... | 23— | Inches | M. | 811 | 65·0 | 66·1 | 67·2 | 68·2 | 69·0 | 69·9 | 70·6 | 71·4 | 72·3 | 73·6 | 74·8 |
| | 51 | ,, | F. | 770 | 58·6 | 59·5 | 60·7 | 61·7 | 62·4 | 63·0 | 63·7 | 64·5 | 65·4 | 66·7 | 68·0 |
| Weight in ordinary | 23— | Pounds | M. | 520 | 121 | 125 | 131 | 135 | 139 | 143 | 147 | 150 | 156 | 165 | 172 |
| indoor clothes | 26 | ,, | F. | 276 | 102 | 105 | 110 | 114 | 118 | 122 | 129 | 132 | 136 | 142 | 149 |
| Vital or Breathing | 23— | Cubic | M. | 212 | 161 | 177 | 187 | 199 | 211 | 219 | 226 | 236 | 248 | 277 | 290 |
| Capacity | 26 | Inches | F. | 277 | 92 | 102 | 115 | 124 | 131 | 138 | 144 | 151 | 164 | 177 | 186 |
| Strength of Pull as | 23— | Pounds | M. | 519 | 56 | 60 | 64 | 68 | 71 | 74 | 77 | 80 | 82 | 89 | 96 |
| archer with bow | 26 | ,, | F. | 276 | 30 | 32 | 34 | 36 | 38 | 40 | 42 | 44 | 47 | 51 | 54 |
| Strength of Squeeze | 23— | Pounds | M. | 519 | 67 | 71 | 76 | 79 | 82 | 85 | 88 | 91 | 95 | 100 | 104 |
| with strongest hand | 26 | ,, | F. | 276 | 36 | 39 | 43 | 47 | 49 | 52 | 55 | 58 | 62 | 67 | 72 |
| Swiftness of Blow | 23— | Feet per | M. | 516 | 13·2 | 14·1 | 15·2 | 16·2 | 17·3 | 18·1 | 19·1 | 20·0 | 20·9 | 22·3 | 23·6 |
| | 26 | second | F. | 271 | 9·2 | 10·1 | 11·3 | 12·1 | 12·8 | 13·4 | 14·0 | 14·5 | 15·1 | 16·3 | 16·9 |
| Keenness of Sight by distance of reading | 23— | Inches | M. | 398 | 13 | 17 | 20 | 22 | 23 | 25 | 26 | 28 | 30 | 32 | 34 |
| diamond type | 26 | ,, | F. | 433 | 10 | 12 | 16 | 19 | 22 | 24 | 26 | 27 | 29 | 31 | 32 |

Taking the table, it can be shown by plotting in reverse directions that the male and female ogive curves, for squeeze of hand say, intersect at 7 °/。, or if we had to select the 100 strongest individuals out of 100 men and 100 women taken at random, we should select 93 men and seven women. On this Galton makes the comment:

"Very powerful women exist, but happily perhaps for the repose of the other sex, such gifted women are rare." (p. 278.)

It is probable that if Galton had taken the fisherfolk of Aberdeen, the peasantry of the Tyrol, or any primitive people, he would have found a considerably greater percentage of women in his 100 strongest individuals. He was largely dealing with a town population. Our experience in the Galton Laboratory shows that there are quite a few characters in which the absolute values reached by the women equal or excel those of the men, notably in Memory of Form, Discrimination of Colour, Steadiness of Hand, and Speed-Accuracy of Hand.

[1] The value that is unreached by $n$ °/。 of any large group of measurements and surpassed by $(100-n)$ °/。 is called the $n$th percentile.

At this point in his paper Galton draws attention to a "Common Error in Statistics[1]," which he says has not hitherto attracted attention. It occurs in a classification, such as "65—" for stature, which is assumed to mean 65 *inches and less than* 66 *inches*. He points out that the centre of this group is usually taken as 65·5 inches, but that this is most often erroneous, and that the true centre of the group (if we arrange the group uniformly over its range) may differ very considerably from this. The true centre depends on the fineness of our readings. If, for example, we read stature to the nearest half-inch only, then all measurements in excess of 64·5 would be included in 65, and all measurements in excess of 65·5 in 66. The former would all appear in "65—" and the latter would not. Thus the centre of "65—" would be actually the centre of the range 64·5 to 65·5 or 65 and not 65·5. In other words, there would be half an inch error in the resulting mean. Most tyros in statistics are now aware of this point, i.e. that the centre of a subrange depends upon the fineness or coarseness of the readings. It is, therefore, startling to think that Galton was the first to point out this pitfall even as late as 1884!

The next paragraphs are of some historical interest. Galton is seeking to find a relation between various characteristics, and considering the old idea of the ratios of anatomical measurements as constants. He notes that in certain cases these ratios are not constant (p. 281). Galton also gives rough growth curves for vital capacity, but has missed the point of contraflexure in the neighbourhood of puberty, which his own data treated more in detail exhibit. These are cases of what we should now term skew regression. Galton, we notice, had not yet reached in 1884 the full conception of correlation. He gives (p. 285) what we should now term a correlation table between vital capacity and strength of squeeze for 522 males of ages 23, 24 and 25. He says that he was surprised to find no close relation between these two characters.

"The importance of a large breathing capacity to a man who expends force rapidly, as to a runner or mountain climber, is undoubted, but for a strain of short duration it seems comparatively non-essential." (p. 284.)

He reaches his conclusion by comparing the means for strength of squeeze corresponding to vital capacities of 150 and 300 cubic inches. He says there is a difference of 17 lbs. But without a measure of the variabilities we could not deduce any measure of the association of the two characters. Now-a-days we should say that the coefficient of correlation was certainly not negligible: it is $\cdot 40 \pm \cdot 02$ from his table; but we are able to express the intensity of the association in this simple manner, only because Galton later taught us how to do it[2].

Some data are given for the relationship of right and left sides. Thus Galton found the left hand to be about 6 °/° weaker than the right. The right and left eyes on the average had equal keenness of sight. Galton

[1] See p. 280 of the present paper. Galton also contributed a paper on the same point to the *Jubilee Volume of the Statistical Society*, June 22–24, 1885, p. 261.

[2] Galton's wording is rather vague, but I think there is more correlation than his phrase would suggest.

investigated whether there was any relation between superior strength of the right or left hand and superior reading power of the right or left eye. Presumably he made a correlation table; he tells us that he found no association, but he considers that an association would be less improbable had he been able to compare the difference in skill of right and left hands with the difference in vision of the right and left eyes.

The paper concludes with the data on highest audible note referred to on our p. 221.

A study of this paper suggests at once the road along which Galton was being carried towards the conception of correlation; we shall see soon how he recognised that the numerical association between two anthropometric variates and the quantitative measure of the intensity of the hereditary factor were one and the same mathematical problem.

While Galton began immediately to use his data for the problem of correlation he did not hesitate to place portions of it before other investigators that they might if they liked reduce it by other methods. Thus in 1889 he published[1] data for 400 of the 518 individuals, extracted by Mr J. H. Young (see our p. 375). These data are for adults aged 23, 24 and 25, and for other individuals than those he himself used in his 1888 Royal Society memoir on *Correlation*. They give Age, Status (married or single), Eye Colour, Birthplace, Occupation, Residence, Vital Capacity, Squeeze of both hands, Span, Sitting Height, Stature and Weight in ordinary indoor clothing. As far as I know they have never been reduced, and would prove rather inadequate in number if allowance were made for birthplace and occupation. For a really full discussion of these matters, it would be needful to return to the much fuller material in the far more numerous original records.

Two further papers remain to be touched on, although of a much later period than the first Anthropometric Laboratory. It will, I think, be wiser to take the later of these first, because it gives us more of the history of his laboratories. It is entitled: "Retrospect of Work done at my Anthropometric Laboratory at South Kensington," and was published in 1891[2].

Galton's Anthropometric Laboratory on the closing of the Health Exhibition in 1885 had been transferred to a piece of vacant ground, which later was taken over by the Imperial Institute. Here the laboratory, under Serjeant Randal as Superintendent, continued its work for five years, and an additional 3678 persons were measured. New measurements and observations were made, including the wonderful collection of finger-prints now in the *Galtoniana*. But Galton after six years' experience had begun to realise that an Anthropometric Laboratory cannot remain stationary either in its methods or instruments. It must always be starting new inquiries, and needs for this purpose a scientific research staff. He had himself used his laboratory for various special researches, such as the Finger-Print inquiry, the question

[1] *Journal of the Anthropological Institute*, Vol. XVIII, pp. 420–30.

[2] *Ibid.* Vol. XXI, pp. 32–5. The Laboratory was dismantled in February 1891, and reopened August 3, 1891. Galton states in his *Retrospect* that the South Kensington Museum authorities had offered to place a larger and better lighted space at his disposal under their own roof.

PLATE LI

Francis Galton's Second Anthropometric Laboratory, South Kensington Museum, 1891–95.

of marks for physical tests, and the problem of the symmetry of the two sides of the body. Such inquiries required an ever available scientific director, and Galton himself was far too busy with the multitude of claims on his energy and executive skill to undertake such duties. He writes:

"In brief, what little has been accomplished at the laboratory during the three years of its existence justifies to my own mind the trouble and expense I have been put to in building, equipping and maintaining it. But it never reached to my ideal. Besides the objects already named, I was almost equally desirous of establishing a place where the keenness of the senses and other faculties in any individual who applied, might be measured with all the accuracy and painstaking that is achieved by the few biologists who occupy themselves seriously in such pursuits. To effect this it would be necessary to secure the occasional services of a skilled experimenter and to ensure at the same time that a sufficiency of persons should come to be measured. The time did not seem to have arrived for such an enlargement of the existing methods, though I hoped and still hope that it may not be far distant, as the utility of the laboratory becomes more widely appreciated. The measurements thus far employed are of a comparatively rude, but not ineffective, character. It would give me pleasure at any time to receive suggestions as to new and useful special inquiries, such as might be carried out and brought to conclusion without a too serious expenditure of time and effort."

But although Galton was very modest about what his laboratory could achieve in the future without a scientific staff, it really had accomplished a great deal.

"Persons of all ranks went to it[1], a knowledge of its existence was extending, and it was becoming increasingly frequented up to the day of its closure. Many correspondents in the United Kingdom, in America, and elsewhere, have more or less adopted its methods, and it was, I may add, a great consolation to me to receive, on the very day that I began to dismantle it, the proof sheets of the register, and other forms in many respects like my own, that are to be used in the laboratory at Dublin, which has been set on foot through the efforts, and will be carried on under the superintendence, of Professors Cunningham and Haddon." (p. 32.)

Immediately following[2] Galton's *Retrospect* is a paper by Cunningham and Haddon entitled "The Anthropometric Laboratory of Ireland." They say that Mr Galton, who has given them every encouragement in their work, proposed that they should give some account of the steps they were taking to introduce anthropometric work into Ireland and their aims in doing so.

"We need hardly explain in the Institute where the important and interesting results obtained by Mr Galton in this field of inquiry have been so largely made known, that it was these that stimulated us to endeavour to do likewise in Ireland." (p. 35.)

Directly or indirectly Galton's Laboratory was the parent of Anthropometric laboratories at Eton, Dublin and Cambridge; indeed of the much later work also of Schuster at Oxford. Just as Galton generally transferred his laboratory to the varying loci of the British Association, so Cunningham and Haddon proposed peregrinations for their laboratory during the Long

[1] Among whom we may note Mr W. E. Gladstone, whose head measurements afterwards were as serviceable to Mr Brock, as those of the Biometric Laboratory on the head of Professor Weldon were to Mr Hope Pinker—both being used for posthumous portraiture. Gladstone was amusingly insistent on the size of his head, saying that hatters often told him that he had an Aberdonian head—a fact which he did not forget to tell his Scottish constituents. It was not, however, of very great circumference and rather low (like Sir Thomas Browne's and Bentham's). It was less than Spottiswoode's, Sharpey's and Galton's own. "Have you ever seen as large a head as mine?" Gladstone said to Galton, on which the latter observed: "Mr Gladstone, you are very unobservant!"

[2] *Journal of the Anthropological Institute*, Vol. XXI, pp. 35–9.

Vacations to special "ethnical islands" in Ireland in the hope of unravelling "the tangled skein of the so-called 'Irish Race.'" The peripatetic principle of Galton was again adopted by Gray and Tocher in their Scottish surveys. If the people will not come to you, you must go to the people, and stand beside the itinerant dentist and the cheapjack in the market-place and measure their crowds.

Looking back on those years during which the Anthropometric Laboratory existed we ask what definitely was achieved? Well, (i) An immense amount of material was collected, which only forty years later is being adequately reduced. (ii) From small portions of it Galton deduced the foundations of the correlational calculus. Here Galton evolved an entirely new principle described in his paper of 1889, yet to be placed before the reader. This was in itself a very big achievement, much bigger in the light of what has followed than it might have appeared to casual observers of that day. It reduced

"all forms of correlation, including hereditary qualities, to one simple law, namely that of the relation between two variables partly dependent on a common set of influences."

(iii) It led to laws of growth and development whose study had hitherto been impossible, and it enabled investigators to take account of social status and of occupations. (iv) The limitations of such systems of personal identification as that of Bertillon became apparent and this prepared the way for Galton's finger-print system combining indexing with identification. (v) It provided material by which many interesting problems could be solved, such as diurnal changes in measurement, fallibility of the measurer, and the limits of change after adult age in various anthropometric "constants."

Lastly the Laboratory gave a most vigorous push to anthropology; it indicated what might be achieved by anthropometry taken seriously and the impetus is yet far from exhausted. There are still anthropologists who believe that great racial problems can be solved by juggling with a few cephalic indices. They do not recognise that a human being is a vast congeries of faculties, and that only certain of these are highly correlated. Others are practically independent and are largely modified by intermixture in each new generation. It is wholly impossible to define an individual, still less a race or associated group of individuals, on the basis of a single character. It requires a great many measurements to describe with moderate accuracy an individual, and quite as many to characterise, or provide the type of, a local group of men. Anthropometry whether physical, psychical or medical has this end in view: the definition of Type—in particular racial type—by the measurement of a fairly representative system of characters. Anthropologists learnt from Galton's Anthropometric Laboratories not only the importance of their task, but how to set about it. They learnt for the first time to what extent characters are correlated and how to measure the degree of their association. That could only result, when the investigator learnt to deal with the whole system of variations and did not occupy himself simply with the averages of characters. The last quarter of the nineteenth century revolutionised anthropology, and Galton was the main mover in those momentous changes.

We now turn to the second memoir, published in 1890, entitled: *Anthropometric Laboratory, Notes and Memoirs*, No. I. This seems to have been printed at Galton's expense and was probably sold (price 3*d.*) at the second Anthropometric Laboratory in the Science Museum, South Kensington. The pamphlet, now unpurchasable, consists of four chapters, thirty-two pages in all. The first chapter bears for title: "Why do we measure mankind[1]?" Galton first answers the supposed reader, who may say, "I do not care for science, why then should I go and be measured?" For this "very cynical but not quite imaginary speaker" Galton writes as follows:

"The cost of being measured has been proved to amount to something between 3*d.* and 1*s.*, and the real question is whether it is worth your while to pay a shilling at a maximum to have yourself or your boys and girls measured."

Galton indicates first the advantage to non-adults; given age, sex and social position, we are able to tell them how they rank among their contemporaries.

The value of testing sight is specially emphasised and it is pointed out how a knowledge of mischief may lead to the removal of its source. Then Galton notes how often colour-blindness has been discovered late in life and after possibly a failure has been made in a profession where it is an absolute bar. He cites the case of a widow bringing a son of 18 years to the laboratory and getting quite angry with him for his supposed stupidity in blundering between reds and greens, quite unconscious that there was such a natural incapacity as colour-blindness!

It is then pointed out how important to an adult is a knowledge of his strength and his vital capacity, and how valuable a warning may be, not only in the selection of an occupation, but in the case of an incapacity arising from unrecognised causes.

"A register of measurements resembles a well-kept account-book. It shows from time to time the exact state of a man's powers, as the account-book shows that of his fortune. Whatever may be whispered by the inner voices of vanity or of envy, no sane and experienced person can doubt the enormous difference between the natural gifts of different men, whether in moral power, in taste, in intellect, or in physical endowments. Those who have frequently pitted themselves fairly against others, doing their very best to succeed, must have often known what it is to be utterly beaten by their natural superiors. It is only those who have kept aloof from contest who can possibly flatter themselves with the belief that their failures are wholly due to circumstance and in no degree to natural incapacity. Such persons will quickly be awakened from their self-conceit by submitting themselves to measurement and thereby ascertaining their exact rank in each several respect. They will be sure to receive a good moral lesson from the results." (p. 8.)

This passage explains to a considerable extent why Galton preferred the method of grades to the use of frequency curves. He liked a system in which the uninitiated could put his finger on a single number and say: "I am worse than so many per cent. of men and better than so many per cent." That is what the method of Anthropometric Percentiles achieves and the fundamental diagram reproduced on our p. 390 appears on p. 23 of this reprint.

[1] It also appears in *Lippincott's Magazine* for February, 1890.

The next advantage of anthropometry Galton finds in its *industrial* value, a matter which some of our moderns believe they have discovered for the first time.

"Employers of labour might often find it helpful to require a list of laboratory measurements when selecting between many candidates who otherwise might be equal in merit. Certainly a man who was thereby shown to be measurably much more highly endowed than the generality of his class with physical efficiency, would have a corresponding high chance of being selected for any post in which physical efficiency of the kind tested was of advantage. I have great hope of seeing a system of moderate marks for physical efficiency introduced into the competitive examinations of candidates for the Army, Navy and Indian Civil Services." (p. 8.)

We have here a point to which we must shortly return—the question of marks for physical proficiency, which Galton strongly advocated. I might be inclined to go somewhat further and suggest that when there is doubt—and there often is—between the intellectual merits of two candidates the *mental* tests of a well-organised anthropometric laboratory would effectively discriminate. The County Council educational authorities are annually in difficulty in the award of their secondary scholarships, not about the boys or girls at the top, but those at the bottom of the selected list, where there are numbers on a nearly dead level, with no adequate examinational differences to guide the judgment. It would be a valuable and justifiable procedure to further separate out the better of these candidates by well-chosen anthropometric tests, which would be certain to appeal to faculties whose differentiation could not be achieved by a written examination.

Another advantage Galton finds in anthropometry is the registration of individuals for identification. Honest men may need identification as well as rogues, and the measurements, especially if finger-prints are included, would suffice to identify any one between the ages of 20 and 60.

Apart from the—shilling's worth of—advantage to the individual, Galton emphasises the scientific ends which can be attained by anthropometric laboratories. He refers to problems such as whether the promises of youth are fulfilled in adult life; if a boy is of high rank among his compeers is it an indication of superior future efficiency in the man? Another problem is that of the influence of education or practice upon both mental and physical characters. Again the influence of environment could be tested as soon as we have precise measures of faculties. Galton notes that even if there be a rapid rise in the efficiency of any factor due to training or environment, it soon reaches a condition in which the daily improvement lessens and at last stands still; the limit of perfectibility has been reached.

"Experiences of this kind on a large enough scale to give trustworthy results would have a direct bearing on the science of education." (p. 10.)

Lastly Galton deals with the educational value that a habit of measurement has in promoting accuracy in ideas and language:

"The present vague way in which men mostly estimate and describe the performances of themselves or others, testifies to much muddleheadedness and to a sad lack of expression.... There is a world of interest hidden from the minds of the great majority of educated men, to whom the conceptions and laws of the higher statistics are unknown. A familiarity with these conceptions would soon be gained by the habit of dealing with human measurements, as by the

PLATE LII

| Taille 1m | Oreille dr. Tête. Longr | Pied g. | Coulr de l'iris g. N° de cl. | Agé de |
|---|---|---|---|---|
| Voûte | Largr | Médius g. | Aurle | né le |
| Enverg 1m | Longr | Auricre g. | Pérle | a |
| Buste 0, | Largr | Coudée g. | Partés | dépt |
| | | | | âge appt |

(Réduction photographique 1/7.)

Mr. Galton 19.4.93

277

| Front. Inclin° | Nez. Racine (cavité) | Oreille droite. Bord. o. s. p. f. | Barbe | Colore (pigre |
|---|---|---|---|---|
| Hautr | Dos Base | Lob. c. a. m. d. | Cheveux | /sange |
| Largr | Hautr Saillie. Largr | A. trg. i. p. r. d. | Car | Ceint. |
| Partés | l l | Pli. f. s. h. E | *Autres traits caractéristiques :* | |
| | Partés | Part. | Sigt dressé par M. | |

**Francis Galton, aged 71, photographed as a criminal on his visit to Bertillon's Criminal Identification Laboratory in Paris, 1893.**

assignment of rank in a class, or by making other deductions that I have not space to refer to here, such as the numerical values by which the nearness of different degrees of kinship may be expressed, or the closeness of correlation between different parts of the body. There is no intrinsic difficulty in grasping the conceptions of which I speak, but they are foreign to present usage and look strange at first sight. They are consequently very difficult to express briefly and intelligibly to those to whom they are wholly new."

It will be seen that as early as 1889 Galton was fully assured that the ideas of a correlational calculus opened a new world of thought, not only to the trained scientist, but to every man of education who could master his natural inertia and endeavour to grasp the new conceptions.

Chapter II of the pamphlet is entitled "Human Variety." This was Galton's final address to the Anthropological Institute at the end of the four years during which he had held the office of president[1]. The paper opens with a paragraph on finger-prints which shows that Galton was working at the subject, and already fully recognised its importance for identification. He especially refers to Sir William Herschel's use of the method in India, and suggests its use in North Borneo for identifying the coolies, and in other cases where there may be fraud from impersonation of pensioners and annuitants. Galton then turns to correlation,

"a very wide subject indeed. It exists wherever the variations of two objects are in part due to common causes; but on this occasion I must only speak of such correlations as have an anthropological interest."

He tells us that the particular problem he first had in view was to ascertain the practical limitations of the ingenious method of anthropometric identification due to M. Alphonse Bertillon, which was then in habitual use in the criminal administration of France. Correlation between the various measurements would obviously be a serious defect of the Bertillon system, and Galton suspected strongly the existence of this source of error[2]. An element of history is now revealed:

"The first results of the inquiry, which is not yet completed, have been to myself a grateful surprise. Not only did it turn out that the measure of correlation between any two variables is exceedingly simple and definite, but it became evident almost from the first that I had unconsciously explored the very same ground before. No sooner did I begin to tabulate the data than I saw that they ran in just the same form as those that referred to family likeness in stature,

[1] Delivered Jan. 22, 1889. See *Journal of the Anthropological Institute*, Vol. XVIII, pp. 401–19.

[2] The first complete analysis of these correlations was given by the late Dr W. R. Macdonell and he indicated how an index could be constructed of artificial functions of the Bertillon measurements in which this difficulty of correlation would be satisfactorily surmounted, *Biometrika*, Vol. I, 1902, pp. 177–227, "On Criminal Anthropometry and the Identification of Criminals." He shows that the correlations of the Bertillon measurements are high, far too high for indexing purposes. Galton first expressed his doubts on this point after Bertillon's discourse on his system before the Anthropological Institute. "There may be room for reasonable doubt among anthropologists whether the precision with which the living body can be measured is quite as great, and whether its dimensions are quite as permanent, as they are considered to be by M. Bertillon; and again there may be some hesitation in believing that a very large collection of measures would admit of being so surely catalogued on the Bertillon system as to be ransacked with a promptitude at all corresponding to that with which a word may be found in a huge dictionary" (*Journal of the Anthropological Institute*, Vol. XX, p. 198). This was one of the early clashes in the contest between the Bertillon and Galton systems, which was to end ultimately with the victory of the latter.

and which were submitted to you two years ago. A very little reflection made it clear that family likeness was nothing more than a particular case of the wide subject of correlation, and that the whole of the reasoning already bestowed upon the special case of family likeness was equally applicable to correlation in its most general aspect." (p. 14.)

This passage justifies the assertion that Galton came first to correlation from the study of heredity; that, when testing Bertillon's system of indexing anthropometric measurements, he discovered correlation afresh and then saw that both associations were capable of identical treatment, and only special cases of a far wider conception.

But there was another point Galton now recognised, in dealing with heredity he had compared the *same* characters; he now saw that in comparing different characters we must take each in its own unit of variation, and that when this is done the relations become strictly reciprocal[1]. In this paper Galton states, probably for the first time, that measured in units of variation, there is always 'regression,' i.e. that in these units the proportion that one average deviation of one character is of a given deviation of a second is always a proper fraction. (p. 15.)

The next section of this chapter is headed "Variety," and Galton laments that, while an immense amount of trouble is taken over measurements, anthropologists devote their inquiries solely to the means of groups, passing over the variety of the individuals in those groups with "contented neglect." The whole section is so thoroughly characteristic of Galton's attitude to anthropology and of his genial sagacity that I cannot refrain from reproducing it almost entirely; there is much still to be learnt from its perusal even forty years after it was written.

"It seems to be a great loss of opportunity when, after observations have been laboriously collected, and been subsequently discussed in order to obtain mean values from them, that the small amount of extra trouble is not taken, which would determine other values whereby to express the variety of all the individuals in those groups. Much experience some years back, and much new experience during the past year, proved to me the ease with which variety may be adequately expressed, and the high importance of taking it into account. There are numerous problems of special interest to anthropologists that deal solely with variety.

There can be little doubt that most persons fail to have adequate conceptions of the orderliness of variability, and think it is useless to pay scientific attention to variety, as being, in their view, a subject wholly beyond the powers of definition. They forget that what is confessedly undefined in the individual may be definite in the group, and that uncertainty as regards the one is in no way incompatible with statistical assurance as regards the other. Almost everybody is familiar now-a-days with the constancy of the average in different samples of the same large group, but they do not often realise the completeness with which a similar statistical constancy permeates the whole of the group. The mean or the average is practically nothing more than the middlemost value in a marshalled series. A constancy analogous to that of the mean characterises the values that occupy any other fractional position that we please to name such as the 10th per cent., or the 20th per cent.; it is not peculiar to the 50th per cent., or middlemost. Still less do they realise the fact that all Variety has a strong family likeness, by approximating more or less closely to the normal type, which is that which mathematicians prove must be the consequence of Variety being due to the aggregate effect of a very large number of small and independent influences.

Greater interest is attached to individuals who occupy positions towards either of the ends of a marshalled series than to those who stand about its middle. An average man is morally

[1] In modern language Galton recognised that with standard deviations as units, the 'regression' is equal to the correlation coefficient.

and intellectually an uninteresting being. The class to which he belongs is bulky and no doubt serves to help the course of social life in action. It also affords, by its inertia, a regulator that, like the fly-wheel to the steam-engine, resists sudden and irregular changes. But the average man is of no direct help towards evolution, which appears to our dim vision to be the goal of all living existence. Evolution is an unresting progression; the nature of the average individual is essentially unprogressive. Consider the interest attached to the Hebrew race, whose average value is little worthy of note, but which is of special importance on account of its variety. Its variability in ancient and modern times seems to have been extraordinarily great. It has been able to supply men, time after time, who have towered high above their fellows, and left enduring marks on the history of the world.

Some thoroughgoing democrats may look with complacency on a mob of mediocrities, but to most other persons they are the reverse of attractive. The absence of heroic gifts among them would be a heavy set off against the freedom from a corresponding number of very degraded forms. The general standard of thought and morals in a mob of mediocrities must necessarily be mediocre, and, what is worse, contentedly so. The lack of living men to afford lofty examples, and to educate the virtue of reverence, leaves an irremediable blank. All men would find themselves at nearly the same dead average level, each being as meanly endowed as his neighbour. These remarks apply with obvious modification to variety in the physical faculties. Peculiar gifts, moreover, afford an especial justification for division of labour, each man doing what he can do best." (pp. 15–16.)

Could any man but Galton have written a sermon like the above to bring out the essential meaning to humanity of—let us say—the standard deviation or the quartile? Could any man have written thus but Galton without laying himself open to the charge of self-conceit? He did not regard himself as one whose faculties gave him rank in the extreme tail of a frequency distribution. He would have recognised such a position for half his friends, before he thought of it for himself. He pictured life statistically, and with the naiveness of a child spoke the truth, forgetting that he was talking to a crowd the bulk of whom must have been sufficiently introspective to doubt whether they were not themselves mediocrities. Perhaps there is little to wonder at, if Galton's words fell on barren ground, and his brother anthropologists have continued for more than thirty years to neglect variation and cultivate mediocrity by ascertaining means.

Galton in his next section, "The Measurement of Variety," proceeds to describe his method of reaching the median $M$ and the quartile $Q$ from *any* two sets of observations, i.e. from any two percentiles, thus generalising his previous results, for directly finding $M$ and $Q$. He suggests that a traveller among savage peoples might test them by finding what percentages could stretch two (or better three) bows, the bows having been previously tested by the numbers of pounds' weight which were required to stretch them to the full[1]. Similar tests might be easily applied to the delicacy of hearing and eyesight, to swiftness in running, accuracy of aim, with spear, arrow, sling, etc.

[1] Let $x_1$ and $x_2$ be the unknown distances from the median $m$, of the two known test values $t_1$ and $t_2$. Then $x_1' = x_1/\sigma = (m - t_1)/\sigma$ and $x_2' = x_2/\sigma = (t_2 - m)/\sigma$, where $\sigma$ is the standard deviation, can be found at once from the tables of the probability integral, taking as standard case, one percentage above and one below the median (50 °/$_\circ$). Solving for $m$ and $\sigma$ we have:

$$m = (x_2' t_1 + x_1' t_2)/(x_1' + x_2') \text{ and } \sigma = (t_2 - t_1)/(x_1' + x_2'),$$

which solve the problem. Galton works with the quartile instead of $\sigma$, but assumes the normal relation of the two.

The method is very suggestive and easy, but it assumes the normal distribution of frequency to be the rule in the variation of anthropometric characters. To demonstrate this we should have to compare our percentiles in each case with the theoretical distribution. There is no doubt that a large number of physical measurements in man follow accurately enough for practical work the normal distribution but the rule is far from universal. Galton found when his deviations for all characters were reduced to their respective quartiles, that the average value of all the deviations at each of the grades in the eighteen series closely corresponded to the normal series, though individually they differed more or less from it, some in one way, some in another. He gives the following table of deviations:

| | 5° | 10° | 20° | 30° | 40° | 50° | 60° | 70° | 80° | 90° | 95° |
|---|---|---|---|---|---|---|---|---|---|---|---|
| Normal Frequency | −2·44 | −1·90 | −1·25 | −0·78 | −0·38 | 0 | 0·38 | 0·78 | 1·25 | 1·90 | 2·44 |
| Average of observed | −2·44 | −1·87 | −1·24 | −0·77 | −0·40 | 0 | 0·38 | 0·75 | 1·21 | 1·92 | 2·47 |

This, while not demonstrating the truth of the normal distribution for every anthropometric measurement, suggests at least that on the average there is no persistent deviation from it.

Galton concludes this chapter by stating that the properties of the law of frequency of error are

"largely available in anthropometric inquiry. They enable us to define the trustworthiness of our results, and to deal with such interesting problems as those of correlation and family resemblance, which cannot be solved without its help. Anthropologists seem to have little idea of the wide fields of inquiry open to them as soon as they are prepared to deal with individual variety and cease to narrow their view to the consideration of the Average." (p. 21.)

### F. MARKS FOR BODILY EFFICIENCY

Galton's Chapter III is entitled: "On the Advisability of Assigning Marks for Bodily Efficiency in the Examination of Candidates for those Public Services in which Bodily Efficiency is of Importance[1]." We reach at this point a topic in which Galton had a great and persistent interest, namely the desirability of giving marks in competitive examinations for physical as well as mental proficiency. The development of anthropometric laboratories with more or less standardised tests rendered such a proposal fairly feasible, and for a time Galton pushed it energetically. It is of course related to his earlier proposals for the selection of eugenic youths for endowment, these being determined by both intellectual and physical examination.

Galton read his paper before the Anthropological Section of the British Association in 1889, and obtained a Report from the Council of that body in favour of the proposal in the following year. He wrote at this time several articles urging its importance and he lectured on the subject. Galton says in his *Memories* that he became convinced that although the proposal had strong

[1] This is part of the British Association memoir of 1889: see *Report*, pp. 471–3.

*à priori* claims to consideration, it did not merit acceptance, and therefore he gave it up. But it is difficult to discover a really strong answer to his papers, and it will undoubtedly be revived some day, when the official world is slightly more enlightened. Probably marks will be given for a complete series of anthropometric tests, mental as well as physical. There can be no better remedy against cramming than examinations of this kind wherein it is quite easy to vary the tests, and to prevent anything but general intelligence and good physique scoring.

In 1878 a Joint Committee of the War Office and Civil Service Commissioners had been appointed to inquire into the question

"whether the present literary examinations for the army should be supplemented by physical competition."

The recommendations of this Joint Committee deal principally with marks to be assigned for athletic performance, and very reasonable objection was taken to the proposals. In a report of 1889 the Civil Service Commissioners stated that the War Office were satisfied with the physique of the young men selected by their examinations, but the Commissioners remark that should any department of the public service be desirous of testing the physical qualifications of its officers more severely than at present, they anticipated no more difficulty in determining the relative capacities of the individual candidates in this respect than is experienced in the literary examination.

"Moreover encouragement would be given generally to candidates to maintain a good state of health while preparing for the literary examinations, and any tendency to overpressure would thereby be diminished." (p. 24; *B. A. R.* p. 472.)

This is the point from which Galton starts, and he remarks that to define bodily efficiency and to measure it both in individuals and races is the special task of the anthropologists, who concern themselves with the practice of human measurements, and devise tests that give warning whether growth and development are or are not proceeding normally. Galton at once states that what he has in view has no relation whatever to the pass-examination now made by medical men to eliminate candidates who are absolutely unfit. These pass-examinations are obviously a necessity. The reform asked for consists in giving additional marks to those youths who are not only fit for service, but exceptionally well fit as far as bodily efficiency is concerned.

"The curious and hardly accountable disregard of bodily efficiency in those examinations through which youths are selected to fill posts in which exceptional bodily gifts happen to be peculiarly desirable, must strike the attention of anthropologists with special force, and they of all people are best able to appreciate how much is sacrificed by its neglect." (p. 24; *B. A. R.* p. 472.)

Galton next cites the reduction by Dr Venn[1] of the data from the Cambridge Anthropometric Laboratory to show there is no significant asso-

[1] *Journal of the Anthropological Institute*, Vol. XVIII, pp. 140–54. Galton provided the instruments for the Cambridge Anthropometric Laboratory and when over a thousand students had been measured Venn reduced the data with the above-mentioned result. Galton supplemented Venn's paper with some discussion of the association of intelligence with size of head (pp. 155–56), and considered that it was possible to show that there was increased intelligence with increased head size. This matter is of such importance that a number of years ago I got

ciation between physique and intellect, and we cannot therefore argue from examinational success to any physical fitness. The high honours men, the low honours men and the poll men were alike in their average bodily efficiency, except for slightly worse vision in the high class honours men.

"The intellectual differences are usually small between the candidates who are placed, according to the present literary examinations, near to the dividing line between success and failure. But their physical differences are as great as among an equal number of other candidates taken at random. It seems then to be most reasonable whenever two candidates are almost on a par intellectually, though one is far superior physically, that the latter should be preferred. This is practically all I propose. I advocate no more at present than the introduction of new marks on a very moderate scale sufficient to save from failure a few very vigorous candidates for the Army, Navy, Indian Civil Service, and certain other Government appointments in which high bodily powers are of service. I would give the places to them that would be occupied under the present system by men who are far their inferiors physically, and very little their superiors intellectually. I am sure that every successful employer of men would assign as much weight as this to bodily efficiency, even among the highest class of those whom he employs, and that Government appointments would be still better adjudged than they now are if considerations of high bodily efficiency were taken into some account." (p. 25; *B. A. R.* p. 473.)

Galton considers that the desirable tests should involve measures of strength, vital capacity, agility or promptness, keenness of eyesight and of hearing. We could now add a number of tests, such as have been applied recently to candidates for the Air Force. The chapter concludes with the remark:

"It would certainly be grateful to many parents who now lament the exclusively bookish character of the examinations, and are wont to protest against a system that gives no better chance to their own vigorous children of entering professions where bodily vigour is of high importance than if they had been physically *just not unfit* to receive an appointment." (p. 26; *B. A. R.* p. 473.)

Chapter IV is entitled: "On the Principle and Methods of Assigning Marks for Bodily Efficiency[1]." Galton starts by saying that we may either give marks for ranks or for achievements, and apparently proposes in each

permission to have copies taken of some of the schedules and worked out the actual correlations for over 1000 cases with the following results:

Length of Head and Intelligence $+ \cdot 111 \pm \cdot 020$,
Breadth of Head and Intelligence $+ \cdot 097 \pm \cdot 021$.

Schoolboys and schoolgirls at age twelve gave similar results. (See "Relationship of Intelligence to Shape and Size of Head," *Biometrika*, Vol. v, p. 120.) Thus Galton was *formally* correct in saying there was association, but the correlation is so low as to be absolutely idle for any individual prognosis.

Galton's interest in the Cambridge Anthropometric Laboratory was very great and in association with Mr (now Sir) Horace Darwin not only new anthropometric instruments were devised, but the old ones improved or modified. The Cambridge Scientific Instrument Company produced "A Descriptive List of Anthropometric Apparatus, consisting of Instruments for Measuring and Testing the Chief Physical Characteristics of the Human Body. Designed under the Direction of Mr Francis Galton." Of this, the fourth issue (May, 1890) lies before me; it is rather a reasoned account of the meaning of an anthropometric laboratory, of the measurements which may be taken and of the methods of taking them, than the price list of a manufacturing firm.

[1] This was also contributed as a memoir to the British Association, see *Report*, 1889, pp. 474–8. The diagrams of the pamphlet fail in this B.A. memoir. They were, however, given in a paper in *Nature*, October 31, 1889, Vol. XL, pp. 650–51. On the other hand the *B.A. Report* has the paper by A. A. Somerville discussing Eton experiments on the relative reliability of physical tests and of literary examinations. A system of marking having been devised for physical tests and for medical fitness in certain directions two medical men examined independently 32

case to mark from the mean or median. It is not quite clear in this case why there should not be negative marks for those who fell short of mediocrity. Galton then provides marks for each rank above 50 °/₀ or again for the deviate. In the latter case the full number of marks is supposed to be reached by a grade 99·99. Next he draws up a table in which the total marks for physical efficiency are assumed to be 10 and are supposed to be assigned in different proportions to rank and to absolute achievement, i.e. to deviate from mediocrity. His table (*B. A. R.* p. 476) is as follows:

| Percentage of marks assigned to | | Rank | | | | | |
|---|---|---|---|---|---|---|---|
| Rank | Deviate | 50° | 55° | 75° | 95° | 99° | 99·99...° |
| °/₀ | °/₀ | | | | | | |
| 100 | 0 | 0 | 1·0 | 5·0 | 9·0 | 9·8 | 10·0 |
| 50 | 50 | 0 | 0·7 | 3·5 | 6·9 | 8·4 | 10·0 |
| 33·3 | 66·6 | 0 | 0·6 | 3·0 | 6·2 | 8·0 | 10·0 |
| 25 | 75 | 0 | 0·5 | 2·7 | 5·8 | 7·7 | 10·0 |
| 0 | 100 | 0 | 0·4 | 2·0 | 4·8 | 7·0 | 10·0 |

It is clear that Galton is compounding what every schoolmaster has to consider, namely: "place in class" with "marks in examination," or indeed "place" and "marks in examination." If the distribution were normal the marks would be readily deducible from "place" in examination. Considering what wide differences occur between the top individuals—for example in the old Cambridge Mathematical Tripos system—and what slight differences between mediocre individuals, one is inclined to doubt the legitimacy of marking by rank. This point was recognised, I think, by Galton later when he endeavoured to estimate the average differences between individuals arranged according to rank. However the table is suggestive and we leave it to the consideration of the educationalist.

Galton supposes the chief physical measurements to have been reduced for the class under examination to percentile scales, such as in the table on our p. 376. He gives a rather clear diagram of the absolute values for males and females at each rank for seven characters on p. 29[1]. In some of the characters it would be desirable for rapid and safe interpolation to insert a few more values of the deviate.

boys. Their average difference in judgment was 8·75 °/₀ of the maximum marks. Nineteen of the boys were subsequently examined in English Essay, and the essays submitted to two independent examiners. The average difference was now 16·7 °/₀ of the maximum marks. The experiments seem to have been undertaken at Galton's suggestion and are used by him as an argument that physical test marks can be as accurately determined as marks for literary work. Some years ago the present writer reported on the marking statistics of the London Matriculation Examination. He was startled to find that the relative personal equation of examiners in history, languages and literature was very greatly larger than the relative personal equation of examiners in branches of science; and that no systematic method of correcting for this personal equation had been adopted. Thus a candidate's chance of passing depended largely on the examiner to whom his paper was allotted! It would appear therefore that the choice of an English essay for comparison of marking differences was rather unfortunate.

[1] See *Nature*, Vol. XL, pp. 650 and 651.

At this point Galton recognises two important matters, namely: (i) that these physical characters are in certain cases highly correlated, and that

GRADES OF RANK, 0° TO 100°

0° 10° 20° 30° 40° 50° 60° 70° 80° 90° 100°

STATURE, WITHOUT SHOES. INCHES.

MALES 63 64 65 66 67 68 69 70 71 72 73

FEMALES 59 60 61 62 63 64 65 66 67 68

HEIGHT SITTING, ABOVE SEAT OF CHAIR. INCHES.

MALES 33 34 35 36 37 38 39

FEMALES 32 33 34 35 36

SPAN OF ARMS BETWEEN OPPOSITE FINGER-TIPS. INCHES.

MALES 64 65 66 67 68 69 70 71 72 73 74 75

FEMALES 58 59 60 61 62 63 64 65 66 67 68

WEIGHT IN USUAL INDOOR CLOTHING. LBS.

MALES 120 130 140 150 160 170

FEMALES 100 110 120 130 140 150

BREATHING CAPACITY. CUBIC INCHES.

MALES 170 180 190 200 210 220 230 240 250 260

FEMALES 90 100 110 120 130 140 150 160 170 180

KEENNESS OF EYESIGHT. DISTANCE OF READING DIAMOND TYPE. INCHES.

MALES 10 15 20 25 30 35

FEMALES 10 15 20 25 30 35

STRENGTH OF GRASP. LBS.

MALES 65 70 75 80 85 90 95 100 105

FEMALES 30 35 40 45 50 55 60 65 70 75

0° 10° 20° 30° 40° 50° 60° 70° 80° 90° 100°

**Diagram of Absolute Values at each Rank.**

accordingly it is scarcely accurate to mark them independently, and (ii) that from the fitness point of view a vital capacity, for example, for a small man may be amply adequate, while the same vital capacity would be insufficient for a big man. If we judged independently by rank for vital capacity and stature, say the big man might be more heavily marked than the small man, although his combined system was far less efficient. To throw light on these matters Galton constructs "mixed" correlation tables in which he gives the rank of one variate and the deviate of the other and draws contour lines for the first. Thus for each value of stature—i.e. for every two inches—he constructs a scale of deviate of vital capacity to each grade. He then combines these to give a single diagram and joins the points of equal vital capacity. He thus obtains what he terms *isograms* at every 20 cubic inches of vital capacity. He gives a similar system of isograms for strength of squeeze and weight. We reproduce one of these diagrams.

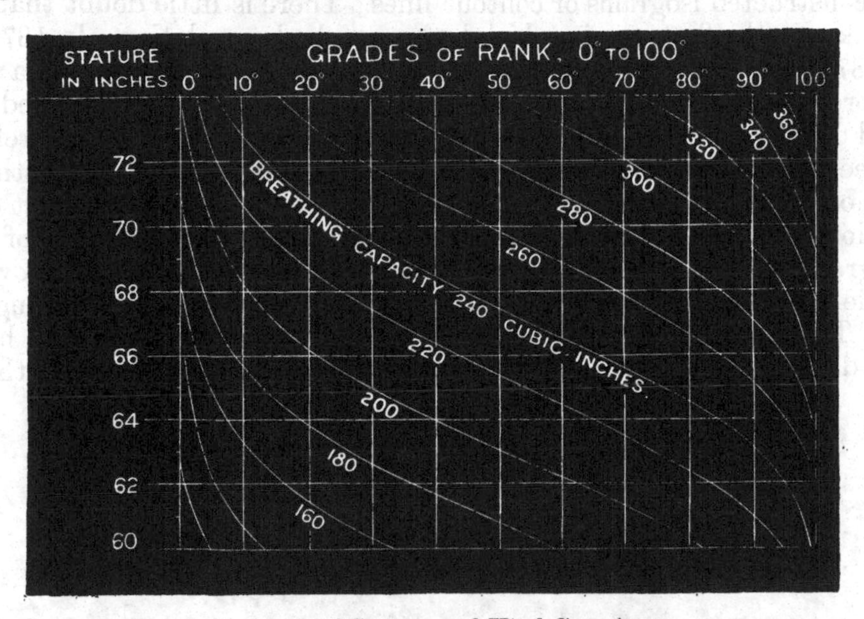


Isograms of Stature and Vital Capacity.

Galton remarks that these and similar methods indicate how marks may be determined for physical measurements. It is, perhaps, needless to say that they would apply also to mental measurements, and even to examinational results. The ordinary process of marking by adding up the marks acquired by the same boys in a variety of subjects is really a fallacious one, as there is usually a high correlation between success in different subjects, and the true difference between two individuals cannot be measured by adding up a series of differences, highly correlated. We might as well repeat papers in one subject until the first of two boys had multiplied his excess marks a hundred times, but this deviate difference would certainly not be the measure of their relative intelligence. Their true ranking would have to be obtained from the multiple correlation surface, allowing correlation between subjects. The theory of examination marking sadly needs scientific treatment[1]. Galton himself concludes as follows his discussion of marking:

[1] Ranking in multiple marking can be thrown back, through multiple correlation, on the $P$ and $\chi^2$ of the 'Goodness of Fit' tables.

"It is now the part of those who have to fix the scales of marks to determine the weight to be given respectively to relative rank and to absolute performance in examinations of each different kind of service." (p. 32; *B. A. R.* p. 477.)

A point may be noted here before we consider Galton's further contributions to the marking problem. In the paper just discussed Galton illustrates by *isograms* his method of exhibiting graphically the correlation of the rank for one character with the deviate for a second associated character. He undertook, however, the construction of a diagram of isograms for the case in which rank of one character was correlated with rank of a second. For all three cases—rank with rank, rank with variate, and variate with variate—Galton constructed isograms or contour lines. There is little doubt that rank with rank was the first way in which he approached correlation. In 1875 (see our p. 187) Galton was dealing with the inheritance of size in sweet-pea seeds, but before he obtained his data for sweet-peas, he appears to have tried what he could do with much smaller seed, apparently that of cress. The correlation of the seed of mother and daughter plants was dealt with by the method of his memoir of 1875, and that is, I think, the probable date of his first crude correlation table, which he obtained from five groupings of each size of seed; the isograms are represented by ink lines on the sheet of glass covering the little compartments which contain the ranked seeds of the daughter-plants. These isograms have been smudged and almost obliterated by the wear and tear of fifty years, but can still be traced. The facts (*a*) that he

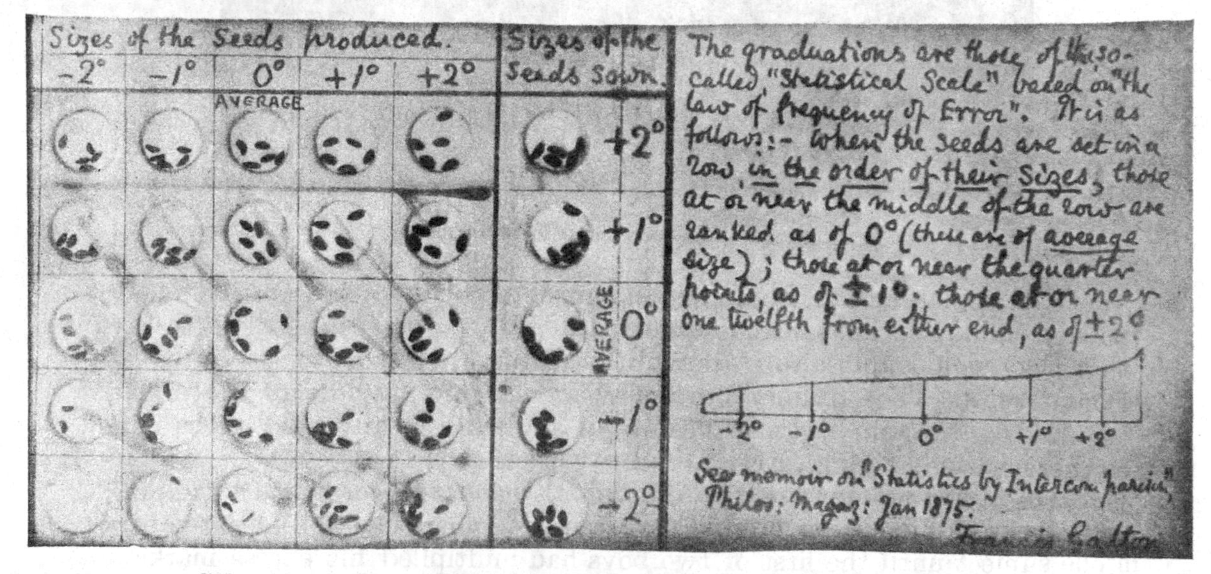


Galton's first illustration of Correlation, *circa* 1875. From the *Galtoniana*.

uses the word *average* when he later used *median*, (*b*) that he divides not into the percentiles of his later work, but into quartile and double-quartile[1]

[1] Galton himself says one *twelfth* from the end of the range, but if the reader looks at Galton's table on p. 42 of his paper on "Statistics by Intercomparison" (see our p. 338), he will find 82 in 1000 given for two quartile units, which Galton takes approximately as $\frac{1}{12}$.

groups suffice to indicate that the model must have been made about the time of his paper of 1875. It is, I think, sufficient evidence that Galton dealt with the correlation of ranks before he even reached the correlation of variates, and the claim that it is a contribution of the psychologists some thirty or forty years later to the conception of correlation does not seem to me valid. Galton, we may with high probability suggest, had satisfied himself that the correlation of ranks was more cumbersome than the correlation of variates, because in the simplest case, that of the normal distribution, it fails to provide linear regression, but gives a non-integrable curve, which can only be plotted by aid of the probability integral table.

Galton undoubtedly first attacked the problem of correlation from the standpoint of ranks, and it naturally did not lead him to any simple expression for the relation between relatives. In his *Memories* (p. 300) he tells us how:

"As these lines are being written, the circumstances under which I first clearly grasped the important generalisation that the laws of heredity were solely concerned with deviations expressed in statistical units [for Galton the quartile values] are vividly recalled to my memory. It was in the grounds of Naworth Castle[1], where an invitation had been given to ramble freely. A temporary shower drove me to seek refuge in a reddish recess in the rock by the side of the pathway. There the idea flashed across me, and I forgot everything else for a moment in my great delight."

That 'recess' deserves a commemorative tablet as the birthplace of the true conception of correlation!

Galton continued his campaign in favour of marks for physical efficiency. He published a paper in the *Nineteenth Century*, Feb. 1889 (pp. 303–8), entitled, "The Sacrifice of Education." The title does not seem well chosen, and I expect it was an editorial choice; Galton himself has written "Tests of Physical Capacity" above the paper and so indexed it. The object of the paper is to show the ease with which certain anthropometric measurements can be made, and the cost of making them. With most of the facts stated in the paper the reader will be already familiar. The tests are described, the time taken in making them, and the cost of running an anthropometric laboratory.

"The problem is to give marks for physical qualifications just as they are now given for intellectual ones, in order to pass those candidates who being a little under par intellectually are far above par bodily; conversely to weed out those other candidates who, not being particularly fit in respect to their brains, are at the same time of decidedly inferior physique. The relative weight to be assigned for intellectual and bodily excellences is a question of detail, most important no doubt, but one that need not be discussed here."

Galton states the various criticisms of the proposal that have been made, but points out that they apply not only to marks for physical capacity, but to all examinations whatsoever. We can never test *all* the faculties even of the mind, and again we can only test the examinee on the special occasion on which he presents himself. It is admitted that the examination of *any* faculty, physical or mental, is a difficult art, and one not to be perfected offhand.

[1] Naworth Castle is in Cumberland, north-east of Carlisle. The Galtons went north on Aug. 12 and lodged at Wetheral in the neighbourhood of Naworth; later, on Sept. 10, they travelled to Newcastle and stayed with the Spence Watsons at Bensham Grove during the B.A. meeting at which Francis Galton read his paper on marks for physical efficiency. "Frank has been busy all the year on problems connected with correlation but has published nothing yet about them." (*L. G's Record* under 1889.)

The next effort of Galton was a lecture before the *Society of Arts* on November 26, 1890[1], entitled: "Physical Tests in Examinations." This lecture tells us a little more of the history of the movement. Galton's proposal to give marks for physical efficiency was sent by the Anthropological section to the Council of the British Association, and the latter drew up a report on the subject which was distinctly favourable, although cautiously expressed; finally the recommendations reached were submitted to various government departments. Not improbably the original draft was due to Galton. The recommendations ran:

(1) That an inquiry should be held as to the best system of assigning marks for physical qualifications, on the double basis of inspection and anthropometry, with a view to its early establishment as a temporary and tentative system.

(2) That the marks to be given under this temporary system should be small, so as to affect the success of those candidates only, who would be ranked by the present examinations very near to the dividing line between success and failure, and whose intellectual performances would consequently be nearly on a par, though they would differ widely in their physical qualifications.

(3) That a determination should be expressed to reconsider the entire question after the experience of a few years.

The replies received from the government departments were more or less of the usual type.

> "The Civil Service Commissioners, moved thereto by the India Office, are now engaged in considering the practicability of the proposals."

They probably would wait to see what force of public opinion was behind them.

Galton states that—subject to such reservations as that training for physical efficiency must not distract the candidate from his books, that it should not exhaust his energy, and that it should not displace any of the usual examination subjects or the medical pass-examination—it was a truism to say that physical efficiency ought to be taken into account in selecting men to fill posts where high physical powers are advantageous. Our lecturer next dealt with the means of testing physical efficiency. Athletic competitions he discarded as in no way suitable; he wanted to test natural capacities and not these capacities after a severe course of training. He then turned to inspection, and suggested that medical examiners might not only pass men, but mark them. Galton appeared to lay some stress on judgment of physical efficiency by inspection and cited the manner in which horse dealers and slave dealers rapidly reach on inspection fairly accurate estimates[2]. Galton next turns to physical tests and mentions those with which this book has rendered the

[1] *Journal of the Society of Arts*, Vol. XXXIX, pp. 19–27.

[2] Galton illustrated this from an experience of his own when travelling in 1846 in the Soudan. "An Egyptian, who possessed little besides a sword, had attached himself to the caravan with which I was travelling. He was on his way to join a slave-raiding expedition on the borders of Abyssinia, and he had, I found out, considerable experience in slave markets. I asked him many questions, from time to time, about the valuing of slaves, and at last begged

reader familiar, also his latest instruments and the need for considering the correlation of characters.

Then Galton passes to objections. It had been stated that many great commanders and strategists had a poor physique, and that a Nelson might be excluded. But the proposal did not exclude anybody, that was for the pass medical examination to do. Really able men would not be excluded by the marks for physical efficiency, which it was proposed should touch only border-line cases.

Another objection was that anthropometric tests did not reach the important quality of energy, which includes pluck, strong will and endurance. Galton admits *lacunae* here, as in mental testing.

"We must be content with what we can get. It is not impossible that practical tests of energy in some of its forms may yet be discovered. It must be associated with physiological signs that we have not yet had the wit to discover." (p. 24.)

The third objection Galton discusses is the supposed untrustworthiness of the examination he proposes. There is, he says, no reason to suppose that it would be more untrustworthy than a literary examination, and he cites for variability of judgment in a literary examination Edgeworth's paper in the *Journal of the Statistical Society* of the same year[1].

Among problems which the lecturer held could be ultimately settled by tests of this physical kind were the following:

(1) Whether the proposed tests of physical efficiency confirmed the results of athletic competitions.

(2) Whether physical efficiency in youth corresponded to achievement in after life.

(3) What type of physique was best suited to tropical climates.

"It is cruel and costly to tempt youths to the tropics who are less constitutionally capable than others of thriving there. If we could distinguish those who are fitted for life in hot countries, we should select them even though in other respects they may be somewhat wanting. The tropical possessions of England are become so large that it is a matter of national importance to investigate this question thoroughly. It may yet be possible to find varieties of our race who are capable of permanently establishing their families in those climates." (p. 25.)

Thus the far-seeing Galton; we are no nearer a settlement of these problems now than we were when he urged their importance, and the reason of it lies in the fact that the data still fail us. One of his reasons for establishing tests of physical efficiency is the stimulus it would give to the collection of trustworthy records.

him as a favour to price myself, just as if I was a light-coloured African; for I was curious to know my worth as an animal. He took evident pains, and I think was fairly honest, though with a bias towards flattery. Having regard to the then high state of the market, he estimated my worth, on the spot, at a number of piastres that was about equal to £20." (p. 20.)

The price of adult negro slaves in 1690 (Davenant) was £26. Sir William Petty after elaborate calculations valued an Englishman at £69, which King some twenty years later reduced to £65. The most recent valuation that I have come across is that of the average American citizen at about £400 (Davenport). It must be admitted therefore that Galton, aged 23, would have been a distinctly cheap bargain at £20.

[1] Vol. LIII, pp. 460–75, 644–63, 1890, and compare Vol. LI, pp. 599–635, 1888, and *Phil. Mag.* 1890, pp. 171–88.

The discussion which followed the lecture was of the usual character and not very profitable. The lecture, however, shows that Galton himself had not weakened in his faith in the advantages of his proposal. That the matter was dropped in 1890, without any published criticism of a damaging kind, will make it more difficult for success to be achieved when the matter comes up for discussion again, as it evidently must; we shall be told that the proposer himself abandoned the scheme in 1890, and why should it be resurrected? The only answer can be that we really do not know why Galton gave up the fight. His reasons given twenty years after the campaign are not conclusive[1]. Who was the high authority of the War Office who wrote a careful minute, and what evidence is there that he possessed the requisite anthropological and physiological knowledge? Galton, as we have said before, was very apt to assume that other men's judgments, whatever their real intelligence, must certainly be better than his own, and the present was probably a case in point.

### G. PRESIDENTIAL ADDRESSES TO THE ANTHROPOLOGICAL INSTITUTE

During the four years 1886–1889 Galton was President of the Anthropological Institute, and gave not only four presidential addresses, but took a considerable part in its proceedings, and worked for its welfare[2]. The first of his presidential addresses (1886) was on the inheritance of stature, and will be dealt with in our next chapter. The second presidential address (1887) is of a mixed character: it describes the then recent progress of anthropology, and it gives some suggestions and thoughts arising therefrom[3]. It is followed by an obituary notice of Dr George Busk, which I happen to know was written by Galton himself[4]. Among other matters he refers to the anthropological collections recently established in London and Oxford, to the projected Imperial Institute and to the foundation of the International

[1] *Memories*, p. 214. Some experiments undertaken at Marlborough and reported by Meyrick and Eve (*Marlborough College Natural History Society*, 1889) seem to me very wide of the point. The actual anthropometric measurements were placed before eleven masters who were asked to mark them according to their own arbitrary opinion having regard to the boys' ages. Probably none of those masters had any idea of the variations of and the correlations between the measurements; it is hardly likely that they had the physiological or medical knowledge adequate for appreciating the relative values of the tests, nor any idea of how the individual boys would rank in a population of a like class. The conclusion that "there is probably greater vagueness in this examination than in most school examinations" was probably perfectly correct as applied to an examination for physical fitness conducted in such a manner.

[2] The *Report* of the Council for 1888 contains the words: "Mr Francis Galton's second term of office, as President, has now expired, and the Council desires to put on record its sense of the valuable services which he has rendered to the Institute and to the cause of Anthropological Science in general, during the past four years. The many ways in which Mr Galton has promoted the interests of the Institute demand, in an exceptional manner, the grateful acknowledgment of the members." *Journal of the Anthropological Institute*, Vol. XVIII, p. 400.

[3] *Journal of the Anthropological Institute*, Vol. XVI, pp. 387–402.

[4] Galton wrote a considerable number of notices of dead friends, Spottiswoode, Marianne North, Herbert Spencer and George Busk, among others. They are graceful tributes to his friends' work.

Statistical Institute by Sir Rawson W. Rawson. Speaking about the new Anthropological Society of Japan and its memoirs, Galton writes:

"No doubt some of the more valuable papers in this journal will hereafter appear in one or other of the chief European languages. The curse of the Tower of Babel, in whatever sense we may employ the phrase, has long pressed heavily upon scientific men in Europe; the contemplation of the additional burden on our descendants of having possibly to learn Japanese, Russian and Chinese as well as the Western European languages can hardly be indulged in with equanimity." (p. 394.)

Galton next turns to the white man in the tropics—a favourite topic with him—and noting that it is the temperature of the living rooms—especially the bedrooms—which is the difficulty, he enters fairly fully into the possibilities, mechanical and economical, of refrigerating apparatus for use in tropical climates. This leads Galton to consider the possibility of an acclimatised strain of Englishmen, and thus prompted he discusses alternate inheritance:

"Much has recently been written on the difficulty of any rare accidental variety of animal or plant establishing itself, when it has unrestricted opportunity of intercrossing with the parent stock. It is urged that the peculiarity would be halved in each successive generation, and would very soon cease to be apparent in the descendants. It seems to me that this argument is sometimes pressed too far. It cannot be a general truth that characteristics blend, else to take a conspicuous example, there would be a growing tendency in every mixed population for the eye-colour to become of a uniform hazel or brown gray tint, through the intermarriage of persons whose eye-colours differ widely. On the contrary I have lately shown by a considerable body of statistics[1] that among the English, the proportions between the eye-colours, as sorted under seven headings, have not changed at all during four generations. The fact is the heritages are only partially liable to be blended together; partially they are mutually exclusive. No case of inheritance probably falls under either of these opposed extreme conditions, but some approximate to one, and others to the other. I am not aware that the respective results of these two extreme conditions have yet been put forward quite as forcibly as they admit of being and deserve to be....Suffice it to conclude that the establishment of a somewhat rare variety, as that of white men naturally suited to thrive and multiply in tropical climates, is not so great an improbability as those anticipate, who lay exclusive stress on the tendency of rare peculiarities to disappear in a very few generations, through free intermarriage with the ordinary members of the stock." (pp. 400–402.)

Galton's presidential address of 1888[2] turns chiefly on Alphonse Bertillon's system of criminal identification. The address begins with a reference to Galton's short course of lectures on "Heredity and Nurture" given at the South Kensington Museum in December, 1887, under the auspices of the *Institute*.

"Their object was to test the reality of a supposed demand for information on such subjects, and so far as it was possible to judge from the results, there seemed to be a widely spread interest in the matter....These lectures have led to at least one tangible result. I took the opportunity to reiterate my often expressed regret that no anthropometric laboratory existed in this country, at which children and adults of both sexes could at small cost have their faculties measured by the best methods known to science, and a record kept for future use. I explained how difficult it would be to maintain such a laboratory, and to make it effective, except under the shelter of some important institution that was daily frequented by the class of persons likely to make use of it." (p. 346.)

As a result of Galton's reiteration he was given the wooden building associated with the South Kensington Museum which formed his second

[1] We shall return to Galton's 1886 paper on this subject in the following chapter.

[2] *Journal of the Anthropological Institute*, Vol. XVII, pp. 346–54.

anthropometric laboratory, and this arrangement lasted till 1890. He insists largely on the value of such anthropometric records for the identification of individuals, and cites the claims made for them by Bertillon, Topinard and Herbette. He describes further the index-system of Bertillon:

"Whether all that was claimed for the power of M. Bertillon's system, on purely theoretical grounds and in his earlier publications, can be sustained, may fairly be questioned; but there can be no doubt that a series of measurements must be of considerable service as supplementary evidence, either that a person is really the man he professes to be, or negatively that he is not the man for whom he is taken. In speaking of these matters it is impossible not to allude to the Tichborne trial, and the enormous waste of money, effort, and anxiety which might have been spared, had Roger Tichborne passed through an anthropometric laboratory before he went abroad. It would be a reasonable precaution for every person about to leave his country for a long time, having regard to the various accidents of good or ill fortune, to be properly measured, and to leave a copy of his measurements in the safe keeping of an anthropometric laboratory." (p. 252.)

"Another and very important question is as to the degree in which the several bodily proportions that are measured may be looked upon as independent variables. The stature is related with the length of the foot, and with that of the forearm, and we should expect a still closer relation to exist between any two of these taken together, and the third. We have yet to learn the proportion between the number of the elements measured and their value for purposes of identification. The supposition that they may be treated as independent variables, which lies at the bottom of some of the earlier estimates, such as that on page 22 of the Conference at Rome[1] headed 'Étendue infinie de la Classification,' cannot be accepted as correct.

The whole subject of 'Personal Identification and Description' forms an important chapter of anthropological research, and it is one on which I hope before long to be in a position to offer some views of my own." (p. 354.)

The careful reader of this passage will note how Galton is beginning to realise the problems of multiple regression, and to see that with a large number of *correlated* variables, there is a limit to the intensity of the multiple correlation coefficient, which cannot be indefinitely increased by increasing the number. We also see how he is studying the correlation of bodily characters and gradually advancing beyond the indexing by such characters to a method of his own—the identification by finger-prints.

Galton's fourth presidential address, that of 1889, we have already discussed (see our p. 383), and the reader who turns back to our account will have some appreciation of how the ideas of the 1888 address had been developed in the interval. Besides working out the fundamental ideas concerning correlation and heredity to which Galton was led by his "Family Records" and the data from his Anthropometric Laboratory, he contributed numerous short papers on anthropometric and statistical topics to various journals during this period and later at intervals.

### H. MISCELLANEOUS STATISTICAL PAPERS

(i) *The Horse.* The horse had always been a favourite animal with him, notwithstanding his experiences with the camel in Syria and the ox in Damaraland, and no less than four of his papers treat of the horse under various conditions in addition to his paper on standard photographs of horses

[1] Louis Herbette and Alphonse Bertillon: "Les Signalements Anthropométriques," *Conférence faite au Congrès Pénitentiaire de Rome*, Masson, Paris, 1886.

for measurement[1]. One of these papers deals with heredity and will be considered in the following chapter. The first of the remaining three belongs to Galton's work on composite photography. It is entitled: "Conventional Representation of the Horse in Motion[2]." In this paper Galton endeavours to construct the conventional attitude of sculpture or of painting from composites of Muybridge's photographs of the "Horse in Motion." The final result ought to represent a mule, for it is certainly a very hybrid structure.

"The first composite shows the hind legs distinctly, the second shows the fore legs distinctly; and if duplicates of the first and second woodcuts are each divided into two halves and the best defined halves of each are united (in a way that might have occurred to Baron Munchausen if a second rider's horse had suffered as his own, and there had been a mistake in piecing them) a result is produced that shows a very fair correspondence with a not uncommon representation in sculpture."

The second paper is entitled: "The American Trotting Horse[3]." The interest of this paper lies in its great value as indicating the influence of long continued selection on a character. Galton, dealing with the statistics of the speed of American trotters, shows that every three years from 1871 to 1880 the speed of the best horse increased about two seconds, or in the nine years there was an improvement from about a mile in 2 mins. 17 secs. to a mile in 2 mins. 11 secs. Perhaps the most noteworthy point is, however, that not only the speed of the fastest horse thus improved, but the *first hundred* horses maintained their *relative* speeds, or all increased by about the same two seconds. Galton's final table is as follows:

*Number of Seconds and Tenths of Seconds in Excess of Two Minutes that are required for Running One Mile by the Horses, whose order in the Rate of Running in each Year is given at the Top of the Column.*

| Year | 100th | 50th | 20th | 10th | Year | 100th | 50th | 20th | 10th |
|---|---|---|---|---|---|---|---|---|---|
| 1874 | 25·1 | 23·4 | 20·5 | 18·8 | 1880 | 20·8 | 19·3 | 17·6 | 16·0 |
| 1875 | 24·1 | 22·5 | 19·9 | 18·2 | 1881 | 20·4 | 18·8 | 17·2 | 15·7 |
| 1876 | 23·5 | 21·6 | 19·5 | 17·7 | 1882 | 19·9 | 18·4 | 17·0 | 15·4 |
| 1877 | 22·9 | 21·0 | 19·0 | 17·4 | | | | | |
| 1878 | 22·1 | 20·2 | 18·5 | 17·0 | Anticipated | | | | |
| 1879 | 21·3 | 19·6 | 18·0 | 16·6 | 1890 | 16·8 | 15·5 | 14·4 | 13·4 |

*Mem.* The first horse runs the mile in about five or six seconds less than the tenth horse.

Galton's anticipation for 1890 is obtained by noticing that the values in the vertical columns are nearly linear and then extrapolating.

"Supposing the conditions to be maintained, I should anticipate that in 1890 there will be about 15 horses that will run a mile in 2 minutes 15 seconds or less, and that the fastest horse of that year will run a mile in about 2 minutes 8 seconds[4]."

[1] See our p. 317 *et seq.* [2] *Nature*, Vol. XXVI, pp. 228–9, July 6, 1882.
[3] *Nature*, Vol. XXVIII, p. 29, May 10, 1883.
[4] Actual best values for 1891, 1900, 1910, and 1922 were: 2′ 8″·25, 2′ 3″·25, 1′ 58″·75 and 1′ 56″·75 respectively, so that Galton's prediction was exact, if we take 1891 for 1890.

The third memoir on horses also related to American trotters, but it is 14 years later, 1897[1]. In this paper Galton draws attention to the value of the speeds of the "trotters" and "pacers" in *Wallace's Year Book of American Trotters* and suggests their importance for heredity. He notes, however, that it is very difficult to obtain the speeds of the parents and grandparents. No horse can theoretically obtain admittance to the trotting register unless it can trot a mile in under 2 mins. 30 secs. The data, however, seem to indicate either that some grace is given to horses who trot the distance of a mile in a little over the limit time, or else that a good many owners do not care to press their horses beyond the limit of admission. In either case Galton was justified in discarding the 2 mins. 29 secs. to 2 mins. 30 secs. entries as not homogeneous with the remainder. Dealing with the remainder Galton fits the distribution by rather rough methods[2] with a *portion* of a normal curve of deviations, and thus is able to determine not only the mean speed but the variation in the speed of American trotters. Thus he obtains for the median and quartile of the series of 982 records for 1893, a median speed of 2 mins. 26 secs. with a quartile of 5 secs. (2 mins. 21 secs.). The paper is of considerable value as it was, I believe, the first occasion on which an attempt had been made to fit incomplete series.

(ii) *The Median.* There are a number of short papers by Galton which are, perhaps, most suitably dealt with in this chapter. A good many of them appeared in the pages of *Nature*, a ready means of attracting immediate attention, but too often at the cost of later oblivion. Several of these papers concern really important points, which have, since their publication, been again and again overlooked.

In the first place we may turn to a group dealing with the median or the mid-character in a series. It is well-known that the median is subject to a larger probable error than the mean and this has discouraged its use in statistical inquiries dealing with carefully recorded observations. But Galton realised that while its chief value in such cases was the rapidity with which it could be ascertained[3], yet there existed certain cases in which the median may be said to be far more reliable than the mean. In a paper of 1907 entitled "One Vote, One Value[4]," Galton draws attention to how misleading a use of the *average* may be. He cites as instances of importance and frequent occurrence: (i) the assessment by a jury of damages, (ii) the determination by the council of a society or by a committee of a sum of money suitable for some particular purpose. Each voter, whether of jury or council, ought to have equal authority with each of his colleagues. How can the right conclusion be drawn from the many individual estimates? Galton

[1] *Royal Soc. Proc.* Vol. LXII, pp. 310–15. See also *Nature*, Vol. LVII, p. 333, Feb. 3, 1898.

[2] More exact methods have since been applied to the data: see *Biometrika*, Vol. II, pp. 2–6. In the 1893 case a mean of 2 mins. 28 secs. and a quartile of 5·96 secs. were found.

[3] That Galton used median and quartiles so frequently even on careful records must, I think, be attributed to his great love of brief analysis. He found arithmetic in itself irksome; he would prefer to interpolate by a graph rather than by a formula, and while his rough approximations were as a rule justified, this was not invariably the case.

[4] *Nature*, Vol. LXXV, p. 414, Feb. 28, 1907.

holds, and surely he is right, that the *middlemost* estimate, the median, is the correct one. Every other estimate has a majority of the voters against it as either too low or too high. The correct estimate cannot be the *average* for, as Galton puts it, the average "gives a voting power to 'cranks' in proportion to their crankiness." The average allows crankiness to swamp reasonable judgment. For such reasons Galton laid considerable stress on the median, and on various contrivances for rapidly determining it.

I have already referred (pp. 336, 385) to the use Galton made of two bows or two weights to test the strength of a group, and how he determined his median from the resulting percentages. This point is more fully dealt with in a paper on "The Median Estimate" read at the Dover meeting of the British Association in 1899[1]. In this paper Galton applies the two weights test to determine the proper damages by a jury or a suitable grant by a committee. Two sums $A$ and $B$, $B$ being greater than $A$, are fixed on and then three shows or counts of hands are taken, (i) for a sum less than $A$, (ii) for a sum between $A$ and $B$, and (iii) for a sum greater than $B$. The individuals have thus not to determine actual amounts, but only inequalities. Galton now assumes the "normal" distribution of judgments and proceeds to determine the median in the manner of our footnote, p. 385[2]. To expedite the determination he published a table of percentiles giving the ordinates in terms of the quartile. This table is also reproduced in a paper of the following year and originally appeared in his book *Natural Inheritance* of 1889. It can still be used although it only gives three significant figures (two decimals), when the quartile is preferred. It has, however, been superseded for most purposes by the table of five significant figures (four decimals) provided by Dr W. F. Sheppard at the suggestion of Galton, who wrote a prefatory note to the table[3]. This table gives the deviate in terms of the standard deviation and proceeds by permilles not percentiles. The prefatory note is a remarkable one considering that Galton was then aged 85; he there broke a last lance for the use of the ogive curve and the median, which he had introduced 40 years earlier. He took his present biographer's data for the intelligence of Cambridge graduates and represented it on a percentile scale and not on the biographer's "normal" scale; and he made a very good defence of his method.

[1] *British Association Report*, 1899, pp. 638–40.

[2] If $b$ and $a$ be the fractions of the total assessors who vote "above $B$" and "below $A$" respectively, then the ordinates of the probability curve corresponding to $b$ and $a$, in terms of the standard deviation as unit, can be found from a table of permilles (see *Tables for Statisticians and Biometricians*, Table I). If these be $\alpha$ and $\beta$ the median will be

$$m = A + \alpha\frac{B-A}{\alpha+\beta} = B - \beta\frac{B-A}{\alpha+\beta}.$$

Here we suppose $a$ and $b$ both less than 50 per cent. of the total number of assessors. This is the better way of determining $m$; a slight modification is needed, if $m$ be greater (or less) than both $A$ and $B$. The values of $a$ and $b$ should correspond to more than 5 per cent. of the assessors for reasonable accuracy.

[3] *Biometrika*, Vol. v, pp. 400–6. "Grades and Deviates (including a Table of Normal Deviates corresponding to each millesimal grade in the length of an array, and a figure)."

He sums up its merits as follows:

"(1°) It establishes a centesimal scale of precedence, into which the order of any individual, in any array of individuals and of any length, can be easily translated, and it gives the normal deviate at the grade which the individual occupies.

(2°) It easily defines the limiting values of successive classes of given numbers in a normal array.

(3°) It classifies objects that can be arrayed by judgment, though not by actual measurement.

(4°) It gives by inspection the value of $\sigma$ [the standard deviation] in a normal series, and that of the probable error in any series, whether normal or not.

(5°) It exhibits processes under their real forms, and so is free from the danger of errors in principle, to which those unpractised in statistics are liable.

(6°) It affords an excellent criterion as to whether an observed array is or is not normal, and of the degree of its departure from normality." (p. 104.)

There is no doubt that Galton's method of grades and deviates will retain a permanent position in statistics, chiefly as a means of illustrating in a simple manner statistical results; but as a fundamental method of tabulation it cannot be used. The ogive curve has no simple mathematical expression and data described in this way do not readily lend themselves to further quantitative discussion. This will be obvious to anyone who endeavours to determine the correlation coefficient from doubly-graded data, instead of from a frequency table[1]. Yet there was something fine about Galton's defence of his first statistical method in his 85th year! It did not, perhaps, convince the younger school, but it made them reconsider, and possibly judge more favourably and use more frequently, Galton's mode of representation.

Some years earlier Galton reduced his method of determining the median to a very simple process[2]. He transformed his "ogive curve" to a straight line by altering its horizontal or percentile scale. He was thus applying to a special case the conception of Lalanne's anamorphic geometry. In Galton's case the scale of percentiles is so chosen that the vertical ordinate up to an arbitrary sloping straight line represents the deviate in terms of the quartile or standard deviation as unit. Percentiles 5° to 95° limit the practical range of working. Any sloping straight line on this chart will be an ogive curve, i.e. correspond to some one or other normal distribution. If we know that $p_a$ per cent. of individuals have a character less than $A$, and $p_b$ per cent. a character less than $B$, we plot $A$ and $B$ upwards at the percentiles $p_a$ and $p_b$ respectively, join by a straight line the tops of these ordinates, and the point in which this line meets the 50° percentile gives by its ordinate the median. I do not know whether Galton ever prepared an accurate chart ("abac") of his ogive transformed to a line—I have not come across it—but it would not be hard to do with considerable accuracy on a large scale, which might then be reduced by photography to reasonable dimensions.

Galton shows that linear interpolation on the ogive itself is very imperfect unless $p_a$ and $p_b$ are equally distant from the 50° point.

[1] Even the criterion suggested in (6°) is one rather of appearance than actual measure of "goodness of fit."

[2] "A Geometric Determination of the Median Value of a System of Normal Variants, from two of its Centiles." *Nature*, Vol. LXI, pp. 102–4, Nov. 30, 1899.

Some good illustrations of the merit and defect of the ogive-median method may be found in a further paper published in 1907.

In "Vox populi[1]" Galton begins by stating that

"in these democratic days any investigation into the trustworthiness and peculiarities of popular judgments is of interest,"

and proceeds to illustrate the "Vox populi" by discussing the 787 answers given in a weight-judging competition at the West of England Annual Fat Stock Show at Plymouth. The judgments turned on what a selected fat ox would weigh after being slaughtered and dressed. Galton considers that the entrance fee of 6*d*. and the hope of a prize deterred practical joking and that the judgments would be largely those of butchers and farmers experienced in the matter.

"The judgments were unbiased by passion and uninfluenced by oratory and the like.... The average competitor was probably as well fitted for making a just estimate of the dressed weight of the ox, as an average voter is to judge the merits of most political issues on which he votes."

Galton gives the following table of results and the diagram on page 404:

*Distribution of the estimates of the dressed weight of a particular living ox, made by 787 different persons.*

| Degrees of the length of Array 0°—100° | Estimates in lbs. | Centiles | | Excess of Observed over Normal | |
|---|---|---|---|---|---|
| | | Observed deviates from 1207 lbs. | Normal p.e. = 37 | Francis Galton | Karl Pearson |
| ° | | | | | |
| 5 | 1074 | −133 | −90 | +43 | −23 |
| 10 | 1109 | −98 | −70 | +28 | −11 |
| 15 | 1126 | −81 | −57 | +24 | −9 |
| 20 | 1148 | −59 | −46 | +13 | +1 |
| $q_1$ 25 | 1162 | −45 | −37 | +8 | +5 |
| 30 | 1174 | −33 | −29 | +4 | +7 |
| 35 | 1181 | −26 | −21 | +5 | +7 |
| 40 | 1188 | −19 | −14 | +5 | +5 |
| 45 | 1197 | −10 | −7 | +3 | +6 |
| $m$ 50 | 1207 | 0 | 0 | 0 | +8 |
| 55 | 1214 | +7 | +7 | 0 | +7 |
| 60 | 1219 | +12 | +14 | −2 | +4 |
| 65 | 1225 | +18 | +21 | −3 | +1 |
| 70 | 1230 | +23 | +29 | −6 | −2 |
| $q_3$ 75 | 1236 | +29 | +37 | −8 | −5 |
| 80 | 1243 | +36 | +46 | −10 | −8 |
| 85 | 1254 | +47 | +57 | −10 | −9 |
| 90 | 1267 | +52 | +70 | −18 | −11 |
| 95 | 1293 | +86 | +90 | −4 | −8 |

$q_1$, $q_3$, the first and third quartiles, stand at 25° and 75° respectively.
$m$, the median or middlemost value, stands at 50°.
The dressed weight proved to be 1198 lbs.

[1] *Nature*, Vol. LXXV, pp. 450–51, March 7, 1907.

*Diagram, from the tabular values.*

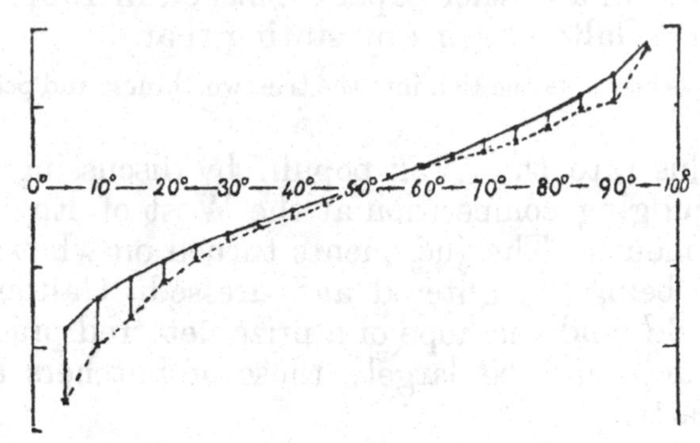


The continuous line is the normal curve with p.e. = 37.
The broken line is drawn from the observations.
The lines connecting them show the differences between the observed and the normal.

According to this method of dealing with the matter the "Vox populi" was only wrong nine pounds (1207 against 1198), or 0·8 per cent. Galton considers that the judgments were not distributed normally and that negative errors were magnified and positive errors minimised by the competitors. But what if Galton be not fitting the best curve to his data? It is not hard to show that the judgment of the middlemost man is not the best median—paradoxical as it may seem! Almost any pair of symmetrical percentiles gives a result with less probable error. For example, the median of the quartiles $\frac{1}{2}(1162 + 1236)$ is 1199, only 1 lb. out. Other medians are:

| 20° and 80° | 30° and 70° | 35° and 65° | 40° and 60° |
|---|---|---|---|
| 1195 lbs. | 1202 lbs. | 1203 lbs. | 1203 lbs. |

—all better than the middlemost value.

Again the 25° and 75° are far from being the best percentiles to obtain the "probable error" from, i.e. the quartile does not give the quartile best, strange as that may appear. If we calculate the quartile from the 15° and 85° percentiles it is $\frac{1}{2}(73 + 55) \times 1/1{\cdot}54 = 41{\cdot}5$ and this is nearly the best position for determining its value, on the assumption of a normal distribution[1]. With median at 1199 and quartile or probable error 41·5, a much more reasonable distribution is found, and there is far less need to assume as Galton did that the individual judgments are abnormally distributed; it is no longer true to say that errors in defect have been exaggerated, although errors in excess are still minimised. Whether the "fit" is a reasonable one it is not possible to determine when the data are thus given in percentiles. I have dwelt on the matter, because Galton's use of the values at 25°, 50° and 75° to determine the median and quartiles is not the best, and may lead, as

[1] Unfortunately the percentile method of tabulation does not permit of very ready determination of the mean and standard deviation and so of getting the best normal distribution. But I find after some labour: mean 1197, standard deviation 61·895, leading to a probable error or quartile value of 41·75. These give a far better fit than Galton's median and quartile values. I have inserted a column on the right of the table giving my results.

in this case, to an erroneous conclusion. The study of popular judgments and their value is an important matter and Galton rightly chose this material to illustrate it. The result, he concludes, is more creditable to the trustworthiness of a democratic judgment than might be expected, and this is more than confirmed, if the material be dealt with by the "average" method, *not* the "middlemost" judgment, the result then being only 1 lb. in 1198 out.

Among other matters which much interested Galton was the verification of theoretical laws of frequency by experiment. He considered that dice were peculiarly suitable for such investigations[1], as easily shaken up and cast. As an instrument for selecting at random there was he held nothing superior to dice[2]. Each die presents 24 equal possibilities, for each face has four edges, and a differential mark can be placed against each edge. If a number of dice, say four, are cast, these can without examination be put, by sense of touch alone, four in a row, and then the marks on the edges facing the experimenter are the random selection. Galton uses another die, if desirable, to determine a plus or minus sign for each of the inscribed values. On the 24 edges of this die he places the possible combinations of plus and minus signs four at a time (16), and of plus and minus signs three at a time (8). Then, when he has copied out in columns his data from the facing edges of the first type of dice, he puts against their values the plus or minus sign according to the facing edge of the sign-die, which gives either three or four lines at a cast. The paper is somewhat difficult reading, and there are a good many pitfalls in the way of those who wish experimentally to test theories of frequency, especially those of small sampling. The importance of distinguishing between hypergeometrical and binomial distributions, between sampling from limited and from unlimited or very large populations, and the question of the returning or not of each individual before drawing the next, are matters which much complicate experimental work with dice.

Galton, however, was not unconscious of the many pitfalls which beset the unwary student of the theory of chance. There is an interesting short paper by him on "A plausible Paradox in Chances," written in 1894[3]. The paradox is as follows: Three coins are tossed. What is the chance that the results are all alike, i.e. all heads or all tails?

"At least two of the coins must turn up alike, and as it is an even chance whether a third coin is heads or tails, therefore the chance of being all alike is 1 to 2 and not 1 to 4."

If the reader can *distinctly* specify off-hand, without putting pen to paper, wherein the fallacy lies, he has had some practice in probability or has a clear head for visualising permutations. We leave the solution to him

[1] Ordinary dice do not follow the rules usually laid down for them in treatises on probability, because the pips are cut out on the faces, and the fives and sixes are thus more frequent than aces or deuces. This point was demonstrated by W. F. R. Weldon in 25,000 throws of 12 ordinary dice. Galton had true cubes of hard ebony made as accurate dice, and these still exist in the *Galtoniana*.

[2] "Dice for Statistical Experiments." *Nature*, Vol. XLII, pp. 13-14, 1890.

[3] *Nature*, Vol. XLIX, pp. 365-6, Feb. 15, 1894.

remarking that Galton confirmed the accuracy of the 1 to 4 result by an experiment with his favourite dice. The paradox is

"a good example of the pitfalls into which persons are apt to fall, who attempt short cuts in the solution of problems of chance instead of adhering to the true and narrow road."

For Galton that "true and narrow road" was the study of the possible permutations, the road followed by the early masters of the doctrine of chances.

We have seen how the "Vox populi" was—at any rate in the judgment of the meat-weight of fatted oxen—not so far from the truth. Galton, in a paper of twelve years earlier, endeavoured to test the "Vox judicum[1]," the reasonableness of the judgments of a presumably educated and trained class of minds. Galton expected that the various terms of imprisonment awarded by judges would fall into a continuous series. He limited his data to sentences on males without option of a fine, and he dealt with 830 sentences for terms of years, 10,540 for terms of months and 43,500 for terms of weeks. All these data give what we now term *J*-curves—i.e. frequency distributions similar to those of cricket scores, of incomes or rents—the shortest sentences in each case being the most numerous, the longest the least frequent. This is probably the nature of criminality in the population—or as Galton would put it of "true penal deserts." But Galton does not lay stress on this remarkable deviation from the normal curve of distribution. He is concerned with another phase of this distribution of criminality, namely that it is extraordinarily irregular; there are marked preferences for certain terms of imprisonment. Thus when sentences are reckoned in months, the maxima occur at 3, 6, 9, 12, 15, 18 and 24 months. In 10,540 sentences in months there are none at 17 months, hardly any at 11 or 13 months. Galton argues that three months or a quarter of a year is a round figure that must commend itself to a judge by its simplicity. He suggests that if our year had been divided into 10 periods, then 2½ periods, the equivalent of 3 months, would not have been used in its place, or the same penal deserts would have been treated differently from what they are now. Again, in the distribution of sentences in years, he draws attention to the emphasis on sentences of 3, 5, 7 and 10 years, showing a tendency at first to a unit of 2 years and then, presumably guided by a habit of decimal notation, a jump from 7 to 10 years. Galton remarks that while there were 7 sentences for 20 years and 6 for 15 years, there were absolutely none for 19, 18, 17 or 16 years. Terms of weeks are distributed with equal irregularity. Galton argues that the powerful cause of disturbance which interferes with the orderly distribution of punishment in conformity with penal deserts lies in the personal fancies of judges for certain series of numbers.

"It would be interesting to tabulate the sentences passed by the several judges since their appointments, to discover their respective peculiarities and personal equations, all who exercise extensive jurisdiction in criminal cases being included under the title of judge."

There is no doubt that the idiosyncrasies of some judges in the matter of sentences are as well recognised in the legal profession as by the habitual criminal himself, but is there not another source of the results observed by

[1] "Terms of Imprisonment." *Nature*, Vol. LII, pp. 174–6, June 20, 1895.

Galton of as great influence as the vagaries of the judicial mind? His material is heterogeneous, in that it covers a great variety of classes of crime from misdemeanours to felonies. For many of these offences a period of imprisonment, or at least a maximum period of imprisonment, has been fixed by the legislature itself, and however much a judge might consider an offender to deserve a longer period of imprisonment he could not inflict it. I think the irregularity which undoubtedly manifests itself in these results is due as much to the 'vox legislatorum' as to the 'vox judicum.' This might be easily ascertained by discussing the returns separated out into individual classes of crime. Whatever may be the exact origin of the anomalies—which are certainly present if we hold that anti-social conduct is a continuous variate—we may safely conclude with Galton:

"by moralising on the large effects upon the durance of a prisoner, that flow from such irrelevant influences as the associations connected with decimal or duodecimal habits and the unconscious favour or disfavour felt for particular numbers. These trifles have been now shown on fairly trustworthy evidence to determine the choice of such widely different sentences as imprisonment for 3 or 5 years, of 5 or 7, and of 7 or 10, for crimes whose penal deserts would otherwise be rated at 4, 6 and 8 or 9 years respectively."

There is a passage in this memoir which would have delighted the heart of the "Passionate Statistician." It runs:

"We test the acquirements of youths by repeated examinations, but do not as yet employ the methods of statistics to test the performances of professional men. Examiners, for example, should themselves be tested in this way[1], and I have a fancy that a discussion of the clinical reports at the various large hospitals might enable a cautious statistician to express with some accuracy the curative capacities of different medical men, in numerical terms. Before putting oneself into the hands of any new professional adviser, it would certainly be a grateful help to know the indices of capacity of those among whom the choice lay, not such as might be inferred from their performances in school and undergraduate days, or by their unchecked professional repute, but as they really are in their mature and practical life." (p. 176.)

What a readjustment of values there would be if those "indices of capacity" were found one morning attached to the brass plates of Harley Street or inscribed in the more sober black and white of the passages in Lincoln's Inn!

Two further contributions of Francis Galton may be just mentioned. On the death of Dr Samuel Haughton he wrote[2] pointing out that amid Haughton's many-sided activities he had introduced the "long drop," as the most painless death by hanging by the neck. Haughton had experimented on the tensile strength of the spine and muscles of the neck and published a formula for the length of drop dependent on the height and weight of the culprit. Galton believed Haughton to have omitted a small factor in the increased section of the muscles of the neck in fat men. The matter, if an unpleasant one, still needed scientific investigation as a death by beheading—which in one case occurred—was not carrying out the sentence of the law.

[1] All universities ought to take periodic stock of their examiners in the manner suggested. I have no hesitation in asserting that in many cases the success or failure of candidates is not a measure of their intelligence, but of their choice of subject and still more of the particular examiner in that subject who has marked their script.

[2] *Nature*, Vol. LVII, p. 79, November 25, 1897.

Those who recognise the relative mercy of the long drop may wonder why capital punishment has not been still further modified in the direction of scarcely less painful, but more seemly, methods of disposing of the socially abhorrent.

Galton further contributed a letter to the discussion on "Corporal Punishment" which developed in the *Times* in 1898. He considered that the writers on the subject had overlooked two important points:

"The first is, the worse the criminal the less sensitive he is to pain, the correlation between the bluntness of the moral feelings and those of the bodily sensations being very marked. The second relates to the connection between the force of the blow and the pain it occasions, which do not vary at the same rate, but approximately, according to Weber's law, four times as heavy a blow only producing about twice as much pain. In a Utopia the business of the Judge would be confined to sentencing the criminal to so many units of pain in such and such a form, leaving it to anthropologists skilled in that branch of their science to make preliminary experiments and to work out tables to determine the amount of whipping or whatever it might be that would produce the desired results. Really these latter considerations might even now be made the subject of a solid scientific paper of no small interest, but they cannot be more than hinted at in a short letter like this, which has to be written in non-technical language."

The unit of pain—quite apart from corporal punishment—seems no more incapable of measurement than a unit of intelligence. The threshold of the sensation of pain might be determined in a number of ways, and then correlated with other mental and physical characters of the individual. Like all Galton's writings, this brief letter suggests unexhausted fields for the persistent and cautious investigator.

Galton's fertility of statistical ideas may be further illustrated by two papers, one belonging to 1894 when he was seventy-two years of age, and the other written fifteen years later when he was 87. The first was contributed to the *Proceedings of the Royal Society*[1] and deals with the important problem of the fertility of marriages according to the ages of father and mother. The fertility is measured by percentage of families which have a child when the husband and wife are of the given ages. If a chart be formed of which one variate is age of father, the second age of mother, and the percental offspring be inscribed for each pair of ages, then Galton proposes to represent the *loci* of equal percentages by contours, which he terms *isogens.* By a fairly simple, if somewhat rough process he constructs these isogens for Körösi's Budapest data, and we reproduce them below.

The diagram indicates that the form of the isogens as long as the husband is older than the wife is very closely a system of straight equidistant and diagonal lines. As a result of this Galton concludes that the fertility of a husband of age $a_H$ and a wife of age $a_W$ will be closely given by

$$p = 93 - a_H - a_W,$$

provided that (i) the wife is not older than the husband, and (ii) she is not less than 23 nor (iii) more than 40 years of age.

[1] Vol. LV, pp. 18–23. "Results derived from the Natality Table of Körösi by employing the method of Contours or Isogens."

*Isogens or Lines of Equal Fertility.*

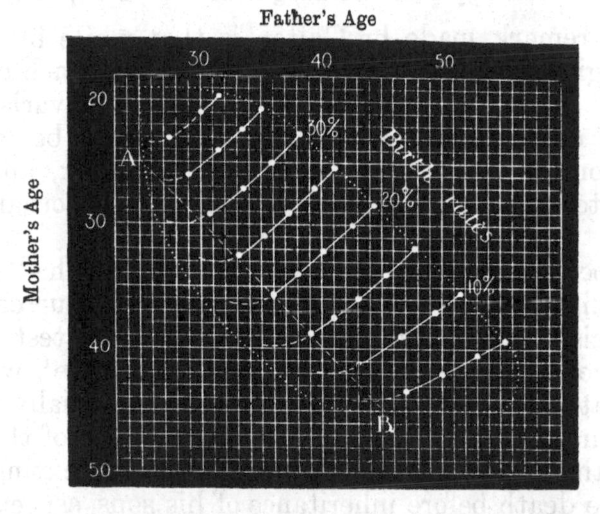


*Illustrations of the accuracy of the Isogens.*

| Wife | Husband | Calculated | Actual |
|---|---|---|---|
| 23 | 27 | 43 °/。 | 42 °/。 |
| 23 | 37 | 33 °/。 | 30 °/。 |
| 27 | 33 | 33 °/。 | 31 °/。 |
| 31 | 35 | 27 °/。 | 26 °/。 |
| 35 | 41 | 17 °/。 | 16 °/。 |
| 39 | 49 | 5 °/。 | 6 °/。 |

Probably a little adjustment would give a better fitting plane to the portion of the surface under consideration, but Galton's rule gives at any rate a first approximation of a quite serviceable kind.

Our author notes the curious change in the direction of the isogens when the wife is older than the husband. This seems to me to indicate that when the wife is older than the husband, the age of the latter is of minor importance. Galton interprets it as follows:

"When she [the wife] is from thirty to thirty-eight she certainly seems to be appreciably more fertile with a husband of her own age or somewhat older than she is [than] with one who is younger[1]. I should hesitate to ascribe this to physiological causes without corroborative

[1] The following results drawn from his data do not seem to confirm Galton's views:

| Age of Wife ... ... ... | 29 | 31 | 33 | 35 | 37 | 39 | 41 | 43 | 45 |
|---|---|---|---|---|---|---|---|---|---|
| Percentage Children: | | | | | | | | | |
| Husband of same age ... | 35 | 22 | 22 | 19 | 17 | 15 | 10 | 6 | 3 |
| Mean for husbands younger | 34 | 25 | 26·5 | 21·3 | 17·3 | 16·3 | 11·7 | 6·3 | 3·3 |

evidence derived from breeders of stock. It is very possible that indifference on the part of young husbands to aging wives may have something to do with it." (p. 22.)

An important remark made by Galton is that while his table and the smoothed isogens give the mean percentile fertilities at each age of husband and wife, they fail to measure the degree of individual variation from this mean; the nature of this variation, he remarks, could be found from the original observations with a moderate amount of work, and it would be of much interest to determine whether it varies in accordance with some definite law.

The second paper deals with a problem which Galton had special delight in handling. Nothing pleased him more than to dispel a current superstition by statistical criticism. We have already seen how he tested in this way the objective efficacy of prayer. In the present instance[1], with Mr Edgar Schuster's aid, he attacked the belief widely spread, especially among Roman Catholics, that Church property sequestrated at the time of the Reformation carried a curse with it; the effect of the curse was to extinguish the line of the owner by the death before inheritance of his sons, especially the eldest son. The phrase "Church property" is applied to estates which were in whole or part ecclesiastical previous to the dissolution of the monasteries under Henry VIII, and "Not Church property" to those that were not.

*Survival of Eldest Sons.*

| Total number of Owners | | Number who were Eldest Sons | |
|---|---|---|---|
| Not Church property | 459 | 241 ... ... | 52·5 °/。 |
| Church property ... | 464 | 240 ... ... | 51·7 °/。 |

*Length of Tenures.*

| | Mean Length |
|---|---|
| Not Church property ... | 27·2 years |
| Church property ... ... | 27·4 years |

There is therefore no appreciable effect produced by the curse in either thwarting the inheritance of eldest sons, or in shortening the tenure of the ownership of Church property. It was found, however, that Church properties changed hands much more frequently than non-Church properties; transmissions by purchase were almost three times as frequent. Galton was inclined to ascribe this to the comparative unsuitability to modern require-

[1] "Sequestrated Church Property." *Nature*, Vol. LXXIX, p. 308, January 14, 1909.

ments of abbeys, etc., as dwelling houses, to their low situations and bad drainage, so that those who had experience of them would be the more ready to sell when the picturesqueness and romance of old buildings created a fictitious market value. Although this, Galton admits, is pure speculation, it may indicate that mundane reasons and not supernatural interferences really account for the excess of transmissions by purchase[1].

I have reserved one statistical paper because of its special importance for a last illustration of Galton's statistical *ingenium*. It appeared in August 1902, when he was more than 80 years old[2]. Its title conveys no idea of its value and would suggest that it dealt only with a minor if interesting point. But Galton's method of proportioning first and second prizes demands a knowledge of the average interval between the first, the second and the third place men in a competition between a number $n$ of individuals. Hitherto statistical frequency had been looked upon as a *continuous* distribution, which could be represented by the equation to a mathematical curve. I know no one before Galton, in 1902, who had proposed to consider a population for what it really is—a *discontinuum* with finite intervals between its individual members when arranged in order of their intensities with regard to a given character. What are the average values of these finite intervals, and how do they vary? What we now term Galton's *Individual Difference Problem* created at once a whole set of new conceptions and suggested questions, a few only of which have as yet been answered. But the explanation of certain well-recognised phenomena flowed at once. Taking only 100 individuals supposed to follow a normal distribution for any character the interval between the 50th and 51st is only one-tenth of the interval between the 99th and 100th, or between the 1st and 2nd. Given very large populations indeed mediocrity is crowded together; the exceptional are

[1] It is clearly impossible to give a summary of *all* Galton's many contributions to *Nature*, or of his letters to the daily press, although so many of them contain thoughts or suggestions, which it is sad should not be put on more permanent record. The curious may care to read *inter alia* (*a*) a long letter to the *Times*, October 6, 1887, entitled "The Proposed Imperial Institute: Geography and Anthropology,"—the word "Statistics" might have been added. We fear the existing Institute, largely deprived by cheeseparing governments of its own buildings, falls far short of Galton's ideal. (*b*) *Nature*, Vol. XXXVI, pp. 155–7 (June 16, 1887) contains a paper on "North American Pictographs." These interested Galton from more than one side. He thought the pictographic calendars of the Indians might be modernised for family records, and for some years was wont to have a small medallion drawn illustrating his own family occurrences in the year. He considered it possible to think by aid of pictographs as Laura Bridgeman had found an adequate basis for the exercise of a considerable amount of reasoning in the unassisted sense of touch. So also dogs may be occasionally "carrying out some real act of thought by aid of imagined and symbolic odours." This will seem fantastic only to those who have not observed what a Pekingese dog does when introduced for the first time to a new house with unknown occupants. (*c*) "A New Step in Statistical Science" (*Nature*, Vol. LI, p. 319, Jan. 31, 1895) deserves mention, as typical of Galton's splendid generosity to the younger generation when it came knocking at the door. Such innate generosity appears also in (*d*) a paper on "Bertillon's System of Identification" (*Nature*, Vol. LIV, pp. 569–70, Oct. 8, 1896), where in the "Signaletic Instructions" there was a not very happy attempt to claim finger-print identification as a French discovery.

[2] "The Most Suitable Proportion between the Values of First and Second Prizes." *Biometrika*, Vol. I, pp. 380–90.

widely separated from each other. I cannot describe better my sense of the importance of Galton's Difference Problem than by citing, with but slight additions, the words I used about it more than twenty years ago[1]:

"Now, of course, the normal distribution in a general sort of way indicates that the differences between modal or what the biologists term 'normal' individuals are very small. But Mr Galton's Difference Problem enables us for the first time to appreciate quantitatively how much wider the differences are between the extreme (biologists' 'abnormal' or atypical individuals) and modal ('normal' or typical) individuals. Now the range of a distribution being somewhat about six times the standard deviation, we see that extreme individuals, even in a population of only 100, may be separated by as much as $\frac{1}{17}$th of the range, while modal individuals have only a difference of $\frac{1}{250}$th of the range and even individuals at the quartile only a difference of $\frac{1}{200}$th of the range. The relative differences become much greater in populations of several millions.

It is not possible to pass over the general bearing of such results on human relations. If we define 'individuality' as difference in character between a man and his immediate compeers, we see how immensely individuality is emphasised as we pass from the average or modal individuals to the exceptional men. Differences in ability, in power to create, to discover, to rule men do not go by uniform stages. We know this by experience—our Shakespeares, our Newtons, our Napoleons have no close compeers in the populations of their own generations—but we see a reason for the gulf which separates the genius from ourselves, the phenomenon flows from a characteristic and familiar chance distribution. We ought not to be surprised, as we frequently are, at the results of competitive examinations, where the difference in marks between the first men is so much greater than occurs between men towards the middle of the list. In the same way the marked individuality of extreme criminality, and the appalling differences in stupidity and imbecility at the lower end of the moral and intellectual scales, receive their due statistical appreciation.

We stand in a better position to discriminate the pathological from the merely exceptional; mere isolation no longer leads us *à priori* to question the position of an outlying observation or of an exceptional individual.

In short Galton's Difference Problem leads us to look upon samples of populations, and even on populations themselves, no longer as arrays of individuals with continuously varying characters, but as systems of discrete units. We see discontinuity in every sample and in every population. We obtain a new and most valuable conception of a normal or standard population. It is one in which each individual is separated from his immediate neighbours,—when the whole is arranged according to any character,—by definite calculable intervals. These intervals are, of course, the *average* intervals which would be found by taking the mean of many such samples or populations, but they are none the less of extreme suggestiveness. Just as the *continuous* representation by a frequency curve is only an ideal representation of the observed facts, so we now reach an ideal representation of the actual *discontinuity* in the given population. As in the case of many physical investigations, so we find in statistical theory both continuous and discontinuous representations of the phenomena equally important and equally valid within the legitimate limits of interpretation."

Did Galton immediately recognise all that flowed from his treatment of the proper proportions of first and second prizes? Possibly not; he took some years to realise all that must eventually flow from his conception of correlation. But is not this failure to grasp immediately all that results from a new standpoint the essential peculiarity of the creative mind, whether it be that of a great scientist or of a great poet? Galton has himself so well described the workings of the exceptional mind that I need not labour this point[2]. The mine discovered by Galton more than twenty years ago is far

[1] *Biometrika*, Vol. I, p. 398.

[2] See the first footnote on p. 236 of this volume. The inspiration is the product of the subconscious mind. The man who has reached a truth knows it to be true, wrote Spinoza truly

from being exhausted, indeed its treasures are hardly yet touched; and probably will not be until some laborious German breaks by accident into its vein of wealth[1].

Admitted that Galton did not see the whole bearing of his conception, yet as pioneer he blazed the track. His method of approach was simple as usual, a rough but adequate approximation for the purpose he had in hand. He took as his average distribution of $n$ individuals the normal curve divided into $n$ equal areas, and he took as the character value of his individuals that corresponding to the areal bisector of each of these equal areas. He might have taken the means of those equal areas. Neither is in accordance with the more accurate method of approach which results from a fuller application of the theory of probability. But the difference is not very important unless the number of individuals be small; thus for a sample of 100 we find:

| In terms of Standard Deviation | Method of Median (Galton) | Method of Mean | Accurate Theory |
|---|---|---|---|
| Distance between first and second | ·406 | ·488 | ·360 |
| Distance between second and third | ·210 | ·215 | ·202 |

The ratios of first to second prizes given by the three methods are respectively: Median 74·6 to 25·4, Mean 76·6 to 23·4, Actual 73·6 to 26·5.

Thus the three methods do not lead to widely different results as far as the prize ratio is concerned; the first prize may, as Galton observes, be adequately taken as three times the second. The value for 10 competitors is 71·9 to 28·1 and for 1000 is 74·1 to 25·9. Hence the greater the number of competitors the more accurate is Galton's ratio of 3 to 1. But, as he remarks, the number in most competitions is limited by the fact that many individuals know it would be futile to enter, so that it is reasonable to proportion the prizes on numbers far larger than are actually entered[2]. Galton's inspiration certainly divined that a better result could be obtained by using the median rather than the mean in this case!

As far as I know no attention has been paid to Galton's ratio in settling the monetary value of prizes in athletic competitions[3], agricultural shows and scholastic awards. Its importance is obscured by the fact that it is merely an *average* value, and we are only to-day discovering how the system of differences—for example, between the first and second individuals—is

but really undemonstrably. What he did not write was that the applications of a truth are rarely fully recognised by its first discoverer. It is just because the exceptional mind does not and cannot exhaust the utility of its own creations that lesser minds value its teachings and recognise a master-builder.

[1] A prophecy which has come unexpectedly true! See von Bortkiewicz's recent paper on the range or distance between first and last individuals in a series.

[2] The first and second horses in the Derby are not merely the best out of many horses that actually run, but out of the very much larger number of potential runners.

[3] I am informed that the 3 : 1 ratio is in use in at least one public school, but the origin of it is obscure.

distributed *round* the mean; we are not even clear as to the frequency of brackets in competitions of different sizes. There is still much work to be done on the lines of Galton's Difference Problem.

One conclusion, however, may be safely drawn, namely, that in this contribution of the octogenarian Galton to the theory of statistics there was no sign of a failure of that fertile suggestiveness which had led the sexagenarian Galton twenty years earlier to develop his far-reaching ideas on correlation.

Readers of this chapter will have observed that I have avoided almost entirely any reference to Galton's work on correlation, which is so essential a part of his contributions to statistical theory. This has been done purposely because it would not only have overweighted an already lengthy chapter, but its development belongs peculiarly to Galton's statistical studies of heredity. Both topics will form the subject of chapters in the remaining volume of his *Life, Letters and Labours.* It suffices in this chapter to see Galton deeply interested in almost every branch of statistics, but especially in their bearing on anthropometry. We have seen him pass from geography to ethnology, from ethnology to anthropometry, and from anthropometry to statistical theory. In the course of a long and crowded life, his contemporaries recognised him as a master-builder and as a pioneer in one branch of science after another. Space does not permit of our citing the innumerable questions and problems propounded to Galton by scientific correspondents from all quarters of the world. Galton's replies would indeed be a *repertorium* of information and suggestion, but in the majority of cases the recipients are now dead—for Galton outlived his generation—and I found the quest for his own letters a hopeless task. One such quest was, however, fruitful, and forms a fitting theme with which to close this chapter and volume.

### I. THE PROPOSED PROFESSORSHIP OF APPLIED STATISTICS

Florence Nightingale has been usually estimated by that gracious phase of her life which appealed to the emotional sympathies of a little-instructed public. For that public she is the "Lady of the Lamp." Sympathy with suffering is, however, of small avail—no more so than charity—unless it be accompanied by administrative insight, and this side of Florence Nightingale's character has been too often overlooked. She was a great administrator, and to reach excellence here is impossible without being an ardent student of statistics. Florence Nightingale has been rightly termed the "Passionate Statistician." Her statistics were more than a study, they were indeed her religion. For her, Quetelet was the hero as scientist, and the presentation copy of his *Physique Sociale* is annotated by her on every page[1]. Florence Nightingale believed—and in all the actions of her life acted upon that belief—that the administrator could only be successful if he were guided by

[1] Presented to the Galton Laboratory by Miss Nightingale's niece, Mrs Vaughan Nash, and now placed beside Darwin's gift of the *Origin of Species* to Galton and Tyndal's gift of his Belfast Address to Herbert Spencer.

PLATE LIII

Francis Galton at Haslemere in 1907.

"When the desired fullness of information shall have been acquired, then and not till then, will be the fit moment to proclaim a 'Jehad,' or Holy War, against customs and prejudices that impair the physical and moral qualities of our race."

statistical knowledge. The legislator—to say nothing of the politician—too often failed for want of this knowledge. Nay, she went further: she held that the universe—including human communities—was evolving in accordance with a divine plan; that it was man's business to endeavour to understand this plan and guide his actions in sympathy with it. But to understand God's thoughts, she held we must study statistics, for these are the measure of his purpose. Thus the study of statistics was for her a religious duty.

Those who have drawn from the earlier chapters of this volume some idea of Galton's religion, will realise how close must have been the sympathy of ideas. For Galton the world was developing; at present under stern forces a mentally and physically superior human type was being evolved, and it was the religious duty of man to assist these changes, but for effective action we must study the laws of evolution, we must know and *statistically* know before the pace could be hastened.

"When the desired fullness of information shall have been acquired, then and not till then, will be the fit moment to proclaim a 'Jehad,' or Holy War, against customs and prejudices that impair the physical and moral qualities of our race[1]."

And again:

"The ideas have long held my fancy that we men may be the chief, and perhaps the only executives on earth. That we are detached on active service with it may be only illusory powers of free-will. Also that we are in some way accountable for our success or failure to further certain obscure ends to be guessed as best we can. That though our instructions are obscure they are sufficiently clear to justify our interference with the pitiless course of Nature, whenever it seems possible to attain the goal towards which it moves, by gentler and kindlier ways[2]."

Thus it came about that for Galton, and for Florence Nightingale, the end and the means were the same: men must study the obscure purpose of an unknown power,—the tendency behind the universe; and the manner of our study must be statistical. Therein, according to Francis Galton, lay the way to that unsolved riddle of "the infinite ocean of being"; therein, according to Florence Nightingale, lay the cipher by which we may read "the thoughts of God." Men of the twentieth century may fail to appreciate the doctrine of either great Victorian, but of one thing they may be sure, the belief in both of them amounted to a religion. And what was a religion to both became at once in both a motive for action. Galton was not content with the office of teacher, he devoted a large portion of his fortune to the foundation of a school of eugenics on the basis of probability, that is of the modern theory of statistics. Statistics were to be applied in gleaning information as to Nature's immediate purpose, which undoubtedly lies in the evolution of man's mind, body, and character towards increased energy, and more efficient co-adaptation. Florence Nightingale strove—possibly with less scientific insight, but with a wider administrative experience and with no less religious earnestness—towards the like end. She sought, before Galton, but with smaller economic resources, to establish a university chair of "Applied Statistics." I have often wondered how far the final form of

[1] *Probability the Foundation of Eugenics:* Herbert Spencer Lecture. Oxford, 1907. p. 30.
[2] *Ibid.* p. 9.

Galton's foundation was influenced by his correspondence with Florence Nightingale concerning this chair of applied statistics. When I sought a name in 1911 for the new department which should combine the Biometric and Galton Eugenics Laboratories, no fitter and more historically worthy name occurred to me than that of "Applied Statistics." Were I a man of wealth I would see that Florence Nightingale was commemorated, not only by the activities symbolised by the "Lady of the Lamp," but by the activities of the "Passionate Statistician." I would found a Nightingale Chair of Applied Statistics to carry out the ideal expressed in the letters below.

The first reference of Florence Nightingale to Francis Galton occurs in a very characteristic letter of hers to Captain (later Sir) Douglas Galton. It is dated August 7, 1867. In this letter she refers to a Standing Committee which was being appointed to consider contrivances for dealing with the wounded after a battle. The keynote to her letter is that appliances kill: Do away with all huts and marquees, give the wounded plenty of air and tend them on the battlefield. For every man that dies of his wound five or six die of the doctors and the removing; as to medicines, make the doctors swallow them all, all that is wanted is a little brandy and a great deal of water. But if the wounded are to be tended in extemporised shelters on the battlefield somebody must be on the committee who understands rough shelters.

"The only person who has written anything worth having on travelling apparatus is Mr Francis Galton, a cousin of yours I believe; I should put *him* on the Standing Committee, if possible."

It is not till twenty-four years later that Florence Nightingale again seeks the advice of Francis Galton; he was then 69 and she over 70. She was reviving one of the great dreams of her younger days and he, with no sign yet of age, was then actively contributing not a little towards its realisation.

10 South Street, Park Lane, W. *Feb.* 7, '91.

*Scheme of Social Physics Teaching.*

Dear Sir, Sir Douglas Galton has given me your most kind message; saying that if I will explain in writing to you what I think needs doing, you will be so good as to give it the experienced attention without which it would be worthless. By your kind leave, it is this:

A scheme from someone of high authority as to what should be the work and subjects in teaching Social Physics and their practical application in the event of our being able to obtain a Statistical Professorship or Readership at the University of Oxford.

I am not thinking so much of Hygiene and Sanitary work, because these and their statistics have been more closely studied in England than probably any other branch of statistics, though much remains to be desired: as e.g. the result of the food and cooking of the poor as seen in the children of the Infant Schools and those of somewhat higher ages. But I would—subject always to your criticism and only for the sake of illustration—mention a few of the other branches in which we appear hardly to know anything, e.g.

*A.* The results of Forster's Act, now 20 years old. We sweep annually into our Elementary Schools hundreds of thousands of children, spending millions of money. Do we know:

(i) What proportion of children forget their whole education after leaving school; whether all they have been taught is *waste*? The almost accidental statistics of Guards' recruits would point to a large proportion.

(ii) What are the results upon the lives and conduct of children in after life who don't forget all they have been taught?

(iii) What are the methods and what are the results, for example in Night Schools and Secondary Schools, in preventing primary education from being a *waste*?

If we know not what are the effects upon our national life of Forster's Act is not this a strange gap in reasonable England's knowledge?

*B* (1). The results of legal punishments—i.e. the deterrent or encouraging effects upon crime of being in gaol. Some excellent and hardworking reformers tell us: Whatever you do keep a boy out of gaol—work the First Offenders' Act—once in gaol, always in gaol—gaol is the cradle of crime. Other equally zealous and active reformers say—a boy must be in gaol once at least to learn its hardships before he can be rescued. Is it again not strange in practical England that we know no more about this?

*B* (2). Is the career of a criminal from his first committal—and for what action—to his last, whether (*a*) to the gallows, or (*b*) to rehabilitation, recorded? It is stated by trustworthy persons that no such statistics exist, and that we can only learn the criminal's career from himself in friendly confidence—what it has been from being in gaol, say for stealing a turnip for a boys' feast, or for breaking his schoolroom window in a temper because he has been turned out of school for making a noise—to murder or to morality.

In how many cases must all our legislation be experiment, not experience! Any experience must be thrown away.

*B* (3). What effect has education on crime?

(*a*) Some people answer unhesitatingly: As education increases crime decreases. (*b*) Others as unhesitatingly: Education only teaches to escape conviction, or to steal better when released. (*c*) Others again: Education has nothing to do with it either way.

*C*. We spend millions in rates in putting people into Workhouses, and millions in charity in taking them out. What is the proportion of names which from generation to generation appear the same in Workhouse records? What is the proportion of children de-pauperised or pauperised by the Workhouse? Does the large Union School, or the small, or 'boarding out' return more pauper children to honest independent life? On girls what is the result of the training of the large Union Schools in fitting them for honest little domestic places—and what proportion of them falling into vice have to return to the Workhouse? Upon all such subjects how should the use of statistics be taught?

*D*. India with its 250 millions—200 millions being our fellow-subjects, I suppose—enters so little into practical English public life that many scarcely know where this small country is. It forms scarcely an element in our calculations, though we have piles of Indian statistics. [As to India the problems are:]

(i) Whether the peoples there are growing richer or poorer, better or worse fed and clothed?

(ii) Whether their physical powers are deteriorating or not?

(iii) Whether fever not only kills less or more, but whether it incapacitates from labour for fewer or more months in the year?

(iv) What are the native manufactures and productions, needed by the greatest customer in the world, the Government of India, which could be had as good and cheap in India, as those to be had from England?

(v) Whether the native trades and handicrafts are being ruined or being encouraged under our rule?

(vi) What is the result of Sir C. Wood's (1853) Education Act in India?

These are only a very few of the Indian things which—I will not say are hotly contested, for few care either in the House of Commons or out, but—have their opposites asserted with equal positiveness.

I have no time to make my letter any shorter, although these are but a very few instances. What is wanted is that so high an authority as Mr Francis Galton should jot down other great branches upon which he would wish for statistics, and for *some teaching how to use these statistics in order to legislate for and to administer our national life* with more precision and experience.

One authority was consulted and he answered: "That we have statistics and that Government must do it." Surely the answering question is: The Government does not use the statistics which it has in administering and legislating—except indeed to "deal damnation" across the floor of the H. of C. at the Opposition and *vice versâ*. Why? Because though the great majority of Cabinet Ministers, of the Army, of the Executive, of both Houses of Parliament have received a university education, what has that university education taught them of the practical application of statistics? Many of the Government Offices have splendid statistics. What use do they make of them? One of the last words Dr Farr of the General Register Office said to me was: "Yes, you must get an Oxford Professorship; don't let it drop."

M. Quetelet gave me his *Physique Sociale* and his *Anthropometrie*. He said almost like Sir Isaac Newton: "These are only a few pebbles picked up on the vast seashore of the ocean to be explored. Let the explorations be carried out."

You know how Quetelet reduced the most apparently accidental carelessness to ever recurring facts, so that as long as the same conditions exist, the same "accidents" will recur with absolutely unfailing regularity[1].

You remember what Quetelet wrote—and Sir J. Herschel enforced the advice—"Put down what you expect from such and such legislation; after —— years, see where it has given you what you expected, and where it has failed. But you change your laws and your administering of them so fast, and without inquiry after results past or present, that it is all experiment, see-saw, doctrinaire, a shuttlecock between two battledores."

Might I ask from your kindness—if not deterred by this long scrawl—for your answer in writing as to heads of subjects for the scheme? Then to give me some little time, and that you would then make an appointment some afternoon, as you kindly proposed, to talk it over, to teach, and to advise me? Pray believe me, Yours most faithfully, FLORENCE NIGHTINGALE.

I confess—but then I am a prejudiced person, for the prophetess was proclaiming my own creed—that this letter appears to me one of the finest that Florence Nightingale ever wrote. What is more it is almost as true to-day as it was thirty years ago. We are only just beginning to study social problems—medical, educational, commercial—by adequate statistical methods, and that study has at present done very little to influence legislation. What is more the requisite statistical teaching on which real knowledge must be based has hardly yet spread throughout our universities. The time has yet to come, when the want of a chair of statistical theory and practice in any great university will be considered as much an anomaly as the absence of a chair of mathematics. The logic of the former is as fundamental in all branches of scientific inquiry as the symbolic analysis of the latter.

To many it may seem as if we had here a proposal which Galton would welcome at once, none would doubt that he would give the closest consideration to it. But to those who knew Galton well three points of hesitation would suggest themselves: (i) He had no faith in a man simply because he was a professor; the men who in his day had made the most important contributions to science—Darwin, Wallace, Lubbock—and such personal friends as Spencer and Groves—were not professors (ii) He did not till long after this lose his faith in working by committees, or in some form of cooperative work[2]. (iii) He believed that small monetary prizes would produce excellent research work by able young men, and overlooked the fact that a stiff preliminary training is needful (then wholly lacking in statistics) if

[1] *Footnote*. I presume that no one now but understands, however vaguely, that if we change the conditions for the better, the evils will diminish accordingly.

[2] He had experienced the great value of the Meteorological Council and the Kew Observatory Committee, but these were largely homogeneous groups of highly trained men. Some of Galton's later committees, selected with the most catholic spirit, were such heterogeneous teams that he might as well have harnessed a thoroughbred, a mule, an ass and a camel to his wain, and have hoped for reasonable progress!

even the highest ability is to be productive of really scientific research. Without realising these characteristic elements of Galton's mind—phases of which I saw closely in after years—the reader cannot appreciate the trend of the present correspondence.

Galton at once thought most carefully over the matter. I have before me the rough notes he drafted before he replied. A few extracts from these notes emphasising my points may be made.

*Professorship of Social Statistics.*

Need of a professor of the theory of statistical methods and the application of them to definite social problems.

Higher laws of statistics, a mathematical head required....Professor will never get a class. Query as to Oxford at all? and query as to a professor? He will draw a big salary for certain but that is all you can be sure of....

Great difficulty of interrogating Nature etc. aright. One wants a committee for discussion. Great loss of time by false lines of quest....A man like Moltke to plan campaign; not necessary that he himself should work out results...that might be done by special grants....The hardest task is to frame questions. To obtain men who shall be masters not slaves of statistics and whose hearts shall be set on the solution of social problems....Each problem is a separate and severe problem to be attacked in its own way by such facts as are available....

This is all true, but it is not the whole truth. I take it that the kernel of Florence Nightingale's proposal was the foundation of a school of higher statistics, and the production by it of minds keen on applying novel methods to social problems. It was not till much later that Galton fully realised this, giving up his faith in committees and in work to be done by small grants for essays. The chief factor in that change was, I believe, his friendship with a professor of the best type, W. F. R. Weldon, whose energy, idealism and enthusiasm, showed Galton how much could be achieved by the right academic spirit.

With these precursory remarks I give Galton's reply.

42 Rutland Gate. *Feb.* 10, '91.

Dear Miss Nightingale, I think most progress may be made if I send the general ideas that your letter suggested to me, rather than by delaying to make a list of subjects suitable for inquiry; the reason why will be seen directly.

In the first place your object of obtaining a supply of men well versed in the appropriate methods of statistics, who shall apply them to the social problems of the day, seems to me a *most worthy* one, and well deserving a *great effort.*

In addition to the problems you specify, such may be mentioned as:

(1) Number of hours' work, and corresponding amount and value of output in different occupations, whether purely mechanical, partly mental or aesthetic.

(2) The effect of town life on the offspring, on their number and on their health.

(3) What are the contributions of the several classes—as to social position and as to residence—to the population of the next generation? Who in short are the proletariat?

In pursuance of what I have said above I will not multiply instances.

The real difficulty in treating these and similar subjects is to specify exactly what is aimed at in a way free from all ambiguity, and again in a way to which the statistics that are available will give an answer also free from ambiguity. The difficulty of the physicist is to interrogate Nature by framing searching questions to her, and it is by this method, the applications of which seem so simple, after some philosopher has had the ingenuity to think of them, that all physical science is forwarded. But there are very few men capable of interrogating Nature aright; those who are become the great men of science. Now the difficulty in social statistics is of exactly the same kind, but *greater*. Therefore by no straightforward and expeditious method can the problems in which you—and I may be permitted to add myself—are so much interested, be solved. Each is a separate and difficult undertaking requiring a vast deal of thought and planning, just like planning a campaign. Quetelet's own history is an example of this. His promises and hopes and his achievements in 1835–6 remained *in statu quo* up to the last edition of his work (*Physique Sociale*) in 1869. He achieved nothing hardly of real value in all those 33 years[1]. So again Buckle, who started with a flourish of trumpets in the first chapter of his *History of Civilisation*, did next to nothing beyond a few flashy applications that have rarely stood after-criticism.

The way in which your object might best be attained requires, I think:

(1) A man (or men) conversant with the methods, and especially the *higher methods*, of statistics.

(2) Conversant with the existing statistical data.

(3) With his heart directed towards the solution, one by one, of such parts of such of your problems as he can, after much thought, see his way to attack successfully.

(4) Proportioning his labour so as to stop short when he has reached a fairly near approximative result, and not to waste himself in figures in order to procure a slightly closer approximation. In short he must be the master and not the slave of his statistics. The waste of effort by statisticians seems appalling. (I know it is so in meteorological statistics.)

How to get all this? I gather that you have in view the establishment of a Professorship or Readership at Oxford. Before you *fix* your mind in that direction or in that of Cambridge, I should like to tell you by way of warning the experience our Geographical Society has had in doing the same for Geography in the two Universities. I happen to have been closely connected with the movement and am indeed going down to Cambridge next week to see if the dismal want of success of our Reader there can be obviated. The result of very much inquiry has been, that unless the subject on which a Professor lectures has a place in the examinations he will get *no class* at all. His position will be that of a salaried sinecurist, which is proverbially not conducive to activity. Still, he would have leisure and personally would have interest in his work, and if only a Reader, is removable after 5 years. A professor is permanent. He would live in much isolation at Oxford as far as his own subject is concerned, for all the main interests of the place are scholastic, and many of them are rather petty. It occurs to me that perhaps as good a way as any might be to found a professorship at the Royal Institution in London, and to require a yearly course of lectures. The Royal Institution audience is just the sort to stimulate on the one hand and to curb the vagaries of the inquirer on the other. It is a mixture of some of the ablest philosophers, of many persons of wide social interests and of the general public. The existing professors are all men of the highest ability in their several lines: Lord Rayleigh in Physics, Dewar in Chemistry, Victor Horsley in Biology. If a Professorship in Social Statistics could be established there on the same basis as those mentioned, it would have to be nominally renewed each year up to five years' (I think) tenure. Then the re-election is for another (practically) 5 years. The cost is, *I think*, about £300 to £500 a year, not more. Pray excuse my impertinence if you think it such, in venturing to suggest, but my only object is to show what seems to me to be the best *direction* of action. I think London would be by far the best residence for an inquirer into social statistics. Believe me, Very faithfully yours, FRANCIS GALTON.

Looking back after thirty years one is compelled to think that Florence Nightingale's scheme, if it could have been carried out, was essentially better than Francis Galton's. How could a school of trained applied statisticians have been created by six lectures a year at the Royal Institution; that institute has a most valuable platform for announcing in a popular way the results of recent research, but it is not an academic centre for training enthusiastic young

[1] I venture to think that this is far too sweeping, it overlooks not only what Quetelet achieved in organising official statistics in Belgium, but his great work in unifying international statistics.

minds to a new departure in science. The very names mentioned by Galton are those of men who had become famous for research in their own lines before they became "professors" at the Royal Institution, of men, whose means of support did not depend on that institution. Looking round the possible field of candidates in 1891 what man was there who would have fulfilled the same conditions in statistical science as these men in their respective branches? There was only one man—Galton himself—and it is quite certain that he had not that man in view. And also that, if the endowment had been made, and others had suggested him, he would have refused the post. It could have contributed nothing to his influence or research activity, and would have curtailed his freedom in a way wholly distasteful to him. There is small doubt that Florence Nightingale's plan of a professorship round which a school of young enthusiasts might be developed was the wiser, if less showy policy[1]. Between Galton's letters of February 10 and March 15, a brief note written by him on February 19—

"it would give me pleasure to call and talk over the scheme when you feel disposed. The more I think of it the more important it strikes me to be"—

indicates that the discussion had been continued by interview. During this or a later meeting Florence Nightingale must have emphasised the importance of a "school of youngish men." But Galton did not surrender his Royal Institution lectureship, or his advisory committee, or his essayists. Writing on March 15, 1891, he says:

With reference to your scheme, I have not been idle but have made some few inquiries; of course withholding your name. I think the net result is this:

(*a*) Lectureship or Professorship at the Royal Institution with the duty of giving at least six lectures a year and writing a paper.

(*b*) A studentship, prize or scholarship at Oxford or Cambridge.

(*c*) A regular Professorship somewhere. Query in London.

(*d*) Endowment of a Course of Annual Lectures—like the Hibbert Lectures—at some great centre. Query in London.

The selection between these would depend much on the funds disposable eventually.

There is no doubt that a small body of youngish men inspired with a common enthusiasm would do incomparably more than any endowment can ensure. One is often in despair at the thought of how little money can secure in the way of original work. The enthusiasm I mean is not that which is fed by public notice or high patronage, but by the intelligent kindly interest and prompt appreciation of a very few capable and honoured people like yourself of whatever really good work may be done. In short one wants a school of inquirers, having a nucleus of a few able and single-minded persons, not distracted by too many other interests, to originate and maintain the enthusiasm of their fellows and co-adjutors.

Then again some journal suitable for receiving such memoirs, long or short as the case may be, is a desideratum, as well as means of discussing them. This raises the question whether the Statistical Society might not appropriately be the body, in whose hands the endowment might be placed, in order to forward your object under the best attainable safeguards. Most statisticians belong to it, and a suitable committee of them might be trusted.

I will take *the chance* of finding you at home about 5 to-morrow (Monday) unless I receive a card to the contrary, Very faithfully yours, FRANCIS GALTON.

[1] An energetic professor would very soon have compelled even Oxford or Cambridge to grant degrees on the basis of "schools" or triposes in statistics, and I do not despair of such a future after the full admission of women to Oxford, and the extreme difficulty, even for a Cambridge don, to detect any feminist push in a proposal to graduate in statistics!

It is strange that Galton did not grasp that of his four alternatives (*c*), Florence Nightingale's own suggestion, was the sole one that could lead to a school of "enthusiastic youngish men," and that even such young men could not do their work in the spare moments of other employments; it was not only that they would need a leader, but they would need a livelihood. How strangely different the development of modern statistics might have been, had Galton confined himself solely to Florence Nightingale's proposal of a professorship and the creation of a school of social statistics or as she later headed some of her letters "Applied Statistics"! Boldly to have said we need £50,000 or £60,000 to carry out a real scheme would have been the wisest policy. Can we between us and with the aid of others who realise our standpoint induce the public to see the importance of the whole matter, and aid in such an endowment? Instead of appealing to the enthusiasm that a big scheme might have raised, Galton drew up a memorandum to be sent round to a number of prominent statisticians asking their advice as to the disposal of a sum of £4000 available to further the scientific study of social problems from a statistical point of view. He stated that a plan had provisionally commended itself for the distribution of three hundred pounds in honoraria of £50 each to a few selected writers, who should severally draw up a list of what seemed to them to be the most feasible problems in the branch of inquiry with which they were familiar.

"It would be their part to think out and to draw up reasonable plans of campaign, specifying the available data now in existence, and such other data as would be required, and which at the same time might be procured without serious difficulty."

The simultaneous direction of these six highly competent persons to different branches of the same scheme would, Galton thought, greatly assist in its inauguration and drawing public attention to its importance.

The fundamental suggestion then made for the remainder of the endowment was that of the Royal Institution lectureship. There is no evidence that this memorandum was ever issued, or received Florence Nightingale's approval. Indeed some of the sentences in later letters seem to suggest that it did not. She writes in a letter of April 19 (1891) with regard to the subjects of the essays:

"I would only suggest that the statistics on business which the Statistical Society so often and so wisely publishes are not quite the sort of thing, nor are Hygiene and Sanitation proper, for which also there is already much large machinery, official and unofficial. And I would ask: Would 'the matters that affect a large part of the community' include such subjects as so press on my mind, and to which you have so generously given a home?"

and then she reiterates the headings of the suggested topics of her first communication. Again, in a letter of a few days earlier evidently referring to the leaders of the Statistical Society whom Galton proposed to consult:

"Mr Giffen, I suppose, is a bright particular star, but not in my line of business—that of moral sanitation. Nor Sir J. Farrer. Also they are not your 'youngish men' whom you so wisely and so well propose to collect and educate."

It is not of importance for us now to know how far Galton's proposals failed to satisfy Florence Nightingale, or how far further examination of economic possibilities on her own side cooled her ardour.

In 1891 a Demographic Congress was to meet in London and Galton suggested that a memoir should be read before the Congress, if possible under Miss Nightingale's name, urging the more systematic collection and utilisation of demographic statistics, with a view to applying them to the solution of social problems. It is not clear from the remaining letters whether Florence Nightingale, while approving of this scheme, was unwilling that her name should be associated with it; still she was very desirous that her three or four problems should be especially mentioned, and remained willing to subscribe towards honoraria for the proposed essays. On April 21 Galton sent another letter, enclosing a memorandum, which was to be circulated to "half a dozen or so eminent authorities" asking about precise subjects and persons. This memorandum runs:

It is desired to promote Statistical Inquiry into the efficiency of legislative acts, intended to promote the well-being of large classes.

With this object in view it is proposed during the present year to offer £50 to £75 in remuneration for each of two or three essays, severally referring to selected branches of any of the following subjects: (i) Board School Education, (ii) Treatment of the Criminal Classes, especially of boy-offenders, (iii) Effect of Poor-Law and Workhouses, whether de-pauperising or not.

A statement or discussion is desired in each essay of the nature and value of the statistical information now accessible, and of such other information as exists in an unpublished form, and again of such as has not yet been collected but which might apparently be procured without serious difficulty. It is then expected that the writer would discuss the ways in which these data should be treated so as to lead to sound and practically important conclusions with the minimum of difficulty.

Should the results of this first attempt be encouraging, it is proposed to follow it up by further action in future years, perhaps of a wider character.

Galton's letter appears to have remained a month unanswered. The original proposal[1] had shrunk to comparative insignificance, and it is little wonder that there was no enthusiasm for it in its final form. On May 23, 1891, Florence Nightingale wrote apologising for her delay—"I can only sum up my apologies in: how good you have been and how bad I." She returned Galton's memorandum initialed and asked him to send it to the eminent authorities he might select. Five days later Galton replied that to his sorrow he must say that the season was too far advanced for him to attempt to carry through the preliminaries with hope of success:

"You would necessarily and naturally have to be consulted at each important stage, financial arrangements would have to be made and there is not now time for doing all this before the vacation begins and people, especially those of the Universities, scatter. I therefore am obliged to desist for the present at least....The more I think of it the more convinced I am that the assurance in some form of a continuation of these awards or other form of endowment would be an important element of success."

[1] As drawn up by Francis Galton and corrected by Florence Nightingale evidently at their first interview this ran:

——— is desirous of founding a professorship of statistics, to be called by the name of the ——— Professorship of Statistics, for promoting by means of lectures or otherwise the cultivation and improvement of statistical science, and especially its practical application to social problems.

A last brief letter closes the correspondence:

10 SOUTH STREET, PARK LANE, W. *June* 13, '91.

*Statistical Inquiry Essays.*

MY DEAR SIR, I sorrowfully acknowledge your just award that the season is now too far advanced for you to attempt to carry out the preliminaries. I can only hope that when the vacations are over I may still appeal to your wisdom. You have been more than kind. And no one could do for the matter what you would. I trust your Demography is making favourable progress. I am ever yours gratefully, FLORENCE NIGHTINGALE.

One can but regret this conclusion to what might have been a great success, the realisation of an ideal common to two of the most remarkable minds of the nineteenth century. They were both "passionate statisticians," both saw a great need—a need which still largely exists—and both had shown themselves capable of carrying great enterprises to successful conclusions. Yet somehow Francis Galton seemed to overlook the very kernel of Florence Nightingale's scheme, and the whole vanished in a trivial essay project. Yet the correspondence was, I believe, not without influence on Galton himself, and probably contributed not a little to guide him consciously or unconsciously when he came to make his own foundation in linking it up with a school of statistical training. An additional twenty years demonstrated to him not only the futility of advisory committees, but how little in the way of research could be achieved by the offer of small monetary prizes.

Something would certainly have failed in this chapter, if we had been unable to show even this slender link between the master builder of the modern theory of statistics and the "Passionate Statistician" whose mind had been so deeply stirred by his greatest forerunner, Quetelet:

"I might have done it for you. So it seems:
Perhaps not. All is as God over-rules.
Besides, incentives come from the soul's self,
The rest avail not."

PLATE LIV

SIR FRANCIS GALTON
1910
From a sketch by his niece, Mrs. Ellis

*Note illustrating Francis Galton's Views on Religion.*

I found the following remarks in Francis Galton's handwriting among material collected for a new edition of *Inquiries into Human Faculty*. Its bearing on what has been said on pp. 257, 282, and in the footnote, p. 102 will be obvious. The date of the manuscript must be about 1892.

Probably every one has at some time had the feeling that if a dearly loved parent were taken from him, the grief and loneliness after the loss would be insupportable; yet parents die, and their children, after a burst of poignant grief, recover themselves and survive, and most persons of middle age are orphans, leading happy lives full of interests, and mellowed rather than saddened by recollections of the past. The early loves of men and women are intense; they are wholly bound up in one another and the words 'for ever' and the like are the stock expressions of their phraseology, but how transient in many cases are these dispositions. The mind is not wholly dependent on its anchorage to any one given sentiment; if it be cut adrift, at least in early life, after a short while new interests will arise, to which it will moor itself as securely as before. The sense of necessary dependence on any given sentiment may be very strong, but its reality is belied by the experience of what daily occurs around us. Thus if a suspicion were lodged in the mind of a fervent Roman Catholic that the Virgin Mary exercised no protective power over him, the dread lest that suspicion should grow into a conviction would be a far worse terror to him than the anticipation of any earthly orphanage; yet Protestants holding that view lead lives as calm as those of the Catholics. Similarly, the thought to the Christian of being orphaned of Christ is no less horrible; but Jews and Unitarians, some of high position in society, and others, philosophers and men of letters, having no belief in the incarnation and intercessory powers of Christ, live and die as contentedly as Christians. So again, the thought of being orphaned of the paternal guidance of a being having the peculiar attributes of the Jewish Jehovah, would give a terrible shock to many, yet it is notorious that the majority of thoughtful Germans and numerous English Agnostics, whose views on other subjects are treated with general respect and who lead well balanced and contented lives, do not entertain that belief. It is astonishing how devoid of sympathetic intelligence most men are. They are afraid to face the fact that good and able men disagree fundamentally on the elements of religious doctrine, and that therefore no certainty can be claimed for any one of these doctrines. At the best they are only persuasions.